Volume 9

Ceramic Transactions

NUCLEAR WASTE MANAGEMENT III

Edited by G. B. Mellinger

The American Ceramic Society, Inc.
Westerville, Ohio 1990

Proceedings of the Fourth International Symposium on Ceramics in Nuclear Waste Management held in Indianapolis, Indiana, April 24–26, 1989.

Library of Congress Cataloging-in-Publication Data

Nuclear waste management III / edited by G. B. Mellinger.
p. cm. — (Ceramic transactions ; v. 9)
Proceedings of the Fourth International Symposium on Ceramics in Nuclear Waste Management held in Indianapolis, Indiana, April 24–26, 1989.
ISBN 0-944904-24-6
1. Radioactive waste disposal—United States—Congresses. 2. Radioactive waste disposal—New York (State)—West Valley—Congresses. 3. Glass waste—Environmental aspects—United States—Congresses. 4. Ceramics—Congresses. I. Mellinger, G. B. II. International Symposium on Ceramics in Nuclear Waste Management (4th : 1989 : Indianapolis, Ind.) III. Series.
TP898.118.N84 1990 90-622
621.48'38—dc20 CIP

ISBN 0-944904-24-6

Printed in the United States of America

1 2 3 4 5—94 93 92 91 90

Preface

The Fourth International Symposium on Ceramics in Nuclear Waste Management was held April 24–26, 1989, in Indianapolis, Indiana. The symposium was held in conjunction with the 91st annual meeting of The American Ceramic Society and was sponsored by the Nuclear Division of The American Ceramic Society in cooperation with the Society's Glass, Cements, and Basic Sciences Divisions. As the symposium title indicates, this was the fourth in a series of symposia; the first was held in 1979.

The continued interest in the materials aspects of nuclear waste management was evidenced by the significant number of registrants. The symposium participants included international representatives from industry, national laboratories, universities, and government agencies. Seven countries were represented, including Australia, Canada, England, France, Germany, Japan, and the United States. Forty-five papers were presented in seven technical sessions. Technical sessions were devoted to ceramic waste forms and radiation damage effects; transuranic and low-level waste; leaching modeling and mechanisms; systems and field testing; spent fuel characterization and testing; waste form development, characterization, and qualification; and glass waste form processing.

Many people contributed to the success of the symposium and the issuance of the proceedings. In particular, I would like to acknowledge the contributions of the session chairmen, A. B. Harker, L. R. Dole, C. M. Jantzen, G. G. Wicks, J. E. Mendel, and C. C. Chapman. In addition to chairing their sessions, these individuals also arranged for the technical reviews of the papers associated with their sessions. This contributed greatly to the quality of the proceedings. I am also grateful for the assistance of W. A. Ross, Nuclear Division Program Chair for his assistance in organizing and implementing the symposium. I would also like to express my appreciation to J. A. Graham and P. Achter of The American Ceramic Society for their help related to program and meeting arrangements and publication of the symposium proceedings. Finally, the dedicated efforts of Billie Neth during the planning of the symposium, the symposium proper, and the issuance of this proceedings are gratefully acknowledged.

G. B. Mellinger

__Ceramic Transactions__ is a new proceedings series designed to meet two needs: high quality content and rapid publication. Volumes in the series come from meetings, symposia, and forums. Each paper is reviewed by two peers, and final manuscripts are prepared by authors in a camera-ready format. The volumes in this series would not be possible without the hard work, dedication, and cooperation of editors, reviewers, and authors, who all deserve a great deal of thanks.

Your comments, questions, and suggestions for future __Ceramic Transactions__ volumes are welcomed and should be addressed to the Director of Publications, The American Ceramic Society, Inc.

Contents

Section II. TRU and LLW

Section III. Leaching-Modeling and Mechanisms

Section IV. Systems and Field Testing

Section V. Spent Fuel Characterization and Testing

Section VI. Waste Form Development, Characterization, and Qualification

Section VII. Glass Waste Form Processing

Section I. Ceramic Waste Forms and Radiation Damage Effects

TAILORED CERAMIC CONSOLIDATION FORMS FOR ICPP WASTE COMPOSITIONS

Alan B. Harker and John F. Flintoff
Rockwell International Science Center
Thousand Oaks, CA 91362

ABSTRACT

A polyphase tailored ceramic has been developed for the consolidation of simulated ICPP-type high Zr content high-level waste (HLW) calcines. The ceramic is specifically designed to provide chemically stable host phases for each species present in the HLW and to maximize waste volume reduction through high loadings and form density. The ceramic is designed for a 73 wt% waste loading with a density of $3.35 \pm 0.05 (g/cm^3)$. The major phase in the ceramic is a high-silica glass, which contains the neutron poison boron as well as the majority of the nonrefractory species in the waste. The primary crystalline phases are calcium fluoride, calcium-yttrium stabilized cubic zirconia, a hexagonal apatite type silicate containing the plutonium simulant Ce, and a Cd metal phase. Minor phases include zircon, zirconolite, and a sphene-type. Leach testing and microscopic analysis shows the ceramic form to be chemically durable, with only the glass phase showing any detectable dissolution in deionized water at 90°C.

INTRODUCTION

A research study is being conducted to assess the use of polyphase ceramic forms for the consolidation of ICPP high-level waste (HLW) fluid bed calcines. Earlier efforts have shown that significant volume reduction improvement can be obtained over a fully vitreous waste form by this approach for several simulated ICPP-HLW compositions since waste loadings of 70 to 80 wt% can be accommodated with a final form density of over 3.3 g/cm^3.[1,2] Such polyphase ceramics also offer the potential benefit of chemically binding the long-lived actinide species in highly durable crystalline phases whose properties can be predicted from the geologic record of mineral analogues.

Several potential compositions for future ICPP wastes have been examined in this study since its inception, with a number of additives being considered to assist the waste in forming desirable crystalline phases mixed uniformly with a durable boro-silicate glass matrix.[1,2] Even with the high refractory content of the ICPP calcines, a glass phase is necessary in the ceramic form to accommodate the neutron poison boron, and the sodium present as a processing by-product.

The simulated HLW composition given in Table 1 was used in the present work to represent future ICPP type HLW. The waste contains significant Na and has levels of both boron and cadmium as neutron poisons. Hexavalent Cr is present in the mixture as well as nonradioactive isotopes of Cs and Sr. Depleted uranium and Ce are used to simulate the chemistry of the actinides.

Table 1. Simulated ICPP High-Level Waste Composition Used in This Study

Solid Component	Weight %
CaF_2	52.21
ZrO_2	21.41
Al_3O_3	7.14
CdO	5.58
B_2O_3	4.15
Na_2O	3.76
CaO	2.58
$CaSO_4$	1.01
K_2O	0.55
Ce_2O_3	0.5
U_3O_8	0.5
Cs_2O	0.25
SrO	0.25
Fe_2O_3	0.11
Cr_2O_3	0.54

Given that the simulated waste is composed of both refractory and nonrefractory species, the waste form was designed to have a major glass phase. The crystalline phases forming from the refractory species are then controlled by the use of additives, redox control, and consolidation conditions. The ceramics produced in this effort have been characterized as to microstructure, the distribution of the individual species among the phases, and the leaching performance under static MCC-1 conditions.[3] Based upon these results the ceramic composition and processing have been optimized.

EXPERIMENTAL

The simulated HLW was prepared in calcine form by firing nitrate and sulfate precursors in air to 500°C. All chemical tailoring additives were added to the simulated HLW calcines as finely powdered oxides or metals by ball milling for periods of 15 to 20 minutes. Additives were not incorporated with the original HLW composition in the calcining step in order to more appropriately simulate actual processing conditions. The mixed waste and additives were cold pressed into cylindrical stainless steel canisters and then evacuated under heating to above 400°C. When a base vacuum below 13 Pa was obtained the canisters were crimped and weld sealed. Consolidation was carried out by hot isostatic pressing (HIP) at 1000°C and about 100 MPa.

After consolidation the waste forms were sectioned and polished for MCC-1 leach testing, examination by analytical scanning electron microscopy, x-ray diffraction analysis, and density determination. All solution analyses were carried out by inductively coupled plasma emission spectroscopy.

RESULTS AND DISCUSSION

Glass Phase Optimization:

It has been demonstrated in the development of boro-silicate glass waste forms that there is a small compositional regime in which the glass can be optimized for minimum silica release rate in aqueous solutions.[4] This can be expressed in the form of a triangular composition diagram in which silica is put at one corner, boron + alkali at the second, and other oxides at the third. This is shown in Fig. 1. The optimum glass phase performance occurs at the high silica end of the diagram with a silica to alkali + boron weight ratio of about 2 to 1 and a contribution of about 20 to 25% from other oxides in the waste.

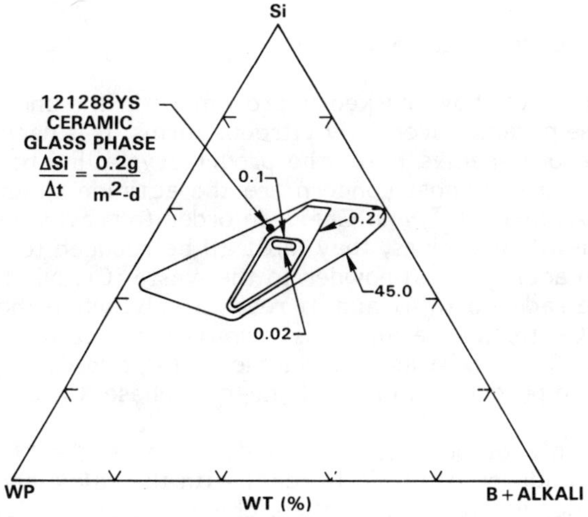

Fig. 1 Glass stability diagram.

Applying this diagram to the glass phase in the ceramic requires some assumptions. First it is assumed that all the boron, alkali, calcia, alumina, sulfates, iron oxide, and chromium form a glass phase. Then enough silica must be added to optimize the dissolution behavior. From Table 1 it can be seen that the sum of the glass forming elements in the waste is approximately 19.8 wt%, with 8.7 wt% of the waste being the sum of the oxides of boron and alkali metals. At the most simple level, the diagram would prescribe the addition of 17.4 grams of silica per 100 grams of waste, producing a waste loading of about 85 wt%. This, however, leaves the glass modifiers at a level of about 30% in the glass phase, well above the optimum region. Adding a higher level of silica to reduce the glass modifiers is the obvious approach and this was undertaken in an empirical manner, increasing the level of silica and measuring the Si release rate from the consolidated ceramic. In these studies the near optimum performance in the Si release rate from the ceramic (about 0.2 g/m^2-day at 28 days) was obtained at a waste loading of 73 wt% with 20.5 wt% silica added to form the glass phase. Some of added silica did, however, react with the other waste elements (Ca, Zr, and Y) to produce an apatite-type crystalline phase in the ceramic ($Ca_4Y_6(SiO_4)_6$) as well as minor zircon ($ZrSiO_4$) and a sphene-type phase ($Ca(Cr,Ti)SiO_5$). Assuming that the significant apatite phase consumed the majority of the added yttria and cerium, the residual silica and waste oxides can be estimated to place the glass phase composition at about 58 wt% silica, 22 wt% waste oxides, and 20 wt% boron + alkali. This agrees quite well with the predicted leach behavior from the diagram, but places the glass phase composition in a region where increased waste loading or higher silica content would be deleterious. Hence, if any further changes in composition were expected, a reduction in waste loading and the addition of some further alkali would be desirable.

Crystalline Phase Optimization:

In order to show marked improvement in the chemical durability of the polyphase ceramic over fully vitreous forms, it is necessary to assure that the refractory species enter the proper crystalline host phases. The elements which are of most concern are the actinide simulants U and Ce and the toxic species Cd^{2+}, and Cr^{6+}. In order to provide crystalline hosts for these multi-valent cations, they must all be reduced to a lower valence state. Through adding metal powder to the waste (Cr, Ni, Ti), the uranium and cerium are reduced to 4+ and 3+ respectively, while the Cr is reduced to 4+ and the Cd to the metal. This permits the U to enter the dominant zirconia phase, Cd to exist as an insoluble metal, and Cr to react with the Ti^{4+} and silica to produce a minor sphene-type phase, $Ca(Cr,Ti)O_5$.

The only other required modification to the chemistry of the waste, was the addition of yttria to react with the calcia and silica to form an apatite, analogue with a nominal stoichiometry of $Ca_4Y_6O(SiO_4)_6$. This apatite-type phase accepts the Pu simulant Ce as well as a small amount of residual Cd^{2+}.

The resulting crystalline phase content of a ceramic made up of 73 wt% waste, 20.5 wt% silica, 4.2 wt% Y_2O_3, and 2.3 wt% mixed metal powder is relatively simple as shown in the x-ray diffraction pattern in Fig. 2. Calcium fluoride, formed during the waste calcining, remains the dominant crystalline phase, though it is not a major radionuclide host. Zirconia generally retains the calcium stabilized cubic structure found in the waste calcines, though the monoclinic form is also present. The lesser apatite-type phase is just observable, while the minor sphene phase and trace zircon are below the sensitivity of the instrument. The individual phases are more easily identified by their morphology and chemical content using analytical scanning electron microscopy.

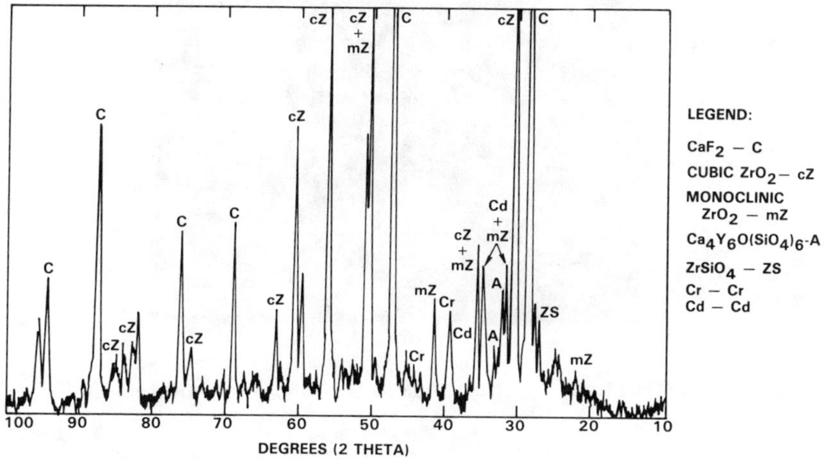

Fig. 2 XRD pattern of 121288YS ICPP ceramic.

Figure 3 shows the microstructure of the ceramic. The individual crystalline phases can be recognized by their morphology and contrast in the glass matrix. In Fig. 3 the Cd metal phase nucleates around particles of the added Ni powder and appears as light spheres and aggregates of spheres, 2 to 10 μm in size. The apatite-type phase appears as light grey rectangular crystallites from 2 to 5 μm in length. The zirconia forms clusters of submicron scale spheres, and the calcium fluorite is present primarily as large grey contrast spheres, 5 to 10 μm in diameter.

This ceramic form, labeled 121288YS, is grey-green at a waste loading of 73 wt% and has a density of 3.35 ± 0.05 g/cm³. The calculated approximate phase composition and observed elemental distribution for the ceramic are given in Table 2.

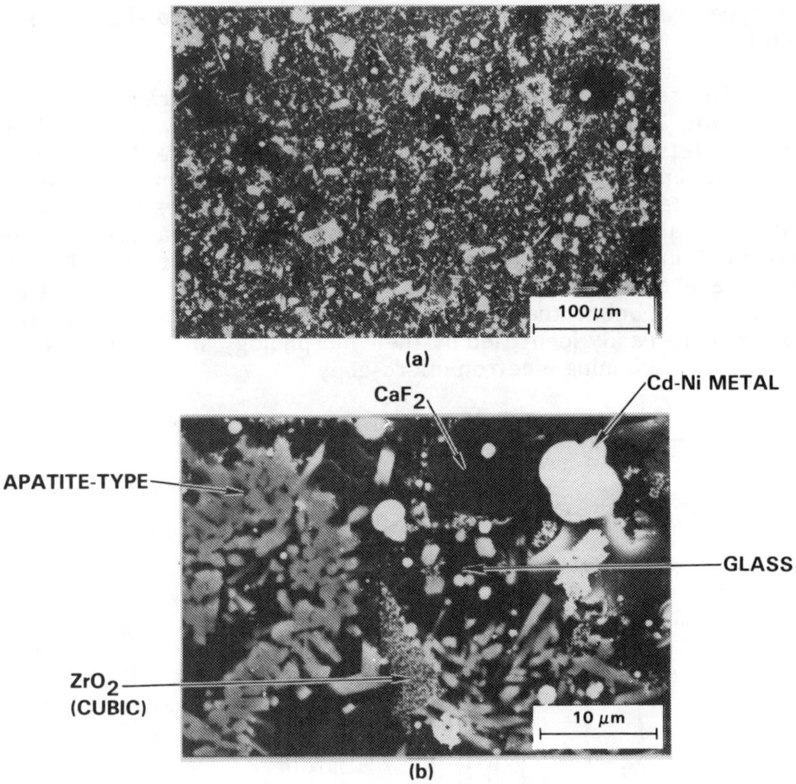

Fig. 3 Microstructure of polyphase 121288YS simulated ICPP HLW ceramic.

Chemical Durability:

Cut and polished samples of the 121288YS ceramic were subjected to MCC-1 static deionized water exposure tests.[3] The samples were immersed at 90°C for periods of 1, 14, 21, and 56 days. When the tests were completed the pH of the solutions was determined at room temperature and solution concentrations of individual species were determined by inductively coupled plasma spectroscopy. The average elemental release rates and solution concentrations for the individual species are given in Table 3. The solution analysis procedures are not sensitive to uranium and cesium and the solution concentration data is not currently available for these species.

The solution data show that there is an initial release of most of the species present in the glass, followed by decreasing release rates for all elements. The 56 day data also shows a solution concentration decrease for

Table 2. Composition and Phase Content of the 121288YS Science Center Ceramic Waste Form. The elemental content information was derived from EDXS mapping.

Composition:	73.0 wt% Science Center Reference Simulated Waste 20.5 wt% Silica 4.2 wt% Yttria 1.0 wt% Ti metal 0.8 wt% Ni metal 0.5 wt% Cr metal			
Density:	3.35 ± 0.05 g/cm3			
Phase		Volume %	Estimated Specific Gravity	Elemental Content
Glass		40%	2.4	Si, Al, B, Na, Ca, Cs, Sr, K, Fe Ti, Cr, S, trace U
CaF_2		38%	3.18	Ca, F, Sr, Si
ZrO_2 (cubic + monoclinic)		10%	5.6	Zr, Ca, Y, U
$Ca_4Y_6O(SiO_4)_6$		7%	4.07	Y, Si, Ca, Ce, Cd
Metal		2%	8.6	Cd, Ni
Minor Phases		3%	~ 5	
$ZrSiO_4$				Zr, Si, U, Y, Ce
$Ca(Cr,Ti)SiO_5$(sphene-type)				Ca, Cr, Ti, Si
calculate density = 3.34 g/cm³				

Table 3. MCC-1 Leach Test Data on the 121288YS Ceramic Conditions: Initially Deionized Water, 90°C, SA/V=10 cm^{-1}

Element	7 Days	14 Days	21 Days	56 Days	
	Average Leach Rate (g/m2-day) Measured Solution Concentration (ppm)				
pH	7.8	7.9	7.8	7.8	
Ca	0.20 3.45	0.11 3.63	0.09 4.19	0.037 4.79	(g/m2-day) (ppm)
Zr	< 0.0008 < 0.007	< 0.0004 < 0.007	< 0.0003 < 0.007	< 0.0001 < 0.007	(ppm)
Al	0.30 0.66	0.24 1.0	0.35 2.13	0.052 0.88	(ppm)
Cd	0.23 0.66	0.16 0.87	0.10 0.78	0.063 1.38	(ppm)
B	0.48 0.36	0.33 0.48	0.26 0.53	0.084 0.49	(ppm)
Na	0.56 0.91	0.38 1.19	0.31 1.39	0.10 1.27	(ppm)
Sr	0.68 0.08	0.44 0.1	0.37 0.12	0.12 0.11	(ppm)
Fe	< 0.1 < 0.004	< 0.05 < 0.004	< 0.04 < 0.004	< 0.01 < 0.004	(ppm)
Cr	< 0.04 < 0.006	< 0.02 < 0.006	< 0.01 < 0.006	< 0.005 < 0.006	(ppm)
Si	0.43 3.34	0.32 4.81	0.28 5.90	0.069 4.13	(ppm)
Ce	< 0.4 < 0.1	< 0.2 < 0.1	< 0.15 < 0.1	< 0.05 < 0.2	(ppm)
Y	< 0.004 < 0.01	< 0.002 < 0.01	< 0.001 < 0.01	< 0.0005 < 0.01	(ppm)

Cs and U were not measured. Cs is normally about twice the measured Na release rate, U is normally below 0.005 g/m2-day after 28 days.

sodium, silicon, and aluminum, indicating that precipitation reaction is occurring in the static system.

Figure 4 shows two scanning electron micrograph views of the 56 day leached surface. Some general surface dissolution of the glass matrix has occurred which brings the crystallites into relief on the initially polished surface; however, no obvious reaction products or "gel-layer" are visible. The higher magnification micrographs in Fig. 5 show that there is enhanced glass phase reaction at the boundaries with the Cd-metal and zirconia phases. There is no apparent enhanced dissolution at the interfaces of the apatite, sphene, and calcium fluoride phases. None of the crystalline phases show any morphology changes with the dionized water exposure.

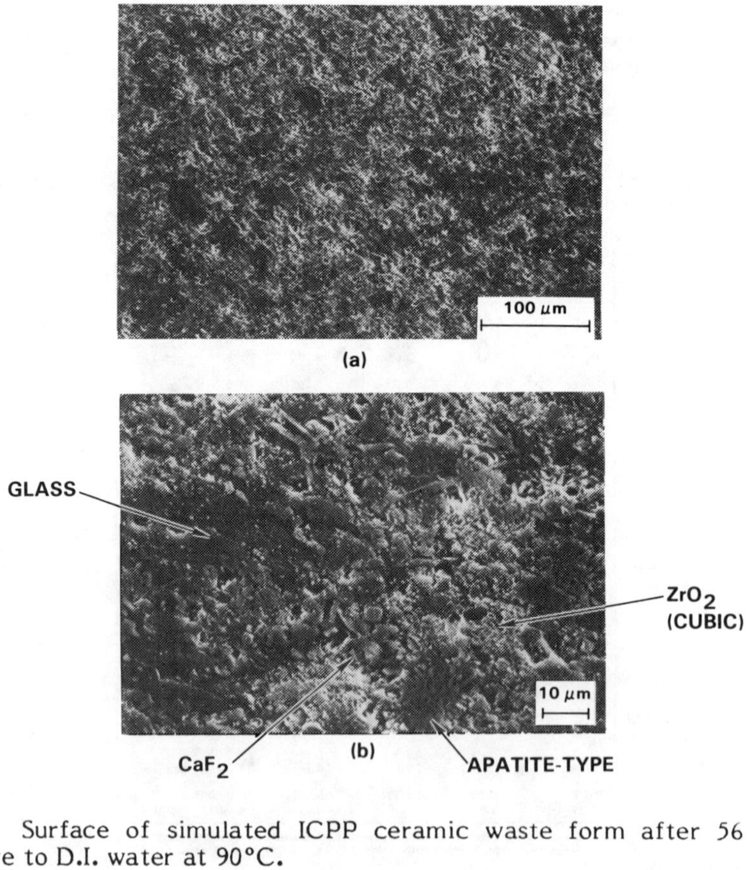

Fig. 4 Surface of simulated ICPP ceramic waste form after 56 days exposure to D.I. water at 90°C.

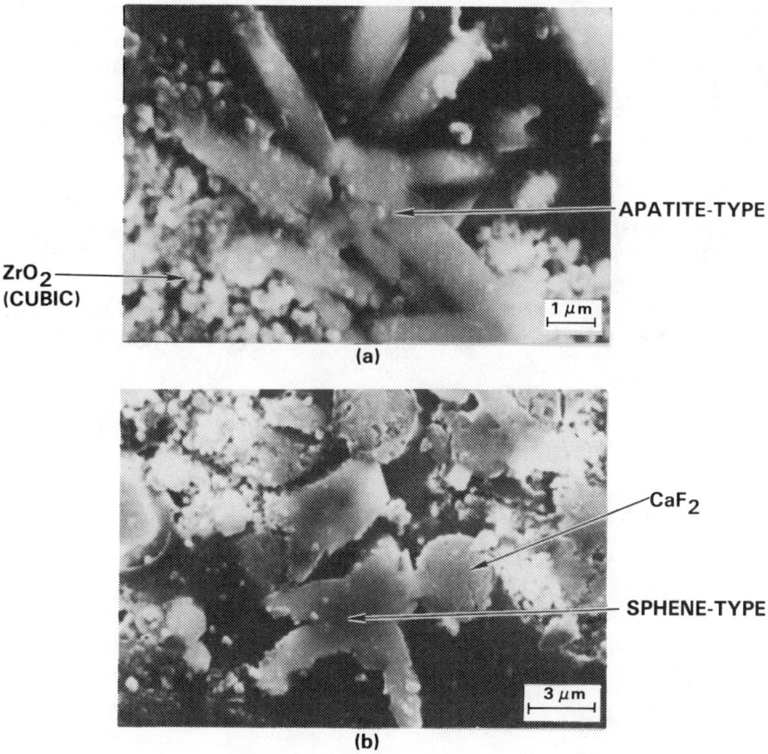

Fig. 5 Uniform glass phase dissolution near crystalline silicate phases.

Summary:

An experimental polyphase tailored ceramic form has been designed, fabricated, and characterized for the consolidation of a simulated ICPP-type high-level nuclear waste. The ceramic has a nominal 73 wt% HLW loading, is consolidated in a sealed stainless steel canister with silica, yttria, and metal powders as tailoring additives, and has a density of 3.35 ± 0.05 g/cm^3. The waste form contains a designed boro-silicate glass phase and three major crystalline phases which act as hosts for the actinide simulants. The ceramic shows good chemical durability in MCC-1 static testing with no evidence of pH run-away or crystalline phase dissolution in 56 days.

Acknowledgement

This work was sponsored by The United States Department of Energy through Westinghouse Idaho Nuclear Co., Inc.

REFERENCES

1. A.B. Harker and J.F. Flintoff, "Polyphase Ceramic and Glass-Ceramic Forms for Immobilizing ICPP High-Level Nuclear Waste," Scientific Basis for Nuclear Waste Management, Vol. 7, ed., G.L. McVay, Elsevier, NY, 513-520 (1984).

2. A.B. Harker and J.F. Flintoff, "Crystalline-Phase Formation in Hot Isostatic Pressing of Nuclear Waste Ceramics with High Zirconia Content," J. Am. Ceram. Soc. 68, 159-165 (1985).

3. J. Mendel, "Materials Characterization Center Test Methods," Battelle Pacific Northwest Laboratories Publication PNL-3990 (1986).

4. A.A. Hench and L.L. Hench, J. Nuclear Chem. Waste Management 4(4), 231 (1983).

A GLASS-CERAMIC COMPOSITION TO IMMOBILIZE ICPP HLW*

ROSEANNE S. BAKER AND JULIUS R. BERRETH
Westinghouse Idaho Nuclear Company, Inc.
Idaho National Engineering Laboratory
P. O. Box 4000
Idaho Falls, ID 83403

ABSTRACT

Candidate glass-ceramic waste forms have been prepared by Hot Isostatic Pressing (HIPing) simulated calcined Idaho Chemical Processing Plant (ICPP) high-level waste (HLW) and glass-ceramic forming additives which include SiO_2, B_2O_3, Na_2O, and Li_2O. Titanium dioxide is added to enhance titanate phase formation to host cadmium, and Ni or Ti powder is added as a reducing agent. Durable products with waste loadings up to 75 wt% and product densities up to 3.3 g/cm^3 have been formed. The glass-ceramic waste form has the potential to reduce final waste volume by up to 60% compared to glass. Preliminary leach rates of the products determined using the 14 and 28 day MCC-1 leach tests are less than 1 g/m^2-day and are comparable with HLW glass products for the glass matrix elements of Si, B, Li, and Na and for the waste components of Cd, Cr, Ca, and Al.

INTRODUCTION

The acidic high-level liquid waste (HLLW) generated at the Idaho Chemical Processing Plant (ICPP) consists mainly of aluminum, zirconium, sodium, nitrate and fluoride ions[1,2,3] which are routinely solidified as granular calcined waste. While a decision has not been made on the disposition of ICPP calcine, evaluations of a potential glass-ceramic waste form to reduce final waste volumes for repository disposal are in progress. Compared to a glass waste form,[4] the glass-ceramic product has

* Work supported by the U.S. Department of Energy Assistant Secretary for Defense Programs, under Contract No. DE-AC07-84 ID 12435.

the potential to reduce immobilized waste volume by up to 60%. Candidate glass-ceramic products have been prepared with calcined waste loadings of 50 to 80 wt% at densities of 3.0 to 3.3 g/cm^3. The glass-ceramic is formed by Hot Isostatic Pressing (HIPing) a mixture of calcined waste, glass-ceramic forming additives, and a reducing agent at pressures between 69 and 138 MPa at temperatures from 950^oC to 1050^oC. The product is formed at near theoretical density and consists of several crystalline phases and an amorphous phase, both of which serve to host radionuclides.

EXPERIMENTAL PROCEDURES

Prior to mixing, the simulated calcined waste is heat-treated in air at 600^oC for 4 to 6 hours to remove volatiles such as nitrates. Table 1 gives the composition of a heat-treated simulated calcined waste. The additives, including SiO_2, B_2O_3, Na_2O and Li_2O are fused at 1200^oC, ground to <100 mesh and then mixed with the heat-treated waste. Crystalline phase formation enhancers, such as TiO_2, and a reducing agent, such as Ni or Ti powder are added. The waste-additive mixture is pre-compressed at 34.5 MPa (5000 psi) into a 2.5 cm (1 inch) O.D., 7.6 cm (3 inch) high 304 stainless steel can, then evacuated to <100 millitorr and seal welded. It is HIPed at 950^oC to 1050^oC at 69 to 138 MPa (10,000 to 20,000 psi) for 1 hour using argon as the pressurizing gas. The heat-up and cool-down times are 1.5 hours each.

After HIPing, the experimental waste form is separated from the can and sectioned using a low-speed diamond blade saw. The samples are measured for density by the Archimedes method, and analyzed for phase composition by X-Ray Diffraction (XRD) and Scanning Electron Microscopy with Energy Dispersive X-Ray Spectroscopy (SEM-EDS). Resistance to leaching is measured by the MCC-1[5] static leach test at 3, 14, and 28 days, and the 3 day Soxhlet[6] leach tests.

Table 1. Composition of Heat-Treated Simulated Sodium-Blended Fluorinel Calcine[a]

Component	Wt%	Component	Wt%
Al_2O_3	14.0	Cs_2O	0.5
B_2O_3	3.8	Fe_2O_3	0.6
CaF_2	31.3	K_2O	1.0
CaO	12.8	Na_2O	5.2
CdO	5.2	$SO_4^{=}$	1.8
CeO_2	1.1	SrO	0.8
Cl^-	0.1	ZrO_2	17.7
Cr_2O_3	1.7	Misc[b]	2.4

[a] 4:1 weight ratio of bed particles to fines.
[b] includes SnO, $PO_4^{=}$, NiO and MnO_2.

EXPERIMENTAL FORMULATIONS

Numerous glass-ceramic formulations have been generated by statistical methods with the waste composition given in Table 1. For some of the experiments, a self-directing optimization (SDO) technique[7,8] was used to generate successive formulations. The SDO technique ranked the products by total and normalized Cr Soxhlet mass loss rates to optimize the waste form composition. In addition to the SDO generated compositions, several compositions were prepared with Ni reductant contents ranging from 1 to 4 wt% to evaluate the effect of increasing reductant content on Cr leachability.

Table 2 lists the formulations which were chosen for MCC-1 leach testing. These formulations were prepared with the glass-ceramic forming additives, D-19 and D-23. The compositions of these two additives, given in weight percent, are: D-19: SiO_2 - 70.1, B_2O_3 - 4.6, Na_2O - 0.7, Li_2O - 1.1, and TiO_2 - 23.5; and D-23: 93.9 -SiO_2, B_2O_3 - 4.0 , Na_2O - 0.4, and Li_2O - 1.6. In the formulations prepared with additive D-23 and TiO_2, the TiO_2 was added separately in the amounts specified in Table 2.

The purpose of adding TiO_2 to the waste form was to form perovskite ($CaTiO_3$) and/or zirconolite ($CaZrTi_2O_7$) with Cd substituting for Ca thereby forming a host phase for Cd. However, as Table 3 shows, sphene ($CaTiSiO_5$) appears to form more readily than perovskite and zirconolite. Therefore, three formulations were prepared using the additive D-23 without TiO_2 to determine if eliminating the TiO_2 would affect the Cd leachability and what phase would be formed by the CaO. A composition containing the additive D-23, TiO_2 and Ti as a reductant, was also prepared to determine if the Ti would enhance the formation of the CdO-TiO_2 phases. These compositions are given in Table 2.

Table 2. Glass-Ceramic Waste Form Compositions

		Wt%						
Formulation Number	Waste Loading	SiO_2	B_2O_3	Na_2O	Li_2O	TiO_2	Ni	Ti
HD-19	71.9	19.7	1.3	0.2	0.3	6.6	0.0	0.0
HD-19-N	70.9	19.7	1.3	0.2	0.3	6.6	1.0	0.0
HD-19-4N	67.9	19.7	1.3	0.2	0.3	6.6	4.0	0.0
HD-23H-2N	72.0	18.5	0.8	0.1	0.3	6.3	2.0	0.0
HD-23-2T	75.0	20.7	0.9	0.1	0.3	1.0	0.0	2.0
HD-23A-3N	70.8	24.7	1.0	0.1	0.4	0.0	3.0	0.0
HD-23LA-3N	63.5	31.5	1.4	0.1	0.5	0.0	3.0	0.0
HD-23HA-3N	75.9	19.8	0.9	0.1	0.3	0.0	3.0	0.0

RESULTS AND DISCUSSION

As seen in Table 3, all of the waste forms contain CaF_2 and ZrO_2 as major crystalline phases. These two phases are formed during the calcination of the HLLW and remain as crystalline components of the glass-ceramic waste form. Sphene formed in all of the waste forms in which TiO_2 was added. Titania was also found in these same waste forms, which indicates that the form of TiO_2 added may not have been very reactive. In the formulation where 0 or 1 wt% TiO_2 was added, the CaO reacted with SiO_2 and ZrO_2 to form $Ca_3ZrSi_2O_9$ as a major crystalline component of the waste form. Calcium zirconium silicate is a more desirable phase than sphene because it requires less SiO_2 per mole of CaO and therefore removes less SiO_2 from the glass

Table 3. Crystalline Phases Present in Glass-Ceramic Waste Forms

	Formulation Number							
Phase	HD-19	HD-19-N	HD-19-4N	HD-23H-2N	HD-23-2T	HD-23A-3N	HD-23LA-3N	HD-23HA-3A
CaF_2	M	M	M	M	M	M	M	M
ZrO_2	M	M	M	M	M	M	M	M
$Ca_{0.15}Zr_{0.85}O_{1.85}$			m					
$ZrSiO_4$				m	m	m	m	m
TiO_2	m	m	m	m				
$Ca_xCd_{1-x}ZrTi_2O_7$	m	m						
$CaTiSiO_5$	m	m	m	m	m			
$Ca_3ZrSi_2O_9$			M	m	M	M	M	M
SiO_2							m	
CdS	m				m			

M - Major phase
m - Minor phase

phase and it does not require the addition of TiO_2 which may decrease the waste loading in the glass-ceramic waste form.

Table 4 gives the 3, 14, and 28 day MCC-1 leach test results for the formulations given in Table 2. Figure 1 shows the decreasing trend in Cr leachability with increasing Ni reductant content in the HD-19 series formulations. It does not appear that the leachability changes with the elimination of TiO_2 and therefore it would appear that the addition of TiO_2 is not necessary. However, the addition of Ti as a reductant instead of Ni significantly reduces the Cr leachability. Formulation HD-23-2T, which has Ti as a reductant rather than Ni, has the lowest Cr leachability.

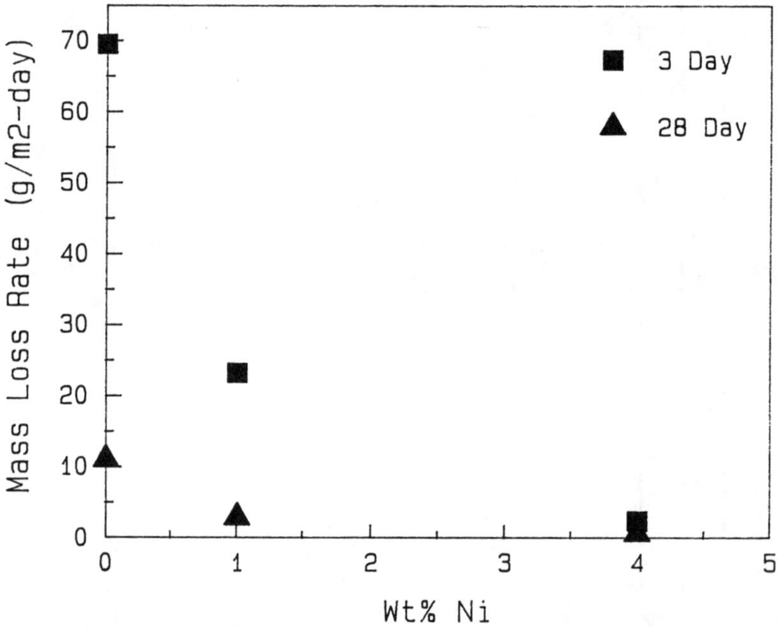

Figure 1 MCC-1 normalized Cr leach rates

Figure 2, a SEM micrograph (2000X) of glass-ceramic waste form HD-23-2T, shows the major phases of CaF_2, ZrO_2, $Ca_3ZrSi_2O_9$ and an amorphous phase. This micrograph also shows CdS and Cr metal. The presence of Cr metal is due to the reduction of Cr^{+6} by Ti. Tin and nickel were also seen by SEM analysis and therefore these two waste cations were also reduced by the Ti.

Table 4. MCC-1 Leach Test Results

Formulation Number	Mass Loss Rates, g/m^2-day[a]								
	Total	Al	B	Ca	Cd	Cr	Li	Na	Si	Sr
3 Day										
HD-19	5.6	6.9	30.0	3.1	0.1	69.4	35.2	22.9	0.9	NA[b]
HD-19-N	3.7	5.9	22.7	2.7	0.3	23.1	21.6	17.4	1.3	NA
HD-19-4N	1.7	4.2	8.9	1.5	0.1	2.3	8.2	7.3	1.9	NA
HD-23H-2N	2.4	4.7	12.8	1.8	0.1	10.0	10.8	9.9	1.5	NA
14 Day										
HD-23-2T	0.5	1.1	1.7	0.5	0.01	0.04	1.8	1.2	0.4	1.0
HD-23A-3N	0.6	1.6	2.5	0.6	0.02	2.1	1.8	1.7	0.3	1.5
HD-23LA-3N	0.5	1.6	1.4	0.6	0.04	2.2	2.1	1.0	0.2	1.9
HD-23HA-3N	0.7	1.9	4.0	0.7	<0.001	3.0	3.0	3.6	0.3	1.2
28 Day										
HD-19	0.8	1.5	6.4	0.6	<0.01	10.9	6.3	14.0	0.2	NA
HD-19-N	0.5	1.1	4.8	0.5	<0.01	2.8	3.9	8.4	0.3	NA
HD-19-4N	0.3	0.8	2.2	0.3	<0.01	0.6	1.8	4.3	0.3	NA
HD-23H-2N	0.4	0.8	2.4	0.3	<0.01	1.6	1.9	6.1	0.3	NA

[a] Determined by the mass-loss response of a monolith to 3, 14 and 28 day MCC-1 leach tests at $90^{o}C$ in deionized water.

[b] Analysis not performed.

Figure 2 SEM micrograph (2000X) of glass-ceramic waste form, HD-23-2T

Table 5 gives a range of expected glass phase compositions for the glass-ceramic formulation, HD-23-2T. This range was determined by semi-quantative EDS and by a spreadsheet computer program developed to calculate glass-ceramic formulations based on waste composition and percent of specific waste components assumed to be amorphous in the formed product. Even though the actual glass phase composition has not yet been defined, the range is sufficient to show that glass phase is high in silica and the weight ratios of alkali oxides to boron oxide (1.4:1 to 1.9:1) and silica to boron oxide plus alkali oxides (2.0:1 to 3.4:1) are in the range which forms a durable glass.

Table 5. Range of Glass Phase Composition in Glass-Ceramic Waste Form HD-23-2T

Component	Range (wt%)	Component	Range (wt%)
Al_2O_3	21.0 - 27.3	Fe_2O_3	0.0 - 0.1
B_2O_3	5.0 - 9.0	K_2O	2.0 - 2.9
CaF_2	5.6 - 6.1	Li_2O	0.0 - 0.4
CdO	0.0 - 3.2	Na_2O	5.2 - 10.1
CeO_2	0.0 - 1.0	SiO_2	44.4 - 49.5
Cs_2O	0.9 - 1.1	TiO_2	0.0 - 2.1
Cl	0.0 - 0.1	ZrO_2	0.0 - 2.9

CONCLUSION

Durable experimental glass-ceramic waste forms have been prepared with waste loadings up to 75 wt%. These products all have 14 and 28 day MCC-1 total mass loss rates of less than 1 g/m^2-day. Normalized mass loss rates for Ca, Cd, and Si for the compositions in Table 2 subjected to 14 and 28 day MCC-1 leach tests were less than 1 g/m^2-day. Formulation HD-23-2T (75 wt% waste, 1 wt% TiO_2, 2 wt% Ti, 20.7 wt% SiO_2, 0.9 wt% B_2O_3, 0.1 wt% Na_2O, and 0.3 wt% Li_2O) is the most durable composition based on the normalized elemental 14 day MCC-1 mass loss rates. The 28-day MCC-1 leach rates for this composition are expected to approach less than 1 g/m^2-day for all elements.

Increasing the amount of Ni reductant in the waste form reduces the Cr leachability. However, it appears that Ti is a more effective reductant since it reduces the normalized Cr mass loss rate by an order of magnitude. Therefore, future compositional studies will include the use of Ti as a reductant.

REFERENCES

1. C. M. Slansky, "High-Level Radioactive Waste Management Program at the National Reactor Testing Station," Adv. Chem. Ser., 153, 31-53 (1976).

2. R. A. Kirkbride, Inventory of Calcined Wastes Stored at ICPP as of September 1979, ENICO-1044, Idaho National Engineering Laboratory (September 1980).

3. J. R. Berreth, Inventories and Properties of ICPP Calcined High-Level Waste, WINCO-1050, Idaho National Engineering Laboratory (February 1988).

4. B. A. Staples, D. A. Pavlica, and H. S. Cole, Properties of Formula 127 Glass Prepared with Radioactive Zirconia Calcine, ENICO-1120, Idaho National Engineering Laboratory (September 1982).

5. J. E. Mendel, "MCC-1P Static Leach Test Method," Nuclear Waste Materials Handbook Test Methods, DOE/TIC 11400 (September 1983).

6. E. D. Hespe, "Leach Testing of Immobilized Radioactive Waste Solids - a Proposal for a Standard Method," Atomic Energy Review, 9, 195-207 (1971).

7. C. D. Hendrix, "Through the Response Surface with Test Tube and Pipe Wrench," Chem. Tech., 10, (8) 488-497 (1980).

8. R. S. Baker, B. A. Staples, and H. C. Wood, Development of a Ceramic-Based Waste Form to Immobilize ICPP HLW, WINCO-M-10077, Idaho National Engineering Laboratory (September 1986).

IRRADIATION DAMAGE TO THE SIMULATED IDAHO CHEMICAL PROCESSING PLANT (ICPP) NUCLEAR WASTE

H. W. Chan and D. G. Howitt
Department of Mechanical Engineering
University of California
Davis, CA 95616

INTRODUCTION

The simulated ICPP waste forms consist of the high liquid waste and an additive composed of different oxides additives used in this study are either yttria-silica or l silica based, and the consolidated waste forms are best des as glass ceramics. The glass phase is basically a silicate the crystalline phases are calcium fluoride (CaF_2), zir (ZrO_2) and zircon ($ZrSiO_4$) [1].

Since the simulated ICPP waste forms consist of amorphous and crystalline phases, the damage from irradi experiments can be compared to the results derived from experiments with the individual crystalline phases and w glass forms studied elsewhere. For example, the amorphous ph are expected to behave similarly to the glass waste f produced by the Savannah River Laboratories (SRL) and, in after gamma irradiation, bubble defect formation was obse throughout the amorphous phases. There were, however, obvi differences in the formation kinetics and morphologies result from the chemical inhomogeneity at the microstructural level, it can be dramatic when the amorphous phase actually underg phase separation due to phase immiscibility. The crystall phases are generally very resistant to radiation damage, and it is the bubble formation in the amorphous phases that is like to be the determining factor in predicting the stability of t waste forms for long term storage.

Previous studies of the radiation damage of alkaline ear halides and the γ and α damage of zirconia and zircon can be use to supplement the simulations of irradiation damage to th simulated ICPP waste forms. The microstructural changes to th ICPP forms were evaluated after both ionization and displacemen damage using a combination of electron, lead ion and γ-

irradiation experiments. Lead ions are in fact produced in nuclear waste as fission products, but the ion implantation is used here to simulate the displacement damage from the α-decay.

There are three types of chemical modifications that can occur in ion implantation experiments [2]. Abstraction, the replacement of an existing atom with the implanted one, prevails when the energy of formation of the new bond is greater than that of the old one. Alternatively, polyvalent atoms such as carbon and nitrogen can insert themselves into the bonds of target atoms, and there is the effect that is observed here where the projectile atom collides with atoms of the material followed by atomic substitution and rearrangement in the solid. This cascade type of displacement mixing in the target lattices prevails as the major source of damage in the crystalline phases but in the glass phase it is the ionization component that produces the greatest effect. The ionization damage introduced by ion collisions can be locally more severe than that by β rays because the energy deposition is larger, and it has been observed that the α decay processes account for 80% of the damage seen in nuclear waste forms such as borosilicates [3]. Displacement damage, unlike that introduced by γ irradiation, is from atomic collisions and occurs whenever the energy of an ion is greater than the threshold displacement energy of an atom in the material. Such damage has been shown to alter the leaching behavior of waste forms [4], but it is unclear whether these effects are truly significant or simply due to the chemical effects which are introduced by ion implantation.

Ionization damage can be induced by γ-irradiation at doses of approximately 10^6 rad/hour using a Co 60 source. Although this flux is substantially less than the ionization dose rate of 10^{10} rad/hour which accompanies the lead ion irradiation, there is almost no corresponding displacement damage introduced by the γ-rays. The displacement and ionization effects can therefore be distinguished in a combination of Pb ion and γ irradiation experiments.

Although the β rays can be simulated by the electrons within the electron microscope, there are difficulties associated with the sample geometry as well as the very high fluxes involved [5]. Depending upon the energy of the beam, electron irradiation can introduce various proportions of ionization and displacement, although the former will always dominate, and the flux dependence of the ionization processes in nuclear wastes can be substantial.

EXPERIMENTAL

High energy β irradiation of the ICPP waste forms were simulated with the electrons produced by conventional (TEM) and high voltage electron microscopy (HVEM). TEM (100-120 keV) can

be used to simulate the ionization damage process while electrons from the HVEM (0.5 → 1.5 MeV) have enough potential to do both knock-on and ionization damage. Bulk samples of the waste form were also irradiated in a cobalt 60 gamma pond at a dose rate of 2.5 x 10^6 rad/hour and then made into TEM foils for comparison. The specimens were irradiated to fluences of 2 x 10^8, 4 x 10^8, 5 x 10^8, and 1.3 x 10^9 rads which approximates about 10 years storage of defense waste based on an anticipated dose rate of approximately 3 x 10^4 rad/hour.

The α and α-recoil damage was simulated with Pb ion implantations into the thin TEM foils before ion milling but after dimpling. The irradiation doses ranged from 10^{11} to 10^{14} Pb ion/cm^2 which is equivalent to a deposited density of about 10^{17} to 10^{20} α/cm^3 [4]. The specimens were glued to a metal holder and placed in front of the ion source and irradiated at about 200°C. With 250 kV Pb ions there is only a short penetration distance of about 600Å in this material. The doses performed were determined from previous studies on zircon particles, and zircon was expected to turn amorphous at the highest dose of 10^{14} Pb/cm^2 [4]. After irradiation, the specimens were cleaned in acetone and ion milled to perforation with the unirradiated side facing the milling gun. Both the gamma and Pb ion damage samples were then carbon coated and observed in the TEM for any microstructural changes.

RESULTS AND DISCUSSION

Two major types of amorphous silicate phases have been observed in the waste forms, and one appears as single phase while the other shows definite phase separation. X-ray microanalysis of these glassy regions indicates that volume defects (bubbles) form much easier when the silicon content is reduced, but as the silicon content increases, the glass becomes more refractory and stable to irradiation. Some of the glass phases in the lithia based waste form can be easily decomposed to form large oxygen bubbles which can grow to more than two microns in diameter in less than 30 minutes of electron irradiation. These radiation induced bubbles also consistently grow much faster in the glass phases adjacent to the crystalline phases of zircon presumably because these regions are depleted in silica due to the formation of zircon through the following reaction,

$$ZrO_2 + SiO_2 \rightarrow ZrSiO_4$$

The radiation damage to the glass phases brought about by the lead ion irradiation after an equivalent dose of 1 x 10^{20} α/cm^3 is shown in Figure 1. The oxygen bubble formation is evenly distributed in the microstructure and so different from that obtained from electron irradiation but similar to those induced by γ irradiation. This irradiation dose is equivalent to an ionization dose of about 4 x 10^{10} rads and all of the irradiated samples show bubble formation in the glass phases, although the phase separation that was commonly observed in the γ experiments is generally absent.

Previous electron radiation experiments have already shown that bubble growth in the nuclear waste glass occurs in several stages [6,7]. The first stage is the nucleation of fine bubbles which becomes observable at about 100,000x magnification. The bubbles will then grow under the electron beam and finally regress after prolonged electron irradiation. The first set of graphs (figure 2) illustrate the response of the phase separated glass after electron ionization damage. The number of bubbles, the average size and the total volume are plotted against the ionization doses in rads, and these results represent the typical behavior observed.

These electron irradiation experiments suggest that the total number of bubbles actually decreases with respect to the radiation doses in the phase separated glass. It appears that the bubbles actually grow together under the electron beam which must involve oxygen diffusion which is not normally observed in nuclear waste glasses. The average size of the bubbles increases at the expense of the number of smaller bubbles, and this radiation induced diffusion continues until all of the small bubbles have disappeared. After an electron dose of about 10^{24} rads, the total number of observable bubbles decreases to about 12 bubbles per 100 square microns. At this stage, the size of the bubbles also reaches a maximum diameter of about 200A. Further irradiation results in the regression of the bubbles since the oxygen generation source has depleted, and the irradiation induced diffusion continues to drive the oxygen molecules to diffuse and desorb from the free surfaces.

Figure 3 compares the electron irradiation damage data of the glass phase from the yttria-silica additive ICPP waste forms to the Savannah River Laboratory (SRL-165) glass. The extrapolated saturation dose at 3 x 10^{23} rads for bubble formation in the SRL-165 glass is about the same. The similar threshold dose at about 10^{21} rads may be misleading since the average dose rate for TEM imaging is about 10^{21} rad/s, and so the threshold dose may be substantially less than 10^{21} rads for electron irradiation. Despite the similarities, the bubble volume fraction for the SRL glass is about 3 times higher than that for the yttria based ICPP waste form. When the average size of the bubbles from both wastes is about the same, the SRL glass has 3 times as many bubbles as the simulated ICPP waste and presumably reflects the higher soda concentration in the SRL-165 glass.

Micrographs from the γ irradiated glass phases suggest that bubble growth can occur at doses below 2 x 10^8 rads. The bubbles appear to grow to about 200A, and further γ irradiation does not have any effect on the maximum bubble size although a greater number of smaller bubbles form.

Figure 4 is a plot of the total volume of bubbles as it varies with the γ irradiation dose to the maximum fluence of 1.3 x 109 rads and the results for the SRL-165 waste glass are plotted for comparison. The threshold dose for bubble formation in the glass phase both SRL-165 and simulated ICPP occurs at a similar value between 10^8 to 10^9 rads although the fluence for the SRL-165 glass is slight lower. The bubble saturation dose has not been measured for the glass phase of the simulated ICPP waste form, however, if the saturation occurs at a bubble volume fraction of 0.1, similar to the value obtained from the SRL-165 glass, the extrapolated saturation dose will occur at about 1.6 x 10^9 rads.

Fast lead ions are known to induce displacement damage in crystalline zircon although the displacement thresholds have not been precisely determined. Defect clusters with an average radius of about 30A have been also reported as precursors to the metamictization (complete loss of crystallinity) of this material which occurs at about 3.5 x 10^{13} Pb ions/cm^3. We have also observed defect clusters in the monoclinic zirconia particles, and in addition, the twin bands that are associated with the tetragonal to monoclinic transformation disappear from the microstructures. We have neither been able to induce the metamict transformation to the zircon phase nor the disappearance of the planar defects in the zirconia using the high voltage electron microscope which is also capable of inducing displacement damage but not the accompanying implantation.

CONCLUSION

The simulated Idaho Chemical Processing Plant (ICPP) nuclear wastes are composed of both crystalline and amorphous ceramic phases. The crystalline phases of the simulated ICPP waste forms are very resistant to both displacement and ionization damage. At doses approaching zircon metamictization, the zirconia and zircon in the simulated wastes retain their crystallinity although displacement defects can be observed in them. The CaF_2 remains crystalline after irradiation by an equivalent of 10^{20} α/cm^3 lead ions. The amorphous phases do suffer oxygen bubble formation from ionization damage. Some glassy regions also undergo phase separation from γ-irradiation, and such separation is constant with the albite-mullite phase separation found in the soda-alumina-silica system.

REFERENCES

1. H. W. Chan, D. G. Howitt, and A. B. Harker, "Crystalline Phases of the Simulated ICPP Nuclear Waste", Proceedings of the 22nd Annual Microbeam Analysis (MAS) Meeting, July 13-17, 1987.
2. K. Rossler, "Chemical Modification by Implantation", Rad. Eff., **99**, 21-70 (1986).
3. H. Holland, "Radiation Damage and Its USe in Age Determination", Nuclear Technology, p. 178, Faul, H. Ed., John Wiley and Sons, New York (1954).
4. G. W. Arnold, "Radiation Damage Effects in Nuclear Waste Glasses", Radiation Effects, **77** [1-4] 151-159 (1983).
5. J. F. DeNatale and D. G. Howitt, "Importance of Ionization Damage to Nuclear Waste Storage in Glass", Am. Ceram. Soc. Bull., **66** [9] 1393-1396 (1987).
6. J. F. DeNatale, "Radiation Damage in Silicate Glass", Doctor of Philosophy Thesis, University of California, Davis (1985).
7. J. F. DeNatale and D. G. Howitt, "A Mechanism for Radiation Damage in Silicate Glasses", Nucl. Instrum. Methods, **B1**, 489-497 (1984).

ACKNOWLEDGMENT

This work was supported by the U. S. Department of Energy under Contract No. DE-AC07-761D01570.

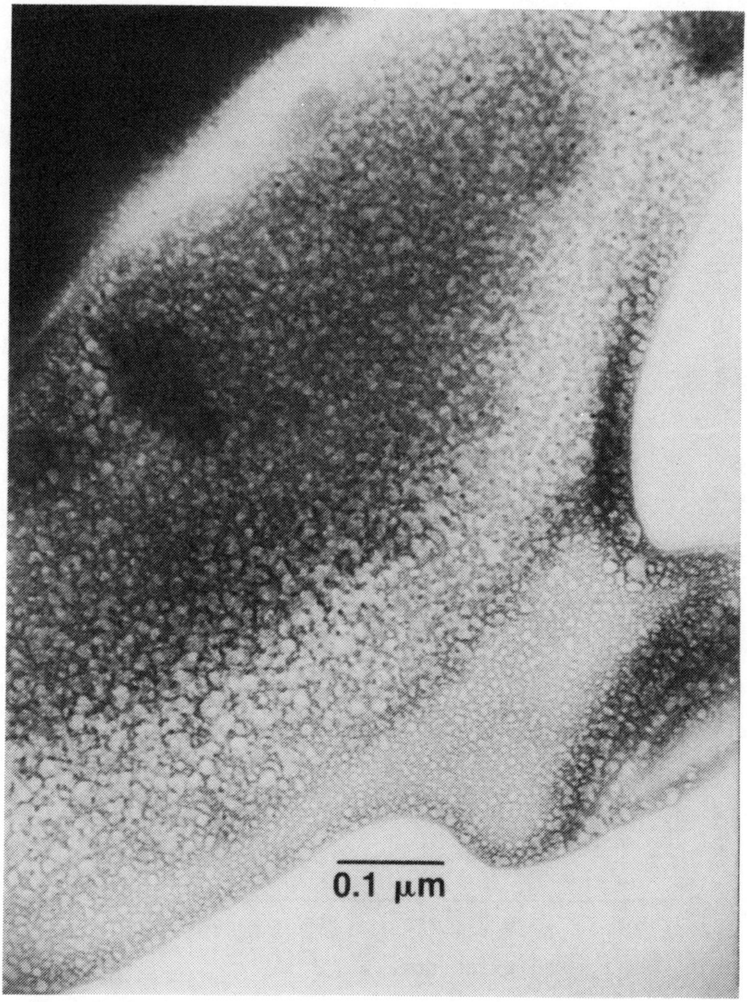

Figure 1: Example of bubble formation in the Pb ion irradiated glass phase

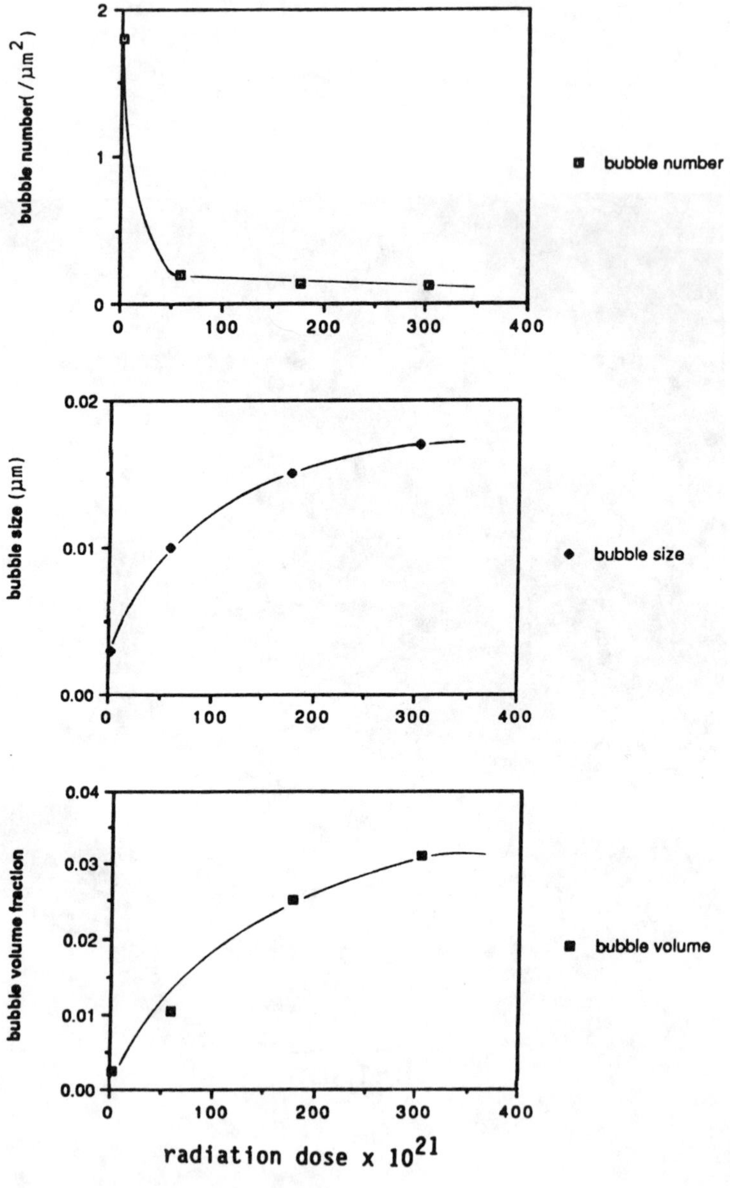

Figure 2: The changes in bubble size, contration and total volume as they vary with γ-fluence

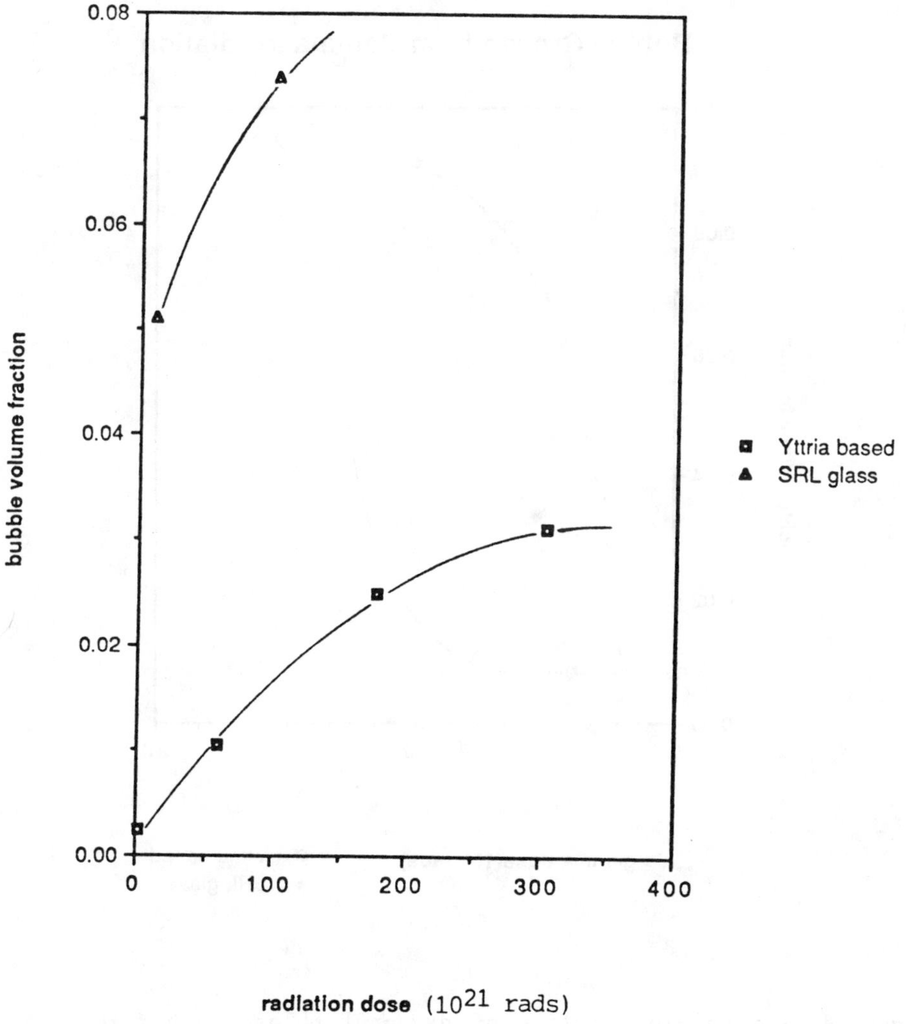

Figure 3: Comparison of the volume of oxygen bubbles as they vary with electron fluence for the Savannah River waste glass and the phase separated glass from the yttria-silica simulated ICPP waste form

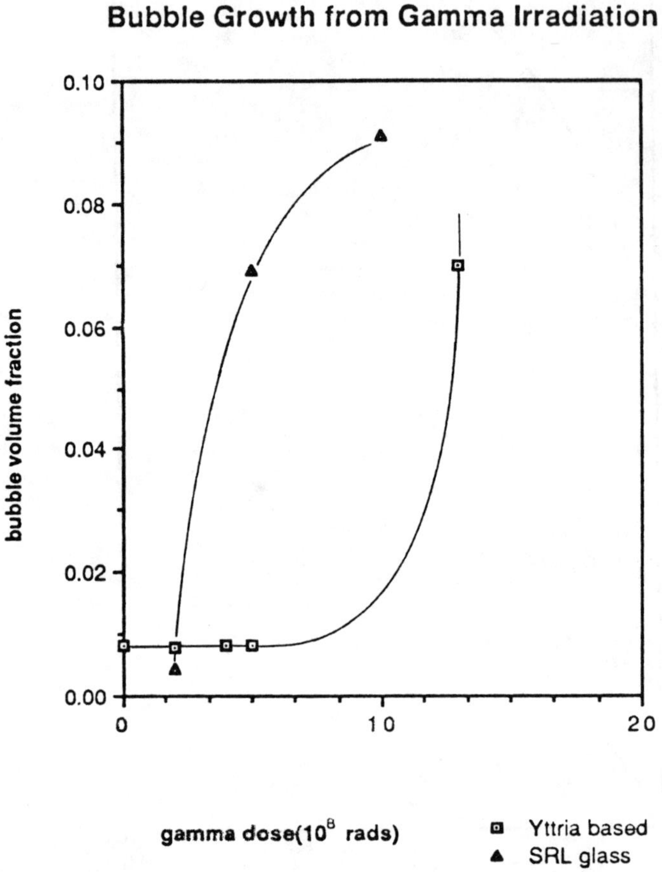

Figure 4: The volume fraction of oxygen bubbles which form from γ-irradiation

RADIATION DAMAGE EFFECTS: COMPARISON OF BOROSILICATE GLASS TO SYNROC PHASES

R. C. Ewing, Dept. of Geology & Ctr. for Micro-Engineered Ceramics, University of New Mexico, Albuquerque, New Mexico 87131, USA;
W. Lutze, Hahn-Meitner Inst., 1000 Berlin 39, Fed. Rep. of Germany

INTRODUCTION

This paper briefly summarizes and contrasts radiation effects that are to be expected in crystalline ceramic (e.g., Synroc) and glass nuclear waste forms. Information from naturally occuring, U and Th bearing minerals are used to extrapolate behavior in crystalline actinide hosts. There is no appropriate natural analogue for radiation effects in nuclear waste glasses, so experimental results are cited. Detailed references and tabulations of data are given in Lutze and Ewing[1].

The chief sources of radiation in high-level nuclear waste forms are the beta decay of fission products (e.g., ^{137}Cs and ^{90}Sr) and the alpha-decay of actinide elements (e.g., U, Np, Pu, Am and Cm). There are minor contributions to the damage process from spontaneous fission events of the actinides, and alpha-neutron reactions are sources of fission fragments and neutrons, but the production rates are low. Beta-decay is the primary source of radiation during the first 500 years of storage, as it originates from the shorter lived fission products (e.g., the half-life of ^{137}Cs is 30.2 years and the half-life of ^{90}Sr is 28.1 years). Beta decay of fission products is responsible for the heat generation and the thermal event early in the history of the waste form. Alpha-decay damage becomes and remains important after approximately 1000 years (e.g., the half life of ^{239}Pu is 2.411×10^4 years).

This work was supported by the Department of Energy, Office of Basic Energy Sciences under grant DE-FG04-84ER45099.

The beta and alpha decay events can cause three types of radiation damage in the waste form: (1) collision cascades which cause atomic displacements and the creation of Frenkel defect pairs due to elastic collisions between nuclear particles (e.g., alpha-particles or the alpha-recoil nuclei); (2) ionization effects associated with the beta particles or the alpha-particle during the highest velocity portion of its trajectory; (3) the transmutation of radioactive parent nuclei into different elements. Of these three effects, the most important are processes which cause atomic displacements, as these are responsible for the atomic-scale rearrangement of the structure and hence lead to the greatest change in physical and chemical properties. Each alpha-decay event results in 1500 to 2000 displacements while a beta-decay event results in only 0.15 to 0.10 atomic displacements. There is no compelling evidence that ionization events result in important structural changes in the material, and thus this discussion will focus on alpha-decay event damage.

Alpha-decay event damage will result in changes in atomic-scale structure which cause changes in the molar volume, corrosion rate, stored energy, mechanical properties and microstructure. The magnitude, and therefore the importance, of these changes will be fundamentally different in glasses as compared to single phase or polyphase crystalline ceramics. Of the ceramic, nuclear waste forms, the data are most complete for borosilicate glass and Synroc; therefore, the focus of this comparison will be on these two waste forms. This discussion is probably equally applicable to a comparison of radiation effects in any glass (e.g., sintered-glass) or crystalline (e.g., monazite) waste form.

ALPHA-DECAY DAMAGE

Alpha-decay damage in nuclear waste forms is the result of the decay of actinides. The damage is caused by two, separate, simultaneous processes associated with the alpha-decay event: (1) An alpha particle (4.5 to 5.0 MeV) with a range of 10,000 nm dissipates most of its energy by ionization; however, at low velocities near the end of its track, it displaces several hundred atoms creating Frenkel defect pairs. (2) The alpha-recoil atom (0.07 to 0.09 MeV) with a range of 10 nm produces several thousand atomic displacements creating "tracks" of disordered material. These two damaged areas are separated by thousands of unit cell distances (e.g., referring to the typical one nanometer unit cell periodicity in phases such as pyrochlore) and will have different effects on the crystalline structure. This has been clearly shown by Weber[2] in a study of unit cell parameter changes in single crystals of $U0_2$ as a function of increasing alpha-particle and alpha-recoil dose. There is a factor of two difference in the change in unit cell parameter that results from alpha-particle irradiation (0.8 %) as compared to alpha-recoil damage (0.4 %) at a saturation dose of one displacement per atom.

Just as important as the damage ingrowth process is the annealing process and the accumulation of stored energy. The annealing kinetics will vary for each of the two types of damage (e.g., the alpha-particle damage vs. alpha-recoil damage). The accumulation of alpha-particle or alpha-recoil damage is dependent on the ability of the material to "repair" the damage as it occurs to the original aperiodic or crystalline state. The energy barrier to the movement of interstitial defects or heterogeneous nucleation at the margins of a recoil track are related to the complexity of the structure and the energy of the chemical bonds that have been broken and reformed. For crystalline materials, structural dependence is apparent when one considers three structure types of increasing structural complexity -- a closest-packed metal alloy, a fluorite structure oxide (e.g., $Pu0_2$ or $U0_2$) and a pyrochlore oxide -- all derivatives of a cubic closest packed structure. The disordered alloy has the lowest energy barrier to recrystallization because of the nondirectional bonding and the equivalence of crystallographic sites. The fluorite structure oxide has the next lowest energy barrier to recrystallization because of the single cation site, the large coordination number of the cation and low bond energy. For pyrochlore structure types, with the highest energy barrier, recrystallization is hampered by directional bonding, multiple cation sites and the larger unit cell. The energy barrier to annealing in aperiodic materials is minimal, as there are no structural constraints on bonding direction or atomic positions. Thus, the annealing process is important because it allows one to anticipate the amount of stored energy, the final state (periodic or aperiodic) of the damaged material, and the structure types that may form over the life-time of the waste form.

COMPARISON OF ALPHA-DECAY DAMAGE IN GLASS AND CRYSTALLINE PHASES

There has been a limited effort to compare alpha-decay damage effects in glass and crystalline waste forms[3]. The summary comparisons predate much of the recent work on alpha-decay damage effects in crystalline phases as observed in natural analogues to waste form phases which occur in polyphase ceramic waste forms (e. g., pyrochlore in Synroc)[4-5]. A detailed review of alpha-decay damage effects in borosilicate glasses and Synroc is given in Lutze and Ewing[1] and data are summarized in Table 1. Both waste forms have been studied extensively; yet, fundamental differences in the atomic structure of these materials (a homogeneous, aperiodic phase vs. a polyphase aggregate) make a direct comparison difficult. As an example, it is more difficult to describe the effect of collision cascades in an aperiodic material than in a crystalline phase because techniques such as x-ray diffraction and analytical electron microscopy provide only limited information on the atomic structure of aperiodic materials. A comparison of simple physical properties, such as volume expansion or fracture toughness, can only be generalizations because measurements include single-phases doped with actinides and polyphase, crystalline aggregates in which void spaces and complex grain boundaries are present. In discussing Synroc properties, we mainly rely on data pertinent to zirconolite and pyrochlore, as these are the principal actinide-bearing phases. Another important difference between the waste forms is the absence of appropriate natural analogues for glasses and the relative abundance of important natural analogues for actinide-bearing, crystalline phases (e.g., zirconolite and pyrochlore).

Physical Properties

The principal difference between the two waste forms is the density change and hence the volume expansion. Borosilicate glasses show only a limited density change ($\pm$ 1.5 %) reaching saturation doses at 10^{24} alpha-decay events/m^3. In contrast, crystalline ceramic waste forms (single phase and polyphase aggregates) expand by as much as 8% reaching saturation doses at 10^{25} - 10^{26} α-decay events/m^3. As an example, ^{238}Pu-doped, polycrystalline aggregates of zirconolite swell by 5% (dilatometric measurement on bulk sample) at a dose of

Table 1. Summary of alpha-decay damage effects.

	Borosilicate glass	Synroc
Methods of Study		
Actinide-doping	yes	yes
Fast neutron irradiation	yes	yes
Neutron induced reactions	yes	no
Fission fragment damage	yes	limited
α-particle irradiation	yes	yes
β-particle irradiation	yes	no
ion-implanation	yes	yes
transmutation effects	yes	no
natural analogues	no	yes
Physical properties		
% change in volume	$\pm$1.5	up to 8
% change in hardness	-15	-30
% change in fracture toughness	+25-80	+30
stored energy	150 J/g	50-200 J/g
He-bubble formation	yes	yes
Annealing kinetics studies	yes	yes
Chemical Durability		
% change corrosion rate	max. 300	1000
selective leaching of alpha-recoil nuclei	?	yes
Saturation dose	$10^{25}\alpha/m^3$	10^{25}- $10^{26}\alpha/m^3$

2 X 10^{25} alpha-decay events/m^3 at ambient temperatures[6]. At higher temperatures (575 K), maximum swelling is 4 % at a saturation dose of 4 X 10^{25} alpha-decay events/m^3. This illustrates the importance of annealing during the damage ingrowth process. At 875 K, swelling is less than 0.5 %. Even at the highest level of volume expansion (5-6%), microcracking was not observed. In a ^{244}Cm-doped, pyrochlore structure type, $Gd_2Ti_2O_7$, unit cell volume expansion was determined by x-ray diffraction for doses of up to 10^{25} alpha-decay events/m^3. The volume expansion is very

anisotropic. The expansion of the monoclinic cell along the c-axis was over five times that along the a-axis. However, even with this large, anisotropic expansion, no microcracking was observed.

The fracture toughness of both glasses and crystalline phases increases with alpha-decay dose[7]. In the case of borosilicate glasses, the increase in fracture toughness is approximately 25%. There are no data available on the change in hardness of a glass with increasing alpha-decay dose, but little change would be expected. For ^{238}Pu-doped zirconolite, there is a signicant change in hardness (30% decrease) and an associated increase in fracture toughness[6]. For a ^{244}Cm-doped apatite structure, $Ca_2Nd_8(Si0_4)_60_2$, a 30% increase in fracture toughness was observed[8]. This is consistent with similar measurements on metamict and annealed euxenite, a mineral which is commonly rendered amorophous by alpha-decay damage and in which a 30% increase in Vicker's hardness was measured on annealing. The increased fracture toughness with increasing alpha-decay dose is associated with crack bowing, crack deflection, internal stresses associated with the coherency of second-phase particles, and differences in elastic properties of phases at different stages of the damage process. The most dramatic example of this process is shown in a complexly zoned zircon, $ZrSi0_4$, which has variable concentrations of U and Th such that the full range of damage microstructures is represented for the crystalline to aperiodic transition[9-10]. Those zones which have experienced the highest alpha-decay dose and are therefore aperiodic (metamict) have no cracks, while the low dose zones contain fractures. This is the type of fracture pattern that might develop in crystalline materials that have complexly zoned actinide concentrations.

Stored Energy

Stored energy measurements are perhaps even more difficult to compare than volume or hardness changes. Most measured values for stored energy in borosilicate glasses are less than 150 J/g[11]. Stored energy values for crystalline ceramics are difficult to assess because of the variety of phases. Most actinide-bearing phases are derivatives of pyrochlore structure types, and Lumpkin et al.[12] have summarized recrystallization data for natural pyrochlores and compared them to data available for ^{238}Pu- and ^{244}Cm-doped pyrochlores and zirconolites (note, zirconolite is a monoclinic derivative of the cubic pyrochlore structure type). The natural pyrochlores (completely metamict prior to annealing) have recrystallization energies in the range of 125 to 210 J/g. In contrast, Pu-doped zirconolite has a value of 50 J/g at 15 X 10^{26} alpha-decay events/m^3; Cm-doped zirconolite has a value of 127 J/g at 2 X 10^{25} alpha-decay events/m^3. Activation energies of recrystallization for natural pyrochlores are between 0.29 and 0.97 eV. These activation energies are intermediate between those measured for a ^{244}Cm-doped glass (0.25 eV) and those for Pu-doped zirconolite (1.22 eV) and Cm-doped zirconolite (5.8 eV). The most interesting feature of these studies is that synthetic samples doped with highly active actinides generally have a lower value of "stored" energy than the natural samples. In considering this difference, one must clearly distinguish between the heat of crystallization (released on annealing of completely disordered materials) and stored energy that is directly related to alpha-particle and alpha-recoil damage. In the latter case, stored energy will be related to the removal of isolated Frenkel defect pairs (created by the alpha-particle) and "track" fading (annealing of the alpha-recoil collision cascade). These results suggest that the stored energy will never be greater than the heat of crystallization which covers only a slightly larger range (50 to 200 J/g) than the stored energy values of borosilicate glasses (less than 150 J/g in most cases).

Chemical Durability

There is no straightfoward comparison that can be made between the corrosion of a single, aperiodic phase and a crystalline, polyphase material. The corrosion process extrapolated over time is one in which critical parameters of the waste form environment change, such as the temperature, flow rate and/or solution composition. The only issue that we can address in this section is whether under some set of specified conditions alpha-decay damage causes fundamental structural modifications which lead to an increased corrosion rate. Although alpha-decay damage has been simulated by a number of techniques (e.g., ion implantation), we restrict ourselves almost entirely to the data derived from actinide-doping experiments and actinide-bearing (U and Th) natural analogue structure types.

For borosilicate glasses, corrosion rate changes in actinide-doped glasses have been compiled by Weber and Roberts[5]. For most of the experiments, alpha-decay doses reached values of 10^{24} to 10^{25} α-events/m^3. Based on these results, the maximum change in bulk corrosion rate (network dissolution) is no more than a factor of three (based on weight loss measurements). One should note, however, that during the early 1980's considerable controversy was generated by ion implantation results which suggested increases in corrosion rates by as much as a factor of fifty. The ion implantation results could never be corroborated by actinide-doping experiments, and the discrepancy in the results between the two techniques could be due to differences in dose rate, thermal annealing, or compositional changes associated with the implantation process.

In crystalline materials, one may expect a more dramatic change in the corrosion rate because of the transition from a stable, periodic atomic array to a metastable, aperiodic atomic array[13]. Not surprisingly, there is really very little in the way of systematic data relating changes in corrosion rate to alpha-decay damage. Bulk Synroc samples have been irradiated by fast neutrons up to a dose of 2.6×10^{24} neutrons/m^2, and based on Cs release, no systematic effect of radiation dose on leach rate was observed. More relevant experiments provide results on actinide-doped, single phases. Weber and coworkers[8,23] have reported results on nuclear waste phases -- $Ca_2Nd_8(SiO_4)_6O_2$, $Gd_2Ti_2O_7$ and $CaZrTi_2O_7$ -- doped with ^{244}Cm and accumulated doses of 10^{25} α-decay events/m^3. Their results show an order of magnitude increase in corrosion rate (normalized elemental mass loss based on at least four elements measured in solution for each sample). Ringwood et al.[14] have raised questions concerning the interpretation of these results, because they have erroneously assumed that the fully-damaged and the fully-recrystallized samples were the same specimens. In fact, the leached damaged and leached recrystallized samples were different, and the results are for a parallel set of experiments (Weber, personal communication). The results are based on five elements, principal among these is Ca, a major constituent of the crystalline phases. These results are also consistent with similar measurements of natural phases in which annealing experiments are not part of the leaching experiments. Still, the change in radionuclide release rate with increasing α-decay dose is a critical question that requires additional, well-designed experiments for actinide-doped single phases.

Finally, two important comments should be made concerning these data: (1) Even with a substantial increase in corrosion rate of Ti-based ceramics due to α-decay damage, the final corrosion rate may still be substantially less than that observed in other waste forms. (2) The change in corrosion rate observed in these experiments is most likely associated with the high degree of disorder and broken bonds in the amorphous state; however, this does not include the effect of increased surface area due to fracture caused by the volume expansion of irradiated crystalline phases. Depending on the corrosion model used, changes in surface area may or may not be of critical importance.

Of particular relevance to the discussion of radiation damage effects in crystalline phases are the results and observations obtained on natural analogues. Perhaps the most important results are those that have been reported on zircon, $ZrSiO_4$. Although zircon is seldom suggested as a waste form phase, results on ^{238}Pu-doped zircons and natural zircons, which commonly contain U and Th, can be compared[9-10,15-20]. These rather detailed results demonstrate that the changes in the physical and chemical properties which occur in ^{238}Pu-doped zircons over a period of 5 to 6 years (10^{26} alpha-decay events/m^3) are similar to the same changes observed in natural zircons which have accumulated equivalent alpha-decay doses in 10^8 years. For natural zircons of variable alpha-decay dose spanning the range from the crystalline to metamict transition, Ewing et al.[21] have reported a two order of magnitude increase in the corrosion rate. These results are consistent with observations concerning the dissolution kinetics of crystalline and metamict zircon and "disturbed" uranium/lead systematics in metamict zircons used in geologic age dating.

One of the most important observations made with natural samples has been the recognition that based on x-ray diffraction analysis and high resolution transmission electron microscopy, the alpha-decay dose at which the transition from the crystalline to metamict state occurs increases with geologic age. The increase in "calculated" dose is attributed to annealing of isolated alpha-recoil tracks back to the original crystalline structure. The mean alpha-recoil track life for phases that become metamict are: (1) pyrochlore, 100 million years; (2) zircon, 400 m.y.; (3) zirconolite, 700 m.y. [4]. In contrast, phases which remain crystalline, despite high alpha-decay doses (e.g., uraninite and monazite), have mean alpha-recoil track lifes on the order of 10^4 years[22]

CONCLUSIONS

The summarized data suggest that both glass and crystalline waste forms may sustain substantial doses of α-decay damage and still retain their durability. Radiation effects in glasses are less pronounced and less complicated than that in single or poly-phase ceramics; thus, the latter category requires careful research and consideration. Perhaps the most important conclusion is that short-term actinide doping experiments in crystalline phases provide a realistic simulation of long-term effects based on the comparison of observed radiation effects in Pu-doped zircon and naturally damaged zircon (there is a 10^7 difference in dose rate). Deviations from the similarity in effect (e.g., saturation dose) may be attributed to low-temperature, long-term annealing effects.

Future research efforts should focus on the following:

- Annealing mechanisms, kinetics and temperature dependence in periodic solids.
- Volume expansion in crystalline phases that leads to micro-fracturing.
- Selective leaching and release of alpha-recoil nuclei from glasses and crystalline phases.
- Radiolysis effects in the solid, and particularly in the leaching solutions.
- Ionization effects (particularly in glasses) that may lead to the formation of bubbles.

The last two effects are associated with ionization effects expected from β- and γ-radiation.

REFERENCES

[1]W. Lutze and R. C. Ewing, "Summary and Evaluation of Nuclear Waste Forms"; pp. 699-740 in Radioactive Waste Forms for the Future. Edited by W. Lutze and R. C. Ewing, North-Holland, Amsterdam, 1988.

[2]W. J. Weber, "Ingrowth of Lattice Defects in Alpha Irradiated UO_2 Single Crystals," *J. Nuclear Materials,* **98** 206 (1981).

[3]W. J. Weber, L. R. Pederson, W. J. Gray and G. L. McVay, "Radiation Effects on Nuclear Waste Storage Materials," *Nuclear Instruments and Methods in Physics Research,* **B1** 527 (1984).

[4]G. R. Lumpkin and R. C. Ewing, "Alpha-Decay Damage in Minerals of the Pyrochlore Group," *Physics and Chemistry of Minerals,* **16** 2 (1988).

[5]W. J. Weber and F. P. Roberts, "A Review of Radiation Effects in Solid Nuclear Waste Forms," *Nuclear Technology,* **60** 178 (1983).

[6]F. W. Clinard, Jr., "Review of Self-Irradiation Effects in Pu-Substituted Zirconolite," *Ceramic Bulletin,* **65** 1181 (1986).

[7]J. L. Routbort, P. Offermann and Hj. Matzke, "Mechanical Stability of Cm-Doped Celsian Glass-Ceramic."; pp. 415-422 in Scientific Basis for Nuclear Waste Management VI. Edited by D. G. Bookins. North-Holland, New York, 1983.

[8]W. J. Weber and Hj. Matzke, "Radiation Effects in Actinide Host Phases," *Radiation Effects,* **98** 259 (1986).

[9]B. C. Chakoumakos, T. Murakami, G. R. Lumpkin, and R. C. Ewing, "Alpha-Decay-Induced Fracturing in Zircon: The Transition from the Crystalline to the Metamict State," *Science,* **236** 1556 (1987).

[10]B. C. Chakoumakos, J. C. McCallum, L. A. Boatner, W. C. Oliver, G. R. Lumpkin and R. C. Ewing, "Microhardness in Natural Zircon as a Function of α-Decay Dose Compared with Synthetic Zircon Damaged by Pb Ion Implantation," abstract in Radiation Effects in Insulators-5, Proceedings Volume, in press.

[11]W. Lutze, "Silicate Glasses,"; pp. 1-159 in Radioactive Waste Forms for the Future. Edited by W. Lutze and R. C. Ewing, North-Holland, New York, 1988.

[12]G. R. Lumpkin, R. C. Ewing, B. C. Chakoumakos, R. B. Greegor, F. W. Lytle, E. M. Foltyn, F. W. Clinard, Jr., L. A. Boatner and M. M. Abraham, "Alpha-Recoil Damage in Zirconolite ($CaZrTi_2O_7$)," *J. Materials Research,* **1** 564 (1986).

[13]R. C. Ewing, B. C. Chakoumakos, G. R. Lumpkin and T. Murakami,

"The Metamict State," *Materials Research Society Bulletin*, **May/June** 58 (1987).

[14]A.E. Ringwood, S. E. Kesson, K.D. Reeve, D.M. Levins and E. J. Ramm, "Synroc,"; pp. 233-334 in Radioactive Waste Forms for the Future. Edited by W. Lutze and R. C. Ewing, North-Holland, Amsterdam, 1988.

[15]W.J. Weber and G. D. Maupin, "Simulation of Radiation Damage in Zircon," *Nuclear Instruments and Methods in Physics Research*, **B32** 512 (1988).

[16]R. C. Ewing, B. C. Chakoumakos, G. R. Lumpkin, T. Murakami, R. B. Greegor and F. W. Lytle, "Metamict Minerals: Natural Analogues for Radiation Damage Effects in Ceramic Nuclear Waste Forms," *Nuclear Instruments and Methods in Physics Research*, **B32** 487 (1988).

[17]T. Murakami, B. C. Chakoumakos and R. C. Ewing, "X-Ray Powder Diffraction Analysis of Alpha-Event Radiation Damage in Zircon ($ZrSiO_4$)," in Nuclear Waste Management II, Advances in Ceramics, vol. 20. Edited by D. E. Clark, W. B. White and A. J. Machiels, The American Ceramic Society, Westerville, Ohio, pp. 745-753 (1986).

[18]T. J. Headley, R. C. Ewing and R. F. Haaker, "TEM Study of the Metamict State," in Physics of Minerals and Ore Microscopy. Edited by R. I. Kostov and B. K. Kamenov, Bulgarian Academy of Science, Sofia, pp. 281-289 (1986).

[19]G. J. Exarhos, "Induced Swelling in Radiation Damaged $ZrSiO_4$," *Nuclear Instruments and Methods in Physics Research*, **B1** 538 (1984).

[20]H.D. Holland and D. Gottfried, "The Effect of Nuclear Radiation on the Structure of Zircon," *Acta Crystallographica*, **8** 291 (1955).

[21]R. C. Ewing, R. F. Haaker and W. Lutze, "Leachability of Zircon as a Function of Alpha Dose," in Scientific Basis for Nuclear Waste Management V. Edited by W. Lutze, North-Holland, New York, pp. 389-397 (1982).

[22]Y. Eyal and R. L. Fleischer, "Preferential Leaching and the Age of Radiation Damage from Alpha Decay in Minerals," *Geochimica et Cosmochimica Acta*, **49** 1155 (1985).

[23]J.W. Wald and W. J. Weber, "Effects of Self-Radiation Damage on the Leachability of Actinide-Host Phases," Advances in Ceramics, vol. 8, Edited by G.G. Wicks and W. A. Ross, American Ceramic Society, Columbus, Ohio, pp. 71-75 (1984).

PRODUCTION OF SYNROC POWDERS BY ALKOXIDE HYDROLYSIS

John R. Bartlett *, James L. Woolfrey and William J. Buykx

Advanced Materials Program, Australian Nuclear Science and Technology Organisation, Private Mail Bag 1, Menai NSW 2234, Australia.

ABSTRACT

Advanced chemical processing has been applied to the large scale production of the multicomponent ceramic, Synroc B. Approximately 2000 kg of material has been prepared at the Australian Nuclear Science and Technology Organisation (ANSTO). The technique, which involves the hydrolysis of a mixture of Ti, Zr, and Al alkoxides, and concurrent sorption of Ba^{2+} and Ca^{2+} cations under basic conditions, is discussed. The effects of processing conditions on the physical properties of the powders are also discussed.

1. INTRODUCTION

Synroc is a complex, multiphase, titanate ceramic, designed to immobilise High Level Waste (HLW) from nuclear reactors [1]. The Synroc concept involves immobilising the elements present in HLW within a series of thermodynamically stable mineral phases (hollandite, zirconolite, and perovskite). These minerals are formed during ceramic processing with a precursor powder containing TiO_2, ZrO_2, Al_2O_3, BaO, and CaO in relative mass abundances of (70.4 to 76.4), (5.5 to 8.5), (3.5 to 5.5), (4.6 to 6.6) and (10.1 to 12.1), respectively.

Synroc precursor powders are routinely produced by advanced chemical methods involving alkoxide hydrolysis and sorption of Ca^{2+} and Ba^{2+} cations [2], since the conventional oxide route does not yield acceptable homogeneity or sufficient sorption capacity for HLW components. In addition, unacceptably high temperatures are required to hot press powders produced by the oxide route (> 1250 °C). This paper discusses selected aspects of the process chemistry, and compares the physical characteristics of precursor powders produced.

2. EXPERIMENTAL

2.1 Powder Preparation

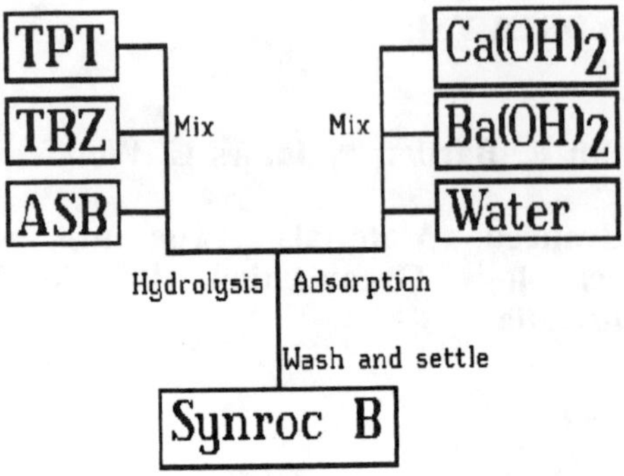

Fig. 1. Hydroxide route used to prepare Synroc B.

Titanium tetra(2-propoxide) (TPT), zirconium tetra(1-butoxide) (TBZ), and aluminium tri(2-butoxide) (ASB), in the required proportions, were dissolved in an equal volume of anhydrous ethanol. The alkoxide mixture was held under a dry nitrogen atmosphere prior to reaction, to prevent premature hydrolysis. The alkoxide solution was then hydrolysed at "ambient temperatures" by a measured amount of water (typically 24 moles H_2O per mole of alkoxide) containing dissolved barium hydroxide, and finely dispersed calcium hydroxide, Fig. 1. A high-shear, dispersion mixer was used to ensure that intimate mixing occurred during hydrolysis. Component additions were made over a 60 minute interval, and the resulting slurry was stirred for an additional 30 minutes to promote ageing of the hydrolysate.

Two processes were examined :

(a) Water-into-alkoxide (W->A) . The alkoxide solution was vigorously agitated, and the aqueous phase injected into the mixing head, either inside or outside the shear zone, where intimate mixing and hydrolysis occurred.

(b) Alkoxide-into-water (A->W) . In this reverse process, the aqueous phase was agitated, and the alkoxide phase was introduced either inside or outside the shear zone.

The precursor was processed at ANSTO in batch sizes up to 100 kg, on an oxide basis [2]. Slurry samples were taken at predetermined intervals during the processing, and were centrifuged at 3500 RPM to separate the solid and liquid phases for further analysis.

2.2 Powder Characterisation.

The specific surface areas, pore surface areas, pore volumes (Micromeritics ASAP 2400), scanning electron micrographs (Jeol), and x-ray diffraction (XRD) patterns (Siemens) of selected dried powders were obtained. Fourier transform infrared (FTIR) spectra (Biorad Digilab FTS-60) of dried powders were acquired using both diffuse reflectance and transmission techniques. Samples examined in diffuse reflectance experiments were diluted in diamond dust (2 to 3 μm particle size) prior to acquisition of spectra, while samples scanned in the transmission mode were diluted in either KBr or CsI. Spectra were scanned using a bandpass of 2 cm^{-1}, and the interferometer was purged with N_2 gas. Conductivities of supernatant liquids were measured at a fixed temperature of 302 K. A nominal cell constant was used to calculate relative conductivities.

3. RESULTS AND DISCUSSION

The physical and chemical properties of Synroc B precursor powders prepared by W->A and A->W processing (*Synroc-WA* and *Synroc-AW* respectively) are summarised in Table 1. Synroc process chemistry has also been examined in earlier studies [2,3]. The key results obtained during these studies are also reviewed in Table 1, and are briefly discussed below.

3.1 Stoichiometry of hydrolysis reactions and polymerisation of Ti/Zr/Al solid matrix.

The hydrolysis of Ti(IV), Zr(IV), and Al(III) alkoxides can be represented by the equations :-

$$Ti\text{-}(OR)_4 + xH_2O \rightleftharpoons Ti(OH)_{2x-4}O_{4-x} + 4R\text{-}OH \quad (1)$$
$$Zr\text{-}(OR)_4 + yH_2O \rightleftharpoons Zr(OH)_{2y-4}O_{4-y} + 4R\text{-}OH \quad (2)$$
$$Al\text{-}(OR)_3 + zH_2O \rightleftharpoons Al(OH)_{2z-3}O_{3-z} + 3R\text{-}OH \quad (3)$$

where x, y, and z denote the minimum quantities of water needed to completely hydrolyse the alkoxides, and $2 \leq x \leq 4$, $2 \leq y \leq 4$, while $1.5 \leq z \leq 3$. It is logical to assume that the hydroxyl groups reside within the surface layer of the hydrolysate, while the oxide species are located both within the surface and within the bulk. The value of x (equation 1) is particularly important in the present study, since titanium is the most abundant element present in the Synroc B hydrolysate (Ti:Al:Zr mole ratio = 1.00:0.12:0.062). It can be shown (see APPENDIX) that

Table 1. Physical properties of Synroc B powders prepared by water-into-alkoxide and alkoxide-into-water processing.

Property	*Synroc-WA*	*Synroc-AW*
Median Particle Size (μm)	5-15	10
Powder Density (g cm^{-3})	3.3±0.2	2.91
IEP (pH)	9.1±0.1	8.15±0.05
Tap Density (g cm^{-3})	1.4±0.1	0.8±0.1
Pore Volume (cm^3 g^{-1})	0.2-0.3	0.63-0.64
Surface Area (m^2 g^{-1})	400-450	430-450
Particle Morphology (SEM)	Raspberry structure	Acicular, spherical
Comments	.No evidence of component segregation. .IEP of slurries increased with processing time. .Slurries settled readily. .Powder surface area decreased following washing and ageing.	.Evidence of component segregation during initial 30 minutes of processing. .IEP of slurries decreased with processing time. .Slurries did not settle readily. .Powder surface area decreased following washing and ageing.

$$x = 0.5F+2 \tag{4}$$

if each surface Ti(IV) site is associated with a single surface hydroxyl group (assuming pH << point of zero charge (PZC)), and

$$x = F + 2 \tag{5}$$

if each surface Ti(IV) site is associated with two hydroxyl species. Equations (4) and (5) are illustrated in Fig. 2 for $F \leq 0.5$.

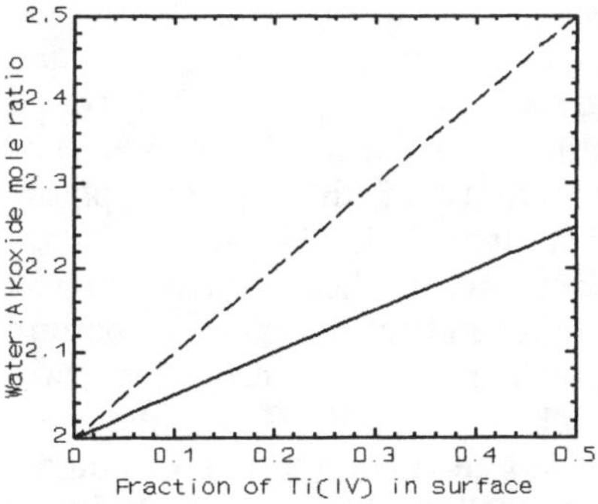

Fig. 2. Relationship between x and the fraction, F, of metal atoms residing within the surface layer of Ti/Zr/Al hydrolysate. The unbroken line refers to equation 4, while the broken line represents equation (5).

3.2 Synroc Process Chemistry

In the W->A method, the aqueous phase was introduced into the alkoxide at a constant rate, over a period of approximately 60 minutes, Fig. 3. A stoichiometric quantity of water was added every 5 to 7 minutes (2-2.5 moles H_2O per mole alkoxide, Section 3.1), resulting in an ultimate ten fold excess of water. Of course, complete hydrolysis may occur at a slower rate, as a consequence of reaction kinetics. This effect was clearly evident in the present study, since FTIR spectroscopy revealed that a small quantity of alkoxide molecules were still sorbed on the hydrolysate after 10 minutes of processing [3], confirming that hydrolytic desorption of the final residual alkoxide groups occurs at a significantly slower rate than the initial formation of the bulk solid phase. FTIR data showed that hydrolysis was complete within the first 20 minutes.

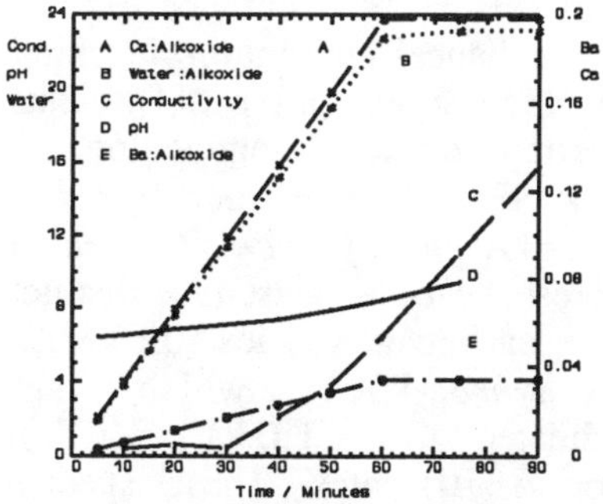

Figure 3. Processing chemistry for Water-into-alkoxide route. (A, B, and E are mole ratios; relative conductivity in μS/cm).

The conductivity of the reaction medium was extremely low during the initial 40 minutes, indicating that all added soluble barium and calcium species were sorbed on the surface of the hydrolysate. The minor increase in conductivity (about 0.04 % of the initial value of the aqueous phase), observed after 40 minutes, may reflect leaching of small quantities of surface species during the final stages of processing.

In the A->W method, the alkoxide was added to the aqueous phase at a uniform rate over a 60 minute interval, Fig. 4, and hydrolysate was formed continuously during this period. The conductivity of the aqueous phase, which was initially about 25000 μS/cm, approached zero after 30 minutes, indicating that all barium and calcium cations had been sorbed when only 50 % of the total hydrolysate had been precipitated. This implies that any hydrolysate subsequently formed will not have sorbed surface barium and calcium species, unless redistribution occurs. Clearly, this could promote segregation of the alkaline earths in the final powder. In addition, the initially high pH of the reacting system (pH = 13) would be expected to enhance aluminium leaching from the solid phase [4], thus promoting additional segregation.

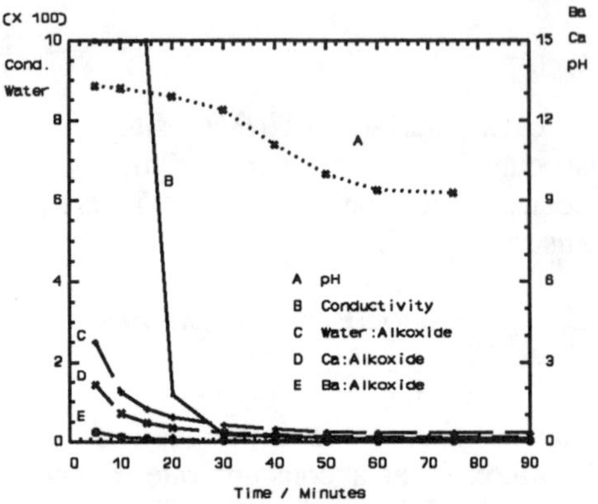

Figure 4. Processing chemistry for Alkoxide-into-water route. (C, D, E are mole ratios; relative conductivity in μS/cm).

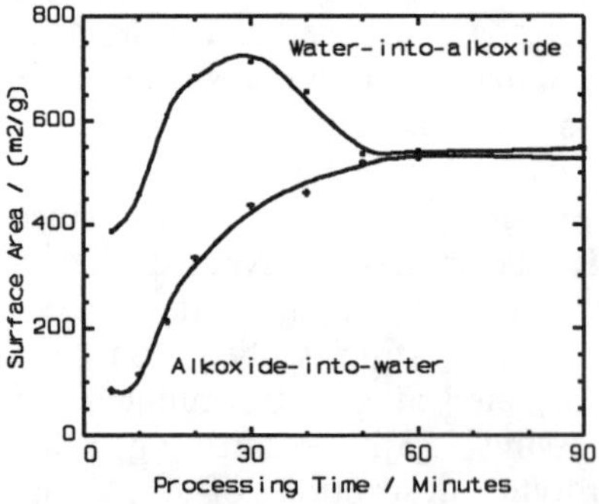

Fig. 5. Variations in total surface area as a function of processing time

FTIR spectroscopy of *Synroc-WA* and *Synroc-AW* revealed the presence of at least two distinct surface carbonate species (0.65 wt% for *Synroc-WA*), which are coordinated to Ti(IV), Zr(IV), and/or Al(III) sites. These species are predominantly formed following uptake of atmospheric CO_2 during processing in basic solutions, although one Ti(IV):$[CO_3]^{2-}$ surface complex appeared to be formed as a result of CO_2 chemisorption on dried powders [5].

3.3 Surface area, pore volume and pore morphology of dried *Synroc-WA* and *Synroc-AW* precursors.

Synroc-WA: The surface area and pore volume vary substantially during processing, Figs. 5 and 6. Both parameters attain a maximum after 30 minutes, and then steadily decline with further processing. The initial increase in both surface area and pore volume is also reflected in the contribution of the open surface pores to the total surface area, since the ratio of pore surface area to total surface area increases from an initial value of 0.28 to a value of 0.98 after 30 minutes of processing, Fig. 7. The apparent agglomerate density also increased significantly during this period. These effects are attributed to an increase in open surface porosity, suggesting that closed pores located near the surface are converted to open pores during the initial processing.

The subsequent reduction in total surface area, pore volume, and (pore surface area)-to-(total surface area) after 30 minutes of processing, Figs. 5,6 and 7, is consistent with a decrease in the volume of the open surface pores. This effect would appear to stem from preferential adsorption of Ca^{2+} and/or Ba^{2+} and $[CO_3]^{2-}$ within the surface pores, rather than pore closure [4].

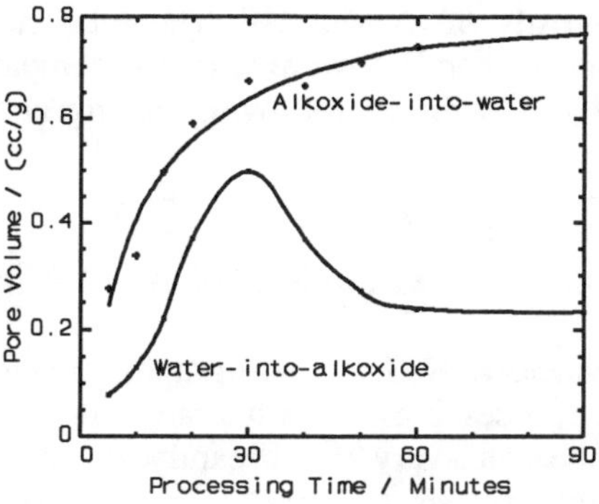

Fig. 6. Variations in pore volume as a function of processing time

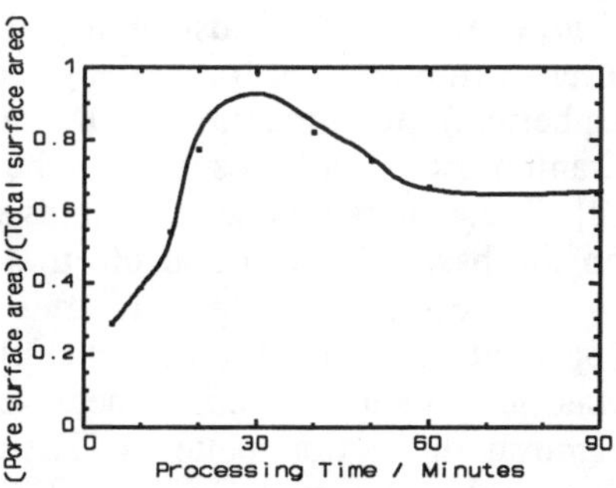

Fig. 7. Variations in (pore surface area):(total surface area) as a function of processing time for Synroc-WA.

Synroc-AW. The variations in surface area and pore volume of *Synroc-AW* during processing were significantly different from those of *Synroc-WA*. Both parameters increased continuously during processing (Figs. 5 and 6), and the pore volume was substantially higher than that of *Synroc-WA* at all times. In contrast to *Synroc-WA*, the pore surface areas of samples taken at various intervals were comparable to the total surface areas, reflecting the greater porosity of *Synroc-AW*. The total surface areas of fully processed *Synroc-AW* and *Synroc-WA* were initially comparable (Fig.5), and subsequently decreased to $\leq$ 450 m^2 g^{-1} following ageing and washing.

3.4 SEM analysis of powder morphology.

Synroc-WA. The micrographs of samples taken at the beginning and ending of processing are illustrated in Fig. 8. The particle morphology did not appear to vary significantly during processing, and the "raspberry-structure" agglomerates (10 μm) were composed of spherical particles with an average diameter of approximately 1 μm. No component segregation was observed by energy dispersive x-ray (EDX) analysis, although XRD revealed the presence of minor quantities of unreacted $Ca(OH)_2$ in samples obtained after 5 minutes of processing.

Synroc-AW. In contrast to *Synroc-WA*, the particle morphology of *Synroc-AW* varied substantially during processing, Fig. 9. Samples obtained after 5 minutes of processing contained both dense (acicular) and porous (spherical) agglomerates. EDX analysis revealed the porous particles to be titanium-rich, with respect to the bulk, indicative of component segregation [2]. These particles were tentatively identified as crystalline calcium titanate, on the basis of XRD and electron diffraction analysis.

Samples obtained after 30, 75 and 90 minutes of processing appeared to contain predominantly acicular agglomerates, Fig. 9, and there was no evidence of component segregation or crystallisation in their XRD or electron diffraction patterns. The XRD pattern of crystalline calcium titanate was only detected in samples taken during the first 15 minutes of processing.

4. CONCLUSIONS

Selected physical and chemical properties of Synroc precursor powders prepared by two chemical routes, involving alkoxide hydrolysis, are discussed. Both W->A and A->W processes yield powders with high surface areas (400 to 450 m^2/g), which are readily densifiable to >99 % of theoretical density. However, the pore volume of fully processed *Synroc-AW* was found to be significantly higher than that of *Synroc-WA*. Component segregation was evident in the case of A->W processing, but no segregation was detected in the case of W->A processing. Both processes yielded homogeneous powders.

Based on these observations, W->A processing has been adopted as the standard processing method for the large scale production of Synroc B

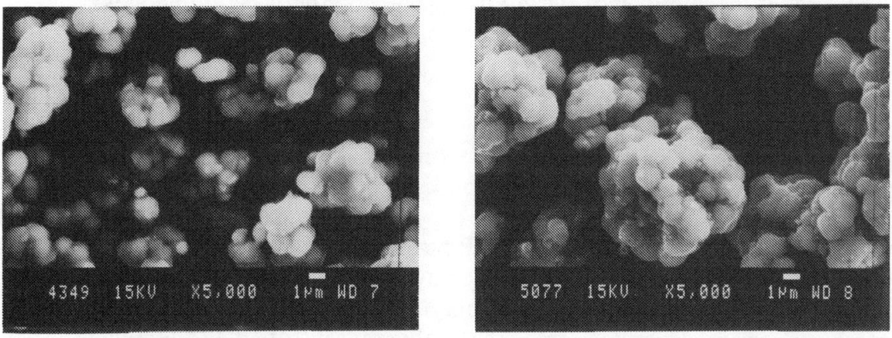

(A) (B)

Fig. 8. Scanning electron micrographs of Synroc B prepared by the W->A route after processing for (A) 5 minutes and (B) 90 minutes

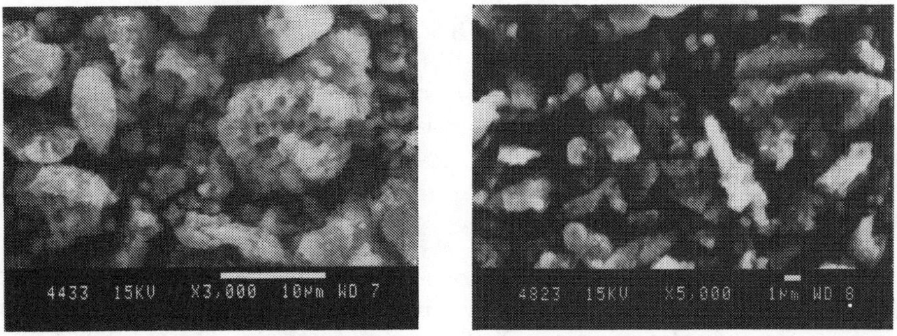

(A) (B)

Fig. 9. Scanning electron micrographs of Synroc B prepared by the A->W route after processing for (A) 5 minutes and (B) 90 minutes.

precursor at ANSTO.

5. APPENDIX

Derivation of relationship between x and the fraction of Ti atoms in the surface layer.

Let s = the number of Ti atoms in the polymeric agglomerates
= the number of alkoxide molecules required to form the agglomerates
F = the fraction of Ti atoms in the surface layer

Then, the number of Ti atoms in the surface layer is given by F.s, while the number of Ti atoms located in the bulk is given by (1-F).s. The s molecules of $Ti(OR)_4$ require sx molecules of water for complete hydrolysis and polymerisation, from equation (1). Protonation of the RO^- alkoxy fragments consumes 4s moles of protons, leaving 2sx-4s moles of protons to form surface hydroxyl species on the polymeric aggregates, and 4s-sx oxygen atoms to be incorporated into the bulk region of the aggregates. If we assume that each surface Ti site is associated with a single hydroxyl, then :

$$F.s = 2sx - 4s$$

Hence $$2sx = F.s + 4s$$
and $$x = 0.5F + 2$$

Similarly, if we assume that each surface Ti site is associated with two hydroxyls, then :

$$F.s = 0.5(2sx - 4s)$$

Hence $$Fs = sx - 2s$$
and $$x = F + 2$$

6. ACKNOWLEDGEMENTS

The authors gratefully acknowledge the assistance of Mr K Watson, Mr P Bendeich, and Mr D Cassidy, in obtaining scanning electron micrographs, nitrogen adsorption and XRD data respectively. The assistance of Mr B Seatonberry and members of the Synroc Feedstock Development group in obtaining the samples characterised in this study are also acknowledged.

7. REFERENCES

[1] A.E. Ringwood, S.E. Kesson, K.D. Reeve, D.M. Levins and E.J. Ramm, "Synroc", ***Radioactive Waste Forms of the Future*** (W. Lutze and R.C. Ewing, Eds.), Elsevier Science Publishers B.V., Chapter 4 (1988).

[2] J.L. Woolfrey, J.R. Bartlett and W.J. Buykx, "Production of Multicomponent Ceramic Powders by Alkoxide Hydrolysis. The Preparation of Synroc", ***Proceedings of the 2nd International Conference on Ceramic Powder Processing Science***, Berchtesgaden, FRG, OCT 12-14 (1988), in Press.

[3] J.R. Bartlett and J.L. Woolfrey, "The Chemistry of Synroc Preparation : A Multicomponent Ceramic", Paper presented at ***The Symposium on Colloidal Processing and Green Forming of Ceramics***, 91st Meeting of the American Ceramics Society, April 1989 (to be published).

[4] J.R. Bartlett and J.L. Woolfrey, Manuscript in preparation.

[5] J.R. Bartlett and J.L. Woolfrey, Manuscript in preparation.

IN-BELLOWS CALCINATION OF ALKOXIDE PRECURSOR TO SYNROC

E.J. RAMM AND E.R. VANCE

Australian Nuclear Science and Technology Organisation,

Lucas Heights Research Laboratories,
MENAI, NSW 2234, AUSTRALIA

ABSTRACT

In-bellows calcination of dried alkoxide Synroc precursor containing simulated high-level waste (HLW) has been investigated for samples of ~ 0.1 m in linear dimensions. Process advantages include elimination of complex calcining equipment, minimal dust carry-over and possible removal of the need to add metallic buffers to the calcine before hot-pressing. Volatilization losses of Ru and Cs were very small. Satisfactory densification and microstructural characteristics, together with satisfactory leach rates, were achieved on subsequent hot uniaxial or isostatic pressing at 1175°C, for material containing 20 wt% simulated HLW.

INTRODUCTION

The nuclear waste form Synroc [1] is a titanate-based, four-phase ceramic formed by hot-pressing intimately mixed Ba-Al-Ca-Ti-Zr oxides, plus waste, at 1100-1200°C. The waste-immobilizing phases and their idealized formulae are barium hollandite ($Ba_{1.14}Al_{2.28}Ti_{5.71}O_{16}$), zirconolite ($CaZrTi_2O_7$) and perovskite ($CaTiO_3$). These phases have chemically durable natural mineral analogues and between them can incorporate in solid solution nearly all the radionuclides from reprocessed nuclear fuel. Titanium oxide is also present.

The flow diagram (Fig. 1) illustrates the current fabrication route [2] in the Synroc Demonstration Plant (SDP) at the Australian Nuclear Science and Technology Organisation (ANSTO). Briefly, Zr, Ti and Al alkoxides are dissolved in ethanol, hydrolysed by the controlled addition in a high-shear mixer of a (Ba + Ca) hydroxide slurry in water, and subjected to three wash/filter cycles to remove the alcohols. The resultant filter cake is slurried in water and mixed with simulated high-level waste (HLW) solution, and fed into a rotary calciner for drying and calcination to remove organics and nitrogen oxides. The calcined powder is then mixed with 2 wt% of fine Ti metal powder as a redox buffer and the mix is hot-pressed uniaxially in stainless steel bellows at 1150°C and 20 MPa.

The overall aim of the SDP project is to demonstrate that non-radioactive Synroc can be fabricated on an engineering scale, the design throughput being 10 kg/h of Synroc, and to collect process data which will be useful in designing a radioactive plant. However the SDP does not necessarily represent the latest concept of Synroc processing and this technology is the subject of intensive research at ANSTO, from the preparation of the precursor through to the final hot-consolidation step.

The aspects of Synroc fabrication technology with which the present paper deals are a concept known as in-bellows calcination (IBC) and hot isostatic pressing (HIP). The work described here was largely on the laboratory-scale, but some semi-quantitative calculations and initial experiments relating to scale-up have been made.

Table 1
Composition of Waste-Free Synroc

Oxide	(wt%)
Al_2O_3	5.4
BaO	5.6
CaO	11.0
TiO_2	71.4
ZrO_2	6.6

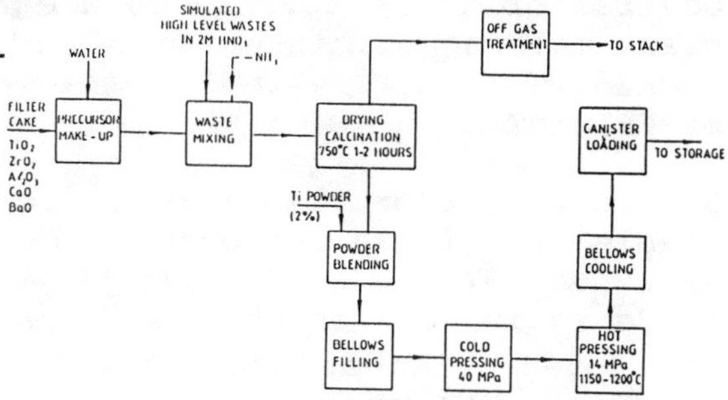

Figure 1

Flow diagram for the Synroc Demonstration Plant

A patent application for the IBC process has been lodged [3]. IBC offers many advantages over other calcination methods, mainly in terms of overall simplicity. The two methods currently used in large radioactive plants are rotary and fluidized-bed calcination. IBC eliminates the drive system, mechanical seals and rabble bar required by the rotary method and should overcome the problem of incomplete denitration which can occur with a fluidized-bed. The major foreseen advantages of IBC are:

- Elimination of powder transport after calcination.
- Dust generation and volatile losses are greatly reduced since the powder is well packed in the bellows and particulate filters are incorporated.
- Avoidance of exposure to atmosphere after calcination may eliminate the need to add powdered metal (e.g. Ti) to maintain reducing conditions during hot-pressing.
- Bellows can be moved straight to the hot-press after calcination.
- A significant reduction in radioactive cell volume can be achieved.

Several of these advantages are also present in in-can glass melting or sintering [4, 5].

EXPERIMENTAL

Synroc precursor powder produced by the alkoxide route (Table 1) [2] was thoroughly mixed with simulated waste solution (Table 2) to give a waste loading of 20 wt% oxide equivalent in the final ceramic. The mixture was then oven-dried at 110°C before being used in calcination and HIP experiments.

X-ray diffraction (XRD) was performed with a Siemens diffractometer fitted with a graphite monochromator. Scanning electron microscopy (SEM) was carried out with a JEOL instrument run at 15 keV and a Tracor Northern energy-dispersive X-ray analyzer was attached for microchemical analysis. Leaching measurements in triplicate were conducted on 2 mm thick, 10 mm diameter polished discs. The MCC-1 protocol was followed and the tests themselves were of 7 days' duration in deionized water at 90°C; subsequent solution analyses were made by inductively-coupled plasma spectroscopy, except for Cs which was determined by atomic absorption spectroscopy. Densities were determined by Archimedes' method, with the samples boiled in water to allow open porosity measurements.

Table 2
Simulated Waste Composition

Oxide	(wt%)	Oxide	(wt%)
Cr_2O_3	0.9	Nd_2O_3	17.3
Fe_2O_3	4.3	Gd_2O_3	4.2
NiO	0.4	ZrO_2	14.0
Cs_2O	9.3	Ag_2O	0.2
SrO	3.0	CdO	0.2
BaO	4.4	MoO_3	14.8
Y_2O_3	1.7	TeO_2	2.1
CeO_2	13.6	P_2O_5	1.8
		RuO_2	7.8

Figure 2

Type 1 convoluted bellows with top cut off, showing filter system.

IBC Experiments

Experiments were conducted using three types of stainless steel 316 bellows, with 0.9 mm wall thickness.

Type 1 (Fig. 2) is a convoluted bellows, 9 cm in diameter x 11 cm high. This type of bellows was used for determining the properties of the calcined powders. The top of the bellows in Fig. 2 has been removed by arc-cutting. A zirconium oxide fibre filter is incorporated in either end to prevent blockage in the inlet and outlet gas lines and to remove particulates in the off gas. The top filter is shown in Fig.2. The lid was welded into the filling hole after the bellows was filled with 600-700 g of dry Synroc + HLW powder.

The Type 2 bellows is also convoluted, 8.5 cm in diameter x 10.8 cm high, and contains in-built filters similar to those in type 1. It was used for in-bellows calcination and subsequent hot isostatic pressing experiments.

Type 3 (Fig. 3-see below) The dumb-bell shape of type 3 is designed for higher eventual disposal canister loadings [6]. Another feature is that should pinhole failure occur on HIP the increase in length of the bellows on fast decompression would be at least a factor of 5 less than for the convoluted bellows. The container incorporates zirconium oxide fibre filters as in types 1 and 2. Its capacity is approximately 250 g of Synroc/waste powder.

During all IBC experiments, 3.5% H_2/N_2 or H_2 was fed through at a flow rate of 2 L/min, and the pressure in the bellows due to the head of water in the wash bottles was 4 KPa. NO_x emission reached a maximum at ~ 500°C, consistent with DTA/TGA results on similar material [7]. After heating to 800°C at ~ 5°C/min holding for ~ 1 h and cooling back to room temperature at ~ 10°C/min, the inlet pipe was crimped and sealed by welding. Subsequent to evacuation to a pressure of 50 millitorr, the exit pipe was crimped and welded.

Hot Uniaxial Pressing (HUP)

After calcination in the Type I bellows using flowing H_2, the powder was extracted and 70 g samples were hot uniaxially pressed in graphite dies of 40 mm bore, to compare the pressing behaviour with that of powders calcined in rotary furnaces. Samples hot pressed at 1175°C/20 MPa were found to have essentially 100% theoretical density (4.48 ± 0.02 g/cm³) and gave leach results (Table 3) which were entirely satisfactory compared to ANSTO's reference-quality Synroc [2]. Although the Ca leach rate was slightly higher than that in reference-quality Synroc [2], the Ba rate was substantially lower. X-ray diffraction showed the presence of the three waste-immobilizing Synroc phases (see above), together with rutile when no Ti metal buffer was added or Magnéli phases (Ti_nO_{2n-1}; $7 < n < 9$) when 2 wt% fine Ti metal was added.

In four separate experiments with different types of container, the amounts of Cs and Ru collected in the exit gas line (by subsequent washing) and the wash bottles were found to represent < 0.05% of the total respective inventories. No visual evidence of any particulate deposits was present.

Hot Isostatic Pressing

Following earlier discussion by Larker [8] on the use of HIP to consolidate radioactive waste forms, Synroc samples designed for Savannah River Plant defense wastes have been successfully densified by HIP at 1100°C/150 MPa, but the compositions differ substantially from those discussed here [9]. Other samples, closer in composition to the present ones, were subjected to HIP at 1250°C/207 MPa, but these generally only reached 90-95% of theoretical density, due apparently to the presence of residual glasses in the precursor powders which were calcined by fluidized bed methods [10]. However Murakami [11] has produced dense Synroc containing simulated HLW by HIP.

The laboratory-type HIP apparatus (Autoclave Engineers,

Erie, PA, USA) used in the present work has a maximum pressure rating of 200 MPa and is located in the Commonwealth Scientific and Industrial Research Organization (CSIRO) Division of Materials Science and Technology in Melbourne, Australia. The furnace is graphite with a maximum operating temperature of 2000°C and the pressurizing gas is argon. Sealed Type 2 bellows were HIPed at 1175°C/56 MPa for 2 h, the full cycle being shown in Fig. 4.

Figure 3

Dumb-bell shaped Type 3 bellows before (right) and after (left) HIPPing.

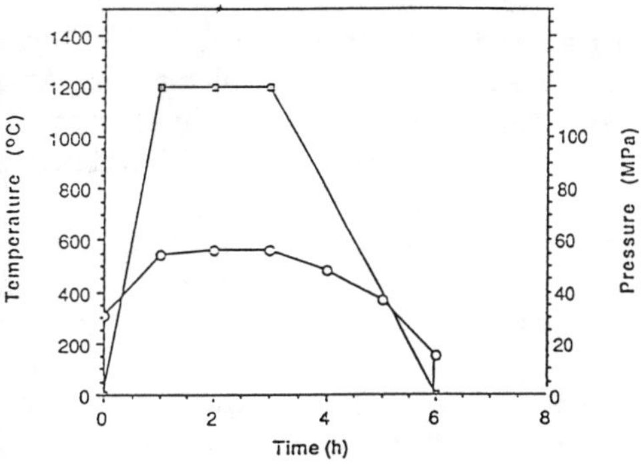

Figure 4

Temperature (squares) - pressure (circles) - time cycle for the HIP

Upon cutting the bellows open, some coarse fracturing was evident near the edges of the Synroc, but 10 mm diameter cores were drilled successfully and sectioned for leach testing and SEM. The density of the material was determined as 4.48 ± 0.02 g/cm^3, i.e. 100% of theoretical density within experimental error, based on experience with equivalent material HUPed under various conditions in graphite dies. XRD showed that the three waste-immobilizing Synroc phases were present, together with rutile (TiO_2). While there is some evidence [4] that the presence of rutile rather than Magnéli phases denotes that redox conditions during hot consolidation are insufficiently reducing, the leach results (Table 3) were entirely in line with those for best quality ANSTO laboratory grade reference Synroc [2]. The lack of rutile reduction may be argued to correspond to the loss of a second line of defence rather than directly increasing the leachability. SEM showed the micro-structure of the HIPed material to be comparable to that of Synroc formed by HUP at 1200°C/20 MPa in graphite dies [2]. Fig. 5 shows examples of bright (Mo, Ru) alloy particles and dark, fine-scale Al segregation. This latter feature is occasioned by the very rapid hydrolysis rate of the Al alkoxide compared to those of Zr and Ti, and is endemic in the alkoxide preparative technique. There is no evidence however that such fine scale segregation is in any way deleterious to the mechanical properties or durability of Synroc. There was no indication of waste segregation.

Similar results were obtained in two experiments using rotary-calcined powder [12] with Type 3 bellows. Leach results on these materials are included in Table 3 and these agreed within a factor of ~ 3 with those of the HIPed oven-dried powder which was treated by the IBC method. Fig 4 shows the Type 3 bellows, before and after HIP.

IBC/Vitrification

Laboratory scale IBC of a dry glass/HLW powder mixture and subsequent vitrification in the bellows was also demonstrated.

SCALE-UP CONSIDERATIONS

Calculations were made for a 30 cm diameter x 60 cm high bellows filled with 50 kg of dried Synroc precursor powder having an equivalent waste loading of 20 wt% and a density of ~ 1.2 g/cm^3 (oxide equivalent) (27% of the theoretical density obtained after hot-consolidation at ~ 1200°C). Such a powder would liberate 350 Mol each of NO_x and steam after complete denitration, assuming that the original HLW solution was 2 Mol/L in HNO_3 and contained 80 g/L HLW oxide equivalent. The STP volume of gas would then be ~ 15 m^3 for a throughput of 10 kg/h Synroc (the current design value for the SDP), and the rate of gas removal

Table 3
Leach Results for Hot-Pressed Synroc Samples

Press Type	Powder Source[++]	Bell-ows Type	Pressure Temperature	Calcin-ation Atmos-phere	Leach results[*] ($g/m^2/d$) Ba	Ca	Cs	Mo	Sr
HUP[+]	IBC	2	1175°C/20 MPa	H_2	.02	.03	.05	.3	.02
HIP	IBC	2	1175°C/56 MPa	H_2	.03	.03	.07	.08	.04
HIP	ACP	3	1185°C/55 MPa	H_2/N_2	.01	.03	.2	.1	.03
HIP	ACP	3	1185°C/55 MPa	H_2	.04	.04	.025	.3	.05
Reference Quality Synroc					.09	.02	.08	.3	.03

[*]Triplicated 7-day MCC-1 test run at 90°C. All Al, Ti and Zr leach results were < .05, < .0002 and < .003 $g/m^2/d$ respectively, as found from solution analyses.

[+]2 wt% Ti metal added after Type 1 in-bellows calcination.

[++]ACP = rotary calcination at engineering scale [12].

from the bellows would need to be ~ 3 m^3/h (50 L/min) at STP.

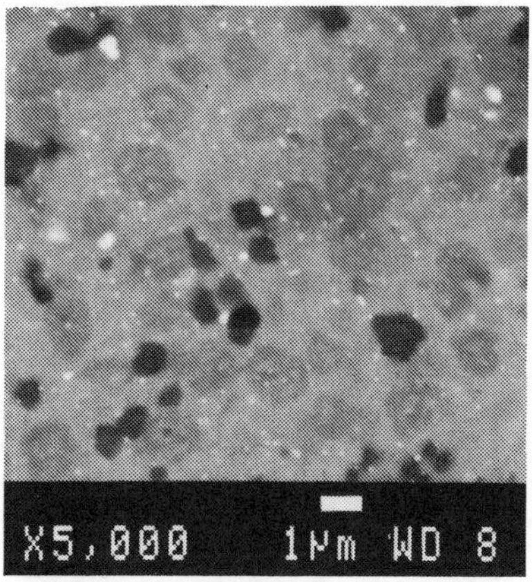

Figure 5

Scanning electron microstructure of HIPped Synroc (x 5000). Material appearing white corresponds to metal alloy particles; waste-immobilizing phases show contrast in this backscattered micrograph.

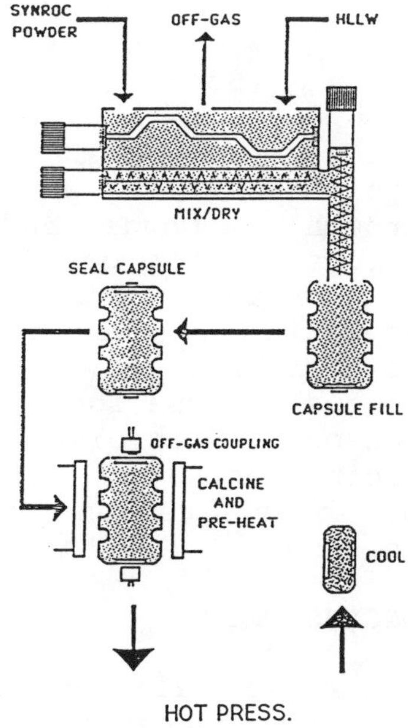

Figure 6

Schematic of in-bellows calcination.

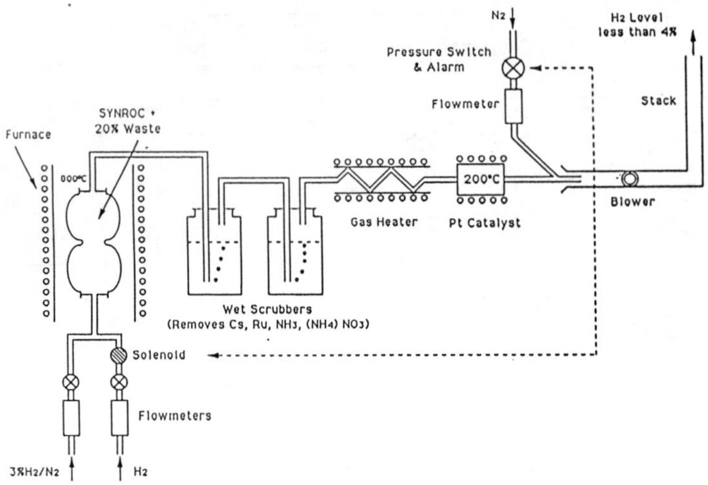

Figure 7

Schematic of in-bellows process for Synroc fabrication

both the vertical and horizontal positions. Gas flow through the powder bed will be examined in both orientations and the gas inlet and outlet systems optimised to give an acceptable gas flow through the bed. A schematic of the in-bellows calcination process is shown in Fig. 6. Fig. 7 is a schematic of the in-bellows process for Synroc fabrication. Minimum powder transport is involved in the process and the in-bellows calcination stage pre-heats the container and powder for densification by hot uniaxial or isostatic pressing.

CONCLUSIONS

In-bellows calcination of alkoxide Synroc precursors has been successfully demonstrated at the ~ 0.1 m scale, the resultant powder being amenable to hot-pressing in the normal pressure-temperature range. Advantages of the method include minimal dust and volatilization losses, elimination of metallic redox buffer additions and minimisation of powder transport

To achieve a suitable temperature vs. time profile to give a steady denitration rate (over the time period allotted to denitration) would require research into heat transfer across the bellows as well as detailed DTA studies, but it is emphasised that since denitration is endothermic [7], thermal runaway and the attendant pressure buildup from a high rate of gas generation could be avoided, as long as the temperature ramp itself was strictly controlled.

The H_2/N_2 atmosphere is used to effect reduction of transition metal ions in the calcination treatment after denitration in the heating cycle. This reduction minimises volatility losses of Ru, Tc, Mo etc. as discussed in a companion paper [7]. All the transition metal ions will tend to be oxidised during denitration and the reduction reactions of particular interest are Cr^{6+}, Fe^{3+}, Ni^{2+}->Cr, Fe, Ni metal; Te^{6+}->Te^{4+}, Mo^{6+}->Mo^{4+}.

Based on these reactions, the stoichiometric H_2 requirement is easily calculated as ~ 2 L/min at STP, corresponding to a flow rate of 60 L/min for 3.5% H_2/N_2. Because of kinetic limitations, these values should be taken as minima for achieving full reduction. The feasibility of this flow-rate for a moderately-well packed powder, and of avoiding channelling and large pressure gradients across the bellows, has not yet been demonstrated, but it is obvious that these possible difficulties could be greatly reduced by the use of H_2 rather than 3.5% H_2/N_2 gas. Moreover the use of H_2 in the denitration stage would render the atmosphere more reducing and thus lead to lower volatility of radionuclides during denitration.

PLANT SCALE UP

A calcination facility has been built for calcining Synroc in 300 mm diameter bellows to a maximum length of 1.5 m. The initial experiments will use a container of 30 cm diameter x 60 cm high with a spiral gas inlet and outlet tube embedded in a zirconium oxide fibre filter at each end of the container. The 36 kW heating furnace is mounted on a trunnion to allow operation in steps. Small-scale (~ 0.1 m) hot isostatic pressing was also successfully demonstrated, using both the in-bellows calcination route and bellows filled with rotary-calcined Synroc/waste powder.

ACKNOWLEDGEMENTS

We wish to thank Miss Jemima Cooper and the CSIRO Division of Materials Science and Technology for assistance with the hot isostatic pressing, K.G. Watson and G.T. Stevens for help with the electron microscopy, F. Jackson for carrying out the leach measurements, D.K. Shirlaw for metallographic sample preparation, B.W. Seatonberry for provision of the waste-doped starting alkoxide powder, P. O'Callaghan, W.E. Perrin and F. Van Luyt for the welding and crimping and K.D. Reeve and A. Jostsons for valuable discussion and comments on the manuscript.

REFERENCES

1. A.E. Ringwood, S.E. Kesson, N.G. Ware, W.O. Hibberson and A. Major, The Synroc process: A geochemical approach to nuclear waste immobilization, Geochem. Journ., 13, 141-65 (1979).

2. A.E. Ringwood, S.E. Kesson, K.D. Reeve, D.M. Levins and E.J. Ramm, Synroc, in Radioactive Waste Forms for the future, eds. W. Lutze and R.C. Ewing, Elsevier, pp 233-334 (1988).

3. E.J. Ramm, In-bellows calcination of Synroc, Australian Patent Application PJ 1156.

4. S. Gahlert and G. Ondracek, Sintered glass, in Radioactive Waste Forms for the future, eds. W. Lutze and R.C. Ewing, Elsevier, pp 161-192 (1988).

5. A.B. Harker and J.F. Flintoff, Hot isostatically pressed ceramic and glass forms for immobilizing Hanford high-level wastes, in Nuclear Waste Management, Advances in Ceramics, Vol 8, eds. G.G. Wicks and W.A. Ross, American Ceramic Society, Columbus, OH, pp 222-33 (1984).

6. E.J. Ramm, Hot-pressing of Particulate Material, Australian Patent Application PJ 0641.

7. E.R. Vance, D.J. Cassidy, K.L. Smith and J.L. Woolfrey, Calcination, Redox Control and Waste Loading in Synroc, this Symposium.

8. H.T. Larker, Hot Isostatic pressing for the consolidation and containment of radioactive waste, in Scientific Basis for Nuclear Waste Management, Vol 1, ed G.J. McCarthy, Plenum, N.Y., pp 207-10 (1979).

9. C. Hoenig, R. Otto and J. Campbell, Densification studies of Synroc D for High-Level Defense Waste, Lawrence Livermore National Laboratory Report UCRL - 53392 (1983).

10. F.J. Ackerman, J.Z. Grens, F.J. Ryerson, C.L. Hoenig, F. Bazan, P.E. Peters, R. Smith and J.G. Campbell, Fluid-bed process for Synroc production, in Nuclear Waste Management, Advances in Ceramics Vol. 8, eds G.G. Wicks and W.A. Ross, American Ceramic Society, Columbus, OH, pp 580-95 (1984).

11. T. Murakami, Microstructure of Synroc, Nucl. Chem. Waste Mgmt., 5, 269-78 (1985).

12. E.J. Ramm, Final Report (ANSTO/C93) on the Advanced Ceramic Processing System (1989).

CALCINATION, REDOX CONTROL AND WASTE LOADING OF SYNROC

E.R. Vance, D.J. Cassidy, K.L. Smith and J.L. Woolfrey

Australian Nuclear Science and Technology Organisation,
New Illawarra Road, Lucas Heights,
Menai, 2234, Australia

ABSTRACT

Differential thermal and thermogravimetric studies were made of the various stages of calcination,viz. drying, denitration and crystallization of intermediate phases, in waste-containing alkoxide precursors to Synroc. Apparent activation energies for the crystallization of the cubic and orthorhombic intermediate phases were found to be ~ 1100 and 600 kJ/mole, respectively. When up to 5 wt% of finely-divided Ti metal was added as a redox buffer to the calcined alkoxide powders prior to hot-pressing, the rutile -> Magnéli phase transformation occurred above ~ 2 wt% Ti,the other waste-immobilising phases apparently being unaffected. The amount of Ti necessary to effect the transformation increased slightly with waste loading. No significant changes were observed in the leach rates when the transformation occurred. The allowable Purex waste loading in Synroc may be as high as 35 wt% oxides.

1. INTRODUCTION

Synroc is a Ti-Zr-Al-Ba-Ca oxide ceramic for nuclear waste disposal [1] in which radionuclides from reprocessing waste are incorporated in solid solution in three principal crystalline phases, which are based on minerals known to be chemically durable in Nature. These phases and their idealized formulae are barium hollandite ($Ba_{1.14}Al_{2.28}Ti_{5.71}O_{16}$), zirconolite ($CaZrTi_2O_7$) and perovskite ($CaTiO_3$). Some rutile ($TiO_2$) or Magnéli phase ($Ti_nO_{2n-1}$; $7<n<9$) is also present, depending on the redox conditions prevailing during hot pressing of the ceramic, together with metal (Pd, Ru, Tc, Mo, Fe...) alloys.

In the current fabrication route [2,3], mixed Ti, Zr and Al alkoxides are dissolved in ethanol and hydrolysed in an aqueous slurry of Ba and Ca hydroxides. Three cycles of filtering and washing remove the alcohols. This precursor slurry is combined with simulated high-level Purex liquid reprocessing waste (PW-4b) and calcined in flowing 3.5% H_2/N_2 for 1 h at 750°C. Two wt%

of finely-divided Ti metal is then added as a redox buffer, and the mixture is hot pressed at ~ 1200°C and 20 MPa in a graphite die or metal bellows.

The present work deals with drying and denitration reactions during the early stages of calcination, crystallization of intermediate phases during calcination, redox conditions prevailing during hot pressing, allowable waste loadings, and finally, possible modifications to the precursor chemistry to adjust the phase assemblage.

2. EXPERIMENTAL

The starting material was obtained from a single batch of waste-free, alkoxide-route precursor which was oven dried at 110°C. The simulated Purex high-level waste (HLW) solution (Table1) was thoroughly mixed with the precursor, as an aqueous slurry (extra water was added for small waste additions), and then either tray-dried or stir-dried (using a hot-plate and magnetic stirrer system) at ~110°C. After rotary calcination at 750°C, the calcine was combined with finely-powdered (-325 mesh) Ti metal, where necessary, and hot pressed at either 1150 or 1200°C, at a pressure of 20 MPa, for 2 hours, in a 40 mm graphite die.

X-ray powder diffraction (XRD),scanning electron microscopy (SEM),density and leaching measurements were conducted as described in a companion paper [4].Transmission electron microscopy (TEM) was performed on ion-thinned samples with a JEOL instrument fitted with a Tracor Northern energy-dispersive X-ray analyzer.

Differential thermal analysis (DTA) and thermogravimetric analysis (TGA) were performed by a Harrop instrument and a Cahn 1000 thermobalance. Alpha-alumina was used as the DTA reference material and the heating rates were in the range 1 to 50°C/min, up to 800°C. The atmosphere was flowing 3.5% H_2/N_2.

3. RESULTS and DISCUSSION

3.1 CALCINATION PROCESSES

Powders heated in the DTA/TGA in the 100 to 600°C range show endothermic behaviour and weight losses due to drying and denitration (Fig 1). The results depend on the HLW loading and the drying method used to incorporate the waste. The relative amounts of absorbed and chemically-bound water were independent of the drying method. Denitration over the broad temperature range 250 to 600°C, which has been reported previously [5], was observed in both materials with an additional large endotherm at 600°C (Fig 1). The DTA peak heights associated with the denitration increased with increasing waste loading.

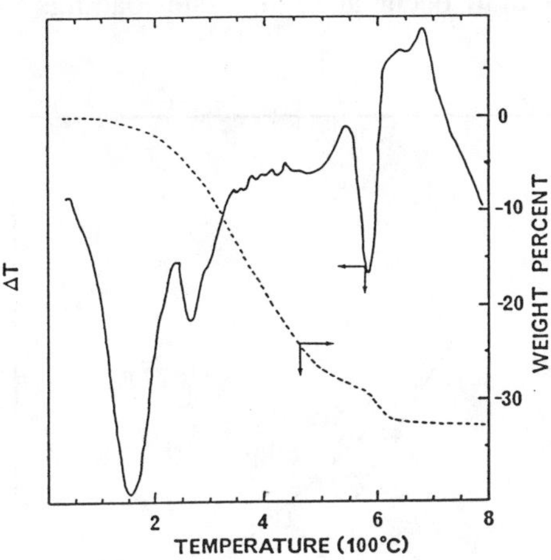

Fig.1 DTA/TGA curves for stir-dried alkoxides incorporating 25 wt% simulated HLW. Heating rate = 10°C/min.

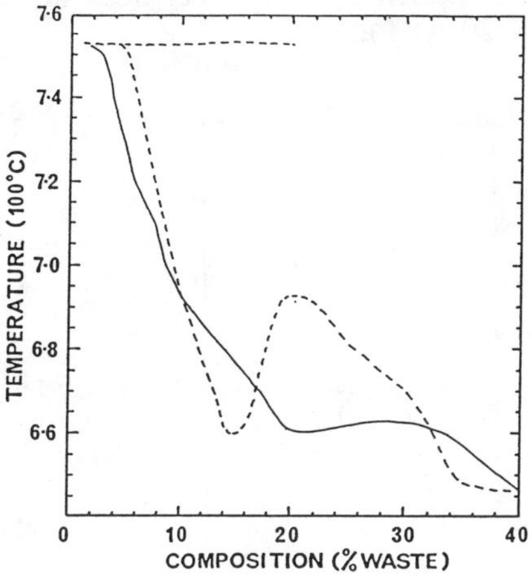

Fig.2 Effect of HLW loading on the DTA exotherm peak temperatures due to crystallization of intermediate phases. Solid and dashed lines refer to stir- and tray-dried materials respectively.

Samples in which the pH had been adjusted to ~ 9 with a strong NH_3 solution during waste addition to promote radwaste sorption [6], showed an additional well defined exothermic peak near 300°C, due to the decomposition of NH_4NO_3, but no endotherm at 600°C.

Mass spectroscopy of the gas atmosphere prevailing during calcination showed a variety of species: NO, NO_2, NH_3, H_2O, H_2, N_2, N_2O [5]. The initial stage up to 200_oC is signified by a weight loss of 20%, Fig 1, corresponding to the evolution of absorbed water. The loss of bound water and denitration, with associated evolution of NO_x reaction products, occurs during the second stage, 200 to 600°C. A reducing atmosphere, either flowing H_2/N_2 or pure H_2, is necessary at all times during calcination to reduce transition metal ions, and thus suppress the volatilisation of Ru, Tc, Mo, etc.

The stir-dried samples, heated at 10°C/min, gave only a single exothermic peak in the ~ 650 to 750°C range due to crystallization of intermediate phases, for all waste loadings up to 40 wt%. The crystallization temperature decreased with increasing HLW loading, Fig 2. However, the less-homogeneous, tray-dried material exhibited two exothermic peaks for HLW loadings up to 20 wt%. The poorer mixing in tray-dried samples possibly causes the sample to consist of waste-deficient and waste-rich zones due to segregation of HLW elements, resulting in the splitting of the exotherm-temperature vs. waste loading plot into two branches. This effect does not appear to occur at high waste loadings ($\geq$ 20 wt%).

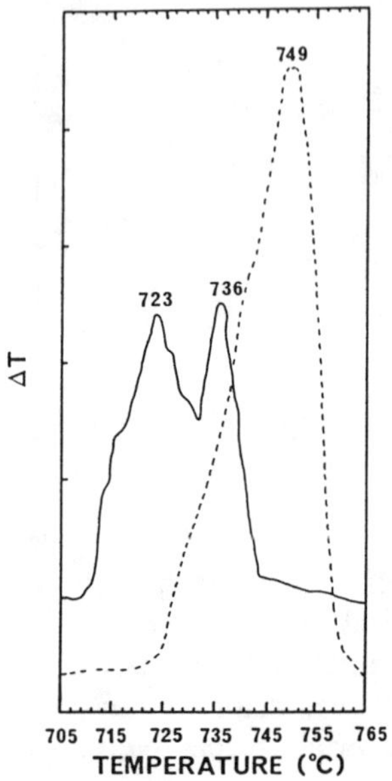

Fig.3 Effect of heating rate (1 and 5°C/min, solid and dashed lines, respectively) on the DTA exothermic peak due to crystallization of intermediate phases, for waste-free alkoxide Synroc precursor.

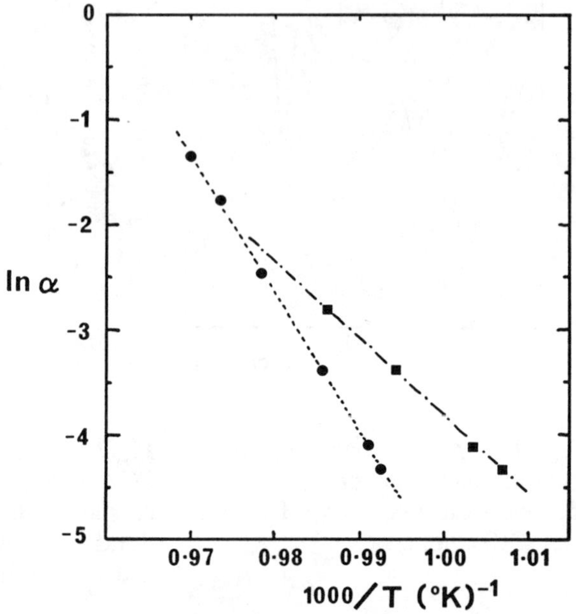

Fig.4 Plot of ln α vs $1/T_p$ for waste-free alkoxide Synroc precursor.

A waste-free sample was studied by DTA at different heating rates to evaluate the activation energy for crystallization of the intermediate phases at ~ 700°C. X-ray and neutron diffraction studies [6] show that the principal metastable intermediate phases are a $(Ca,Ti)(O,OH)_2$ fluorite structure and an orthorhombic phase with possible $CaTi_2(O,OH)_5$ stoichiometry. A single exothermic peak near 750°C was observed for heating rates $\geq$ 5°C/min, but it was split at lower rates (Fig 3). XRD on samples which were 'quenched' by switching off the DTA furnace at different temperatures after heating at 1°C/min in the vicinity of the exotherms confirmed that the orthorhombic and cubic phases gave rise to the exothermic peaks at the lower and higher temperatures, respectively. To obtain the activation energy E for the crystallization of each phase, the Ozawa equation [7]

$$\ln \alpha = -\frac{mE}{nRT_p} + const \quad (1)$$

was used, where α is the heating rate in °K/s, T_p is the temperature of the exothermic peak in °K, R is the gas constant and m and n are constants relating to the crystallization mechanism. For 3-dimensional growth with a constant number of nucleation sites, m = n = 3 [8] so the slope of the line in equation (1) gives E/R. The results are plotted in Fig 4 and the apparent activation energies for the crystallization of the cubic and orthorhombic phases were calculated to be 1110± 50 and 630± 20 kJ/mole ,respectively.

TABLE 1
Simulated Purex PW-4b Waste Composition

Oxide	wt(%)	Oxide	wt(%)
Cr_2O_3	0.85	Nd_2O_3	15.93
Fe_2O_3	3.92	Gd_2O	3.82
NiO	0.32	ZrO_2	12.84
Cs_2O	8.48	Ag_2O	12.39
SrO	2.75	CdO	0.22
BaO	4.04	MoO_3	13.48
Y_2O_3	1.59	TeO_2	1.91
CeO_2	15.77	P_2O_5	1.69

All materials were nitrates, dissolved in 2 Mol/L HNO_3 solution, except Mo, Te and P which were in the form of ammonium molybdate, telluric and phosphoric acids respectively and dissolved in a separate water solution. The mixed solutions had an equivalent concentration of ~ 80 g/L HLW oxides.

TABLE 2
Calculated Phase Assemblage of Synroc in Absence of Ti Metal

HLW (wt%)	0
Ba-hollandite	23
Zirconolite	18
Perovskite	19
Al-rich phases	3
$Ca_3(PO_4)_2$	-
Metal alloy	-
Rutile	37

3.2 REDOX CONDITIONS PREVAILING DURING HOT-PRESSING

The composition of waste-free Synroc, in wt%, is: Al_2O_3 (5.4); BaO (5.6); CaO (11.0); ZrO_2 (6.6); TiO_2 (71.4). Based on the idealized compositions for the three principal phases (see above), and using an ionic model in which the cation valences are Al(+3); Ba(+2); Ca(+2); Ti(+4) and Zr(+4), the phase constitution is readily calculated, assuming all the Ba and Zr are contained in the hollandite and zirconolite, respectively, and that the Al not incorporated in the hollandite phase forms the Al-rich phase, $CaAl_8Ti_3O_{19}$. The calculated phase assemblage (Table 2) is not very sensitive to the precise composition of the Al-rich phase, which is somewhat uncertain, even though the phase has been studied by transmission electron microscopy [3].

Taking a Magnéli phase with a representative value of n = 8, the reaction of rutile with Ti metal can be written as

$$15\ TiO_2 + Ti \rightarrow Ti_8O_{15}\ . \qquad (2)$$

In a closed system, the 37 wt% of rutile should be converted to the Magnéli phase by 1.5 wt% of Ti metal. This agrees reasonably well with the XRD results of the effect of Ti metal loading on the rutile observed in HLW-free Synroc, Fig 5, in spite of possible influences exerted by the graphite die on the redox state.

The corresponding calculations are more difficult when HLW is present. This is due to the large number of possible charge compensation schemes for incorporating HLW ions into the immobilizing phases. Suitable reducing conditions also allow appreciable amounts of Ti^{3+} to form and this can replace Al^{3+} in hollandite [9]. The addition of HLW reduces the amount of TiO_2 formed under oxidising conditions and thus the amount of Ti necessary to offset reaction (2) is reduced. Assuming, as per Table 2, that the rutile content in the absence of Ti metal additions was ~ 20 wt%, a reaction as in equation (2) would require only ~ 1 wt% Ti metal. However, transition-metal ions which have not been fully reduced during calcination will react with Ti according to reactions of the type

$$Mo^{4+} + n\ Ti \rightleftharpoons Mo + n\ Ti^{(4/n)+}\ . \qquad (3)$$

A quantitative study of such reactions is very difficult since no easily-measurable property can be used to define the redox state of a calcine. One possible method would be to measure the equilibrium weight gains on heating at fixed oxygen fugacities. In our laboratory studies of Synroc containing 20% HLW in which the redox reference state was not established after calcination, the rutile -> Magnéli phase transition occurred between 2 and 3 wt% Ti metal addition,Fig 5. Larger Ti additions are evidently needed with higher waste loadings (see below) to suppress rutile formation.

Densification was essentially independent of Ti addition for a hot-pressing temperature of 1200°C for the material containing 20 wt% HLW, but the Ti inhibited densification at 1150°C, as would be expected for the presence of hard agglomerates [10].

Leach rates varied irregularly with Ti content in any given set of samples and no singularity was observed at the Ti content corresponding to the rutile -> Magnéli phase transformation. The absence of a singularity is probably due to the encapsulation of small amounts of leachable material in the matrix and the variability of the leach results due to differences in surface finish, grain pullout and minor open porosity, etc.

TABLE 3
7-Day MCC-1 Leach Results at 90°C on high Waste Loading Samples.

HLW (%)	Drying Method	Leach Results g/m²/d(normalised) Al	Ba	Ca	Cs	Mo	Sr
20	Stir	<0.05	0.1	0.04	0.2	0.9	0.2
30	"	<0.05	0.2	0.06	0.1	1.0	0.2
40	"	5.0	0.1	0.7	300	10.0	2.0
20	Tray	<0.04	0.6	0.1	0.7	1.0	0.3
25	"	<0.05	0.2	0.1	0.9	3.0	0.4
30	"	<0.06	1.0	0.1	0.3	2.0	1.0
35	"	<0.07	1.0	0.2	0.1	2.0	1.0
40	"	<0.07	1.0	0.2	40.0	2.0	1.0
Ref. 10	Stir	<0.02	0.09	0.023	0.08	0.3	0.026

All Ti and Zr leach rates were < 0.0002 and < 0.003 g/m²/d, respectively.

3.3 WASTE LOADING

Previous work suggests that 20 wt% HLW (PW-4b) can be successfully incorporated in Synroc, but there is little explicit work dealing with waste loading limits. Buykx et. al. [11] have found that 25% HLW does not markedly increase the leach rate, relative to Synroc containing 10 wt% HLW.

The densities of calcined tray- and stir-dried powders containing 15 to 40 wt% HLW, hot pressed at 1150 or 1200oC (20 MPa, 2h), are shown in Fig 6. Three wt% Ti metal was added to the calcine before hot pressing. The density generally increases with increasing HLW loading. Considerable segregation of the waste elements was observed in the hot-pressed, tray-dried material with the main features being regions rich in rare earths, Sr and Zr, and Sr and Zr alone. The higher densities observed in these materials, compared to the stir-dried materials hot-pressed at the same temperature (1200°C), possibly arise from the increased occurrence of liquid-phase sintering during the early stages of the pressing cycle.

The leach rates (Table 3) of the stir-dried material containing 20% HLW were somewhat higher than those of reference alkoxide-route Synroc [3]. However, it is clear that there is a sharp dividing line between leach-resistant and leachable materials between 30/35 and 40 wt% HLW, as evidenced from the 7-day leach rates.

XRD of all samples showed the normal development of Synroc phases, but samples containing ≥ 30 wt% HLW also contained rutile, rather than Magnéli phases. Scanning electron microscopy revealed some porosity and complex microstructures, especially in samples with the highest waste loadings. Stir-dried samples were studied in some detail by transmission electron microscopy. The samples containing 20 and 30 wt% HLW had very similar microstructures with a dense matrix of perovskite, hollandite, zirconolite, Magnéli/rutile and metallic alloy phases. All the Cs was in the hollandite and no intergranular films could be seen. Some segregated alumina grains were observed in the 20% specimen. However, the 40% specimen had a very porous microstructure and, as well as the normal phases, contained molybdenum titanate, intergranular films and amorphous material. A high proportion of the Cs was contained in the amorphous material. This would explain the poor leach resistance of this material. Some Cs was also detected in the Mo-Ti phase.

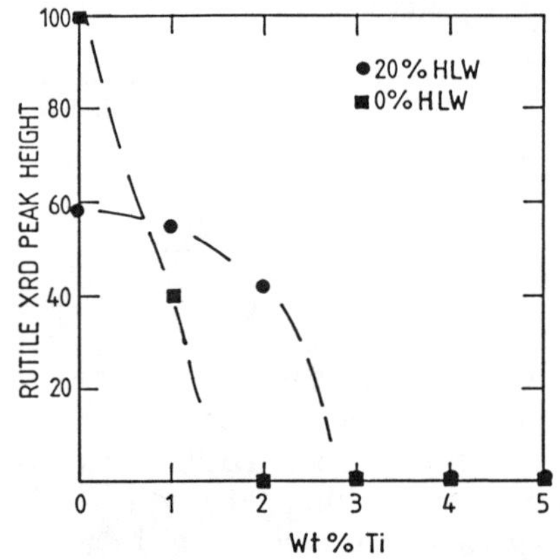

Fig.5 Intensity of rutile reflection (d = 3.25 Å) and sample density as function of Ti content and hot-pressed temperature for 0 and 20 wt% HLW.

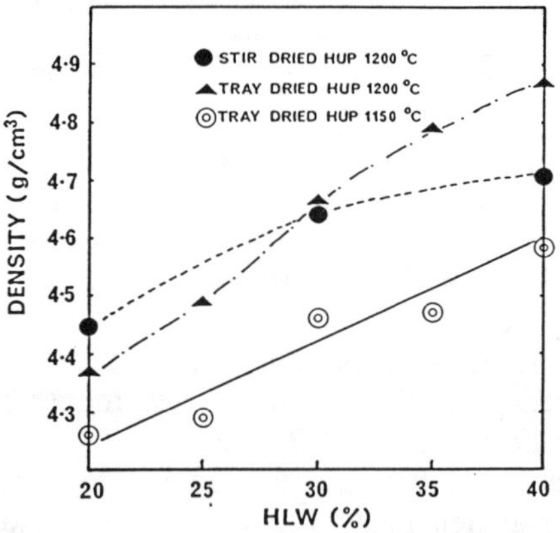

Fig.6 Effect of HLW addition on the density of Synroc hot pressed for 2 h at 1150 and 1200°C.

3.4 PHASE ASSEMBLAGE

More work is necessary to define precisely the waste-loading limits for Synroc, firstly with the existing precursor formulation and then by modifying the formulation. Excess Al^{3+} and Ti are present in the existing formulation in that under suitably reducing conditions, Al^{3+} can be replaced by Ti^{3+} in hollandite [9] and Ti^{3+} can charge-compensate RE^{3+} in perovskite. Perovskite has been found to be the most leachable [12] of the Synroc phases and increased durability might be obtained by decreasing the amount of perovskite present in the Synroc. Perovskite was present in the early formulations solely as the host phase for Sr, but, while Sr preferentially partitions into perovskite [1,2], its solubility in hollandite is considerable [13].

Several attempts to produce a perovskite-free Synroc via the oxide-carbonate route have been unsuccessful. This failure is due to the metastable formation of perovskite as a precursor in the formation of zirconolite [14], where the reactivity of the zirconia powder is a limiting factor, rather than its existence as an equilibrium phase. However alkoxide-route allows the formation of perovskite-free material due to the increased reactivity of the system with the cubic intermediate metastable phase, discussed previously, being the new precursor for zirconolite. A sample having a composition of (wt%) Al_2O_3 (5.5); BaO (5.7); CaO (5.1); ZrO_2 (9.4); TiO_2 (74.4) was prepared and a calculation based on the assumption that all the Ba and Zr occupy the hollandite and zirconolite yielded a "theoretical" phase assemblage of (wt%) hollandite (25); zirconolite (26); Magnéli phase (46); Al-rich phase (1); perovskite (2). This calculated result was in qualitative accord with XRD of the material after hot-pressing at 1200°C/20 MPa in that only peaks due to hollandite, zirconolite and Magnéli phases were observed. The measured density was 4.25 g/cm^3 and open porosity was < 0.2%. Scanning electron microscopy showed good homogeneity in general but there was some unexplained mottling in the backscattered (and secondary electron) image at the micron scale in a sample polished to a 0.25 μm diamond finish. Further studies of waste-doped material containing essentially only hollandite and zirconolite are in progress.

4. CONCLUSIONS AND FINAL REMARKS

The various phenomena occurring during drying, denitration and crystallization of alkoxide Synroc precursors containing 0 to 40 wt% of simulated Purex waste oxides (PW-4b) were studied by differential thermal and thermogravimetric analyses. Activation energies of 1110 and 630 kJ/mole were obtained for the crystallization of the cubic and orthorhombic intermediate phases which form nearly simultaneously at ~ 700°C from the alkoxide precursors. The rutile -> Magnéli phase transformation was suppressed by the addition of ≥ 2wt% Ti metal to the calcines before hot-pressing, with the amount required increasing slightly with increasing waste loading. The maximum allowable waste loading in Synroc appears to be 35 wt% HLW oxides (Purex), but more work,including the adjustment of the basic Synroc formulation, is necessary. Adjustment of the basic formulation to reduce the abundance of perovskite may also improve the chemical durability of Synroc.

5. ACKNOWLEDGEMENTS

We thank D. Shirlaw, R. Warren and J. Warmeant for assistance with preparation of leach and microscopy samples, F. Jackson for carrying out leaching studies, and G. Thorogood for experimental assistance.

6. REFERENCES

1. A.E. Ringwood, S.E. Kesson, N.G. Ware, W.O. Hibberson and A. Major, Geochem Jour. (Japan) 13, 141 (1979).
2. J.L. Woolfrey, W.J. Buykx and J.R. Bartlett, Ceramic Powders from Alkoxide Hydrolysis: The Preparation of Synroc, in Ceramic Developments, (eds. C.C. Sorrell and B. Ben-Nissan), Materials Science Forum, 34-36, p. 583 (1988).
3. A.E. Ringwood, S.E. Kesson, K.D. Reeve, D.M. Levins and E.J. Ramm, Synroc, in Radioactive Waste Forms for the Future (eds W. Lutze and R.C. Ewing) Elsevier, pp. 233-334.
4. E.J. Ramm and E.R. Vance, In-Bellows Calcination of Alkoxide Synroc Precursor, this Symposium.
5. J.L. Woolfrey, D. Cassidy, W.J. Buykx, F. Dickson and T.J. White, The Calcination of Synroc Powders, Proceedings of Bicentennial Conference '88, Institute of Metals and Materials Australasia, Volume 1, pp. 105-11 (1988).
6. C.B. Amphlett, Inorganic ion Exchangers, Elsevier,pp.84-92(1964).
7. E.R. Vance, R.L. Davis and K.D. Hawkins, Intermediate Phases in Synroc Crystallization from Alkoxide Precursors, J. Solid State Chem, submitted for publication.
8. T. Ozawa, A Modified Method for Kinetic Analysis of Thermoanalytical Data, J. Thermal Anal., 9, 369-73 (1976).
9. S.E. Kesson and T.J. White, $[Ba_xCs_y][(Tc,Al)_{2x+y}{}^{3+}\ Ti_{8-2x-y}{}^{4+}]O_{16}$ Synroc-type hollandites: II Structural Chemistry, Proc. Roy. Soc. A408,295-319(1986).
10. G.W. Scherer, Sintering with Rigid Inclusions, J. Amer. Ceram. Soc., 70,719-25 (1987).
11. W.J. Buykx, D.M. Levins, R.St.C. Smart, K.L. Smith, G.T. Stevens, K.G. Watson and T.J. White, Processing impurities and phase assemblage modifiers in titanate nuclear waste ceramics University of

Melbourne Report UM-P-89/05 (1989).
12. T. Kastrissios, M. Stephenson, P.S. Turner and T.J. White, Hydrothermal dissolution of perovskite: implications for Synroc formulation, J. Amer. Ceram. Soc 71, C-144-6.
13. E.R. Vance and D.K. Agrawal, Incorporation of radionuclides in crystalline titanates, Nucl. Chem. Waste Mgmt.3.229-234(1982).
14. W.J. Buykx, D.J. Cassidy, C.E. Webb and J.L. Woolfrey, Fabrication Studies on Perovskite, Zirconolite, Barium Aluminium Titanate and SYNROC-B, Amer. Ceram. Soc. Bull, 60 1284-88 (1981).

SOLIDIFICATION OF TRU-WASTES BY EMBEDDING INTO AN ALUMINIUM-SILICATE BASED CERAMIC MATRIX.

A. LOIDA and L.KAHL
Kernforschungszentrum,Institut fuer
Nukleare Entsorgungstechnik(INE),
P.O.Box 3640, D - 7500 Karlsruhe 1,
Federal Republic of Germany

In the Federal Republic of Germany the solidification of the radioactive wastes, arising in the nuclear fuel cycle will be accomplished by vitrifying the HLLW and cementing the predominant remaining waste streams. Both conditioning methods have been evaluated to permit their use on a production scale.

However, both waste forms suffer significant deterioration of the final products quality,if the relatively low limit of the possible waste loading is exceeded. Heat development and formation of radiolytic gases control the waste loading of the cement matrix. The molybdenum inventory of the waste limits the waste loading of glassy waste forms/1,2/.
Moreover, a high platinum metal's content may rise problems with the melter system/1,2/.

Actual R&D-work at KfK(INE) is aiming to demonstrate a way to reduce the volume of the final waste products, as well as to improve their quality by the development of an alternative waste form, in which the radionuclides are concentrated into a small fraction and incorporated into a stable ceramic matrix of an aluminiumsilicate composition.

By establishing the suitability of ceramic waste forms, specially for immobilizing TRU - wastes, a series of experiments on TRU - wastes ceramitization has been carried out.

An aluminiumsilicate based ceramic has been selected as the radionuclide host matrix, because it is generally accepted, that

- its stability against the attack of aequous (salt)solutions and radiation is very high,
- these ceramic types are very common, even in the conventional ceramic applications.

GENERAL CHARACTERISTICS OF THE ALUMINIUM SILICATE BASED TRU-WASTE BEARING CERAMICS.

The primary method for producing this kind of ceramics is as follows:

- pretreatment of the liquid waste(i.e.concentration, denitration, neutralization),
- homogenization of the radionuclide enriched slurries or powders with the aluminium silicate ceramic raw materials and forming by extrusion and cord cutting,
- heat treatment of the ceramic greenform, drying(110°C) and sintering(Tmax. 1350°C),
- final treatment by embedding the ceramic pellets, thus forming a monolithic block.

The aluminiumsilicate ceramic raw materials should optimally have an $Al_2O_3/Al_2O_3+SiO_2$ ratio of about 0.78. In the binary system Al_2O_3 - SiO_2 this corresponds at the appropriate temperature to a phase composition of mullite($Al_6Si_2O_{13}$) and corundum(Al_2O_3), both of which phases have a high resistance to the attack of aequous solutions. Using clay minerals and reactive corundum in this ratio, phase formation can be described in a simplified form by the following equation:

$$\underset{\text{kaolinite}}{n[Al_2Si_2O_5(OH)_4]}+\underset{\text{corundum}}{m[Al_2O_3]} \longrightarrow \underset{\text{mullite}}{x[Al_6Si_2O_{13}]}+\underset{\text{corundum}}{y[Al_2O_3]}+z[\text{glass}]+H_2O$$

Detailed descriptions of the reactions occuring are given in /3,4/.

If the added waste is low in alkaline and alkaline-earth oxides, the phase composition of the ceramic matrix will not change significantly. The presence of the latter oxides, however, will effect the formation of alkaline, alkaline-earth aluminium silicates and/or -aluminates, as well as a glass phase to a certain extent. These relations have been described by Harker et al. as well, by ceramitization of Savannah River Plant HLW, using aluminium silicate based starting materials/5,6/.

Because of the difference between the ionic radii of the matrix elements Al and Si and those of the TRU-elements and most of the other radionuclides, incorporation into the crystal structure of the matrix phases will not be observed. The radionuclides, especially TRU-elements are encapsulated by the phases of the matrix. Figure 1 demonstrates the basic differences of the microstructures between the considered aluminium silicate based ceramics, the SYNROC ceramics and a glass product, containing HLW oxides.

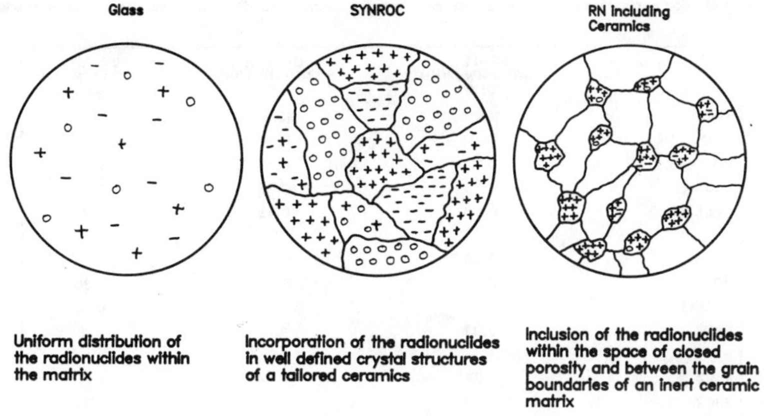

Fig.1: Different types of matrices, immobilizing radionuclides.

CERAMICS BEARING ORIGINAL DISSOLVER RESIDUES(DR).

One kind of TRU-wastes, able to generate problems, when incorporated into cement or glass,are DR from the reprocessing of spent LWR and FBR fuel elements. Their burn ups varied between 31 000 and 74 000 MWd/tHM, with cooling times between 2 and 3,5 years. Therefore, experiments were carried out on their incorporation into the ceramic matrix described above.
The original DR available for the ceramitization experiments were obtained from the laboratory scale reprocessing plant at the "Institute of Hot Chemistry" of the KfK. Because of their self heating rate, all DR samples, were received in small amounts between 6, and 20 gramms, and were in a dry and powdery state. Their actual chemical composition, determined by means of X-ray fluorescence analysis,is evident from table 1, where the chemical composition of the produced aluminium silicate based ceramics loaded with 20 wt% of dissolver residues is listed.

In order to produce the corresponding ceramic pellets the following process steps have been applied (i)homogenizing the DR powder with the appropiate amount of ceramic raw materials,(ii) forming by extrusion and cord cutting after slurrying up with water and (iii) heat treatment(drying and sintering at Tmax 1325 °C for 15-30 minutes holding time).No special cooling programm is quired afterwards.

Table 1: Chemical composition of aluminium silicate based ceramics, containing 20 wt% dissolver residues from different origins

components	DR-FBR 74000 MWd/tHM wt %	DR-FBR 33800 MWd/tHM wt%	DR-LWR 31000 MWd/tHM wt%
Al2O3	62.80	62.80	57.40
SiO2	14.91	14.91	17.80
Fe2O3,TiO2	0.75	0.75	0.50
CaO,MgO	0.75	0.75	3.10
(Na,K)2O	0.65	0.65	1.10
PuO2	8.50	10.91	5.20
UO2	5.62	5.32	5.20
Am2O3	0.26	0.16	0.10
NpO2	0.02	n.d.	n.d.
RuO2	1.07	0.34	4.00
RhO2	0.32	0.05	0.30
Pd	0.31	0.02	0.01
MoO3	1.01	0.67	3.90
(Tc)	(0.35)	(1.13)	(n.d.)
ZrO2	0.24	0.15	1.30
(Nd,Ce)2O3	0.30	n.d.	n.d.
(Cs,Rb)2O (Ba,Sr)O	0.50	n.d.	n.d.
Fe2O3,Cr2O3 NiO	1.44	n.d.	n.d.
spec.α-activity	9.00E8 Bq/g	9.58E8 Bq/g	7.77E8 Bq/g

The volatility of ruthenium during the operation, carried out under normal atmospheric conditions, was much less than 2wt% of the initial ruthenium present. This is in accordance with the results of a series of experiments related to the determination of the volatility rates of ruthenium during the ceramitization as a function of the atmosphere and the chemical form of the ruthenium, when the sintering temperature of 1300 °C is held for 1.5 hours.
Under the same conditions, the volatility of molybdenum is kept below 1.5% by the addition in of an excess proportion of Ca(OH)2 to form CaMoO4(powellite).
The resulting ceramic pellets(Ø~h~10mm) yielded compressive strengths about 100 N/mm2, densities around 2.75 g/cm3 and open porosities between 2 and 5 Vol%.

Their phase composition, determined by means of X-ray powder diffractometry, showed slight variations according to the chemical composition of the individual samples from table 1. The following phases have been determined under the conditions described below:

- corundum(Al2O3) one of the main matrix phases, present in all samples with high X-ray intensities,

- mullite($Al_6Si_2O_{13}$) is present regulary,if the amount of alkaline- and alkaline-earth oxides is kept below 3wt% and if the amount of bentonite in the ceramic raw materials is small,

- anorthite($CaAlSi_2O_8$) occurs instead of mullite, if residual calcium oxide remains to react with the SiO_2 and Al_2O_3,

- ([Pu,U,Am]O_2) represents the main phase related to the waste compounds with high and sharp X-ray reflections;the presence of ternary mixed crystals have been proven by means of SEM methods,

- RuO_2 is detected in all samples; even at low concentrations X-ray peaks were recognizeable; SEM-Methods demonstrated the common appearance of Ru-Rh-Pd compounds,

- powellite($CaMoO_4$) is formed, if the appropriate amount of CaO is added to the starting material to reduce the volatility of the MoO_3,

- baddeleyite(ZrO_2) appears always, if zircaloy chips and fines are part of the waste, which are oxidized during the heat treatment,

- zirkon($ZrSiO_4$) occurs, if available excess SiO_2 of the matrix reacts with the baddeleyite.

Some additional phases are present, but, as their amount is small their detection is limited by X-ray powder diffractometry. An increasing background in the X-ray spectra between 15 and 40° 2-ϑ angle,with a maximum near 25° 2-ϑ, which is originated by a glass phase have not been observed very clearly.It is assumed, based on TEM-investigations using corresponding nonradioactive ceramic specimen, that a glass phase, containing mainly SiO_2 and Al_2O_3 should be an additional phase of the ceramic product/7/.
The waste form microstructure was found to have developed the characteristics of a radionuclide including ceramic, which was demonstrated by autoradiography-, optical- and electron microscopic investigations. Related photographs are shown in figure 2. The size of the corundum crystals of the matrix was found to be 6 µm as a maximum value, down to values under the resolution limit of the equipment. An additional compound is present in the matrix in regions,where association with an aluminiumsilicate phase

could be probable; its crystal size is smaller, than 0.1µm. The waste particles are distributed homogenously and are bonded tightly with the surrounding matrix. Particle size variation is between 700 µm and sub µm.

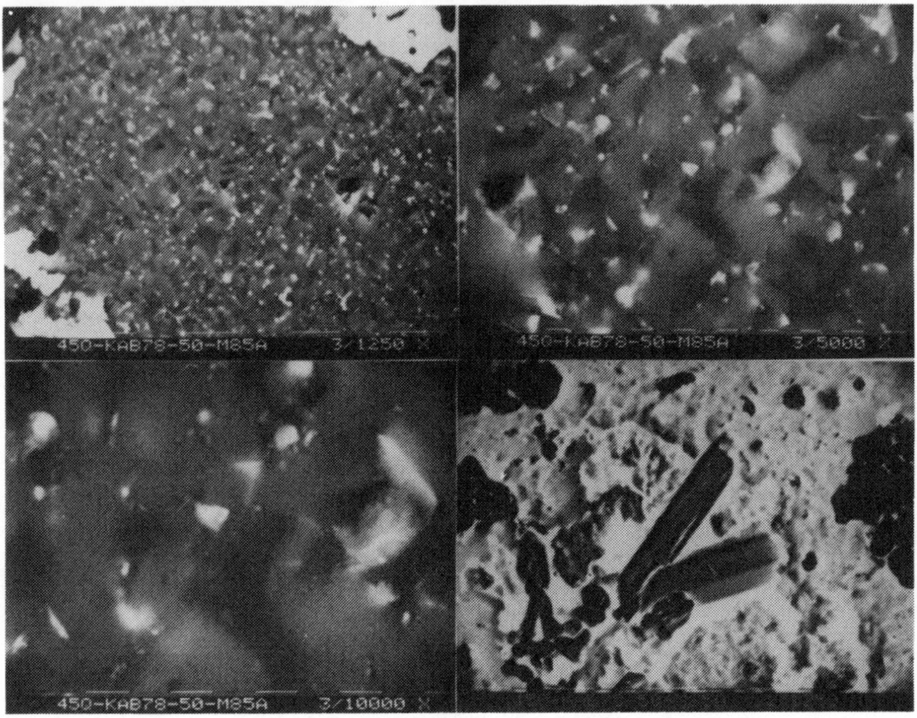

Fig.2. Scanning electron micrographs of an aluminiumsilicate ceramics,loaded with original dissolver residues(FBR);backscattered electrons(1250x,5000x,10000x) and sample current(1250x)

The distribution of the most abundant elements is shown in figure 3. The TRU-elements are concentrated in most parts of the waste particles. Also very clear is the common presence of ruthenium, rhodium, palladium and tellurium in another kind of waste particles.

Leaching tests.

Leaching tests have been performed according the ISO-draft, using a quinary salt brine with the composition 24.8wt% $MgCl_2$, 2.0 $MgSO_4$, 2.4 NaCl, 3.9 KCl, 66.9 H_2O and distilled H_2O as well at 25°C over a total leaching time of more than 6 months.

50ml of solution were in contact with the specimen for a defined period of time, followed by a new period, where the specimen is in contact with a fresh solution. The activity of the solution was then determined by means of α-counting, respectively Pu-241 spectrometry.
The corresponding leach rates were calculated from the activity of the solutions, the specific activity of the sample, the samples surface area and the duration of the leaching period.
Under these conditions the leach rates for actinides from alumosilicate ceramics with dissolver residues were determined to be 1E-8 g/cm2d after more than 6 months of total leaching time. No significant difference was found with the different kinds of leaching mediums.

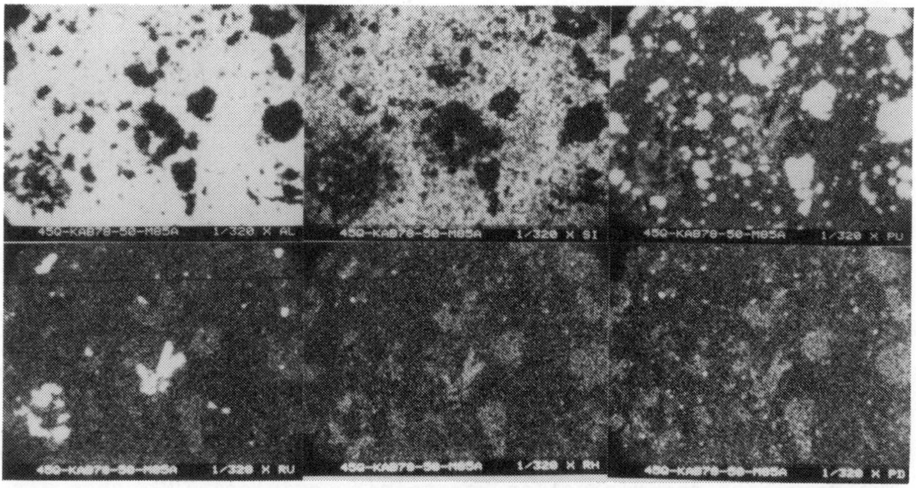

Fig.3. Distribution of the elements Al,Si,Ru,Rh,Pd across a section of an aluminiumsilicate ceramics, loaded with original dissolver residues(FBR); magnification 320x.

CERAMICS BEARING ORIGINAL TRU-ASHES.

The incineration of burnable TRU-wastes may lead to ashes with a TRU-content of more than 10wt%, which requires their incorporation into a matrix with a high long-term stability.

The original TRU-ashes, used for the ceramitization experiments, were generated during the incineration of burnable TRU-wastes from different origins, characterized in more detail by Swennen /8/. They occur in a powdery state, colored brown-, respectively yellowlike. They can contain up to about 20 wt% of carbon and chlorine, their TRU-element content was determined to be 15wt% in maximum,their alkaline oxide content can vary between 5 and near 60 wt%. Their enrichment with the compounds Al_2O_3 and SiO_2 was found to be reverse to the alkaline oxide concentration, from about 40 to 2 wt%. A more detailed chemical analysis from the two different kinds of original TRU-ashes is given in table 2 together with the chemical composition of the corresponding ceramic products.

Table 2: Chemical composition in wt% of two different kinds of incinerator TRU–ashes and the corresponding ceramic products. (waste loading 20 wt%).

component	TRU–ASH EUR 1/1	TRU–ASH EUR 3/1	TRU–ASH CERAMICS EUR 1/1	TRU–ASH CERAMICS EUR 3/1
Al_2O_3	14.23	1.00	67.82	57.60
SiO_2	27.77	1.00	23.40	18.00
CaO	10.32	23.00	2.16	6.90
Fe_2O_3	6.55	tr	1.62	0.30
Na_2O	3.66	43.00	0.87	8.70
K_2O	–	14.00	0.84	3.80
ZnO	2.32	tr	0.48	tr
MgO	1.83	tr	0.39	tr
BaO,NiO,Cr_2O_3, MoO_3,MnO_2	2.00	3.00	0.42	0.60
PuO_2	8.28	14.80	1.74	3.00
Am_2O_3,UO_2	0.38	tr	0.08	0.02
C,Cl–	22.66	tr	–	–
spec.α–activity	7.75 E 8 Bq/g	1.48 E 9 Bq/g	1.24 E 8 Bq/g	2.96 E 8 Bq/g

Sample preparation was carried out, using the same technique as described before, with the main steps being(i) homogenization of the ashes with the desired ceramic raw materials, (ii) forming by extrusion and cord cutting after slurrying up with water and(iii) heat treatment with maximum temperatures of 1300°C and about 15 minutes of holding time.

The resulting ceramic pellets containing original TRU-ashes were characterized by densities around 2.75 g/cm3, compressive strengths in the range from 170 to 365 N/mm2 and open porosities between 3.7 and <2Vol%.
Contrary to the phase formation observed in the ceramics loaded with dissolver residues, where no significant reactions between the waste particles and the matrix compounds were recognized, the greatest part of the nonradioactive ash compounds reacted during the heat treatment with the matrix raw materials in forming new mineral phases. A reaction between the matrix compounds and the TRU-elements however, namely the plutonium, have not been detec-

ted in this case.

The following phases have been observed occurring in the aluminiumsilicate ceramics containing original TRU-ashes by means of X-ray powder diffractometry:

- corundum(Al_2O_3)	is one of the main phases, if the waste load is lower 30wt%; otherwise the alkaline and alkaline-earth oxides react with it by forming alkaline and alkaline-earth aluminates and -aluminiumsilicates,
- nepheline ($[Na,K]AlSiO_4$)	appears as a common waste compound-matrix phase,
- anorthite($CaAlSi_2O_8$)	is present, if the CaO amount is above 5 wt%; plagioclase mixed crystals are also common phases,
- spinell(AB_2O_4) A=Mg,Mn,Ni,Zn B=Al,Fe^{3+},Cr^{3+}	occurs in those kinds of samples, where the ash content exceeds 30 wt%,
- hematite(Fe_2O_3)	can be recognized in high loaded ceramics,
- actinide oxide(PuO_2)	has also been determined to be present as a main phase.

The microstructure of these ceramics is characterized by a fine grained matrix, with crystal sizes below 2 µm and inclusions of TRU-particles, sized up to 250 µm, distributed randomly across the matrix. The characteristics of the microstructure are similar to those of the dissolver residue ceramics. The distribution of the most abundant elements occurring in the ash ceramics is given in figure 4. Aluminium is distributed equally through the matrix, significant enrichments are not evident; silicon is dispersed in this manner as well, but some zones of enrichment are detectable, which are in consistency with calcium and/or sodium enrichment. Plutonium appears mainly in a more or less pure state, but in some special cases a common appearance together with titanium and calcium has been stated, which could be a sign of the formaion of a plutonium zirkonolite($CaPuTi_2O_7$).

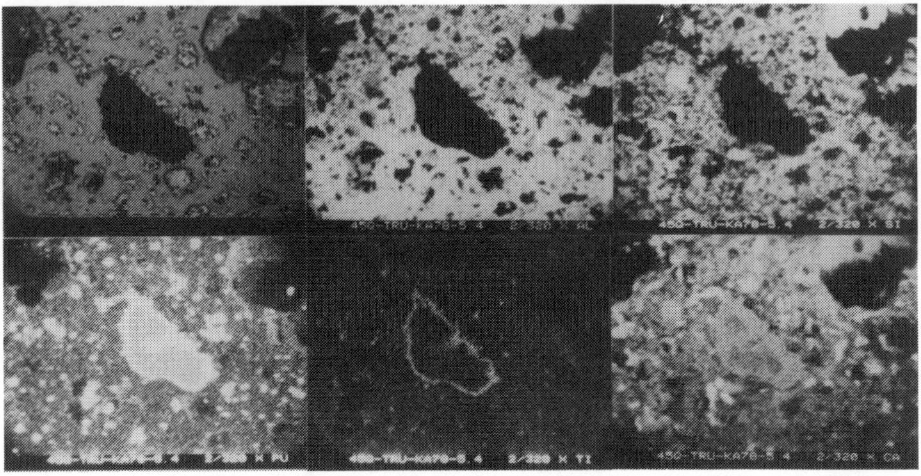

Fig.4. Distribution of the elements Al,Si,Pu,Ti,Ca and sample current photograph across a section of an aluminiumsilicate ceramics, loaded with original TRU-ashes; magnification 320x

Leaching tests.

Leaching tests according to the ISO-draft have also been performed. The conditions were identical to those of, the previous leach tests using dissolver residues ceramics. The average leachrate for plutonium was determined to be in the range between 5E-8 and 1E-7 g/cm2d after more than 6 months of total leaching time, using salt brine and distilled water at room temperature.

CONCLUSION AND OUTLOOK.

The experimental work on the ceramitization of radioactive wastes has demonstrated the suitability of aluminiumsilicate ceramic waste forms specially for TRU-enriched waste concentrates. The properties of these products have been found to be superiour to glassy or cementous waste forms.

However, producing radioactive waste bearing ceramics is more complex compared with the vitrification and/or cementation, be-

cause the process comprises several steps and more moving aggregates. The economic competivity has still to be proven.
The evaluation of a technical process on the ceramization of radioactive wastes is still in an early stage. A nonradioactive half-technical scale pilot plant is under construction and already partly under operation. The transfer of the ceramitization process technique to remote operation is possible.

Anyway, the high potential of the ceramic waste form, esspecially the high degree of the final product's stability of aluminiumsilicate matrices and the potential for the improvement of the technical process is encouraging to enforce the related R&D-works.

The corresponding R&D-works are supported by the Commission of the European Community, Brussels.

References

1 H.Pentinghaus, Process Chemistry of HLLW-Vitrification in a Liquid Fed Ceramic Melter(LFCM), in Spectrum '86, Proceedings of the American Nuclear Society International Meeting, Niagara Falls,NY,USA, Sept.14-18,1986, edited by J.M.Pope et al.,p.1166 - 1173,
2 M.Kelm, H.Pentinghaus, B.Oser, Chemische Vorgaenge bei der Verglasung von HAWC, p.367 - 391, 7.Statusbericht des Projekts Wiederaufarbeitung und Abfallbehandlung des Kernforschungszentrums Karlsruhe, 1988, KfK 4476,
3 K.Okada,N.Otsuka,J.Ossaka,Characterization of the Spinell Phase formed in the Kaoline-Mullite Thermal Sequence,J.Am.Ceram.Soc., 69, C-251 - C-253(1986),
4 M.R.Anseau,M.Deletter,F.Cambier, The separation of Sintering Mechanisms for Clay-Based Ceramics,Trans.J.Br.Ceram.Soc.,80, p.142 - 146,
5 A.B.Harker, C.M.Jantzen, et.al.,Tailored Ceramic Nuclear Waste Forms..,in Scientific Basis for Nuclear Waste Management,Vol 3, Boston 1980, ed J.G.Moore,NY, p.139 - 146.
6 A.B.Harker,P.E.Morgan, et al.,Formulation and Processing of Polyphase Ceramics for High Level Nuclear Waste,in Scientific Basis for Nuclear Waste Management,Vol 6,Boston 1981,ed. S.V. Topp(North Holland,Amsterdam), p.567 - 573,
7 U.Fieger,W.F.Mueller,H.Pentinghaus, TEM-Untersuchungen zur Phasen- und Gefuegeentwicklung in aluminiumsilikatischer Keramik, Fortschr.Mineralogie,65, Beiheft 1, p.47,(1987)
8 R.Swennen, Verification of Measured Quantities of Plutonium in Solid Wastes, IAEA-report Nr.ETR 291, (1976)

Section II. TRU and LLW

GAS GENERATION FROM THE IRRADIATION OF MORTAR*

Michele A. Lewis and David W. Warren
Argonne National Laboratory
9700 South Cass Avenue
Argonne, IL 60439

ABSTRACT

A mortar formulation capable of immobilizing chloride salts with high levels of radioactivity is being developed. As part of the developmental effort, radiation effects are being investigated. The radiolytic generation of gas(es) from irradiated mortar formulations was determined for several formulations with variable salt loadings at 25 and 120°C. The irradiation of a mortar formulation consisting of cement, slag, fly ash, water, and 0 to 10 wt % salt led to the generation of hydrogen. The rate of generation was approximately constant, steady state pressures were not attained, and final pressures were comparatively high. The hydrogen yield was higher for mortar with 10 wt % salt than with 0 or 3 wt % salt at ambient temperature, but it was lower at 120°C. The irradiation of a mortar consisting of cement, fly ash, water, and salt led to the radiolytic generation of both oxygen and hydrogen. The addition of 2 wt % FeS or CaS inhibited oxygen generation and changed the hydrogen production rate.

INTRODUCTION

This report presents the results of mortar irradiation experiments. These experiments were designed to investigate the effects of radiation on various mortar formulations immobilizing chloride waste salts. Previous studies have shown that, during the irradiation of cement-based waste forms, water undergoes radiolysis and hydrogen is generated.[1,2] The radiolytic generation of other gases and the yield of hydrogen were found to depend on the type of salt incorporated into the waste form.

---

*Work supported by the U.S. Department of Energy, Nuclear Fuel Reprocessing and Waste Management Section, under Contract W-31-109-Eng-38.

The effect of gamma radiation on brines (concentrated aqueous solutions of chlorides) has also been investigated. The results of these studies showed that both hydrogen and oxygen are radiolytically generated.[3,4] The effect of gamma radiation on cement-based waste forms containing brine-like concentrations of chloride salts has not been investigated previously. It was, therefore, important to determine the effects of gamma radiation on various formulations of cementitious materials capable of immobilizing chloride salts. The effects of different salt concentrations and different test temperatures were also investigated. Total gas production was monitored during the irradiation, and the composition of the gas phase was determined at the conclusion of the irradiation. The results showed that hydrogen was the only gas radiolytically generated when the irradiated mortar formulation included slag. The rate of gas generation was found to depend on the magnitude of the salt loading, the test temperature, and the dose rate. The gas generation rates, once established, did not fall off for up to 4500 h.

EXPERIMENTAL

Mortar samples were irradiated with ^{60}Co gamma rays at doses varying from 0.1 to 0.9 Mrad/h. The experimental matrix is given in Table I. Each experiment's number, mortar formulation, and test temperature are included. A specific objective for each set of experiments is also given.

Sample Composition/Mortar Formulations

Several formulations were irradiated. The composition of each formulation is given in Table II. The baseline formulation, specified in the text as B, consisted of slag, fly ash, portland cement, and water; and the fly ash formulation, FA, consisted of fly ash, portland cement, and water. The salt content in wt % of the various formulations is specified by the number following the letter identifying the formulation, e.g., B-10 refers to the baseline formulation with 10 wt % salt. The salt is a mixture of KCl-44 wt % LiCl. Both are Fisher ACS certified salts. The slag is Cleveland Ground Granulated Slag from the Standard Slag Company, and the fly ash is Class F fly ash from Commonwealth Edison (Illinois).

Procedure

The mortar samples were poured into quartz sleeves or containers. After curing for approximately 20 hours, the samples were inserted into leaktight stainless steel vessels. A

TABLE I. Matrix of Irradiation Experiments

Experiment Number	Formulation	Temp. (°C)	Purpose
1	T-16	25	Determine gaseous radiolytic products.
2	T-17	25	
3	T-18	25	
4	T-16-1	25	Determine gaseous radiolytic products and the effects of varying the salt loadings in the T-16 and FA formulations.
5	T-16-2	25	
6	T-16-3	25	
7	FA-0	25	
8	FA-3	25	
9	FA-6	25	
10	B-0	25	Determine the yield of gaseous radiolytic products as a function of salt concentration in the B formulation.
11	B-3	25	
12	B-10	25	
13	FA-10	25	Determine effect of different dose rates and a greater salt loading in the FA formulation.
14	FA-10	25	
15	S-10	25	Determine effect of eliminating fly ash from the B formulation.
16	B-3	25	Determine the effect of simulated radiogenic heating for two salt concentrations in the B formulation.
17	B-3	120	
18	B-10	25	
19	B-10	120	
20, 21	B-3	120	Determine the effect of simulated radiogenic heating for two salt loadings in the B formulation.
22	B-3	25	
23, 24	B-10	120	
25	B-10	25	
26	FA-6	25	Determine if CaS or FeS additives inhibit oxygen generation in FA-6.
27	FA-6 + 2 wt % CaS	25	
28	FA-6 + 2 wt % FeS	25	
29	B-10	25	Evaluate Leak-Lok sealant.
30	B-10	25	Evaluate TFE sealant.

TABLE II. Composition of Mortar Formulations (wt %)

	Cement	Slag	Fly Ash	Other	Water	Salt[a]
T-16[b]	19.4	19.4	19.3	4.2 (red clay)	29.0	10
T-17[b]	44.9	0	9.4	19.1 (glass)	24.7	3
T-18[b]	21.4	21.4	21.4	4.7 (red clay)	31.1	0
T-16-1[b]	23.2	23.2	23.2	0	30.3	0
T-16-3[b]	22.2	22.2	22.2	0	30.3	3
T-16-10[b]	19.9	19.9	19.9	0	30.3	10
B-0	17.0	35.0	17.0	0	30.0	0
B-3	16.0	34.0	16.9	0	29.1	3
B-10	15.7	31.5	15.7	0	27.0	10
FA-0	44.3	0	27.1	0	28.6	0
FA-3	42.8	0	26.2	0	27.6	3
FA-6	41.4	0	25.3	0	26.6	6
FA-10	47.3	0	15.7	0	27.0	10
S-10	21.0	42.0	0	0	27.0	10

[a]KCl-44 wt% LiCl.
[b]The numbers 16, 17, and 18 identify a formulation. They do not represent salt concentration.

tee connected the vessel valve, the gas sampling valve, and a pressure gauge. Ashcroft stainless steel gauges, Model #1009, and bellows valves were used.

Dose rates were measured using cobalt glass dosimetry and/or the Fricke dosimeter. Minor corrections for differences between the electron densities of the solution and the mortar were not applied.

After irradiation, the gas phase was sampled using mass spectroscopy. Several samples were taken to assure that the results were representative. The volume of the head space was measured by expanding a known volume of air into the evacuated vessel.

Results

The results and the experimental conditions (temperature, dose rate, etc.) are given in Table III. The results are discussed in terms of three sets of experiments. The first set, experiments 1-9, used several different formulations and provided qualitative information, i.e., what gases were radiolytically generated. The second set, experiments 10-12, 16-25, and 29-30, used the baseline formulation and provided both qualitative and quantitative information. The third set, experiments 13-15 and 26-28, used the fly ash and slag formulations and were designed to provide mechanistic information.

Experiments 1-9

The overall conclusion from experiments 1-9 was that hydrogen was the only radiolytically generated gas when slag was one of the major components of the mortar. In addition, oxygen that was initially in the head space of the vessel prior to the irradiation was consumed. Both hydrogen and oxygen were generated when the mortar formulation did not include slag. No discernible effect of red clay or glass beads was noted, and the salt concentration did not alter the nature of the gaseous radiolytic products.

Experiments 10-12, 16-25, 29-30

These experiments used the baseline formulation exclusively. The results of the gas phase analyses confirmed that hydrogen was the only gas radiolytically generated. Oxygen was not detected in the gas phase indicating that oxygen was either not radiolytically generated or, if it was generated, it reacted immediately upon formation. The oxygen in the head space was consumed as well. The rate of hydrogen generation as a function

TABLE III. Experimental Conditions and Results for Irradiation Experiments

Expt. Number	Formulation	Test Temp.	Dose Rate[a] (MRad/h)	Total Dose (MRad)	Final Pressure (kPa)	Air Volume (cm^3)	Gas Analyses (Volume %) H_2	N_2	O_2
1	T-16	Ambient	Variable	400	435	41	87.6	12.2	<0.01
2	T-17	Ambient	Variable	400	138	40	58.5	25.0	16.2
3	T-18	Ambient	Variable	200	145	48	70.3	29.3	<0.01
4	T-16-0	Ambient	0.7	1300	b	24	97.4	2.2	<0.05
5	T-16-3	Ambient	0.9	1600	b	26	95.4	4.2	<0.05
6	T-16-10	Ambient	0.7	1300	b	22	96.3	3.0	<0.05
7	FA-0	Ambient	0.3	500	b	16	69.2	7.2	23.0
8	FA-3	Ambient	0.3	500	b	22	68.1	24.1	7.1
9	FA-6	Ambient	0.3	500	b	15	70.7	6.7	22.1
10	B-0	Ambient	0.8	450	607	24	90.4	9.3	<0.01
11	B-3	Ambient	0.8	450	566	20	90.4	9.4	<0.01
12	B-10	Ambient	0.9	500	994	21	95.1	4.9	<0.01
13	FA-10	Ambient	0.8	700	414	23	65.2	19.0	13.6
14	FA-10	Ambient	0.3	250	290	21	62.9	17.1	15.4
15	S-10	Ambient	0.4	350	607	24	87.6	9.1	<0.01
16	B-3	Ambient	0.17	180	179	20	78.7	19.9	<0.01
17	B-3	120°C	0.20	210	179	50[c]	50.0	48.3	<0.01
18	B-10	Ambient	0.17	180	421	16	90.3	9.4	<0.01
19	B-10	120°C	0.19	200	124	55	80.6[d]	15.8[d]	<0.01[d]
20	B-3	120°C	0.13	280	338	23	81.4	15.9	≤0.1[e]
21	B-3	120°C	0.13	280	400	26	88.7	9.9	<0.02
22	B-3	Ambient	0.14	300	587	19	89.2	10.1	<0.02

(Continued)

TABLE III. (Continued)

Expt. Number	Formulation	Test Temp.	Dose Rate[a] (MRad/h)	Total Dose (MRad)	Final Pressure (kPa)	Air Volume (cm^3)	Gas Analyses (Volume %)		
							H_2	N_2	O_2
23	B-10	120°C	0.13	280	159	24	62.2	35.8	<0.02
24	B-10	120°C	0.13	280	152	23	63.1	35.4	<0.02
25	B-10	Ambient	0.16	340	711	17	94.4	5.4	<0.02
26	FA-6	Ambient	0.26	210	242	18	68.6	11.8	18.8
27	FA-6 + CaS	Ambient	0.18	140	400	16	84.4	14.4	<0.01
28	FA-6 + FeS	Ambient	0.17	140	136	21	66.4	31.6	0.8
29	B-10	Ambient	0.12	Not terminated					
30[f]	B-10	Ambient	0.14	140	64	51.5	91.2	7.8	>0.02

[a] Dose rate measured using Cobalt glass and/or Fricke dosimetry.
[b] Radiation caused gauges to seize up. No pressure measurements available.
[c] Estimated.
[d] Gas analyses are questionable. The temperature controller used for experiment 19 failed, and the temperature of the vessel reached 350°C. An unusual occurrence was also observed during the gas sampling/analysis of experiment 19. When the valve to the vessel was opened, the gauge pressure increased. In all other experiments, the gauge pressure decreased during this step. It may be that hydrogen absorbed by the mortar was released by the overheating, and that the hydrogen content of the gas sample (experiment 19) was higher than it should have been for comparison with the results from the other experiments. Also, a slight leak between the valves was detected during the MS gas analysis.
[e] Oxygen content higher than normal. Leakage at valve suspected.
[f] Leaks developed after 1030 hours of irradiation.

of salt concentration and test temperature is shown in Figs. 1-4. The yield of hydrogen was calculated in terms of the G values, defined as the number of molecules formed per gram per 100 eV of absorbed radiation.

Figure 1 shows the radiolytic pressurization at room temperature versus irradiation time for the B-0, B-3 and B-10 formulations. The rate observed for the B-10 formulation was significantly higher than that observed for either the B-0 or B-3 formulations. The G values for hydrogen, $G(H_2)$, were 0.17, 0.14 and 0.22 for B-0, B-3, and B-10, respectively.

Figure 2 presents the pressurization data versus irradiation time for the B-10 formulation experiments at 120°C (23 and 24) and at room temperature (25). The rate of radiolytic pressurization was much lower at 120°C than at room temperature for the duration of the experiment. The data in Fig. 2 show that the initial pressure for the heated experiment was higher than for the room temperature experiment. However, this is due to the steam pressure at 120°C rather than to a radiolytic pressurization. The $G(H_2)$ values for the experiments were 0.02 at 120°C and 0.21 at ambient temperature. As shown in Fig. 3, temperature effects were not as large for the B-3 formulation in experiments 20-22. The rates of pressurization were similar for both temperatures during the first 900-1000 hours of irradiation. Again, the initial pressure is higher in the heated experiments because of the steam pressure. A lowering of the pressurization rate at 120°C was observed only at longer times. G values were not calculated for the 120°C experiments (20 and 21) because the rate of hydrogen generation was not constant; however, hydrogen yields were estimated to be slightly lower at 120°C than at 25°C from the pressure-volume data. The results for experiments 16-19 were similar to those of experiments 20-25.

Figure 4 shows the radiolytic pressurization versus time for two time periods, 2000 and 4500 hours. The dose rates for the two experiments were comparable, 0.16 and 0.12 MRad/h; the total doses were 320 and 540 Mrad, respectively. The rates of pressurization were approximately the same, and there was no significant decrease in the rate over the duration of either experiment. This result is probably the most important from this series of experiments.

Experiments 13-15 and 26-28

Experiments 13, 14, and 26-28 used the FA formulation and provided radiolytic mechanistic information. The objectives of these experiments were to confirm that oxygen was a radiolytic

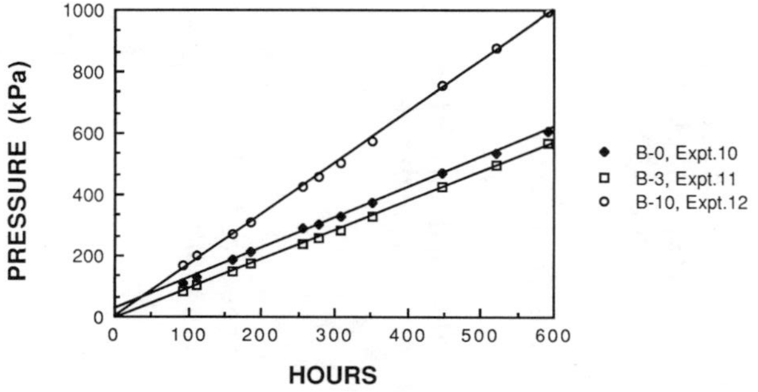

Fig. 1. Rate of Radiolytic Pressurization as a Function of Salt Concentration for Three Baseline Formulations

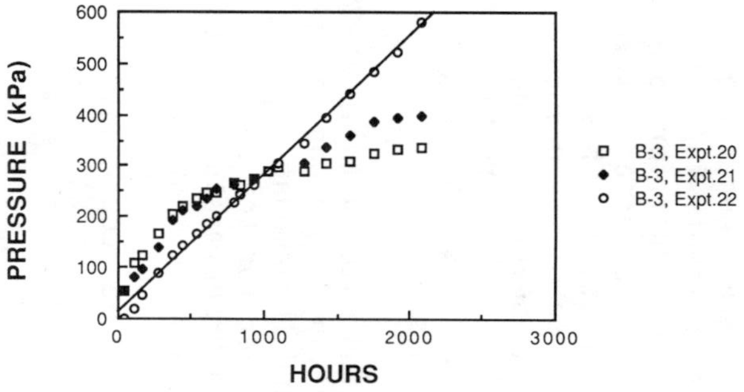

Fig. 3. Rate of Radiolytic Pressurization as a Function of Test Temperature for Baseline Formulation with 3 wt % Salt

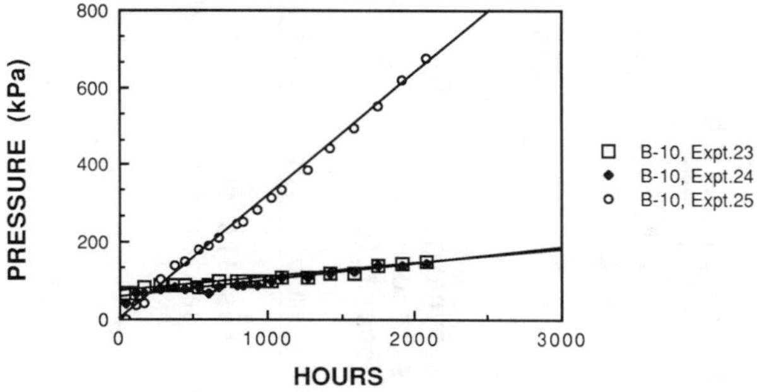

Fig. 2. Rate of Radiolytic Pressurization as a Function of Test Temperature for Baseline Formulation with 10 wt % Salt

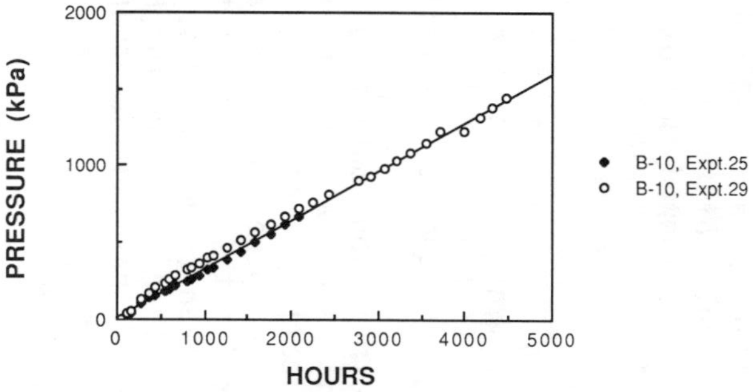

Fig. 4. Rate of Radiolytic Pressurization for 2000 and 4000 Hours of Irradiation of Baseline Formulation with 10 wt % Salt

product for the FA-6 and FA-10 formulations and to determine if additives such as CaS and FeS would inhibit oxygen generation. The results of the gas analyses for these experiments in Table III showed that oxygen was a radiolytic product for both the FA-6 and FA-10 formulations, but it was not detectable when either CaS or FeS was added to the FA-6 formulation. This inhibition of oxygen generation and consumption of the oxygen initially present in the head space was also observed in the experiments with formulations containing slag. These results indicate that the sulfide and possibly the ferrous ions in the slag influence the overall radiolytic mechanism.

A formulation consisting of slag, cement, and water with 10 wt % salt, S-10, was used in experiment 15. Only hydrogen was radiolytically generated. The final pressure and the amount of hydrogen generated (see Table III) were higher for the S-10 experiment (15) than for the parallel FA-10 experiments (13 and 14), even though the dose rate used for S-10 was comparable or lower than that used for FA-10. The lower pressures for the FA formulations indicate that back reactions may be occurring.

Discussion

The most important results obtained from experiments 1-30 are summarized below:

(1) Hydrogen was the only gaseous radiolytic product obtained from the irradiation of slag-containing mortars, and the rate of hydrogen generation did not decrease over the time intervals studied.

(2) The rate of hydrogen generation increased with increasing salt loading of the mortar.

(3) A synergistic effect between the temperature and chloride salt concentration was observed for the rate of hydrogen generation.

Mechanism of Hydrogen Generation

These results can be rationalized using the free radical mechanism for the radiolytic decomposition of water:

$$e^-_{sol} + e^-_{sol} \longrightarrow H_2 \quad (1)$$

$$e^-_{sol} + H \longrightarrow H_2 \quad (2)$$

$$H + H \longrightarrow H_2 \quad (3)$$

Here, hydrogen generation results from the recombination reactions of the solvated electron, e^-_{sol}, and the hydrogen radical, H.[5,6] The findings of oxygen consumption, the non-radiolytic generation of oxygen, and the constant rate of hydrogen generation are consistent with reactions of the hydroxyl radical with the reducing components of slag (e.g., sulfide or ferrous ions) and/or with the chloride ion rather than with molecular hydrogen. Such electron transfer reactions, described generally by

$$OH + Species \longrightarrow Species^+ + OH^- \qquad (4)$$

would compete with the recombination and back reactions,

$$OH + H \longrightarrow H_2O \qquad (5)$$

$$OH + H_2 \longrightarrow H_2O + H \qquad (6)$$

in mortars containing slag and chloride salts.

Effect of Reducing Species in Slag

Whether the hydroxyl radical reacts with the decomposition products of water radiolysis or with reducing species in slag depends on their relative concentration and rate constants. No rate constant data or mechanistic data are available for radiation-induced reactions in mortar or for aqueous solutions with very high pH and ionic strength comparable to the pore water in the mortar matrix. However, rate constant data are available from radiation studies of dilute aqueous systems, and these can be used as guidelines.[7] For example, in dilute aqueous solutions, sulfur species in low oxidation states and ferrous ion react readily with the hydroxyl radical. The rate constants are large, 10^9-10^{10} $\underline{M}^{-1}$ sec^{-1}, which is about two orders of magnitude larger than the rate constant for the back reaction (Eq. 6) of hydrogen and the hydroxyl radical, ~4×10^7 $\underline{M}^{-1}$ sec^{-1}.[5] This back reaction is believed to be responsible for the removal of the hydroxyl radical and, therefore, the non-generation of oxygen and the attainment of equilibrium hydrogen pressure.[1] These rate constant data indicate that, for equal concentrations of reactants, the reactions of hydroxyl radical with reduced sulfur species and ferrous ions would predominate over reactions with hydrogen in aqueous solution and might predominate in the mortar matrix, especially since the concentrations of reducing species in the slag are relatively high (1-4 wt %). The existence of such competing reactions was confirmed in our mechanistic studies using the FA formulation, which indicated that sulfide and possibly ferrous ion inhibited oxygen generation.

The removal of oxygen from the head space was consistently seen in all the slag-containing mortar irradiation experiments. This removal is also consistent with the presence of reducing species in the slag-containing mortar, since ferrous ion, sulfide ion, etc., are readily oxidized by oxygen as well as by hydroxyl radical.

Effect of Chloride Ion Concentration

Several experiments demonstrated that the rate of radiolytic hydrogen generation depended on salt concentration. At room temperature the rate was higher with the B-10 than with the B-3 formulation. These results are consistent with the mechanism described above in which hydroxyl radicals react with species in the mortar such as chloride ion rather than with the water decomposition products. There is no rate constant or mechanistic data that are directly applicable to the reaction of chloride ion and hydroxyl radical in a mortar matrix. However, such reactions have been studied in dilute aqueous solution at various pH's and in concentrated brines.[3,5] The results from these experiments were the following: (1) chloride ion reacts with the hydroxyl radical; (2) the rate of the reaction can be large but is pH dependent; (3) chlorine is formed as an intermediate; and (4) oxygen is a decomposition product when concentrated brines are irradiated. The hydroxyl-chloride reaction is as follows:

$$OH + Cl^- \longrightarrow ClOH^- \tag{7}$$

$$ClOH^- + H^+ \longrightarrow Cl + H_2O \tag{8}$$

$$Cl^- + Cl \longrightarrow Cl_2^- \tag{9}$$

$$2Cl_2^- \longrightarrow Cl_2 + 2Cl^- \tag{10}$$

If the hydroxyl radical-chloride ion reaction occurs in the mortar matrix, an increase in hydrogen yield would follow an increase in the chloride concentration because the recombination reaction of the hydrogen and hydroxyl radicals would be inhibited. This is supported by a comparison between the results of the B-3 and B-10 experiments at room temperature. However, neither oxygen nor chlorine gas was detected in the gas phase in the B-3 and B-10 experiments, indicating that the mortar matrix influences the overall reaction mechanism and the final irradiation products.

Effect of Heat and Salt Concentration

The results of experiments 16-25 with B-3 and B-10 indicated that a test temperature of 120°C rather than ambient decreased the rate of hydrogen generation and the hydrogen yield. This was especially true with the B-10 formulation in which the hydrogen yield decreased tenfold. The effect of the 120°C test temperature with the B-3 formulation was a slight decrease in the rate of hydrogen generation, but this was only observed at times exceeding 1000 irradiation hours. Thus, reactions which decrease the hydrogen yield are enhanced at higher test temperatures and higher chloride concentrations.

A possible explanation for these observations, which is consistent with results from other radiation studies, is that greater amounts of chlorine may be formed from the chloride salts in the mortar at the higher temperature.[9] Scavenging reactions between the chlorine and the hydrogen precursors, the solvated electron and the hydrogen radical, may be enhanced, thereby reducing the yield of hydrogen. This mechanism can be represented by

$$e^-_{sol} + Cl_2 \longrightarrow Cl_2^- \qquad (11)$$

$$e^-_{sol} + Cl_2^- \longrightarrow 2Cl^- \qquad (12)$$

$$H + Cl_2 \longrightarrow H^+ + Cl^- + Cl \qquad (13)$$

The rate constants for these reactions in aqueous solutions are high.[5,8] These reactions can, therefore, account for the lower yields of hydrogen if chlorine formation in the mortar is enhanced at higher temperatures. Preliminary experiments in which solid KCl-44 wt % LiCl salt was irradiated showed this enhancement. About 50% more chlorine was formed interstitially in the solid salt at 150°C than at room temperature.[10] Chlorine formation may also be enhanced in the B-10 formulation because the salt solution used to make the mortar is supersaturated with chloride salts. Thus, the salt may not be incorporated into the mortar matrix as it is in the B-3 formulation.

CONCLUSION

The major finding of our irradiation experiments with slag-containing and chloride salt-containing mortars is that hydrogen is the only gas radiolytically generated. This finding is similar to that reported in other irradiation studies of cement waste forms. The significant difference is that steady-state hydrogen pressures are not attained for a total dose of

540 MRad. Data from our mechanistic studies suggest that the competing reactions for the hydroxyl radical with the reducing species in slag and with the chloride ion may inhibit the recombination reaction of the hydrogen and the hydroxyl radicals and may prevent the back reaction of hydroxyl radical and hydrogen. These reactions increase hydrogen yield and prevent equilibrium conditions from developing. The temperature dependence studies show that other reactions occur at higher temperatures; the rate of hydrogen generation decreases markedly at 120°C for the mortars with 10 wt % salt loading. This may be due to the scavenging of the hydrogen precursors by chlorine or some other species.

ACKNOWLEDGMENT

The authors wish to acknowledge the significant contributions of T. R. Johnson and D. F. Fischer of Argonne National Laboratory, and W. Richmond of Westinghouse/Hanford to this work. Work supported by the U.S. Department of Energy, Nuclear Energy Research and Development Program, under Contract W-31-109-Eng-38.

REFERENCES

1. N. E. Bibler, Radiolytic Gas Production from Concrete Containing Savannah River Plant Waste, Savannah River Laboratory Report No. DP-1464 (1978).

2. P. Colombo and R. M. Neilson, Jr., Properties of Radioactive Wastes and Waste Containers, Brookhaven National Laboratory Report No. NUREG-50957 (1979).

3. W. J. Gray and S. A. Simonson, "Gamma and Alpha Radiolysis of Salt Brines," Scientific Basis for Nuclear Waste Management VIII, C. M. Jantzen, J. A. Stone, and R. C. Ewing, Eds., Materials Research Society, Pittsburgh, PA, pp. 621-630 (1985).

4. W. J. Gray, "Gamma Radiolysis of Groundwaters Found Near Potential Radioactive Waste Repositories," Advances in Ceramics, Vol. 8, G. G. Wicks and W. A. Ross, Eds., The American Ceramic Society, Columbus, OH, pp. 57-61 (1984).

5. J. W. T. Spinks and R. J. Woods, An Introduction to Radiation Chemistry, 2nd Ed., J. Wiley & Sons, New York, NY (1976).

6. I. G. Draganic and Z. D. Draganic, The Radiation Chemistry of Water, Academic Press, New York, NY (1971).

7. L. M. Dorfman and G. E. Adams, "Reactivity of the Hydroxyl Radical in Aqueous Solutions," NSRDS-NBS 46, U.S. Dept. Commerce, National Bureau of Standards, Washington, DC (1973).

8. M. Anbar, M. Bambenk, and A. B. Ross, "Selected Specific Rates of Reactions of Transients from Water in Aqueous Solution. 1. Hydrated Electron, Supplemental Data," NSRDS-NBS 43, U.S. Dept. Commerce, National Bureau of Standards, Washington, DC (1973).

9. P. W. Levy, "Radiation Damage Studies on Natural Rock Salt from Various Geological Localities of Interest to the Radioactive Waste Disposal Program," Nuclear Technology, 60 231-243 (1983).

10. M. A. Lewis and D. W. Warren, unpublished results.

MELTING OF PLUTONIUM-CONTAMINATED RESIDUES BY POWERING MICROWAVE

M. Ohuchi, K. Miyata, J. Ohuchi, E. Inada and N. Tsunoda
Power Reacter and Nuclear Fuel Development Corporation (PNC)
Tokai-mura, Naka-gun, Ibaraki-ken, Japan 319-11

Plutonium-contaminated combustible wastes generated in the mixed oxide (MOX) fuel facilities have been incinerated for volume reduction and immobilizated in the Plutonium contaminated Waste Treatment Facility (PWTF) since late 1987.
The resulting incinerated ash, acid digested residues and oxide slags are melted to be a ceramics-like block in the canister made of the stainless steel by powering microwave.
Characterization is also carried out with these products.

Plutonium-contaminated solid wastes have been generated during MOX fuel fabrication in PNC. These wastes are classified into combustibles, non-combustibles and chlorine-containing organic materials such as PVC and chloroprene. and packed in 200ℓ drums or 1.7 m^3 box containers at the MOX fuel facilities.
As shown in Figure 1, these wastes have been treated for the volume reduction and conditioned them in the PWTF since late 1987. The incinerated ash, acid digested residue and oxide slag are melted without any additives to be a 20~30kg ceramics-like block in the stainless steel made canister by powering microwave. The products are packed in the 200ℓ drums and stored temporarily in the Plutonium-contaminated Waste Storage Facility (PWSF). Characterization is also carried out with these products.

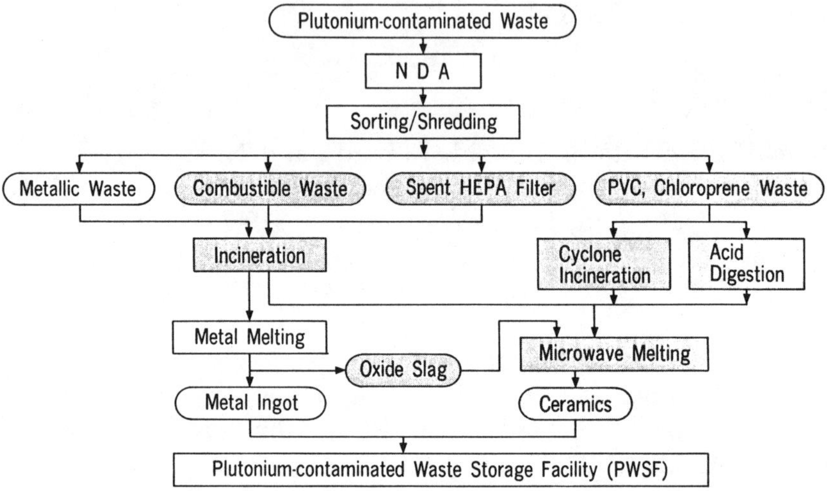

Fig.1. Plutonium-contaminated waste processing flow at PWTF

INCINERATION

Main composition of combustible waste are paper, rags, polyethylene, vinyl acetate and HEPA filter. As shown in Table I, paper bagged combustible waste and cut HEPA filter are incinerated in the primary and secondary chamber at 800~900°C and 900~1000°C, respectively. The ash is crushed and separated from metals, such as nails by sieve, and sent to the microwave melting process by belt conveyer. The shredded waste (less than 5×5mm) such as PVC and chloroprene is burned in the cyclone incinerator at 800°C.

Table I. Operating condition of incineration

Type of Incinerator	Waste Composition	Temperature (°C)	
Conventional	Paper, Rag (Paper Bagged)	Primary Chamber	: 800
		Secondary Chamber	: 900
	HEPA Filter (Cut into 30cm cubic)	High Temp. Filter	: 500~600
Cyclone	PVC, Chloroprene (Cut into 5×5mm)	800	

MELTING

Microwave melting process is based on the dielectric heating

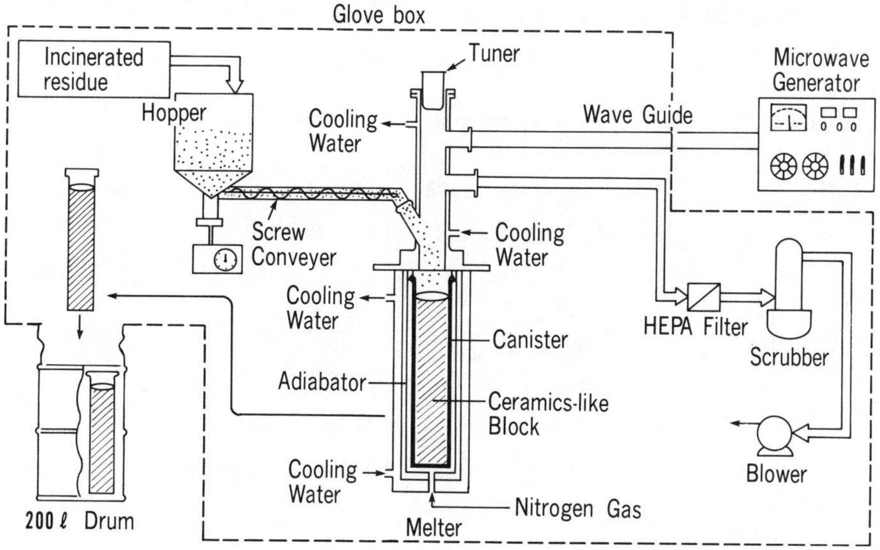

Fig.2. Microwave melting process flow

principle to melt the ash by oscillating the molecules of the residues such as Al_2O_3, SiO_2, MgO and CaO.

As shown in Figure 2, the microwave melting unit consists of melter, ash feeding system, microwave generator and wave guide. The melter and the ash feeding system (Hopper, Screw Conveyer) are installed in a glove box and the microwave is introduced to the melter through wave guide from the microwave generator installed outside of the glove box. The microwave generator has the output of 10KW with the frequency 2450MHz. The melter is equipped with tuner, power monitor, isolator, discharge detector and ITV monitor. The cavity type melter focuses the microwave energy on

Table II. Specification of microwave melter

Items		Specification
Generator	Frequency	: 2450MHz
	Output	: Maximum 10kW
Wave Guide Tube	Size	: 54.6×109.2×14000mm
	Material	: Aluminium
Tuning Method	Head Tuner and Three Stab Tuner	
Melter	Irradiation Mode	: Transverse Electric Mode
	Melting Method	: In-can Melting
	Waste Feed	: Continuously by Screw Feeder
Canister	Size	: 130ϕ ×770mm
	Material	: SUS 304
	Volume	: 10 ℓ
	Cooling	: 0.5 m^3 N_2—gas/h
Throughput	Residues	: 3~4kg/h
		: 30kg/canister

the residues by vertically adjusting the tuner.
Table II shows specification of microwave melter. The ash is fed continuously at the rate of 3~4kg/h and melted in the canister at 1200~1400°C and converted into ceramics-like block. The off-gas is treated by filters and scrubber, and released to inside of glove box. Eight ceramics-like blocks are bagged out for packing in 200ℓ drum, which is stored in PWSF.

EVALUATION

Scanning Electron Microscope, Inductively Coupled Plasma Emission Spectrometer and leaching apparatus have been used for characterization of the ceramics-like block. The evaluation tests include observation of surface, measurement of density and leaching experiment.

OPERATIONAL RESULTS

Generation of residues

Approximately 12 tons of combustible wastes have been incinerated to be 0.9 ton of ashes by Feb. 1989. Main elements of the ash are aluminium, silicon, calucium and magnesium. Size-distribution of the crashed ash is averaged to be 1.8 μm Φ.
Chlorine-containing organic wastes have been also incinerated in cyclone incinerator with oxygen-rich air. Main elements of the analyzed ash are chlorine, calucium, aluminium and silicon. Size-distribution of this ash is averaged to be 2.3~4.0mm Φ.

Microwave melting

Approximately 0.9 ton of ashes have been melted by powering microwave to be 35 canisters of ceramics-like blocks. Volume reduction ratio is approximately 1/150 by incineration and melting, comparing the volume of the containers between untreated wastes and products.
During the melting, electric discharge has been occured because of the pyrolysis and ionization of generated carbon monoxide. It is able to avoid generation of electric discharge by decreasing the

feeding rate of incinerated ash or controling the output of microwave. Table III, Figure 3 and Figure 4 show the operating results, melting surface in the canister and the cross section of the small scale ceramics-like blocks, respectively.

Table III. Operating data

Items	Results
Waste	Incinerated Ash
Feed Rates	3~4 kg/h
Operating Temperature	1200~1400℃
Operating Pressure	−45~−55 mmH_2O
Off-gas	5~6N m^3/h
Density of Ceramics-like Block	3g/cm^3

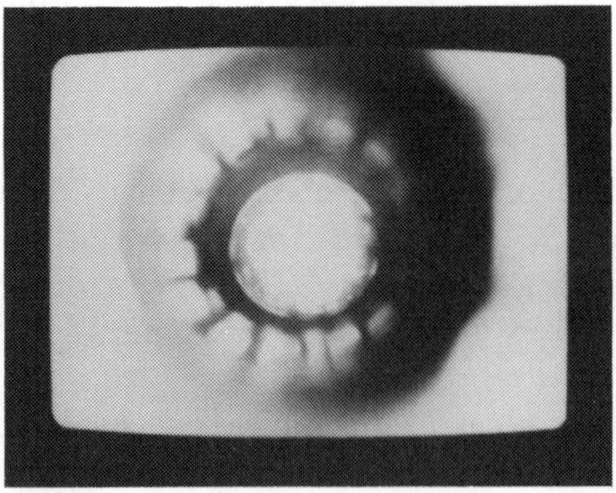

Fig.3. Melting surface in the canister

Characterization of ceramics-like block

Density of the ceramics-like block is averaged to be 3.0g/cm^3. The major mineral phases of the products are forsterite, anorthite and augite.

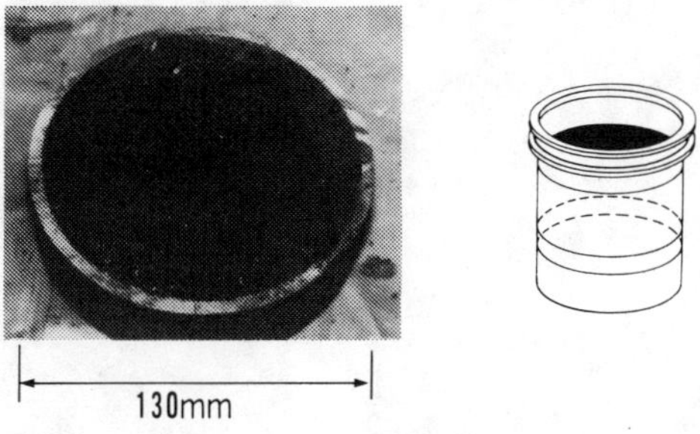

Fig.4. Cross section of ceramics-like block

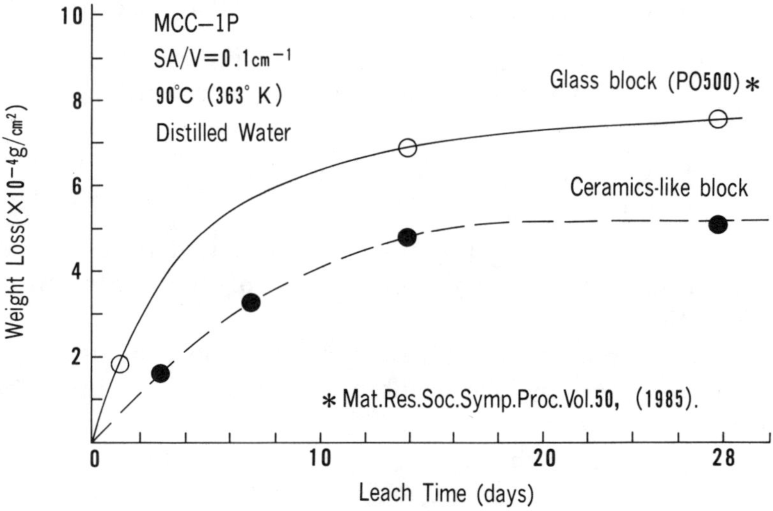

Fig.5. Weight loss in 28days-leaching experiment

Figure 5 shows weight loss in 28days-leaching experiment. Leaching rate of these blocks is on the order of 10^{-5}g/cm^2 · day. Figure 6 shows the cross section of the ceramics-like block before and after the leaching experiment.

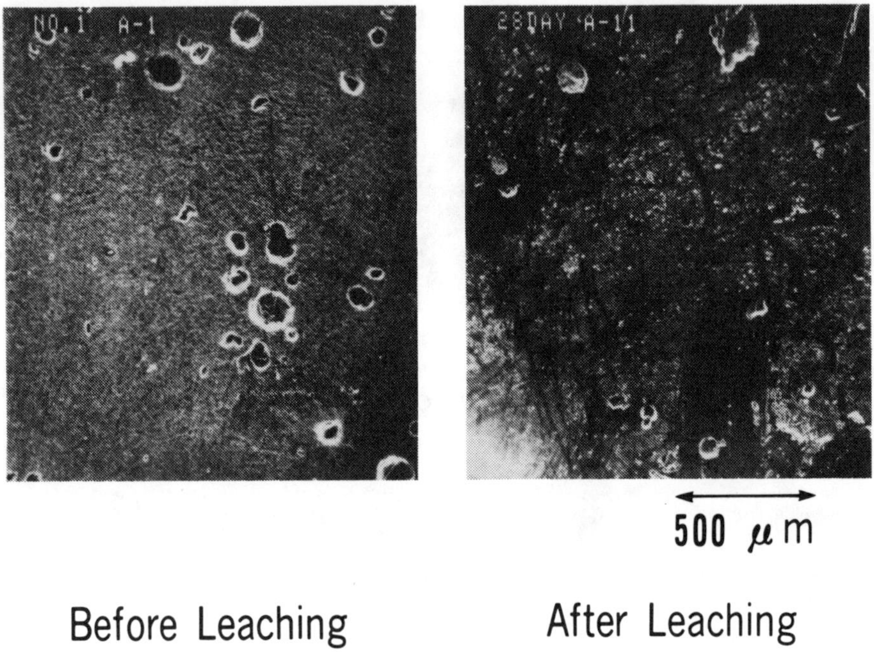

Fig.6. Photographs of ceramics-like block before and after leaching experiment

CONCLUSION

1. Microwave melting of plutonium-contaminated residues has been demonstrated successfully for the volume reduction and immobilization.
2. The volume reduction ratio of 1/150 for the combustible wastes has been attained by the combined process of incineration and microwave melting.
3. The incinerated ash has been conditioned to be the durable waste form without any additives by powering microwave.

4. Density of the ceramics-like block is approximately 3.0 g/cm^3.
5. Leaching rate of ceramics-like block is obtained $\sim 1 \times 10^{-5} g/cm^2 \cdot$ day.

REFERENCES

1. F. Komatsu, Y. Sawada, K. Ohtsuka, J. Ohuchi, "Development of a new solidification method for wastes contaminated by plutonium oxides", Radioactive waste management (IAEA in Seattle, 16-20 May 1983).
2. Y. Ogata, J. Ohuchi, E. Inada and N. Tsunoda, "Processing of plutonium-contaminated waste at PWTF" (IAEA in Stockholm, May 1988).

PRODUCT PERFORMANCE OF THE PILOT-SCALE IN SITU VITRIFICATION OF A SIMULATED WASTE DISPOSAL SITE AT THE OAK RIDGE NATIONAL LABORATORY

S. O. Bates
J. G. Carter
S. S. Koegler

Pacific Northwest Laboratory*
Richland, Washington. 99352
(509) 375-2823.

ABSTRACT

Process feasibility and waste form performance studies have been successfully performed for applying in situ vitrification (ISV) to intermediate-level (low level) waste placed in seepage pits and trenches at Oak Ridge National Laboratory (ORNL). A pilot-scale ISV test 3/8-scale simulation of an ORNL trench was constructed and vitrified at ORNL. The pilot-scale waste form consisted of a glass phase and a multi-crystalline devitrification product. Both these phases had essentially the same elemental composition. The durability of the waste form was evaluated by MCC-1, MCC-3, and EP toxicity tests. The durability of the waste form was found to be similar to or better than two high-level nuclear waste glasses (i.e., PNL-76-78 and SRL-165).

INTRODUCTION

Pacific Northwest Laboratory (PNL) successfully performed laboratory, engineering-scale, and pilot-scale tests of the in situ vitrification (ISV) process on actual Oak Ridge National Laboratory (ORNL) soil and limestone. In situ vitrification is an

---

* Pacific Northwest Laboratory is operated by Battelle Memorial Institute for the U.S. Department of Energy under contract DE-AC06-76RLO 1830

emerging, in-ground thermal treatment technology that fixes radioactive and/or toxic contamination in the soil by converting the soil into a glass and crystalline waste form.

In 1986, and under contract with ORNL, it was decided to test ISV technology using actual soil and limestone from ORNL. Laboratory and engineering-scale tests were performed and reported.(1) A 3/8-scale, pilot-scale ISV test was designed to simulate the processing of the ORNL seepage pits and trenches based on the findings of the laboratory- and engineering-scale tests. The main objective of the testing was to provide technical support for the application of large-scale ISV technology to ORNL intermediate-level waste in seepage pits and trenches. This work included the characterization of the vitrified product for composition, physical properties, volume reduction, crystal structure/devitrification, Cs and Sr retention, and chemical durability.

Process Description

The ISV process involves the insertion of electrodes into the soil, typically in a square array. A starter path for electrical current is established by using a small amount of a graphite and glass frit mixture placed between the electrodes on the soil surface. Dissipation of power through the starter material creates temperatures that are high enough to melt a layer of soil, thus establishing a molten conductive path. This molten zone continues to grow downward and encompasses the contaminated soil.

Figure 1 illustrates the process sequence. At the high temperatures (approximately 1300°C to 1600°C), organic materials pyrolyze, diffuse to the surface, and combust. This movement of gas

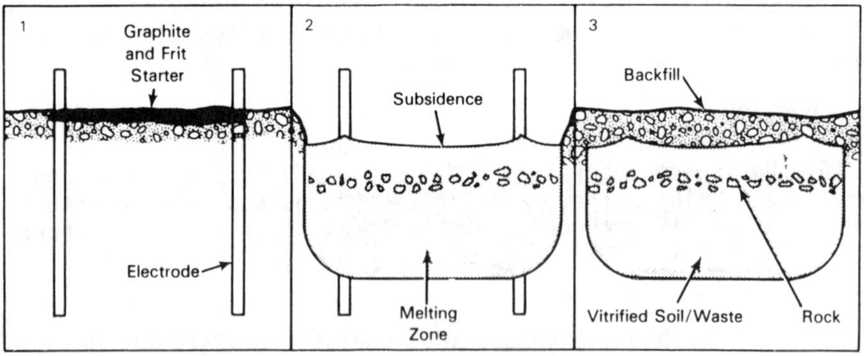

Figure 1. Operating Sequence of In Situ Vitrification System

through the melt may produce some porosity near the surface of the melt. Any off gases from the process are collected, treated, and monitored to ensure their environmentally safe release. Remaining ash, along with other noncombustible material, dissolves or becomes encapsulated in the molten soil. The molten soil cools to a durable glass and crystalline waste form physically resembling natural obsidian.

TEST DESCRIPTIONS

To enable the pilot-scale ISV system to be used, a 3/8-scale model of a well-characterized trench was constructed in a pristine (i.e., uncontaminated) portion of ORNL. The constructed test trench was 1 m wide at the top and tapered to 0.4 m at the bottom. A schematic cross section at the midpoint of the trench is shown in Figure 2. To simulate the contaminated sludge that is present in a typical trench, 526 kg of an 18 wt% Cs- and 82 wt% Sr-carbonate mixture was placed in the central portion of the trench within the square electrode array.

The four molybdenum electrodes (5 cm diameter and 3.7 m long) were placed in graphite sleeves (15 cm O.D.) and placed 1.2 m apart in a square array of augured holes approximately 2.5 m deep. The off-gas hood was then placed over the trench and connected to the off-gas treatment system.

The run time for the ISV process was 110 h and the process used 28.9 MWh. After completion of the ISV run, the vitrified mass was examined by core drilling and excavation. Subsidence from the surface was 0.91 m (36 in.), giving a volume reduction of 43%. The mass of the vitrified monolith was 22.9 x 10^3 kg. The vitrified mass contained nodules of a gray crystalline material. These nodules (up to 2.5 cm in diameter) fused into larger clusters. The largest concentration of the gray material was between the electrodes.

ISV PRODUCT CHARACTERIZATION

The product characterization consisted of the evaluation of the composition and phase behavior properties of the ORNL ISV waste form. Because the primary duty of the ISV waste form is to contain the waste components, the waste form's durability is of prime importance and was also evaluated.

Samples were received from cores 2 through 6 taken from the ISV monolith. The relative position of the cores with respect to the electrodes is shown in Figure 3. Visual observation of these

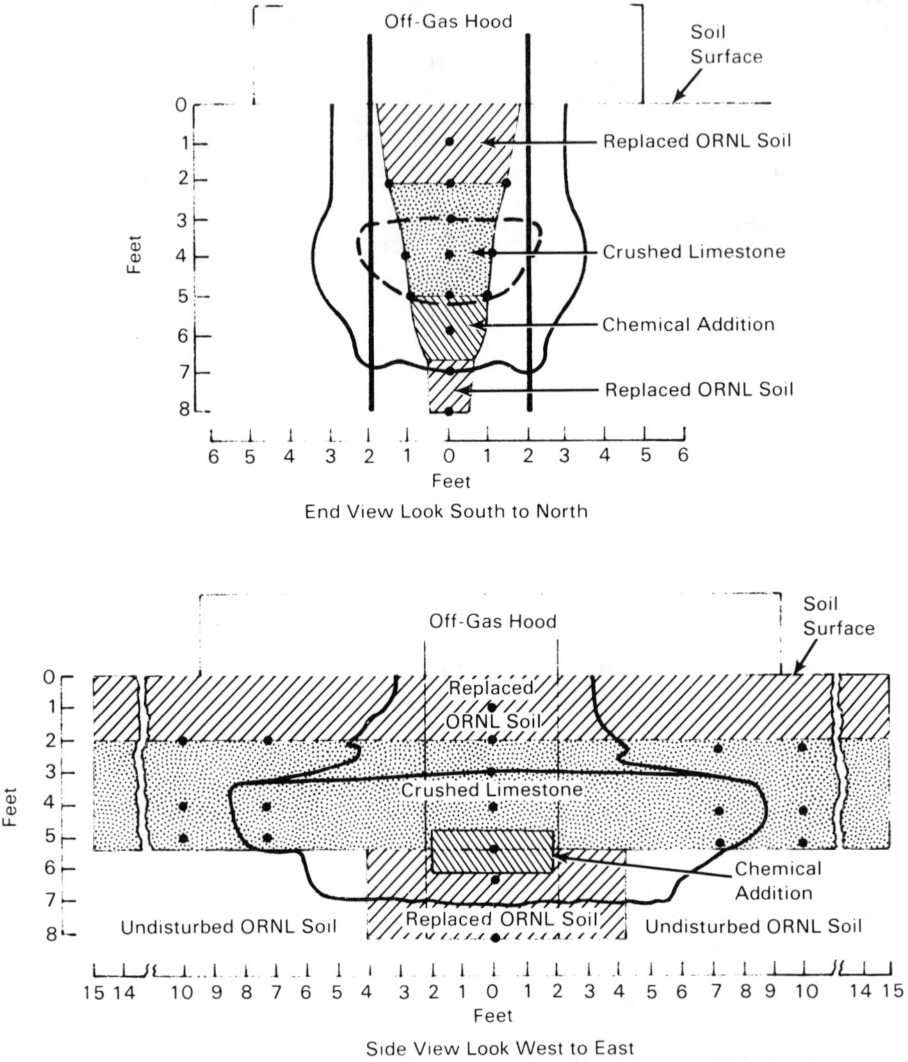

Figure 2. Predicted Shape of the Pilot-Scale ISV Mass (overlayed on the original trench dimensions)

samples showed a frothy, vitreous area at the top of the cores, with the bulk of the core consisting of glass, a gray crystalline (rock) material, or a mixture of both glass and crystalline material. Some soil or calcined soil was also observed at the

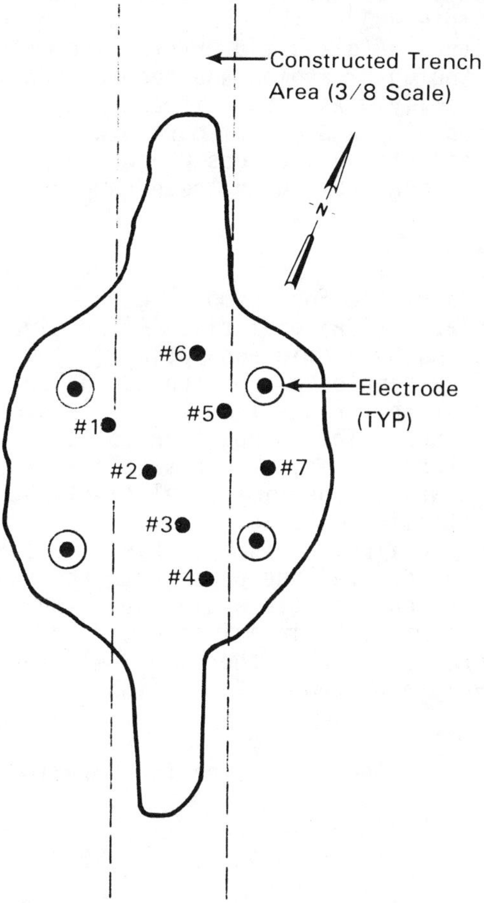

Figure 3. Relative Positions of Core Samples Taken from ORNL

bottom of some of the deeper cores. Core 2, which was taken from near the center of the ISV monolith, was comprised almost entirely of the crystalline material. In contrast, core 4 consisted almost completely of glass. The remaining cores consisted of a mixture of both phases.

Four representative samples were selected for detailed evaluation. Because core 3 was typical of most of the other cores and represented all of the observed phases, three samples from this core were chosen for detailed evaluation. These three samples consisted of a sample of the frothy glass from the surface of the monolith (sample #230), the gray crystalline material (sample #233), and a mixture of glass and crystalline material (sample #232). Because the outer region and the major bulk of the ISV monolith was primarily glass, the vitreous phase of the monolith was also evaluated. A sample from core 4 was also chosen for detailed evaluation of the vitreous phase (sample #231). Table 1 describes the location from which these samples were taken.

Sample Composition

Elemental analyses of the four samples were conducted by inductively coupled plasma (ICP) analysis. After digestion, samples were analyzed for Cs by atomic adsorption (AA). The oxide compositions of the four samples are presented in Table 2, along with the compositions of the original soil and limestone from the pilot-scale test area. The compositions of all the samples, both crystalline and glass phases, were essentially equal, indicating that the monolith was at one time a relatively homogeneous melt and the gray crystalline material resulted from devitrification of the glass upon cooling. This hypothesis is supported by the increased amount of crystalline phase near the center of the monolith where the cooling rate would have been the slowest. Cesium and strontium levels in all samples, both glass and crystalline rock, were essentially the same. All the samples were semi-quantitatively analyzed by X-ray fluorescence (XRF) for Cs

Table 1. Location of Samples Chosen for Detailed Evaluation

Sample #	Core	Position in Core, in. from top	Description of Sample
230	3	0-37, top crust	Frothy, foamed vitreous layer found on the top surface of the monolith
231	4	13-31, top	Glass phase only
232	3	0-37, upper middle	Glass with ≈5% crystalline material
233	3	0-37, bottom	Crystalline phase only
236	3	37-52, middle	Crystalline nodules
240	3	37-52, middle	Glass phase only

Table 2. Oxide Compositions for ORNL ISV Pilot-Scale Test Samples. Values are averages of unnormalized ICP analysis given as oxide weight-percent.

Oxides	Pilot-Scale Test 230	231	232	233	Original Soil	Lime-stone
Al_2O_3	12.8	13.1	13.2	13.2	18.3	0.97
BaO	0.074	0.077	0.078	0.077	0.063	
CaO	17.2	18.6	18.7	18.7	3.09	87.3
Cs_2O	0.33	0.34	0.34	0.37		
Fe_2O_3	4.30	4.41	4.46	4.52	6.32	1.37
K_2O	2.36	2.66	2.34	2.86	2.7	
MgO	2.02	2.15	2.16	2.16	1.83	6.8
MnO_2	0.14	0.14	0.14	0.15	0.12	0.39
Na_2O	1.50	1.38	1.47	1.74	0.77	0.65
SiO_2	51.8	51.5	52.0	52.4	63.1	2.95
SrO	1.30	1.40	1.40	1.39	0.13	0.05
TiO_2	0.58	0.60	0.60	0.60	0.83	
ZrO_2	0.029	0.043	0.035	0.046		
Total	94.5	96.2	96.9	98.2	92.3	100.5

and Sr. The XRF analyses confirmed ICP results that the Sr and Cs were uniformly distributed throughout the monolith.

Phase Behavior

The structure of the waste form was evaluated by x-ray diffraction (XRD) and scanning electron microscopy with energy dispersive analysis by x-ray (SEM/EDAX).

The vitreous samples exhibited no XRD peaks, as expected. All the crystalline samples exhibited the same complex XRD pattern that was the sum of more than one crystalline phase. Substitutions of minor elements into the crystalline phases have likely offset the peaks, making precise identification difficult.

However, it appears that the crystalline material consists predominately of anorthite (feldspar, $CaO-Al_2O_3-SiO_2$) and a wollastonite-like phase ($CaO-SiO_2$), which is consistent with the phase diagram of the $CaO-Al_2O_3-SiO_2$ system(2).

SEM/EDAX was conducted on thin sections of the four major samples to identify the different phases. The glass samples were found to be homogeneous, with no detectable non-vitreous phases. The frothy glass layer found on the top surface of the monolith was also vitreous. X-ray mapping of Cs in the glass samples found that the Cs did not segregate and was uniformly dispersed throughout the glass. Because the levels of Cs in the samples were near the detection limits, the x-ray mapping can only indicate that no concentrations of Cs occurred. X-ray mapping of Sr was not possible due to overlap of the Sr peak with a very strong silicon peak.

The samples consisting of gray crystalline material were similar in morphology. The microstructure is represented by alternate layers of the three crystalline phases (lamellar) typically found when an eutectic composition cools.

SiO_2, CaO, and Al_2O_3 are the major components of ISV product. If only these three components are considered, the normalized composition of the ORNL ISV product is 62 wt% SiO_2, 22.2 wt% CaO, and 15.7 wt% Al_2O_3. This normalized composition was very near a low-melting eutectic and may account for the microstructure of the crystalline phase(3). X-ray mapping of Cs in the gray crystalline phase indicated that even after devitrification, the Cs did not concentrate significantly in any one phase.

CHEMICAL DURABILITY

The durability of the ISV waste form was evaluated using three types of tests: MCC-1(4), MCC-3(5), and the Extraction Procedure (EP) toxicity test(6). To allow the direct comparison of data, the leachate, temperature, and surface area/volume (SA/V) used in testing the ISV waste form were identical to those parameters used to test high-level nuclear waste glasses. The differences in application (repository conditions, groundwater, temperature, etc.) and the nature of the waste form (multiphase ceramic glass versus single-phased glass) makes such comparison difficult. Most high-level nuclear waste glasses are tested at 90°C. By conducting the leach testing of the ORNL ISV waste form at 90°C, the results can be compared with the data base of leaching data on the high-level waste glasses. Leachants were analyzed for cation content only.

To allow the comparison of leach tests results from waste forms with different compositions, the results are given in terms of normalized elemental mass release for the MCC-1 and MCC-3 test.

MCC-1 Testing

The MCC-1 static leach testing was conducted with an SA/V of 10 m^{-1} in deionized water (DIW) at 90°C for 28 days. Because the MCC-1 procedure requires a small monolith of sample, test samples consisted of only the glass or crystalline phase. Testing was conducted on sample numbers 230, 231, 232 (glass phase only), and 233 (crystalline phase). The results from the MCC-1 leaching tests are presented in Table 3. Leaching data from two high-level nuclear waste glasses are also presented for comparison.

For the glass samples (230, 231, and 232), normalized releases for each element did not vary significantly. The normalized releases for the crystalline samples varied slightly, indicating that some incongruent dissolution may have occurred. However, under these test conditions the crystalline phase (sample 233) appears to be more durable than the glass phase (samples 230, 231, or 232), as the normalized releases from the crystalline sample were lower than those of the glass samples. The normalized releases for the ISV product were very comparable to those of the high-level nuclear waste glasses.

Table 3. Comparison of Normalized Release from MCC-1 Leach Testing for ORNL Pilot-Scale ISV Products and Some High-Level Nuclear Waste Glasses[(a)]

	Normalized Elemental Release (g/m^2)					
Element	230	231	232	233	76-68(7)	SRL165(8)
Al	5.9	6.0	5.6	2.8	--	12
Ca	6.9	6.5	6.3	3.8	3.0	--
Si	6.3	6.5	6.2	4.8	26	15
Sr	7.5	7.2	6.9	1.7	3.8	1.7
Cs	9.0	9.0	8.7	8.5	14.7	18
(pH)	(9.3)	(9.2)	(9.1)	(9.1)	--	--

(a) Conditions: MCC-1 in DIW, 28 days, SA/V = 10^{-1}, 90°C.

MCC-3 Testing

The MCC-3 tests were conducted with an SA/V of 2000 m^{-1} in DIW at 90°C for 28 days. Minus 100 mesh + 200 mesh glass powder was used for all tests. Because a powdered sample is used for the MCC-3 tests, a combination of glass and gray crystalline material was also tested. A 50/50 wt% mixture of sample 236 (crystalline nodules) and 240 (glass) was used (see Table 1). All tests were run in duplicate.

The results from the MCC-3 leach tests are presented in Tables 4 and 5. MCC-3 leachants were filtered through 0.45-μm filters, then 18-angstrom filters, to determine the amount of colloidal material in the leachant. The same trends observed in the MCC-1 test results were apparent in the MCC-3 test results. The releases from the combined glass-crystalline sample were between the glass-only and crystalline-only samples, suggesting that no interactions occurred during leaching. The waste form from the ISV test, both glass and crystalline, had releases approximately 0.5 to 10 times lower than those for the SRL-165 under these MCC-3 test conditions.

Table 4. Comparison of Average Normalized Releases from MCC-3 Leach Testing: 0.45-μm Filtrate for ORNL Pilot-Scale ISV Products and Some High-Level Nuclear Waste Glasses[(a)]

	0.45-μm Normalized Elemental Release (g/m^2)					
Element	230P	231P	232P	233P	236/240	SRL165
Al	0.019	0.020	0.019	0.014	0.018	0.27
Ca	0.058	0.048	0.041	0.032	0.037	0.07
Si	0.039	0.039	0.038	0.040	0.039	0.32
Sr	0.046	0.039	0.036	0.020	0.026	0.05
Cs	0.097	0.073	0.072	0.147	0.143	--
(pH)	(9.4)	(9.4)	(9.6)	(9.5)	(9.5)	--

(a) Conditions: MCC-3 in DIW, 28 days, SA/V = 10^{-1}, 90°C.

Table 5. Comparison of Average Normalized Releases from MCC-3 Leach Testing: 18-Angstrom Filtrate for ORNL Pilot-Scale ISV Products and Some High-Level Nuclear Waste Glasses[(a)]

Element	18Å Normalized Elemental Release (g/m^2) 230P	231P	232P	233P	236/240	SRL165
Al	0.018	0.019	0.018	0.013	0.017	0.10
Ca	0.054	0.039	0.038	0.029	0.033	0.02
Si	0.038	0.038	0.037	0.039	0.038	0.26
Sr	0.042	0.033	0.032	0.017	0.022	>0.01
Cs	0.085	0.070	0.070	0.140	0.123	----

(a) Conditions: MCC-3 in DIW, 28 days, SA/V = 2000 m^{-1}, 90°C, 2 replicates

Extraction Procedure (EP) Toxicity Tests

Extraction procedure toxicity tests were conducted to compare the durability of the ORNL ISV waste form with currently established release requirements. The EP toxicity test is designed to evaluate the extraction toxicity of a waste form. The EP toxicity results are presented in Table 6 along with the maximum concentration limits that have been established.

Barium was the only detectable element in the ORNL pilot-scale ISV product for which maximum concentrations limits have been set. The limit for barium is 100.0 mg/L, and the highest concentration detected for any of the samples tested was 0.044 mg/L. Unless the concentrations of any of the elements listed in Table 6 increase substantially in the ISV soil or limestone, the ISV waste form should easily meet any EP toxicity requirements.

CONCLUSIONS

The ORNL pilot-scale ISV waste form was a mixture of vitreous and crystalline phases, with very similar elemental compositions and uniform distributions of cesium and strontium. This indicates that the resulting monolith was once a relatively homogeneous melt, and that the crystalline material formed from devitrification of the glass during slow cooling.

Table 6. Results from EP Toxicity Testing of the Pilot-Scale ISV Test at ORNL

Contaminant	Sample Numbers 230	231	232	233	Maximum Concentration Allowed (mg/L)
Arsenic	<1	<1	<1	<1	5.0
Barium	0.015	0.012	0.016	0.044	100.0
Cadmium	<0.005	<0.005	<0.005	<0.005	1.0
Chromium	<0.02	<0.02	<0.02	<0.02	5.0
Lead	<0.05	<0.05	<0.05	<0.05	5.0
Mercury	<0.001	<0.001	<0.001	<0.001	0.2
Selenium	<1	<1	<1	<1	1.9
Silver	<0.1	<0.1	<0.1	<0.1	5.0

The normalized mass releases at 90°C as determined by MCC-1 and MCC-3 testing of the glass and crystalline samples were approximately equivalent to those for PNL 76-68 and SRL-165 (high-level nuclear waste glasses). In addition, once the ISV vitrified product has cooled, the anticipated storage temperature would be essentially ambient ground temperature. At these temperatures, dissolution rates could be as much as one or two orders of magnitude lower. Based on EP toxicity testing, the pilot-scale waste form, both glass and crystalline phases, is not EP toxic.

These evaluations of the ISV product indicate that the ISV process and waste form are highly suitable for immobilization of ORNL intermediate-level waste.

REFERENCES

(1) J. G. Carter, S. O. Bates, and G. D. Maupin, *In Situ Vitrification of Oak Ridge National Laboratory Soil and Limestone*; PNL-6174, Pacific Northwest Laboratory, Richland, WA, 1987.

(2) E. M. Levin, C. R. Robins, and H. F. McMurdie. *Phase Diagram for Ceramists*; Volume 1, pp. 219. American Ceramic Society, Inc., Columbus, OH, 1964.

(3) E. F. Osborn, and A. Muan, Revised and Redrawn, "Phase Equilibrium Diagrams of Oxide Systems," Plate 1, published by the American Ceramic Society and the Edward Orton, Jr., Ceramic Foundation, 1960.

(4) Materials Characterization Center, Rev. 7, 1986. "MCC-1P Static Leach Test Method." In *Nuclear Waste Materials Handbook. Test Methods*. DOE/TIC-11400. Technical Information Center, Springfield, Virginia.

(5) Materials Characterization Center, Rev. 7, 1986. "MCC-3S Agitated Powder Leach Test Method." In *Nuclear Waste Materials Handbook. Test Methods*. DOE/TIC-11400. Technical Information Center, Springfield, Virginia.

(6) U.S. Environmental Protection Agency. 1986. "Method 1310: Extraction Procedure (EP) Toxicity Test Method and Structural Integrity Test." In *Test Methods for Evaluating Solid Waste*, SW-846, Vol. 1, Section C, Part 2 (Chapter 8). U.S. Environmental Protection Agency, Washington, D.C.

(7) C. H. Kindle, and M. R. Kreiter, *Comprehensive Data Base of High-Level Nuclear Waste Glasses - September 1987 STatus Report. Volume 1 - Discussion and Glass Durability Data*; PNL-6353 Vol 1, Pacific Northwest Laboratory, Richland, WA, 1987.

(8) G. T. Chandler, G. G. Wicks, and R. M. Wallace, *Chemical Durability of SRP Waste Glass - Saturation Effects and Influence of SA/V*; DP-MS-84-37, E. I. du Pont de Nemours & Co., Savannah River Plant, Aiken, SC, 1984.

FORMULATION OF DURABLE WASTEFORMS FOR UK REPROCESSING WASTES

John D Palmer
British Nuclear Fuels
Research and Development Department
Sellafield, UK

ABSTRACT

Inorganic cement formulations have been adopted as encapsulation matrices for Intermediate Level Wastes arising at Sellafield. An extensive product evaluation programme has been carried out to produce data on the properties of the encapsulated waste. The results from the studies are described and show durable wasteforms can be produced using blended cements containing ground granulated blastfurnace slag or pulverised fly ash mixed with ordinary Portland cement.

INTRODUCTION

In the UK, radioactive wastes are divided into three categories: low-level, intermediate level and high level wastes[1].

Low level wastes (LLW) are defined as containing radioactivity at levels greater than those allowed for municipal disposal but less than 4 GBq/tonne alpha and 12 GBq/tonne of beta/gamma activity.

Intermediate level wastes (ILW) contain radioactivity levels exceeding the upper boundaries for LLW's, but do not require radiogenic heating to be taken into account in the design of storage or disposal facilities.

High level wastes (HLW) are wastes in which the temperature may rise significantly as a result of their radioactivity, so this factor must be taken into account in the design of storage and disposal facilities.

The Sellafield reprocessing plant operated by British Nuclear Fuels (BNFL) has historically sent LLW for shallow land burial (SLB) whereas ILW and HLW streams have been stored on site in engineered tanks or silos. As part of BNFL's waste management strategy existing and future waste arisings will be treated to facilitate transport, storage and disposal [2].

LLW will be volume reduced and containerised prior to disposal in engineered trenches.

HLW will be immobilised by vitrification and then stored for 50 years to allow the radioactivity to decline before disposal.

ILW's will be encapsulated in an inorganic cement matrix and sent for storage until such time as a suitable disposal route is available.

This paper details the selection and testing criteria applied by BNFL which have led to the development of cement based systems for ILW encapsulation. It particularly discusses how cement blends based on ground granulated blastfurnace slag (BFS) with ordinary portland cement (OPC) and pulverised fly ash (PFA) with OPC satisfying product requirements for storage and disposal.

INTERMEDIATE LEVEL WASTES AT SELLAFIELD

At present there is approximately 30,000m^3 of ILW stored in an unconditioned form on the Sellafield site. By the year 2000, some 60,000m^3 of ILW will have been produced comprising over 150 individual streams [3]. The ILW can be divided into four main types, namely:

i) Fuel element cladding from Magnox and Oxide Fuel.

ii) Slurries and flocs resulting from reprocessing and the treatment of effluents.

iii) Technological wastes, which comprise of a wide range of solid items produced from plant operations and maintenance.

iv) Plutonium contaminated material (PCM) from glove box operations.

Apart from their physical diversity, ILW streams also vary enormously in specific activity. For example, magnox fuel cladding can consist of upto 1000 TBq/m^3 beta/gamma and

10 TBq/m^3 alpha, flocs may contain upto 10 TBq/m^3 alpha, whilst spent ion exchange resins may contain upto 2000 TBq/m^3 beta/gamma activity. The dose rates from bulk wastes can be several 10's of Gy/hour.

MATRIX SELECTION AND PRODUCT EVALUATION STUDIES

The diverse nature of ILW streams will have a significant effect on the matrix selection and subsequent properties of the encapsulated ILW. Consequently, in 1982 a major R&D programme was initiated by BNFL (with partial funding from the UK Department of Environment) to investigate the properties of a variety of matrix/waste formulations [4]. The objectives of the programme were:

a) to identify potential encapsulation matrices for each waste stream and formally select the preferred option, and

b) to evaluate the product properties of the encapsulated waste and establish a database to support BNFL's waste management strategy.

The programme was sub-divided into three areas, namely:

Small scale non-active
Small scale active
Full scale non-active

Comparison between small scale active and non-active samples enables the effects of radiation to be assessed. The effect of scaling up from small scale to full scale was then investigated inactively. Therefore by combining these two effects the properties of full scale active products can be predicted.

Each major wastestream has been systematically studied in four consecutive phases:

Phase 1

- Waste definition, characterisation and simulation.

- Identification of special characteristics.

- Review of 8 possible encapsulation matrices and selection of at least three potentially acceptable matrices.

Phase 2

- Identification of waste pre-treatment, if required.

- Comparison of the potential matrices by a limited experimental consideration, on small scale samples, of physical and thermal stability, leach testing, mechanical strength and radiation stability in order to identify the preferred 'reference' matrix for Phase 3 studies. Multi-attribute analysis is used to select the reference matrix by evaluating performance against 'essential' and 'desirable' criteria.

Phase 3

- Extensive practical studies of the reference matrix in six areas:

  - Physical properties
  - Mechanical properties/Chemical stability
  - Thermal properties
  - Radiation stability
  - Leaching behaviour

At the completion of the Phase 3 work, adequate information is available to underwrite a basic plant process and to secure approval from the Authorising Departments that construction can proceed.

Phase 4

The work in earlier phases concentrated on a single optimum formulation (eg a particular waste:cement ratio) and work in Phase 4 is designed to identify the extent of the formulation range which still provides an acceptable product. Such information is required for detailed plant design and operation. All information generated during Phase 4 enhances the quality assurance database which will be the foundation of plant safety cases and operations.

This report addresses the formulation selection and testing criteria applied to wasteforms by BNFL. Discussions are limited to phases 1-3 of the programme as Phase 4 deals with more general issues on the effect of variations in wastestreams and process limits on product quality.

Of the 8 possible encapsulation matrices reviewed in Phase 1 of the programme, only inorganic cements, polymers and polymer modified cements were considered suitable for further investigation at Phase 2.

The Phase 2 studies assess the performance of the alternative matrices against a number of essential and desirable criteria for each process stage ie encapsulation, storage, transport and disposal. Details of the assessment criteria applied are given in Table 1.

Table 1. Criteria for Assessing Product Acceptability

Stage of Waste Management	Essential Criteria	Desirable Criteria
1 Processing/ Encapsulation	i) Operationally Safe	vi) Simple Process
	ii) Consistent Product	vii) Economic
	iii) Flow Characteristics (1)	viii) Minimum Secondary Waste
	iv) Pumpability (1)	
	v) Working Life	
2 Storage	i) Essentially Monolithic for at least 50 years	v) Minimum Volume
		vi) High Thermal Stability
	ii) Adequate Strength for Handling	vii) Good Leaching Characteristics
	iii) No Unacceptable Matrix/Container Interaction	viii) Strength/Impact Resistance
	iv) No Unacceptable Fire/Explosion Hazard	

Table 1. Continued

Stage of Waste Management	Essential Criteria	Desirable Criteria
3 Transport	i) Satisfy Transport Regulations	ii) Self Shielding + 2(v) - (viii)
4 Disposal	i) No Criteria Specified since Package will be one Compartment of a Multi-Barrier System	ii) No Enhancement of Isotope Mobility iii) Matrix/Container/ Waste Compatibility iv) Monolithic for Long Times v) Low Permeability vi) Well Understood Properties vii) Minimise Equilibrium Concentration of Long Lived Isotopes

(1) For grout only

Of the 20 major wastestreams evaluated to-date, inorganic cements based on BFS/OPC and PFA/OPC blends have been identified as the most suitable encapsulation matrix for each ILW.

The important benefits associated with the use of cement when compared with alternative matrices include:

i) The processing plant is relatively simple.

ii) Cement is suitable for encapsulating wastes ranging in composition from dry solids to wet sludges.

iii) Cement is composed of thermodynamically stable hydrated species with well documented physical and chemical properties.

iv) Cement has excellent thermal stability due to the low thermal conductivity and non-flammable nature of the material.

v) Because of its high pH and sorptive capacity cement has excellent properties for actinide retention.

As described earlier, the Phase 3 stage of the programme involves a detailed assessment of the full scale properties of encapsulated simulant wastes. The test methods applied are summarised in Table 2 and are tailored specifically for the evaluation of cement based matrices.

Table 2. Summary of Phase 3 Testing Programme

Programme Area		Test Regime
Physical Properties	:	Homogeneity Density Micro-voidage (1mm diameter) Porosity Pore size distribution Liquid Permeability Gas Permeability
Mechanical Properties	:	Compressive Strength Tensile Strength Ultrasonic Pulse Velocity Elastic Modulus Strain Measurements

Table 2. Continued

Programme Area		Test Regime
Impact Resistance		
Chemical Stability	:	Corrosion Monitoring of Embedded Metal Wastes and Drums Phase Formation Monitoring Porewater Chemistry
Thermal Stability	:	Hydration Exotherm Monitoring Thermal Conductivity Freeze/Thaw Cycling Fire Tests Thermal Analysis
Radiation Stability	:	Alpha Gamma
Leaching Behaviour	:	Dynamic Equilibrium

Although it is not appropriate to give a comprehensive list of wasteform properties, the following section gives examples of certain properties as a function of different cement types and comments on their relevance to performance of ILW's.

Physical Properties

Table 3 shows examples of the total porosity of cement pastes as a function of time as measured by helium pycnometry

The results show that slag cements hydrate more slowly than OPC pastes and therefore have a higher initial porosity. However, the continued long-term hydration of BFS/OPC systems results in significantly lower ultimate porosities than for equivalent OPC pastes. This is mainly due to the difference in morphology of BFS/OPC hydration products and the absence of significant quantities of calcium hydroxide in the set paste. (Figure 1) [4]

Table 3. Porosity of Cement Grouts

Cement Formulation		Total Porosity %		
		2 days	28 days	90 days
OPC	w/c = 0.31	26.3	23.4	20.5
3:1 BFS:OPC	w/c = 0.31	34.8	24.5	10.4
9:1 BFS:OPC	w/c = 0.31	39.7	19.5	13.3

The pore size distribution for OPC and OPC/BFS systems as determined by mercury intrusion porosimetry (MIP) is shown in Figure 2. This shows the size of pores progressively reduces with increasing BFS content. The reduction in pore size and total pore volume both limit mass transport within the paste and therefore improve retention of radionuclides within the system.

Mechanical Properties

Compressive strength data for grouts based on BFS/OPC and PFA/OPC are compared in Table 4.

Table 4. Compressive Strength of Cement Grouts

Cement Formulation		Compressive Strength MPa (psi)	
		28 days	90 days
OPC	w/c = 0.31	80 (11600)	91 (13200)
3:1 BFS:OPC	w/c = 0.31	61 (8850)	74 (10700)
3:1 PFA:OPC	w/c = 0.31	19 (2800)	37 (5400)

Although replacement of OPC by BFS or PFA at the 75% replacement level, leads to an overall reduction in compressive strength it should be noted that all the above strengths are adequate for

ILW immobilisation. In the UK there is no minimum strength requirement currently specified for encapsulated ILW's. However, practical experience has shown compressive strengths as low as 0.7 MPa (105 psi) are adequate for handling and storage. Impact tests from 8 metres on the wasteforms have shown similar performance irrespective of strength with a maximum release fraction for respirable dust of less than 10^{-5} of the package (Note - this assumes a complete loss of drum integrity).

Consequently, cement systems containing high replacement levels of BFS or PFA have been demonstrated to have adequate mechanical properties for ILW immobilisation.

Chemical Stability

The chemical stability of encapsulated ILW products are dependent on the potential for detrimental interactions between the cement phase and the embedded waste. Metallic wastes may corrode leading to the formation of expansive corrosion products. Factors that will effect the corrosion rate of metals in a cement matrix include the reactivity of the metal, the pH of the system and the availability and mobility of reactants (eg water, chloride ions and oxygen). The porewater chemistry for a number of cement systems is given in Table 5.

Table 5. Pore Solution Chemistry of Different Cement Grouts (90 Days Cured).

Cement Formulation		Ionic Concentration (mols/m^3)					Redox * Pot.
		Na^+	K^+	Ca^{2+}	OH^-	SO_4^{2-}	mV
OPC	w/c = 0.33	210	1100	2	1230	31	-
3:1 BFS:OPC	w/c = 0.33	140	85	2	275	1	-240
9:1 BFS:OPC	w/c = 0.33	80	90	2	160	2	-400
3:1 PFA:OPC	w/c = 0.42	61	151	1	200	2	-180

* measured against saturated calomel.

All the above grouts are highly alkaline and will tend to passivate the corrosion of steels. Amphoteric metals such as aluminium will tend to corrode slower in BFS/OPC systems due to the lower pH and lower mobility of water. The lower potassium, sodium and hydroxide concentrations of BFS/OPC systems and PFA/OPC systems is also beneficial in limiting attack on potentially reactive forms of silica present in the waste (ie alkali-silica reaction).

The absence of significant quantities of free lime in BFS/OPC and PFA/OPC systems combined with their lower triacalcium aluminate content (C_3A) also gives improved chemical durability to attack by sulphates, weak acids and other aggressive species.

Thermal Stability

The reaction between cement and water, being exothermic, can result in significant temperature rises occurring in full scale cemented ILW products. OPC for example, liberates in the order of 400 KJ per kilogram of cement and this can lead to temperatures in excess of 120°C being reached in the centre of 0.220 cubic metre (45 gallon) products. As BNFL's standard package size for ILW's is 0.560 cubic metres (125 gallon), the control of exotherm temperatures is particularly important. By using cement replacements such as BFS and PFA, exotherms can be controlled to under 100°C therefore eliminating steam pressurisation and preventing thermal degradation.

Wasteforms based on BFS/OPC and PFA/OPC have been found to be resistant to degradation by upto 600 thermal cycles between -6 and +20°C. They also show only minor degradation of the outer few centimetres of the product after exposure to fire for 2 hours at between 800°C and 1000°C.

Radiation Stability

Experimental studies have been undertaken to determine the effects of radiation on the matrices to be used for waste encapsulation. Since the level and type of radiation will vary with the particular waste stream, differing irradiation techniques have been used and long term effects have been simulated by enhancement techniques. Gamma radiation doses up to 10 MGy have been produced by exposure to radiation fields within a fuel storage pond. Enhanced alpha doses have been studied by incorporating plutonium 238 into the samples during manufacture.

Measurements on simulated waste forms immobilised in BFS/OPC under gamma irradiation, have been made on a number of waste streams. The values of $G(H_2)$ vary between 0.05 and 0.15 but oxygen releases are low due to the readsorption by the matrix. Even up to 10 MGy gamma dose the specimens all have remained dimensionally stable and the compressive strength has not been impaired.

Leaching Behaviour

Two types of leach test are commonly applied, short-term dynamic leach testing to determine release rates of soluble radioisotopes and long term equilibrium leach tests to determine equilibrium concentrations of key long-lived radionuclides.

Short-term leach tests are deemed to be of limited value by BNFL although they can be used for comparison of alternative matrices and indicate potential activity release under accident conditions involving immersion in water. Results from leach tests for caesium-137 have indicated reductions in leach rates of upto two orders in magnitude for BFS/OPC systems when compared to equivalent OPC based formulations [7].

The UK disposal strategy for ILW involves burial in a deep repository situated in a low groundwater flow environment. Under these conditions the equilibrium concentration of leached radionuclides will be the factor limiting release. Some examples of equilibrium leach test values for typical BNFL wastes are given in Table 6 [8].

Table 6. Leach Testing of BNFL Intermediate Level Wastes[a] Encapsulated in 3:1 BFS:OPC

Waste	$[^{14}C]$/Mol	$[^{99}Tc]$/Mol	$[^{237}Np]$/Mol	$[^{239}Pu]$/Mol
Magnox swarf	10^{-10}	10^{-9}	10^{-11}	10^{-12}
AGR cladding	10^{-10}	10^{-8}	10^{-11}	10^{-11}
Limiting concentration[b]	10^{-9}	10^{-7}	2×10^{-11}	10^{-11}

a) Experiments were carried out in a carbon dioxide free, air atmosphere using a cement matrix and backfill.

b) The limiting concentration is the concentration in water which would give an adult a whole body dose of 1mSv for each isotope assuming an annual water consumption of 712 l.

The results show the equilibrium concentration for all the species investigated are below the limiting concentration for each radioisotope. This is due to a combination of the high pH and reducing environment afforded by the BFS/OPC matrix and associated cement backfill.

CONCLUSIONS

Inorganic cements have been selected as the best encapsulation matrix for ILW streams arising during reprocessing operations at Sellafield.

Cement systems based on BFS/OPC and PFA/OPC have demonstrated improved performance for waste immobilisation over conventional portland cements.

Comprehensive full scale testing programmes have consistently demonstrated durable wasteforms can be produced for a wide range of encapsulated wastes.

REFERENCES

1	F S Feates	UK Policy on Radioactive Wastes, Radioactive Waste Management, Vol 9 (1-3) pp 3-17, 1987.
2	W Heafield	Encapsulation of Intermediate Level Wastes Arising from Reprocessing Operations at BNFL, Sellafield, Radioactive Waste Management, Vol 9 (1-3) pp 19-25, 1987.
3	P Barlow, D M Beaumont G A Fairhall	Characterisation of Raw and Encapsulated Reprocessing Wastes in the UK. Paper presented to Waste Management 1987, Tucson, Arizona, 1987.
4	G Matthews	The Product Evaluation Development Programme in Support of Treatment Processes for Intermediate Level Waste Arisings from Reprocessing Operations at Sellafield. Radioactive Waste Management, Proceedings of BNES Conference London, pp 165-170. 27-29 November, 1984.
5	A M Neville J J Brooks	Corrosion of Reinforcement pp 275-282, Concrete Technology, 1st Edition, John Wiley and Sons, New York, 1987.
6	C R Wilding G McHugh	The Hydration of Blastfurnace Slag Cements, United Kingdom Atomic Energy Authority, Harwell. AERE R12297, 1986.
7	J D Palmer D L G Smith	The Incorporation of Low and Medium Level Radioactive Wastes (Solids and Liquids) in Cement. United Kingdom Atomic Energy Authority, Winfrith AEEW R1951, 1985.
8	J D Wilkins	The Influence of Organic Materials on the Near Field on an Intermediate Level Radioactive Waste Repository. pp 29-49. Organics in Radwaste. Editor N K Taylor. United Kingdom Atomic Energy Agency, Harwell. AERE R13079, 1988.

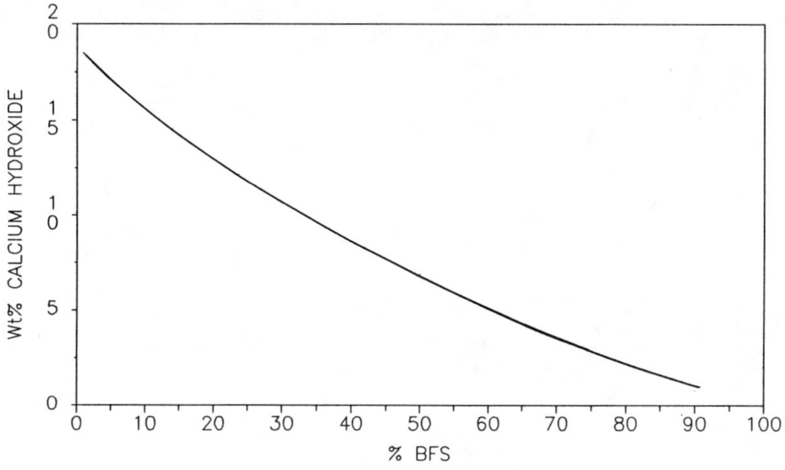

FIGURE 1
ESTIMATED $Ca(OH)_2$ CONTENT OF BFS/OPC
COMPOSITIONS AFTER 1 YEARS HYDRATION

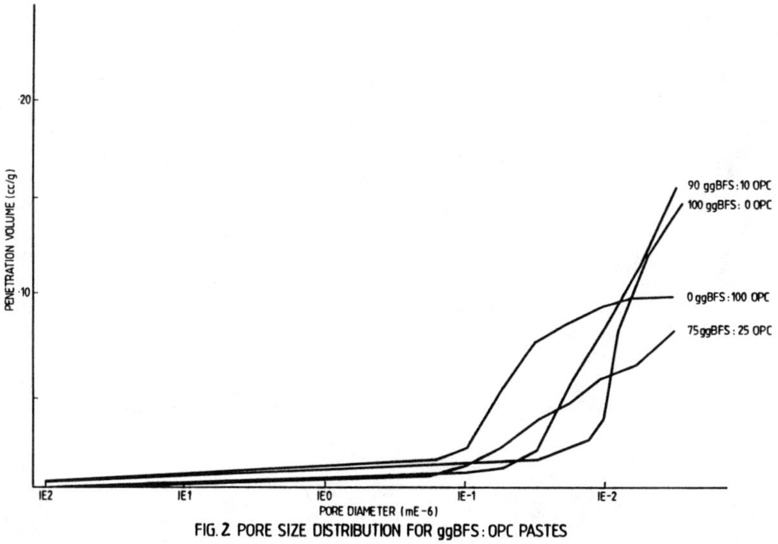

FIG. 2 PORE SIZE DISTRIBUTION FOR ggBFS : OPC PASTES

Section III. Leaching-Modeling and Mechanisms

THE INFLUENCE OF PENETRATING GAMMA RADIATION ON THE REACTION OF SIMULATED NUCLEAR WASTE GLASS IN TUFF GROUNDWATER

W. L. Ebert, J. K. Bates, T. A. Abrajano, Jr., and T. J. Gerding

Chemical Technology Division, Argonne National Laboratory, 9700 South Cass Avenue, Argonne, IL 60439

ABSTRACT

Static leaching experiments have been performed to determine the influence of penetrating gamma radiation on the reaction of simulated nuclear waste glass in tuff groundwater at 90°C. Both the leachates and the reacted glass monoliths were analyzed to characterize the reaction. Radiation was seen to acidify the leachates, but the high bicarbonate content of the groundwater prevented the pH values from dropping below 6.4. The glass reaction tended to raise the pH. Glasses based on SRL 165 black frit and PNL 76-68 glass compositions were leached. The SRL 165 type glasses were quite durable (as measured by the elemental mass loss after constant reaction times) and were unaffected by radiation. The PNL 76-68 glasses were much less durable, with the durability decreasing (after constant reaction times) as the exposure rate was increased. The primary effect of radiation is a lowering of the leachate pH which then affects the glass leaching rate.

INTRODUCTION

An extensive series of experiments has been performed to determine the influence of penetrating gamma radiation on the reaction of simulated nuclear waste glass in tuff groundwater. The Yucca Mountain Project (YMP) is studying the suitability of the volcanic tuff beds at Yucca Mountain, Nevada, as a repository site for the isolation of nuclear waste. While the horizon under study has been characterized as hydrologically unsaturated [1], the waste package performance must be understood under the unlikely but possible condition of liquid water intrusion into the repository. The high post-closure temperature will prevent liquid water from entering the repository for several hundred years, though water vapor may exist. Water which may eventually enter the repository will be subjected to weak gamma radiation fields generated by the waste. The air in the repository will have been irradiated at high doses for many years and will also affect the groundwater chemistry. See [2] and references therein for a detailed discussion of the effect of radiation on the waste package environment for an unsaturated repository. The behavior of the waste in such groundwater must be understood to project the long-term performance of the repository.

Experiments have been performed over a four-year period to measure the effect of radiation on the leaching behavior of simulated nuclear waste glass in a saturated environment relevant to the tuff repository [3-6]. In this report we provide a comparison of the results and discuss the influence of radiation dose on glass reaction.

EXPERIMENTAL

The experiments were similar to the MCC-1 static leach tests [7] and were performed at 90°C under gamma exposures of 2 x 10^5 R/h, 1 x 10^4 R/h, 1 x 10^3 R/h, or without radiation for up to 278 days. The exposure rate was determined by dosimetry performed prior to the experiments. SRL 165 black frit doped with uranium, referred to as SRL U glass, or with uranium, neptunium, plutonium, and americium, referred to as SRL A glass, and PNL 76-68 based glasses ATM-1c and ATM-8, which contains uranium, neptunium, and plutonium, were reacted. While these glasses do not represent actual compositions that will be used to encapsulate waste, they provide a diverse range of glass reactivities which are useful in understanding the effects of radiation on leaching. Repository reference groundwater from well J-13 was prereacted with pulverized tuff rock at 90°C to produce the leachant, which is referred to as the EJ-13 solution. The glass and leachant compositions have been presented elsewhere [3-6]. The experiments were performed in 304L stainless steel vessels having a volume near 22 mL. An amount of leachant was added to attain a glass surface area/leachant volume ratio (SA/V) near 0.3 cm^{-1} and an air/leachant volume ratio near 0.3. Polished tuff wafers were added to some vessels as well.

The final leachates were analyzed for pH, anions (ion chromatography), cations (ICP, AA), and transuranics (α-counting). The reacted glass samples were measured for weight loss and the surfaces analyzed using SEM with EDS and SIMS. Some of the results are presented below, with an emphasis on the affects of radiation on the glass reaction.

RESULTS

The final leachate pH values are plotted vs. the reaction time for experiments with SRL A and ATM-8 glasses at the various exposure rates in Figure 1. The pH values of experiments with SRL U glass were identical to those with SRL A glass, within experimental error. The pH values of the nonirradiated experiments are higher than that of the initial leachant for both glass types. This is due to the production of hydroxide ions accompanying the alkali release from the glass. Irradiated experiments are acidified relative to the nonirradiated experiments due to the radiolytic production of nitric acid in moist air [8]. This was evidenced by a decrease in the pH and an increase in the concentration of fixed nitrogen. Burns et al. have proposed an expression for the nitric acid production in a similar two phase vapor/liquid system in which the amount of acid produced is proportional to the exposure [9]. Thus, more acid is expected to be produced at higher exposure rates for the same reaction time. This is observed in Figure 1 where the rate of acidification decreases as 2×10^5 R/h > 1×10^4 R/h > 1×10^3 R/h for both glass types.

Blank experiments (without glass) were acidified to pHs below 7 at all exposure rates, though the pH was prevented from dropping below 6.4 by the high bicarbonate concentration in the groundwater (~120 ppm). The presence of glass results in slightly higher pH values as the glass reaction produces sufficient hydroxide to partially neutralize the acid. The pH, therefore, reflects the balance between acidification due to radiolysis and basification due to the glass reaction. At long reaction times, the pH appears to approach a common value for all exposure rates, near 7 for SRL A glass and 7.5 for ATM-8 glass. The higher "final" pH of the experiments with ATM-8 glass indicates a greater extent of glass reaction, since the nitric acid production is the same in experiments with both glass types.

The release of various species from the glass into the leachate is indicative of the extent of reaction. The normalized elemental mass losses of boron, silicon, and sodium are shown plotted vs. the reaction time for SRL A and ATM-8 glasses in Figures 2a-2f at the various exposure rates. The release of these species from SRL A glass is not affected by irradiation, with $NL(Na) > NL(B) \cong NL(Si)$. Analysis of the reacted SRL U glasses showed the altered surfaces to be depleted in alkalis relative to the bulk glass, while the silicon concentration remained nearly constant [6].

The normalized elemental mass loss of all elements is greater from ATM-8 glass than from SRL A glass, and the mass loss after a given reaction time is influenced by irradiation. The release rate of all elements increases with the exposure rate as
nonirradiated $< 1 \times 10^3$ R/h $< 2 \times 10^5$ R/h; and the elemental releases decrease as NL(Na) > NL(B) > NL(Si) at all exposures.

The release of transuranic elements was measured for SRL A (Am, Np, Pu) and ATM-8 (Np, Pu) glass in unfiltered and filtered (50 Å filter) leachate samples, and an acid strip fraction of the vessel. Neptunium was found in both the filtered and unfiltered fractions, but no neptunium was plated onto the stainless steel. The concentration of neptunium measured to be in the leachate is plotted against the reaction time for different exposure rates in Figures 3a and 3b for SRL A and ATM-8 glasses. The neptunium concentration is low in experiments with SRL A glass and the results of duplicate experiments show relatively large deviations. Nevertheless, the neptunium release rate does not vary significantly with the exposure rate for SRL A glass. ATM-8 glass, on the other hand, showed marked differences in the neptunium concentration at different exposure rates at constant reaction time, with the concentration increasing with the exposure rate. The plutonium concentration in the filtered fraction is shown in Figures 4a and 4b for SRL A and ATM-8 glasses. The plutonium solubility limit is sensitive to the pH, and only experiments irradiated at 2×10^5 R/h, which reached the lowest pH values, have a significant concentration of plutonium in the aqueous phase. Plutonium is preferentially plated onto the stainless steel vessel surface in other experiments with both glass types.

DISCUSSION

The primary effect of gamma radiation in these experiments is through acidification of the leachate by the radiolytic production of nitric acid. An equation describing the production of nitric acid in a moist air/liquid water system has been given by Burns et al. [9], which takes the following form at low exposures:

$$N = 2.9E\text{-}5 \bullet C_o \; R \; G \; Dt \qquad (1)$$

where N is the molar concentration of nitric acid produced, C_o is the initial concentration of nitrogen, R is the air/liquid volume ratio, G is the radiolytic yield, D is the dose rate, in Mrad d^{-1}, and t is the exposure time, in days. Using the results of the experiments irradiated at

2×10^5 R/h, a G value for the production of HNO_3 of 2.4 molecules/100 eV was obtained. Substituting the experimental values C_o = 3.6E-2M , R = 0.3, and G = 2.4, Eq. (1) is reduced to

$$N(M) = 5.95E\text{-}7\ Dt \tag{2}$$

The nitric acid production is therefore predicted to occur at 2.86 μM day^{-1} at an exposure rate of 2×10^5 R/h, 0.14 μM day^{-1} at 1×10^4 R/h, and 0.01 μM day^{-1} at 1×10^3 R/h. The amount of nitric acid formed is not limited by the available gases. For example, after 100 days at 2×10^5 R/h, which is about twice the maximum exposure attained, only about 0.3 mM of N_2 and O_2, or about 1% of the initial N_2 and 3% of the initial O_2 are consumed. (At greater exposures, the HNO_3 yield may decrease as the oxygen partial pressure decreases [8].) Bicarbonate in the tuff groundwater inhibits acidification below about 6.4 (which is the pK_a of the bicarbonate/carbonic acid equilibrium at 90°C [2]), and so the pH difference between irradiated and nonirradiated experiments is limited to about 2 pH units (the difference between the pH of the starting water and 6.4). This small pH difference may be sufficient to alter the solubility limit of some species and so perhaps alter the reaction rate. The leachates are predicted to be acidified to pH 6.4 after less than 1 day, about 3 days, or after 29 days at exposure rates of 2×10^5 R/h, 1×10^4 R/h, or 1×10^3 R/h, respectively. Figure 1a shows the experiments with SRL A glass to be acidified to near pH 6.5 after 28 days at 2×10^5 R/h. Experiments with ATM-8 glass are acidified to a pH near 6.6 after only seven days at 2×10^5 R/h, but are acidified less at longer reaction times, probably because of the glass reaction.

The final pH is less acidic than predicted by the nitric acid production also because the glass reaction raises the pH. This is seen in the nonirradiated experiments which reach pH values above 9 for both SRL A and ATM-8 glass. The glass reaction partially neutralizes the nitric acid produced in the irradiated experiments. Hydroxide is produced with the release of alkalis but may be consumed by network dissolution reactions. The glass reaction appears to proceed at a maximum rate early and slow with time while the radiolytic production of nitric acid proceeds at a nearly constant rate over the duration of these experiments. Early in the reaction, the glass reaction may overwhelm the nitric acid production and raise the pH. This is seen to occur in the ATM-8 experiments irradiated at 1×10^3 R/h after 28 days. At longer times the glass reaction slows and the pH drops. Eventually the two rates may become similar and the pH attain an apparent steady-state value. This appears to happen in experiments with SRL A glass near a pH of 7 and with ATM-8 glass at a pH near 7.5 over the reaction times tested. Nonirradiated experiments are expected to show increasing pH values until either the reaction stops or the pH becomes buffered by secondary phase formation.

The rate of the SRL A reaction is not influenced by irradiation or by the resulting changes in the leachate pH, as measured by elemental release rates. This reflects the high durability of this glass composition in tuff groundwater at a near-neutral pH. Acidification of the leachate below about 7 is prevented by the bicarbonate buffer. The rate of nitric acid production (that is, the exposure rate) does not influence the rate of the glass reaction. The rate of release of species from ATM-8 glass is increased upon irradiation. These results show the leachate pH to only drop to about 7.5 upon irradiation at all exposure rates. Because more acid is produced at higher exposure rates, more glass must react to maintain the pH near 7.5. The higher exposure rates require greater glass reaction rates to maintain a pH value near 7.5, and so the glass reaction is accelerated by radiation through the pH drop. The glass appears to buffer the leachate to a pH value near 7.5. It is generally found that silica glasses are more reactive at higher pH values than at near neutral pHs due to the greater solubility of silica above pH $\sim$ 9.5. However, the glass reactions may also be inhibited by the formation of surface layers which may be more soluble in neutral or acidic solutions. The SRL A and ATM-8 samples are undergoing further characterization using analytical electron microscopy to determine how the secondary phases that make up the reacted layer are affected by the different irradiation conditions. The present results imply that a clear difference in secondary phase formation will exist for the two glass types.

Irradiation was also seen to produce nitrite ions in the solution at all exposure rates [3,5,6] which indicates the solution Eh is reduced from that of the starting leachant. Experiments irradiated at 2×10^5 R/h generated both nitrate and nitrite ions, but experiments irradiated at the lower exposure rates showed a decrease in the nitrate concentration with a corresponding increase in the nitrite concentration. Equation (2) predicts an amount of nitrate near the detection limit ($\sim$0.1 ppm) will be produced in experiments irradiated at 1×10^4 R/h or 1×10^3 R/h over the duration of these experiments, which is negligible compared to the initial leachant nitrate concentration of about 7.6 ppm. Figure 5 shows the nitrite/nitrate concentration ratio plotted vs. the total exposure for experiments without glass. The presence of glass increases the ratio slightly. The solution is seen to be reduced at low total exposures but to approach the Eh of the starting leachant (which corresponds to a ratio of zero) at higher exposures. This may indicate a transient (or nonequilibrium) nature of the reduction or an influence of the solution pH on the ratio.

The lack of effect of radiation on the SRL 165 glass reaction in the present experimental conditions may be due to the low reactivity of the SRL 165 type glasses, or may be the result of different processes controlling the reaction compared to the ATM-8 glass. Further studies are in progress to assess the effect of radiation on glass reaction at more repository-relevant R values (smaller volumes of water).

CONCLUSIONS

Static leach experiments using simulated nuclear waste glasses based on the composition of SRL 165 black frit and PNL 76-68 were performed under gamma radiation fields of 2 x 10^5 R/h, 1 x 10^4 R/h, and 1 x 10^3 R/h, and in the absence of radiation. The SRL 165 type glasses reacted similarly under all radiation fields while the PNL 76-68 type glasses reacted to a greater extent in the high radiation fields. Radiolysis of the moist air produced nitric and nitrous acids which acidified the leachates relative to the nonirradiated experiments. The leachate pHs remained between 7 and 9 for the SRL glasses and between 7.5 and 9.5 for the PNL glasses. The high bicarbonate ion concentration buffered the pH in experiments without glass and in experiments with SRL glasses. Experiments with the PNL 76-68 glasses were apparently buffered to a pH near 7.5 by the glass reaction. A greater extent of glass reaction was required to maintain a pH of 7.5 at higher exposure rates because of the increased nitric acid production. The behavior of the PNL 76-68 type glasses is contrary to that expected from the silica solubility as a function of leachate pH, and may point towards a secondary influence of radiation.

REFERENCES

1. Site Characterization Plan, U.S. Department of Energy, Office of Civilian Radioactive Waste Management, Volume III, Part A, Chapter 7, pp. 7-8 - 7-12 (1988).

2. R. A. Van Konynenburg, "Radiation Chemical Effects in Experiments to Study the Reaction of Glass in an Environment of Glass-Irradiated Air, Groundwater, and Tuff," Lawrence Livermore National Laboratory report UCRL-53719 (1986).

3. J. K. Bates, D. F. Fischer, and T. J. Gerding, "The Reaction of Glass during Gamma Irradiation in a Saturated Tuff Environment, Part 1: SRL 165 Glass," Argonne National Laboratory report ANL-85-62 (1985).

4. J. K. Bates, T. J. Gerding, D. F. Fischer, and W. L. Ebert, "The Reaction of Glass in a Gamma Irradiated Saturated Tuff Environment, Part II Data Package for ATM-1c and ATM-8 Glasses," Lawrence Livermore National Laboratory report UCRL-15991 (1987).

5. T. A. Abrajano, Jr., J. K. Bates, T. J. Gerding, and W. L. Ebert, "The Reaction of Glass during Gamma Irradiation in a Saturated Tuff Environment, Part 3: Long-Term Experiments at 1×10^4 Rad/Hour," Argonne National Laboratory report ANL-88-14 (1988).

6. W. L. Ebert, J. K. Bates, and T. J. Gerding, Argonne National Laboratory, unpublished information (1989).

7. Materials Characterization Center, Nuclear Waste Materials Handbook.

8. A. R. Jones, "Radiation-Induced Reactions in the N_2-O_2-H_2O System," Rad. Res. 10, 655 (1959).

9. W. G. Burns, W. R. Marsh, and W. S. Walters, "The Gamma Irradiation Enhanced Corrosion of Stainless and Mild Steels by Water in the Presence of Air, Argon, and Hydrogen," Radiat. Phys. Chem. 21, 259 (1983).

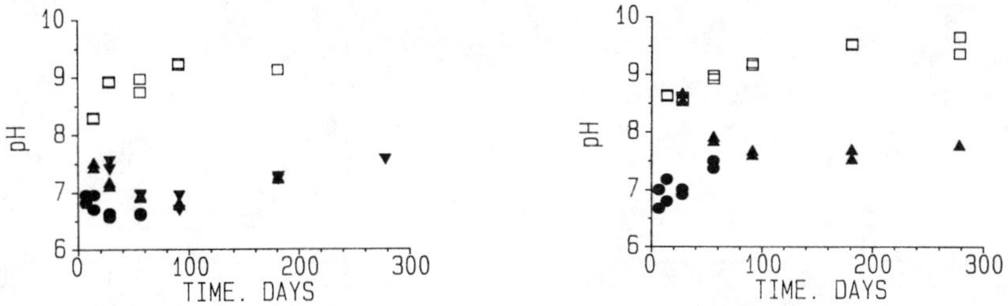

Figure 1. Final Leachate pH for Experiments with (a) SRL A Glass, or (b) ATM-8 Glass Irradiated at (●) 2 x 10^5 R/h, (▲) 1 x 10^4 R/h, (▼) 1 x 10^3 R/h, or (□) Nonirradiated vs. the Reaction Time.

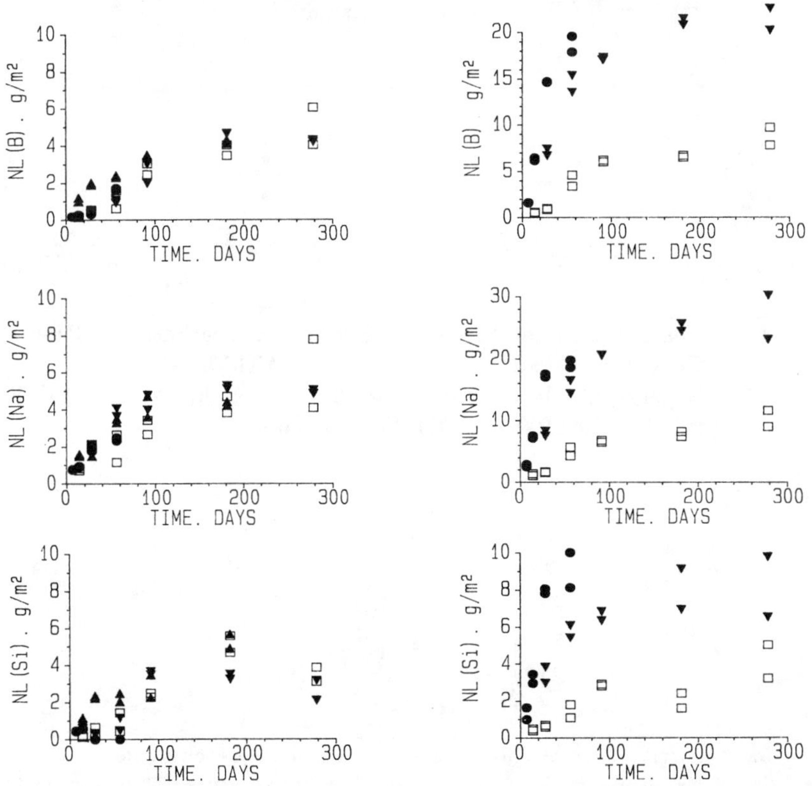

Figure 2. Normalized Elemental Mass Loss for Experiments Irradiated at (●) 2 x 10^5 R/h, (▲) 1 x 10^4 R/h, (▼) 1 x 10^3 R/h, or (□) Nonirradiated vs. the Reaction Time: (a) Boron from SRL A, (b) Boron from ATM-8, (c) Silicon from SRL A, (d) Silicon from ATM-8, (e) Sodium from SRL A, and (f) Sodium from ATM-8.

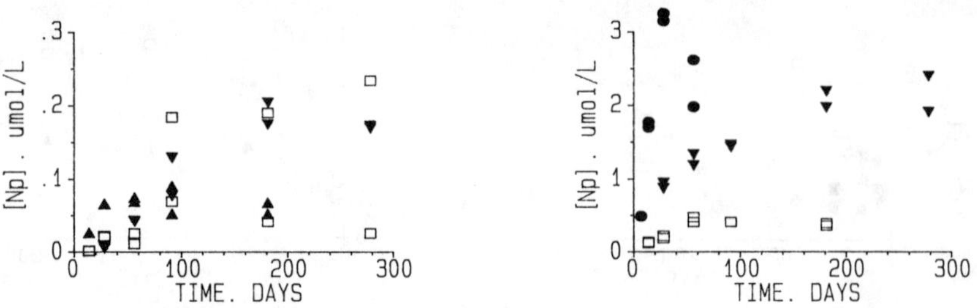

Figure 3. Concentration of Neptunium-237 in the Leachate vs. Reaction Time for (a) SRL A Glass, or (b) ATM-8 Glass in Experiments Irradiated at (●) 2 x 10^5 R/h, (▲) 1 x 10^4 R/h, (▼) 1 x 10^3 R/h, or (□) Nonirradiated.

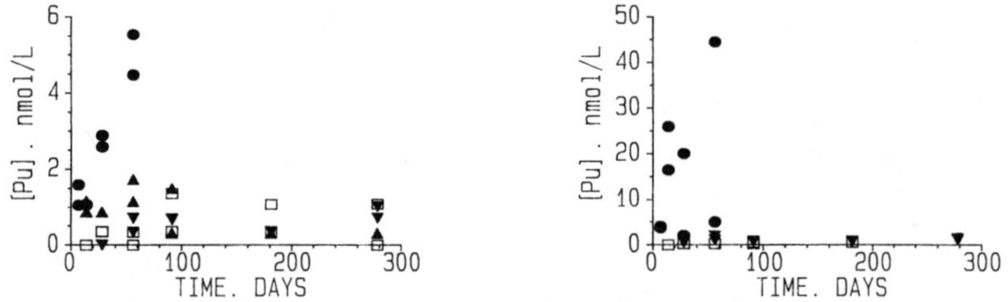

Figure 4. Concentration of Plutonium-239 in the Leachate vs. Reaction Time for (a) SRL A Glass, or (b) ATM-8 Glass in Experiments Irradiated at (●) 2 x 10^5 R/h, (▲) 1 x 10^4 R/h, (▼) 1 x 10^3 R/h, or (□) Nonirradiated.

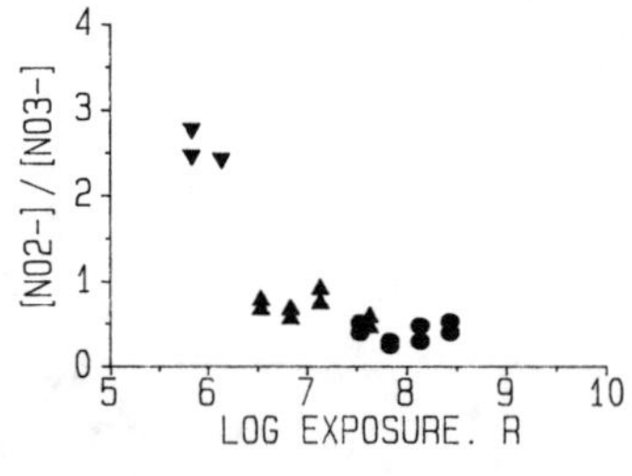

Figure 5. Nitrite/Nitrate Concentration Ratio in Leachate for Experiments Without Glass, Irradiated at (●) 2 x 10^5 R/h, (▲) 1 x 10^4 R/h, and (▼) 1 x 10^3 R/h vs. Reaction Time.

REDOX EFFECTS ON THE DURABILITY AND VISCOSITY OF NUCLEAR WASTE GLASSES

X. FENG, I. L. PEGG, E. SAAD*, S. CUCINELL, Aa. BARKATT

THE CATHOLIC UNIVERSITY OF AMERICA
THE VITREOUS STATE LABORATORY
WASHINGTON, DC 20064

ABSTRACT

Iron oxide is one of the major components in West Valley nuclear waste glasses (about 12%). We report studies of the effect of the redox state of iron oxide on the viscosities and chemical durability of West Valley waste glasses that were prepared with Fe(II)/ΣFe ratios between 0.0 and 0.8 by melting in a controlled atmosphere of CO and CO_2. Durabilities were determined using modified MCC-3 tests and viscosities were measured between 950°C and 1150°C. It is found that the redox effects on durability are strongly dependent on glass composition; e.g. the leach rate differs by a factor of 12 from the most oxidized to most reduced glass at low alumina compositions, while this factor falls to below 2 for similar, but more durable, glass compositions with higher alumina contents. The viscosities of these glasses decrease by about 30% over the range of redox states investigated.

---

* Mobay Co., Baltimore, MD

INTRODUCTION

Iron oxide comprises nearly half of the combined West Valley high-level commercial waste [1] and the resulting West Valley waste glasses are expected to contain about 12% iron oxide [2]. As a major component in the waste glasses, it is clear that not only the amount of iron oxide but also the redox state of the oxide in the glasses will have significant effects on both chemical durability and other physical properties such as the viscosities of the melts. Here we report studies of such redox effects on three groups of West Valley waste glasses: the preliminary reference glass WV205 group, the WV205-Al group, and the WVCM50 group. The compositions of these glasses are shown in Table I. The ratio of ferrous to total iron (Fe(II)/ΣFe) in the glasses studied varies from 0.0 to 0.83.

Table I. Glass Composition (wt%)

Oxide	WVCM50	WV205	WV205-Al
Al_2O_3	9.86	3.30	5.66
B_2O_3	12.27	9.96	10.24
BaO	0.19	0.6	0.14
CaO	0.82	0.61	0.52
Cr_2O_3	0.14	0.22	0.30
Fe_2O_3	11.93	11.80	10.07
K_2O	1.60	3.50	5.20
Li_2O	2.22	3.10	2.72
MgO	0.79	1.30	1.19
MnO_2	1.21	1.70	1.20
Na_2O	9.79	11.00	10.05
NiO	0.30	0.70	0.53
P_2O_5	2.47	2.50	2.42
SO_3	0.22	0.13	0.49
SiO_2	39.60	45.22	44.29
ThO_2	3.52	0.00	0.00
TiO_2	0.82	1.00	1.00
UO_2	0.61	0.00	0.00
ZrO_2	0.39	3.10	2.42

EXPERIMENTAL

Glass Preparation

The three groups of glasses with different redox states were made from the three parent glasses listed in Table I. The WV205 group and the WV205-Al group were made at West Valley Nuclear Services Inc's facility [3,4] under different oxidation conditions. The Fe(II)/ΣFe ratios were determined by colorimetry following glass dissolution [5] at West Valley Nuclear Service Inc. The WVCM50 group glasses were made at Alfred University [6] by remelting VSL WVCM50 glass in a controlled-atmosphere furnace using a mixture of CO and CO_2. The time of reduction, ratio of CO to CO_2, and the weight of the glass were varied to obtain the reductions shown in Table II. The Fe(II)/ΣFe ratios of the WVCM50 set were determined by Mossbauer spectroscopy at Alfred University [6] and also by wet chemical analysis done at Corning Glass Works, Corning NY [6]; the two sets of results agreed to within about 10%. However, Jantzen [7] has recently reviewed analytical procedures for determination of Fe(II)/ΣFe ratios and pointed out that the results can be somewhat method-dependent.

Chemical Durability Measurements

The chemical durabilities of these redox glasses were determined using the MCC-3 powder test procedure [8] modified as detailed in ref.[9]; the principle changes are exclusion of atmospheric CO_2, no filtration of the leachate, no prewashing of the glass powders, and the provision of a means of repeated sampling. A quantity of 4.00 grams of glass powder (74 to 149 microns) was immersed in 40.0 mL de-ionized water in a precleaned Teflon PFA container (Savillex #560 plus #501-37). The container was sealed and enclosed in a large bottle in the presence of Ascarite II (Thomas Scientific) to prevent atmospheric CO_2 from buffering the leachates. The bottle was kept in a 90°C convection oven and was thoroughly shaken once every day. The sampling times were 7, 28, 56, and 180 days and the leachates were analyzed by DC plasma and AA spectrometry. All tests were carried out in triplicate. The precision of the determinations of elements such as boron, silicon, and sodium was about 5%; precision of the pH measurements was about 0.1 pH unit.

Viscosity Measurements

The viscosities of these glasses were measured with a Brookfield viscometer. The precision and accuracy of the measurements, as determined from their reproducibility and measurements on the NBS reference glass SRM 711 [10], respectively, are estimated to be about 10%.

Table II. Fe(II)/ΣFe Ratios and MCC-3 Results (normalized, mg/L)

Glass	Fe(II)/ΣFe	7 day		28 day		56 day		180 day	
		B	pH	B	pH	B	pH	B	pH
WVCM50 GROUP:									
WVCM50-A	0.00	4.68	9.67	6.73	9.60	7.96	9.35		
WVCM50-B	0.22	5.09	9.70	8.11	9.68	10.3	9.45		
WVCM50-C	0.36	5.69	9.80	8.60	9.74	9.82	9.51		
WVCM50-D	0.48	6.10	9.86	8.72	9.88	9.84	9.67		
WVCM50-E	0.76	7.43	9.92	9.13	10.1	10.9	9.85		
WV205-Al GROUP:									
WV205-Al-A	0.01	4.65	9.81	5.45	9.70	6.29	9.05	11.3	9.02
WV205-Al-B	0.02	4.52	9.91	6.21	9.78	8.15	9.02	17.5	9.07
WV205-Al-C	0.09	5.96	9.52	8.99	9.65	12.7	9.12	26.5	9.35
WV205-Al-D	0.19	5.82	10.2	9.03	10.1	12.9	9.51	26.6	9.19
WV205-AL-E	0.30	5.16	10.3	7.92	9.81	11.5	9.35	26.9	9.11
WV205 GROUP:									
WV205-A	0.01	8.84	9.80	13.0	9.66	18.6	9.40	24.6	9.20
WV205-B	0.12	39.4	10.2	172	10.3	224	10.1	251	9.85
WV205-C	0.19	74.6	10.3	256	10.3	269	10.1	308	9.91

RESULTS AND DISCUSSION

The results in Table II clearly demonstrate that the redox state of iron oxide in these glasses can have a large effect on chemical durabilities of the glasses. Fig.1 shows that the normalized concentrations of sodium (concentrations in mg/L divided by the mass abundance of sodium in wt% in the glass) in the leachates of the WVCM50 group of glasses increases as the glass becomes more reduced, i.e., the Fe(II)/ΣFe ratios of the glasses change from 0.0 to 0.76.

The normalized boron concentrations of the WVCM50 group of glasses show the same trend, as illustrated in Fig.2. The large increase in leachability, or decrease in chemical durability, for the WV205 group of glasses with progressive reduction is illustrated in Fig.3 and Fig.4.

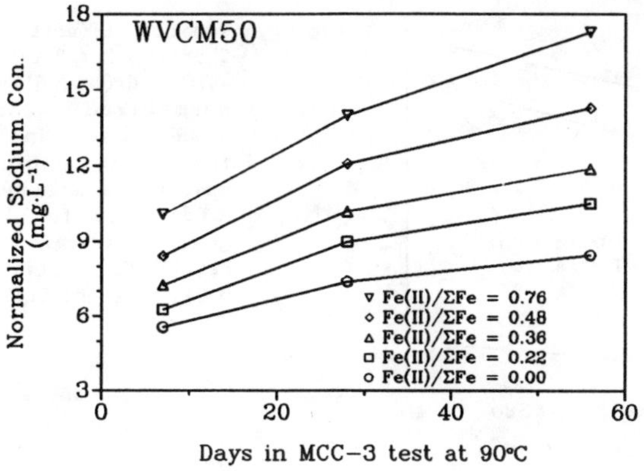

Fig 1. Redox Effects on Chemical Durability of WVCM50 Group Glasses: Normalized sodium concentration in MCC3 tests are plotted against the durations of MCC3 tests for WVCM50 group glasses with Fe(II)/ΣFe ratio of 0.0, 0.22, 0.36, 0.48 and 0.76.

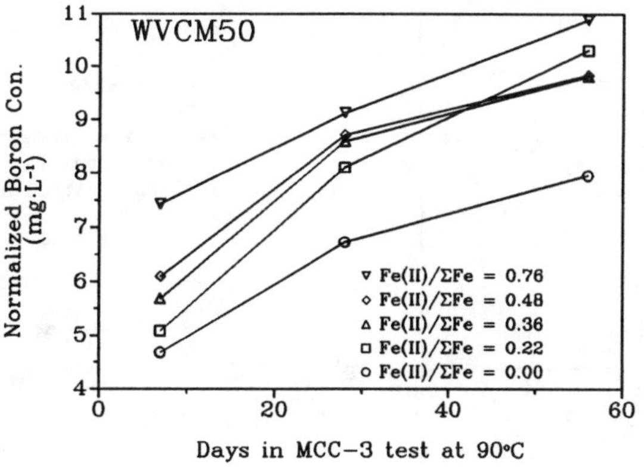

Fig 2. Redox Effect on the Chemical Durability of WVCM50: Normalized Boron concentration in MCC3 test vs the days in MCC3 test for WVCM50 group glasses with Fe(II)/ΣFe ratio of 0.0, 0.22, 0.36, 0.48 and 0.76.

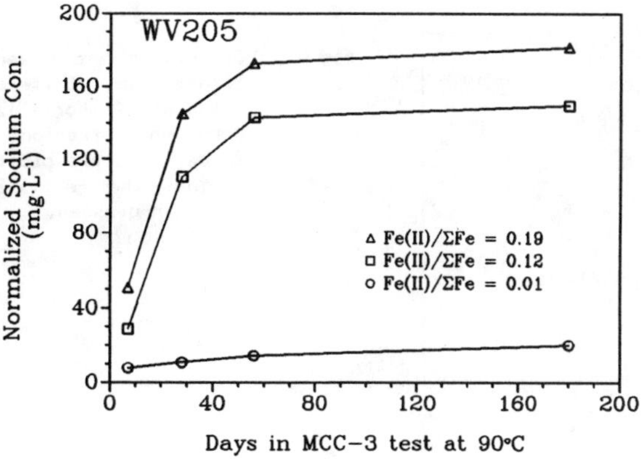

Fig 3. Redox Effects on Chemical Durability of WV205 Group Glasses: Normalized Sodium concentration in MCC3 tests are plotted against the durations of MCC3 tests for WV205 group glasses with Fe(II)/ΣFe ratio of 0.01, 0.12 and 0.19.

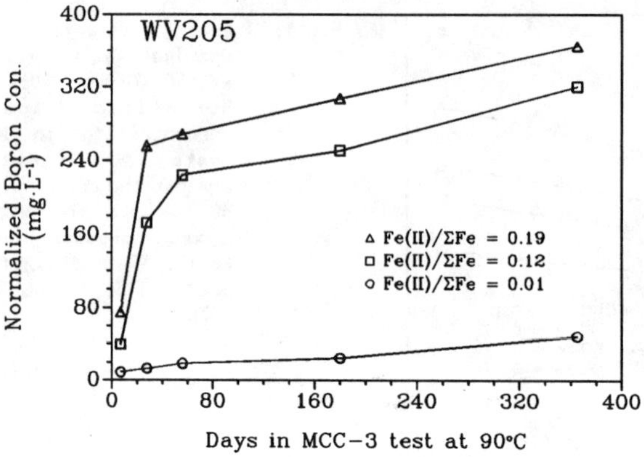

Fig 4. Redox Effects on Chemical Durability of WV205 Group Glasses: Normalized Sodium concentration in MCC3 tests are plotted against the durations of MCC3 tests for WV205 group glasses with Fe(II)/ΣFe ratio of 0.01, 0.12 and 0.19.

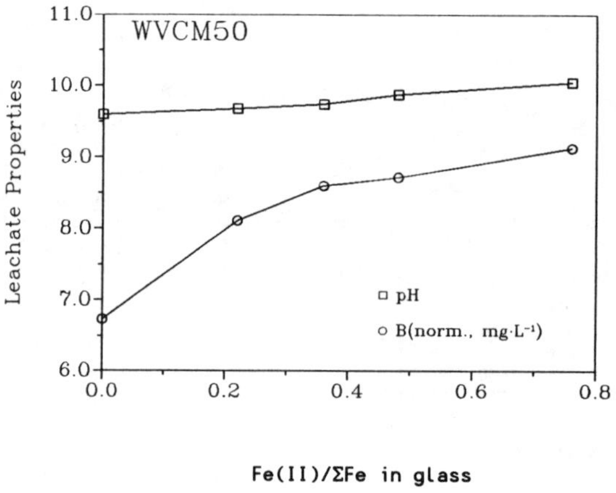

Fig 5. Redox Effect on Chemical Durability of WVCM50 Group Glasses: The properties of MCC3 leachates at 28-days are plotted against the Fe(II)/ΣFe ratio in the glasses.

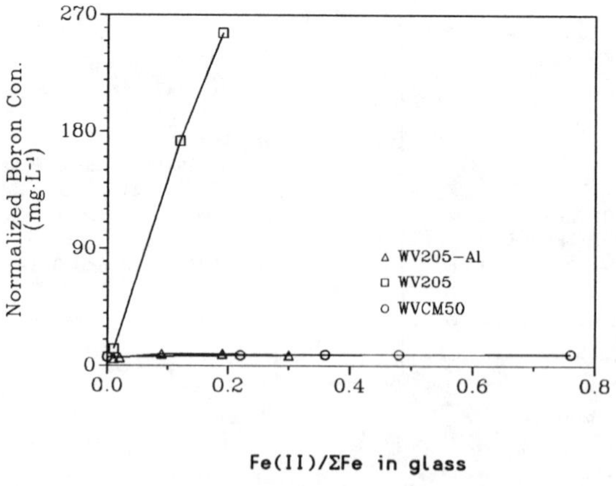

Fig 6. Comparison of Redox Effects on Different Glasses: Normalized Boron concentrations of 28-day MCC3 are plotted against the Fe(II)/ΣFe ratio in glasses.

As the glasses become more reduced the pH's of the leachates are also increased as demonstrated in Fig.5, where the pH of the 28-day MCC-3 leachate is 9.60 for WVCM50-A which has a Fe(II)/ΣFe ratio of 0.00 while the pH's are 9.68, 9.74, 9.88 and 10.05 for WVCM50-B to WVCM50-E whose Fe(II)/ΣFe ratios are 0.22, 0.36, 0.48, and 0.76 respectively.

While the redox effect on chemical durability is very minor on the high alumina glasses (WVCM50 group and WV205-Al group), this effect is much more pronounced for the "less durable" WV205 group glasses, (see Fig.6). Specifically, when WV205 was reduced from an Fe(II)/ΣFe ratio of 0.01 to 0.19, the corresponding leaching of the glasses in terms of normalized boron concentration jumped from 12.98 mg/L to 256 mg/L at 28-days in MCC-3 test, an increase by a factor of 20. In contrast, the WV205-Al groups, shows an increase of a factor of less than two, i.e., from 5.45 mg/L to 9.03 mg/L for a similar reduction from Fe(II)/ΣFe = 0.01 to 0.19. Furthermore, when the higher alumina WVCM50 group was reduced even to the extent of Fe(II)/ΣFe = 0.76 the increase in leaching is still less than a factor of two, i.e., a normalized boron concentration of 6.73 for Fe(II)/ΣFe =0.0 and 9.13 for Fe(II)/ΣFe = 0.76.

The large difference in the effect of the redox state of iron oxide between the "durable" WVCM50 and WV205-Al groups and the "less durable" WV205 group can be qualitatively understood from a consideration of the structural role of iron oxides in the glasses, though the latter is still a matter of some controversy in the literature [11-13]. Some authors consider Fe_2O_3 as a network-former [11] and some do not [12-13]. From the present studies one might reason that in a glass like WV205 which is deficient in network-formers, Fe_2O_3 serves, at least partially, as a network-former. Any reduction of the network-former type of Fe_2O_3 to the non-network-former FeO will decrease the durability and viscosity of the glass. However, in the "more durable" glass region, other network-formers, mainly Al_2O_3, ZrO_2 and SiO_2, are present in a sufficient amount to form a continuous network. In such cases Fe_2O_3 plays a lesser role in network building. The glass is therefore able to tolerate a larger degree of reduction before becoming a "non-durable" type of glass. We have also found that such durable glasses as the WVCM50 group and WV205-Al group are also relatively insensitive to certain ranges of change of other components (such as silica, boron, alkali, etc.)[14]. This is a very desirable feature for nuclear waste glass and the explanation may have similar roots to the redox effect described above.

An indication of the extent of the tolerance to reduction that addition of Al_2O_3 affords on WV205 is shown in Table III. A severe reduction of the glass WV205 plus 3.5% Al_2O_3 to an Fe(II)/ΣFe ratio of 0.83 does push this glass off the "durable-glass plateau" and its leach rate increases from 5.71 mg/L in terms of normalized boron concentration in a 28 day MCC-3 test to 72 mg/L, an increase by a factor of more than 12. These results imply that some of the Fe_2O_3 is still serving as a network-former even in a glass consisting of WV205 plus 3.5% Al_2O_3. Schreiber [15] has made extensive studies of redox equilibria in a variety of glass-forming melts. This work suggests that even for this severe reduction by far the major effect is due to the Fe(II)/Fe(III) couple. The next most abundant redox species, Mn and Ti, are present in these glasses only at around the 1 wt% level and these couples are insignificantly perturbed over the range of oxidation states considered here.

The viscosity data of the WVCM50 group glasses compiled in Table IV, suggest that as the glasses become more reduced they become less viscous. Unfortunately, our data are not yet extensive enough to determine whether the redox effect on viscosity is as strongly dependent on glass composition as is the chemical durability.

Table III. REDOX EFFECTS ON DURABILITY (MCC-3) OF GLASSES (normalized, mg/L)

GLASS	Fe(II)/ΣFe	7-DAY		28-DAY	
		pH	BORON	pH	BORON
WV205+3.5%Al_2O_3	0.00	9.35	5.71	9.01	6.5
WV205+3.5%Al_2O_3	0.83	10.3	67.4	10.0	72.0

Table IV. VISCOSITIES OF WVCM50 SET GLASSES

Glass	Fe(II)/ΣFe	Viscosities (poise)		
		950°C	1000°C	1100°C
WVCM50-A	0.00	515	281	102
WVCM50-B	0.22			
WVCM50-C	0.36	433	237	89
WVCM50-D	0.48	392	211	80
WVCM50-E	0.76	293	177	71

CONCLUSIONS

The redox state of a glass influences both its chemical durability and its viscosity. The general trend of this effect in the composition range studied here is that both the chemical durability and the melt viscosity are decreased as the glass is reduced. However, the extent of this effect is highly composition dependent: "durable" compositions, which are usually sufficient in network-formers, are generally insensitive to relatively large reductions of the glass, while "non-durable" compositions, which are usually deficient in network-formers, are very sensitive to even small reductions. Thus, one might relate the degree of tolerance of the chemical durability to reduction of the glass to the level of excess network-formers. We hope to present such an analysis in a future publication.

ACKNOWLEDGMENTS

Drs. D. Pye and I. Joseph of Alfred university kindly provided us with the WVCM50 group of redox glasses and the redox state data on these glasses. This work is supported in part by West Valley Nuclear Services Inc. under contract number DE-AC07-81-NE 44139.

REFERENCE

[1] L. A. Chick, W. M. Browen, R. O. Lokken, J. W. Wald, L. R. Bunnell, D. M. Strachan, West Valley High-level Nuclear Waste Glass Development: A statistically Designed Mixture Study. PNL-4992, Pacific Northwest Laboratory, Richland, Washington, 1984

[2] L. R. Eisenstatt, C. C. Chapman, R. L. Bogart, "A Method for Showing Compliance with High-Level Waste Acceptance Specifications", in Waste Management'86, p513, R.G. Post, Ed., University of Arizona, Tucson, Arizona.

[3] W. Bowan, T.A.#52, West Valley Nuclear Services Co. Inc., August 21, 1986

[4] L. R. Eisenstatt, T.A.#61, West Valley Nuclear Services Co. Inc., December 8, 1986

[5] D. R. Jones, W. C. Jansheski, and D. S. Goldman, "Spectrophotometric Determination of Reduced and Total Iron in Glass with 1,10-Phenanthroline", Anal. Chem. 53, 923(1981)

[6] A. R. Report #64, 5-13-88, Alfred University

[7] C. M. Jantzen,"Verification and Standardization of Redox Measurement for DWPF", DPST-89-222, Westinghouse Savannah River Co., Aiken, SC

[8] J. E. Mendel, in Test Methods Submitted for Nuclear Waste Materials Handbook (Material Characterization Center, PNL), PNL-3990, 1984

[9] X. Feng, Aa. Barkatt, T. Jiang, "Systematic Composition Studies on the Durability of Waste Glass WV205", in Scientific Basis for Nuclear Waste Management XI, Ed by M. J. Apted and R. E. Westerman, 673(1988)

[10] E. Saad, Ph. D thesis, Physical Properties of Multicomponent Borosilicate Glasses: A Study of Viscosity, Processing And Durability, The Catholic University of America, 1988

[11] J. M. F. Navarro, and R. Brukner, "Structural Incorporation of Iron in Oxidic Glasses", Glastechn. Ber. 49, 82(1976)

[12] Q. Zhong, S. Yu, and C. Tiang, Acta. Phys. Sinica, 21, 618(1965)

[13] T. Nishida, T. Hirai, & Y. Takashima, "Mossbauer Spectroscopic Study of Nonbridging Oxygen Atoms in Potassium Borosilicate Glasses", Phys. Chem. Glasses, 22, 94(1981)

[14] X. Feng, I. L. Pegg, Aa. Barkatt, P. B. Macedo, S. Cucinell and S. Lai, "Correlation between Composition Effects on Glass Durability and the Structural Role of the Constituent Oxides", to be published in June, 1989 in Nuclear Technology.

[15] H. D. Schreiber, J. Geophys. Res., 92, 9225(1987) and refs therein.

EFFECTS OF THE SA/V RATIO ON THE LONG-TERM CORROSION KINETICS OF R7-T7 GLASS

E. Vernaz, T. Advocat and J.L. Dussossoy
CEN - Valrhô, BP 171
30205 Bagnols-sur-Cèze, France

INTRODUCTION

Fission products from reprocessed commercial reactor fuel have been vitrified in France for over a decade. Many laboratory tests have shown that the "R7-T7" glass selected for vitrifying LWR solutions is a stable matrix with very low crystallinity regardless of the heat treatment, and with good resistance to aqueous corrosion. The composition of this glass has been widely published[1].

The ultimate objective of all the leaching investigations is to identify the principal alteration mechanisms in order to develop a "glass source term" model capable of predicting glass behavior over periods of 10 000 years or more.

Most authors agree that the determining factor in dissolution of the glass matrix is the H_4SiO_4 activity in solution[2,3,4,5]. This is generally described by a first-order relation such as the following[6]:

$$r_m = r_o(1 - a_{Si}/a_{sat})$$

where r_m is the glass matrix dissolution rate, r_o is the rate constant for the forward reaction, a_{Si} is the actual activity of orthosilicic acid in solution, and a_{sat} is the corresponding activity at saturation. Most recent models[2,3,5,6] allow for dissolved silica transfer through a gel layer of increasing thickness; a_{Si} is thus the orthosilicic acid activity at the unaltered glass/gel interface, calculated with allowance for the apparent H_4SiO_4 diffusion coefficient in the gel, $\tilde{D}$.

Finally, since the corrosion rates observed in most experiments do not diminish to zero even over the long term, Grambow's model[6] included a corrosion rate under saturation conditions, designated r_{fin} and assumed constant.

Irrespective of coupling this type of model with geochemical codes, which will largely depend on the assumed nature of the precipitating minerals, the dissolution kinetics of a nuclear waste glass at a given temperature are then described by four main parameters:

- the initial dissolution rate r_o
- the final dissolution rate r_{fin}
- the silica saturation limit a_{sat}
- the silicic acid diffusion coefficient in the gel $\tilde{D}$

Although the initial dissolution rate of R7-T7 glass in pure water is known to within a factor of 2, considerable uncertainty remains on the value of r_{fin}[7]. An attempt has already been made to measure this corrosion rate with R7-T7 glass under saturating conditions, but with only limited results[1]: at the highest SA/V ratio tested (350 m^{-1}) it was uncertain whether silicon saturation conditions were obtained, and the scattering of the experimental points made it difficult to measure the final corrosion rate with acceptable accuracy.

A new series of static experiments at higher SA/V ratios has therefore been undertaken with R7-T7 glass* using an improved experimental protocol. Four SA/V values will be investigated to test the hypothesis that the final corrosion rate is constant under saturation conditions. This paper discusses the work done at 90°C, for which a large number of results are already available. The tests will continue at higher temperatures.

EXPERIMENTAL CONDITIONS

Materials

Four high SA/V ratios (400 m^{-1}, 2000 m^{-1}, 8000 m^{-1} and 20000 m^{-1}) were selected to obtain saturation conditions rapidly. Ten experimental durations ranging from 7 days to 1 year were defined for each SA/V value (7, 14, 28, 56, 84, 112, 140, 182, 273 & 365 days) and each experiment was conducted in triplicate.

The 400 m^{-1} tests use rectangular prismatic glass samples sawn into comb-shaped specimens with five "teeth" 1 mm thick and 3.5 mm apart. The specimen surface finish is saw cut, and the total surface area about 20 cm^2.

The tests at higher SA/V ratios are conducted with glass powder in suitable granular ranges: 400-500 μm (2000 m^{-1}), 100-125 μm (8000 m^{-1}), and 40-50 μm (20000 m^{-1}). Ball-milled glass powder batches were washed repeatedly in acetone until the supernatant fraction was clear. Optical microscopic examination confirmed that virtually all the powder and fines were eliminated. Each test involves 2 g ± 1 mg of powder, with surface areas of 100 cm^2, 400 cm^2 and 1000 cm^2 for the 3 SA/V ratios calculated by assimilating the powder grains with spheres equal in diameter to the mean granular value.

---

* *R7-T7 glass is virtually identical with the "JSS-A" glass described by Grambow[6,7].*

The small (7 ml) teflon leaching containers were washed according to the MCC-1 procedure. The leachate is *Volvic* mineral water with the composition shown in Table I. This leachate simulating a granitic water was selected in preference to distilled water so that any effect due to certain anions (Cl^-, $SO_4^=$, HCO_3^-, etc.) will be included in the limiting equilibrium found under saturating conditions.

Table I - Chemical Composition of *Volvic* Water ($mg \cdot l^{-1}$)

pH	Si	Ca	Na	Mg	K	Fe	Zn	Cl^-	$SO_4^=$	HCO_3^-	F^-	NO_3^-	CO_2
7.2	11	9.8	9.2	5.4	5.5	<.005	<.03	7.0	7.2	66	.21	1.0	20

Test Procedure

Two grams of glass powder (or a monolithic specimen) and 5 ml of *Volvic* water were placed in a teflon container, which was then weighed to allow accurate determination of the water loss on completion of the test. In order to limit the water loss to negligible amounts after one year, the three containers in each series (i.e. same SA/V and same duration) were placed in a pyrex tube containing a few drops of water to equalize the water vapor partial pressures inside and outside the teflon containers. The pyrex tube was then sealed with a torch.

The tubes containing powdered glass were placed in an oven regulated at 90°C ± 1°C, lying horizontally on rollers spinning at about 20 rpm to ensure continuous agitation. The tubes containing the monolithic specimens were placed vertically in the oven and were not agitated.

Termination and Analysis

On completion of the test, the leachate pH is measured before and after cooling to room temperature. The container is weighed to determine the water loss, and the result is disregarded if the loss exceeds 10%. A 2 ml supernatant leachate sample is taken by pipette and diluted in 15 ml of acidified deionized water for ICP analysis. For the 20000 m^{-1} experiments, a 2 ml supernatant sample was taken as above, and 1 ml was ultrafiltered to approximately 1.8 nm to separate the colloid or particulate fraction from the soluble fraction. Ultrafiltration was carried out at 25°C, but an ultrafiltration test conducted at 90°C in one case showed no appreciable difference.

EXPERIMENTAL RESULTS

Leachate pH

The leachate pH stabilized from the first series (7 days) and remained appreciably constant up to 364 days; the mean values at 90°C and 25°C are shown in Table II.

Table II - Mean Leachate pH and Standard Deviation

SA/V	400 m^{-1}	2000 m^{-1}	8000 m^{-1}	20000 m^{-1}
pH at 90°C	8.35 ± 0.23	8.66 ± 0.18	9.04 ± 0.13	9.42 ± 0.13
pH at 25°C	8.91 ± 0.18	9.28 ± 0.13	9.61 ± 0.10	9.77 ± 0.05

The pH thus increases with the SA/V ratio. This phenomenon is not attributable to the progression of the dissolution reaction, since the pH after 10 days at SA/V = 20000 m^{-1} did not correspond to the 100-day value for SA/V = 2000 m^{-1}. The systematic difference observed from the beginning of the experiment is probably due to alkaline $\rightleftharpoons H^+$ interdiffusion. In the same ionization layer the quantity of sodium released, and therefore the pH, is proportional to the exchange surface area. As the pH significantly affects the silica solubility limit, the experiments cannot be compared directly with one another since for the same SA/V × t product the pH values are different.

Silicon and Boron Concentrations

Silicon and boron are of particular interest in understanding dissolution mechanisms. The silicon concentration indicates the difference from saturation, while the boron concentration is a good glass corrosion indicator: at 90°C boron is generally not present in any of the glass alteration products.

- **SA/V = 400 m^{-1}**

The silicon and boron concentration variations for this experiment are shown in Figure 1. A steady-state silicon concentration of about 50 ppm was observed after only 14 days, and the boron concentration leveled off as well, indicating that corrosion virtually ceased. After 112 days the silicon concentration rose slightly to 55 ppm while the boron concentration also increased, indicating renewed corrosion at a rate of about 25×10^{-4} $g \cdot m^{-2} d^{-1}$.

- **SA/V = 2000 m^{-1}**

Between 14 and 28 days the silicon concentration (Figure 2) stabilized at about 76 ppm and the boron concentration remained constant. After 56 days the total silicon concentration sharply increased to about 100 ± 10 ppm, while the boron concentration increased steadily, indicating a mean corrosion rate of 6×10^{-4} $g \cdot m^{-2} d^{-1}$ between 56 and 364 days.

- **SA/V = 8000 m^{-1}**

The mean total silicon concentration in solution (Figure 3) reached 200 ± 10 ppm after 56 days. The steady increase in the boron concentration between 56 and 273 days corresponded to a mean corrosion rate of 6×10^{-4} g·m^{-2}d^{-1} under saturating conditions.

- **SA/V = 20000 m^{-1}**

Variations in the silicon and boron concentrations for this experiment are shown in Figure 4. The total silicon concentration in solution rose to 350 ppm after 28 days, then diminished. The dissolved silicon concentration (after 1.8 nm ultrafiltration) is indicated by the broken line in the figure: it remained virtually constant at 120 ± 10 ppm throughout the experiment. The boron concentration increased up to 112 days, showing that corrosion initially progressed despite apparent silicon saturation in solution. Beyond 112 days the concentration of boron and other mobile elements (Na, Li) in solution no longer increased, indicating that the residual corrosion rate was nil (below 1.5×10^{-4} g·m^{-2}d^{-1}). Thus no subsequent increase in corrosion was observed at any time due to precipitation of any phases, despite significant silica supersaturation.

SILICA SATURATION CONCENTRATION

Saturation of the silicon concentration appeared to be reached quickly in all of these experiments. The slight increase in slope that can be observed for long durations at SA/V ratios of 400 m^{-1}, 2000 m^{-1} and 8000 m^{-1} is probably related to the formation of colloidal silica particles. This is confirmed by the 20000 m^{-1} experiment, which showed that after ultrafiltration the silicon concentration was constant at about 120 ± 10 ppm. The sharp rise followed by a slow decline in the total silicon concentration is probably related to the formation, then floculation and decanting of silicated colloids, but not to precipitation of a new silicated phase, which would result in significantly increased corrosion.

Although the dissolved silicon fraction was not measured for the three lowest SA/V ratios, it may be estimated from the following observation. At the three highest SA/V ratios, very high concentrations of relatively insoluble elements (Fe, Al, Zn, etc.) were measured in solution: the values ranging from a few ppm to over 20 ppm were well above the corresponding hydroxide solubility limits. This suggests that these cations were adsorbed or contributed to the formation of silica colloids. Figures 5 and 6 show a linear relation between the concentrations of these elements and the total silicon concentration, as previously observed by Shade and Strachan at high SA/V values[8].

A good estimate of the dissolved silicon concentration is then obtained when no more colloids are present, i.e. when the Fe or Zn curve is extrapolated to zero (or < 0.1 ppm). The extrapolated value was in excellent agreement with the value measured after ultrafiltration at SA/V = 20000 m^{-1}. This gives the following glass silica saturation values:

Table III - Glass Silica Saturation Values

SA/V (m^{-1})	(M)easured or (E)xtrapolated C*	Calculated C* ($\log K_{sat} = -2.80$)
400	52 ± 3 (M)	52
2000	60 ± 3 (E)	62
8000	81 ± 5 (E)	95
20000	116 ± 6 (E) 122 ± 15 (M)	201

The right-hand column represents the total dissolved silicon concentration calculated from an empirical equilibrium constant K_{sat} ($SiO_2 + 2H_2O \rightleftharpoons H_4SiO_4$) and allowing for hydrolysis of silicic acid with the following constants:

$$H_4SiO_4 \rightleftharpoons H_3SiO_4^- + H^+ \ (\log K_1 = -9.09 \text{ at } 90°C)$$

$$H_4SiO_4 \rightleftharpoons H_2SiO_4^= + 2H^+ \ (\log K_2 = -9.60 \text{ at } 90°C)$$

A good fit is obtained with the experimental results with $\log K_{sat} = -2.80$ for the first two SA/V ratios. The deviation for the last two SA/V ratios may be due to uncertainty on the K_1 and K_2 values at 90°C, but may also be attributed to the fact that for highly concentrated solutions the activity cannot be assimilated with the concentration. The approximate $\log K_{sat}$ value of −2.80 proposed here is near but slightly higher than the figure advanced by Grambow ($\log K_{sat} = -2.93$)[7].

FINAL REACTION RATE

Table IV summarizes the final corrosion rates observed after 1 year at 90°C for each of the SA/V ratios tested.

Table IV - Final Reaction Rate after 1 Year at 90°C

SA/V (m^{-1})	r_{fin} (10^{-4} $g \cdot m^{-2} d^{-1}$)	Remarks
50	83	distilled water[9]
50	133	*Volvic* water[9]
400	45	distilled water[10]
400	25	*Volvic* water (this paper)
2000	6	
8000	6	
20000	< 1.0	

The residual dissolution rate is clearly not constant, but diminishes as the product of SA/V × t increases. The significant slowdown of the final rate is not due to diffusion of leached species through a growing surface layer, since the altered glass thickness diminishes as the SA/V ratio increases. In fact, for all these experiments, diffusion phenomena through layers a few tenths of a micron thick were absolutely negligible.

CONCLUSION

A glass silica solubility limit near the value reported by Grambow was observed for R7-T7 glass.

A constant residual corrosion rate was not observed, however. The final rate r_{fin} cannot be considered as a characteristic glass property which can be used in long-term behavior models. As the final rate r_{fin} dropped to less than 10^{-4} $g \cdot m^{-2} d^{-1}$ for the higher SA/V ratios, it may be assumed that there is no glass "residual affinity".

Finally, it may be noted that the very low corrosion rates detected in distilled or "crystalline" water are not directly applicable to complex storage situations where environmental materials, notably clay particles[11], can prevent saturating conditions from being obtained.

ACKNOWLEDGMENTS

Most of the information in this paper was obtained under Contract FI 1W-0030 as part of the "Research Program on Management and Disposal of Radioactive Waste" sponsored by the European Communities.

REFERENCES

[1] Fillet, S., Vernaz, E., Nogues, J.L., and Jacquet-Francillon, N. "Corrosion Rate of Nuclear Glass in Saturated Media". 3rd International Symposium on Nuclear Waste Management, Chicago (1986). Ceramic Advances, 20.

[2] Wallace, R.M., and Wicks, G.G. "Leaching Chemistry of Defense Borosilicate Glass". *Scientific Basis for Nuclear Waste Management VI*, ed. D.G. Brookins. North Holland, NY, pp 23-28 (1983).

[3] Strachan, D.M. "The Effect of Flow Rate on the Leaching of Nuclear Waste Glass". *Advances in Ceramics - Nuclear Waste Management* vol 8, eds. G.G. Wicks and W.A. Ross. American Ceramic Society, Columbus, Ohio, pp 12-18 (1984).

[4] Grambow, B. "A General Rate Equation for Nuclear Waste Glass Corrosion". *Scientific Basis for Nuclear Waste Management VIII*, eds. C.M. Jantzen, J.A. Stone and R.C. Ewing. Materials Research Society, Pittsburgh, PA, pp 15-27 (1985).

[5] Mouche, E., and Vernaz, E. "Aqueous Corrosion of the French LWR Solution Reference Glass: First-Generation Model". *Scientific Basis for Nuclear Waste Management XI*, eds. M.J. Apted and R.E. Westerman. Materials Research Society, Pittsburgh, PA, pp 703-712 (1988).

[6] Grambow, B. "JSS-A Glass Dissolution: Mechanism, Model and Experiments". SKB Technical Report 87-02. Swedish Nuclear Fuel and Waste Management Co., Stockholm (1987).

[7] Grambow, B., and Strachan, D.M. "A Comparison of the Performance of Nuclear Waste Glass by Modeling". PNL 6698. Pacific Northwest Laboratory, Richland, WA (Dec 1988).

[8] Shade, J.W., and Strachan, D.M. "Effect of High Surface Area to Solution Volume Ratio on Waste Glass Leaching". Ceramic Bulletin, 65 No 12 (1986).

[9] Vernaz, E., and Dussossoy, J.L. "Basic Mechanisms of Aqueous Corrosion of Waste Glasses". CEC Contract FI 1W-0030(F). 1988 Annual Report.

[10] Fillet, S. "Mécanismes de corrosion et comportement des actinides dans le verre nucléaire R7-T7" (Corrosion Mechanisms and Actinide Behavior in R7-T7 Nuclear Glass). PhD Thesis, University of Montpellier, France (1987).

[11] Godon, N., Vernaz, E., Thomassin, J.H., and Touret, J.C. "Effect of Environmental Materials on Aqueous Corrosion of R7-T7 Glass". Presented at the 12th International Symposium on the Scientific Basis for Nuclear Waste Management. MRS, Berlin (1988).

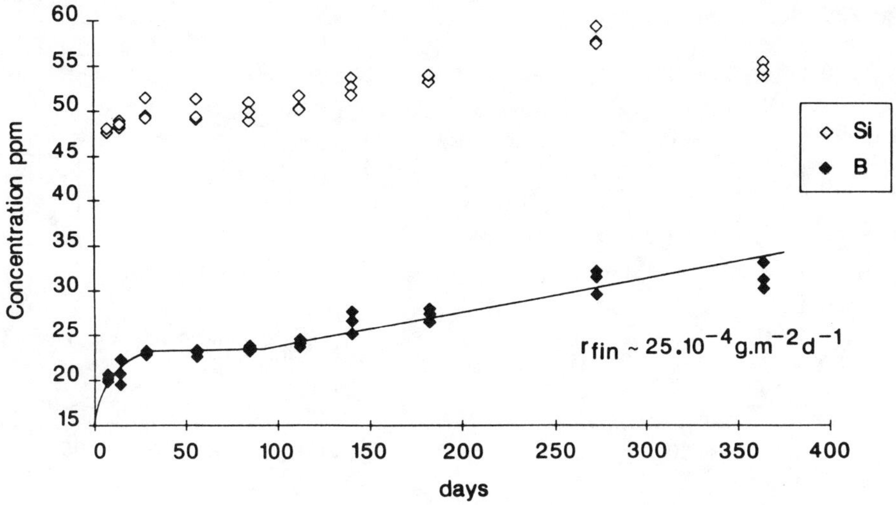

Figure 1 : Silicon and Boron Concentrations versus Time for SA/V = $400m^{-1}$

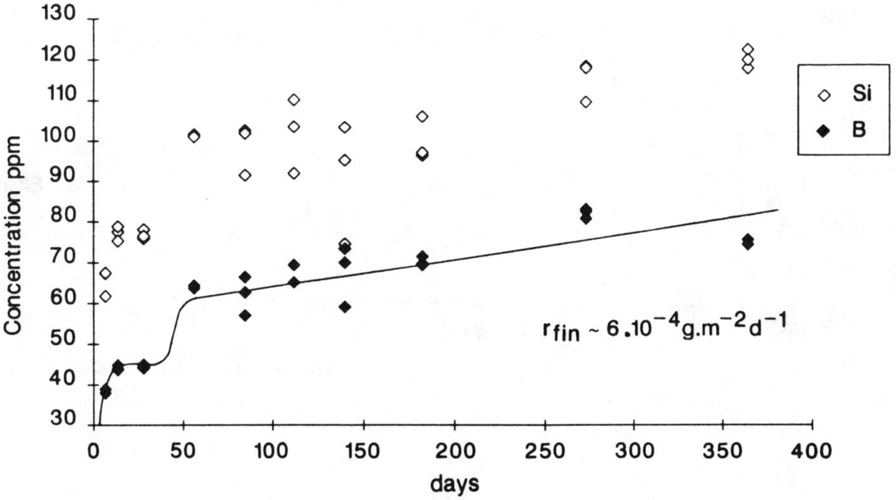

Figure 2 : Silicon and Boron Concentrations versus Time for SA/V = $2000m^{-1}$

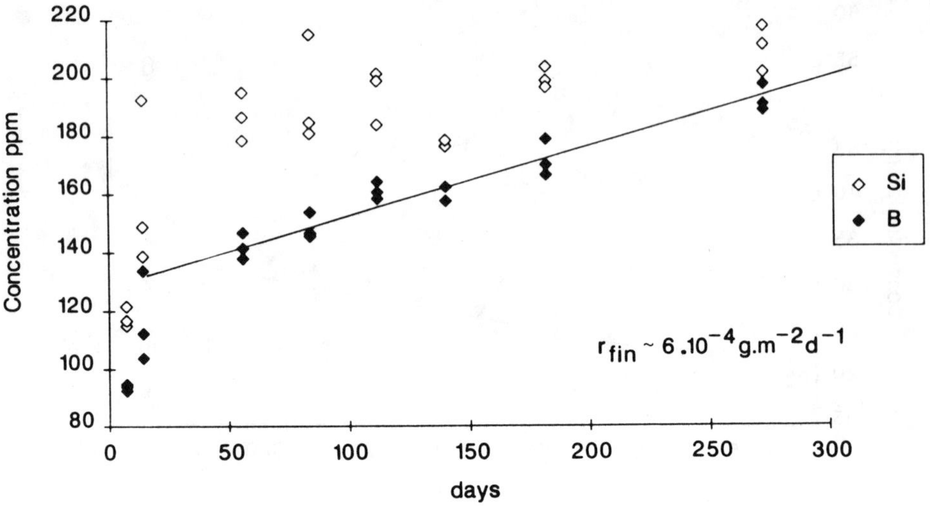

Figure 3 : Silicon and Boron Concentrations versus Time for SA/V = 8000m^{-1}

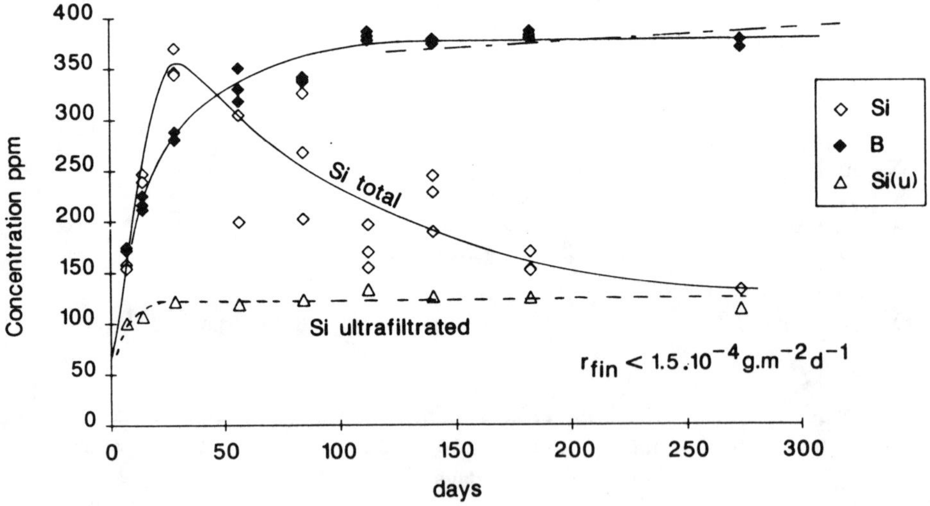

Figure 4 : Silicon and Boron Concentrations versus Time for SA/V = 20000m^{-1}

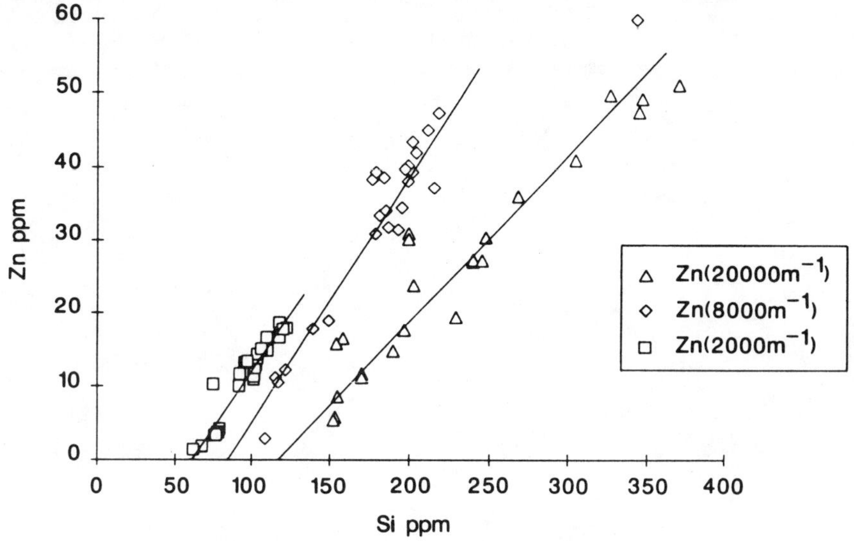

Figure 5 : Leachate Zinc Concentration versus Silicon Content

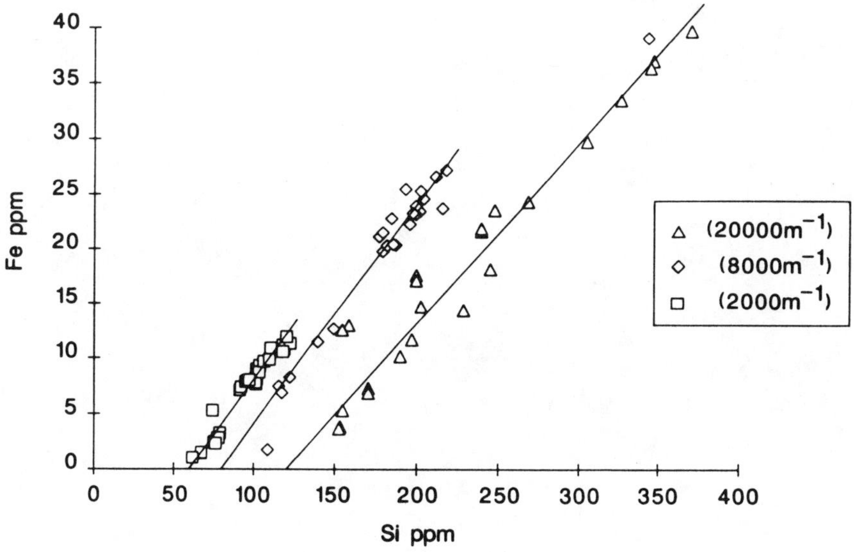

Figure 6 : Leachate Iron Concentration versus Silicon Content

DISSOLUTION MODEL FOR A GLASS HAVING AN ADHERENT INSOLUBLE SURFACE LAYER

K.B. Harvey and C.A.B. Larocque
(Geochemistry and Applied Chemistry Branch)
Atomic Energy of Canada Limited
Whiteshell Nuclear Research Establishment
Pinawa, Manitoba ROE 1LO Canada

ABSTRACT

Waste form glasses that contain quantities of iron, manganese and aluminum oxides, can form a thick, hydrated surface layer when placed in contact with water. Previous attempts to describe the dissolution, based on the "Savannah River Model" of Wallace and Wicks, assume that all of the material dissolved from the glass enters solution, whereas it is observed that substantial quantities of material are retained in the surface layer. An alternative model, presented here, contains a mass balance equation that accounts for material both in solution, and in the surface layer. The model is fitted to experimental data. Dissolution behaviour observed under saturated solution conditions is consistent with a surface area-to-volume effect.

INTRODUCTION

The Savannah River Laboratory glass SRL TDS-131 forms a surface film when placed in contact with water. Barkatt et al.[1] and Macedo et al.[2] have published data on the dissolution of SRL TDS-131 taken under conditions of high specimen surface area-to-solution volume (SA/V) ratio, and slow solution replenishment. In a previous paper[3], it was shown that the data could be fitted by the "Savannah River Model"[4] if a time-dependent diffusion coefficient was assumed for species moving through the surface layer. The assumption of a time-dependent diffusion coefficient is consistent with the observation that the microscopic structure of the layers can change across their thickness[5].

The Savannah River Model assumes that all the material dissolved from the glass passes into solution, whereas, in the case of SRL TDS-131 at least, quantities of the glass matrix elements, Si, Al, Fe, and Mn, were observed to be incorporated into the surface layer[5,6]. This present work suggests an alternative dissolution model, containing a mass balance equation to account for the quantities of the glass matrix species that are incorporated into the surface layer. It is shown that the model can describe experimental data with the assumption of a constant diffusion coefficient for species in the surface layer.

DISSOLUTION MODEL

In Figure 1, a semi-infinite block of glass has a plane surface area, SA, in contact with a volume, V, of water under conditions of constant temperature and pH. The surface of the glass is covered by a layer of the same composition as the glass, whose thickness increases with time, and whose properties allow free access of water to the glass/layer interface, and the diffusion of glass dissolution products away from the interface into the water. For unit area of the glass surface, the flux of glass dissolution products entering solution is dQ'/dt, and the total mass per unit area to have entered solution is Q'. The volume V of solution can be replenished such that a fraction n of solution is replaced with fresh water in unit time. The concentration of dissolved species in solution is c at time t. It is assumed that the glass is dissolving as a single substance in the water at the glass/layer interface. At the interface, the solution is saturated with respect to glass so that the rate of dissolution of the glass is constant, at

$$R = k^{+} - k^{-} \cdot c_s \tag{1}$$

where k^+ is the forward (dissolution) rate constant, k^- is the reverse (precipitation) rate constant, and c_s is the saturation concentration of glass in solution[7].

Following Wallace and Wicks[4], it is assumed that the dissolved species are diffusing away from the interface through the surface layer, and that the concentration gradient of the species through the layer is linear. If dQ'/dt is the flux of species through the layer, then

$$\frac{dQ'}{dt} = \frac{D}{L} \cdot (c_s - c) \tag{2}$$

where D is the effective diffusion coefficient of the species travelling through the layer, c is the concentration of species in the bulk solution and, L is the thickness of the layer.

If a proportion of the species dissolving from the glass is used in the formation of the surface layer, then the mass balance for the dissolution is

$$R = \frac{dQ'}{dt} + \rho \cdot \frac{dL}{dt} \qquad (3)$$

where ρ is the density of the layer.

Integrating Equation (3) with respect to time:

$$R \cdot t = Q' + \rho \cdot L \qquad (4)$$

and eliminating L between Equations (2) and (4):

$$\frac{dQ'}{dt} = \frac{D \cdot \rho \cdot (c_s - c)}{R \cdot t - Q'} \qquad (5)$$

APPLICATION TO EXPERIMENTAL SITUATIONS

Infinite Solution Volume

If the solution volume is infinite, and well mixed, then c = 0 at all times, and Equation (5) can be rewritten

$$\frac{dt}{dQ'} - X \cdot t = - Y \cdot Q' \qquad (6)$$

where

$$X = R/D \cdot \rho \cdot c_s$$

$$Y = 1/D \cdot \rho \cdot c_s$$

Equation (6) is a standard form that can be integrated to give

$$t = Y \cdot (X \cdot Q' + 1)/X^2 + K \cdot e^{X \cdot Q'} \qquad (7)$$

where K is a constant of integration.

Finite Solution Volume

If an area SA of glass is in contact with a fixed volume V of solution, then $c = Q' \cdot SA/V$. Equation (5) can then be rewritten to give

$$\frac{dt}{dQ'} - \frac{X \cdot t}{(1 - Z \cdot Q')} = \frac{Y \cdot Q'}{(1 - Z \cdot Q')} \qquad (8)$$

where $Z = SA/V \cdot c_s$.

Equation (8) is a standard form that can be integrated to give

$$t = Y \cdot \left[\frac{1}{X \cdot Z} - \frac{1}{Z(X + Z)} \cdot (1 - Z \cdot Q')\right] + K' \cdot (1 - Z \cdot Q')^{-(X/Z)} \quad (9)$$

where K′ is a constant of integration.

Finite Solution Volume, Flowing or Replenished Solution

If a glass is dissolving into a volume of water, which is being replenished with frequency n, then

$$\frac{dc}{dt} = \frac{SA}{V} \cdot \frac{dQ'}{dt} - n \cdot c \quad (10)$$

where dc/dt is the net change of concentration of a dissolved species in the water. SA/V is the surface area-to-volume ratio of the glass and the water. dQ′/dt is the flux of the dissolving species into the water per unit area of glass.

Equation (10), together with Equation (5), describe dissolution in a flowing or replenished solution. An analytical solution to this case has not yet been found.

EXPERIMENTAL VALIDATION OF THE DISSOLUTION MODEL

(a) Application to the results of Macedo et al.

The dissolution models have been applied to data published by Macedo et al.[2] for the dissolution of SRL TDS-131 under replenished conditions, and to the dissolution of AECL glass 871, under static conditions. The application to the results of Macedo et al. has been described in detail elsewhere[8], and will be summarized here.

SRL TDS-131 glass was placed in contact with deionized water (DIW) at 70°C, at SA/V ratios of 9.7 cm^{-1} and 19.4 cm^{-1}, respectively. Aliquots of solution were removed and replaced by fresh water sucho that the entire volume of solution would be replaced in 112 days (replenishment frequency, n = 0.0089 d^{-1}). The solution pH stabilized at a value of 11.2 to 11.4. The aliquots were analyzed for Si, B, and Na, and the concentrations were normalized by dividing the actual concentration found in solution by the fractional concentration of that element in the glass. The normalized concentrations were found to coincide within the limits of experimental error, indicating that the dissolution of the glass could be regarded as being congruent.

In Figure 2, the rectangular boxes indicate the reported range of normalized concentration (c) for Si, plotted as a function of time (t). Macedo et al. regard the data taken after 220 d as indicating a "steady state" in which the surface layer is dissolving into solution at the same rate as the glass. This situation is not addressed here. Equation (5) can be rewritten as:

$$\frac{t}{Q'} = \frac{1}{R} + \frac{\rho \cdot D}{R} \cdot \frac{(c_s - c)}{Q' \cdot dQ'/dt} \qquad (11)$$

In previous work[3], a series of values for c, dQ'/dt and Q' were calculated that were consistent with the data of Macedo et al. for the 9.7 cm^{-1} SA/V ratio experiment. Assuming, as in reference (3), that $c_s = 6.25 \times 10^{-4}$ g/cm^3, values for t/Q' and $(c_s - c)/Q' \cdot dQ'/dt$ were calculated for values of t between 20 and 220 days. A straight line fit ($\sigma^2 = 0.996$) to these values, using Equation (11) gave

$$R = 7.81 \times 10^{-7} \text{ g/cm}^2 \cdot \text{d}$$

and

$$\rho \cdot D = 1.58 \times 10^{-8} \text{ g/cm} \cdot \text{d.}$$

These values were then used as the starting points for a computer search for other combinations of values of R, $\rho \cdot D$ and c_s that would also satisfy the criteria that the calculated solution concentration-time curve must pass through the published experimental results at 28, 85, and 220 d, that is, the two ends and the peak of the concentration-time curve.

For each set of values of R, $\rho \cdot D$, and c_s, a numerical integration using Equations (5) and (10) was made, and a table of solution concentration (c) as a function of time (t) was established. A total of 464 solutions which fulfilled the criteria were found. These 464 solutions were then re-examined to see if any of them simultaneously satisfied the equivalent criteria (28 d, 55 d, and 220 d) for the experiment conducted at SA/V = 19.4 cm^{-1}. A group of ten of the sets of values were found to satisfy both criteria.

Choosing one set of values in the centre of the group as an example, where $\rho \cdot D = 2 \times 10^{-9}$ (g/cm.d), $c_s = 2.4 \times 10^{-3}$ (g/cm^3), and $R = 7.8 \times 10^{-7}$ (g/cm^2.d), solution concentration-time curves were calculated for the SA/V ratios of 9.7 cm^{-1} and 19.4 cm^{-1}. These solution concentration-time curves are shown in Figure 2, together with the fit obtained using the previous model[3]. The normalized value of $c_s = 2.4 \times 10^{-3}$ (g/cm^2) corresponds to an actual value of 1.05×10^{-3} g/cm^3, which can be compared to the value of ~ 1×10^{-3} g/cm^3 suggested by Okamoto et al.[9] for the saturation solubility of silica at 70°C, at pH 11.

(b) Application to the dissolution of Glass 871

An AECL glass composition, 871, contains quantities of Al_2O_3, CaO, and ZnO, and might be anticipated to form surface films of insoluble material. A series of static and replenished dissolution experiments were performed using this glass, at 100°C, in DIW buffered to pH 9 with a HYDRION buffer. Static experiments were conducted at SA/V ratios of 1:5, 1:10, and 1:20. The length of the static dissolution experiments, 364 days, were such that, for the majority of the experiment, the solutions should be saturated by dissolution products from the glass, and any surface layers should be well-developed. Four specimens also had the solutions completely replaced every day for 30 days, conditions under which no buildup of dissolution products in solution or any formation of a surface layer would be anticipated; the behaviour of these specimens was used to determine the forward dissolution constant k^+. The progress of dissolution was followed by direct weighing of the specimens. Specimens were weighed every 3 days to day 42, and every 7 days thereafter. A subsidiary experiment was conducted at an SA/V ratio of 1:10, but with weighing only one-third as often. No difference in the unit weight loss/time curves for the two weighing frequencies at this SA/V ratio was observed, indicating that the effectiveness of the surface layer, if present, was not compromised by the operation of weighing.

Plots of the data points for Glasses 871K (SA/V = 1:10) and 871M (SA/V = 1:20) are shown in Figure 3(a,b) as examples of the dissolution behaviour observed. The general characteristics of all of the dissolution curves at all SA/V ratios are an initial dissolution at a declining rate, (OA, OA′) followed by an increasing rate (AB, A′B′), leading to a period of apparently linear rise (BC, B′C′), (Figure 3a), followed by a decline over a 'nose' (CD, CD′), and then a further period of relatively linear dissolution (DE, D′E′) (Figure 3b).

The experimental data will be interpreted in terms of glass dissolution leading to supersaturation of dissolution products, precipitation from solution, and then continued dissolution of the glass under saturated solution conditions, with formation of a surface layer.

The dissolution of specimens of 871K for the first 12 days (OA), and of 871M for the first 39 days (OA′), are interpreted as the dissolution of a 'simple' glass, with no surface layer present. Using methods described previously[7]:

k^+ (the dissolution rate constant) = 4.075×10^{-4} g/cm^2.d.

k^- (the precipitation rate constant) = 1.72 cm/d.

The corresponding 'best fit' dissolution curves calculated for specimens 871K and 871M are shown in Figure 3(a) as OAF and OA′F′, respectively.

From the point A (at 12 d) to the point B (at 24 d) for Glass 871K, and from A′ (at 39 d) to B′ (at 70 d) for Glass 871M, the dissolution rates increase until they apparently stabilize and the dissolution curves become linear (BC, B′C′). This is interpreted as being associated with a reduction in solution concentrations as the result of precipitation from supersaturated solution, and will be discussed later.

From B and B′ onwards, dissolution of Glasses 871K and 871M continues into saturated solutions, from which material is precipitating. The dissolution curves BE and B′E′ are interpreted as being associated with the formation of a layer on the surface of the glass, that retards the release of dissolution products into solution.

Equation (7) can be modified to describe the dissolution of a glass in a solution that contains a constant concentration of dissolution products, c, by the substitution of the term $(c_s - c)$ for c_s in the parameters X and Y. Figure 3(b) shows the best fits using the modified Equation (7) to the 24 to 364 day data for Glass 871K and the 70 to 364 day data for Glass 871M.

For the initial slope of the curves, BC, B′C′, $dQ'/dt = X/Y$. From the initial slopes of the curves for Glasses 871K and 871M, and from the known values of k^+ and k^-, solution concentrations corresponding to the observed dissolution rates were calculated to be 1.55×10^{-4} g/cm^3 and 0.60×10^{-4} g/cm^3, respectively.

For Glass 871K, at the point A on the dissolution curve, the solution concentration, (calculated from $c = Q' \cdot SA/V$) is 2.24×10^{-4} g/cm^3, while it is 1.55×10^{-4} g/cm^3 at point B. For Glass 871M, at A′ and B′, the figures are 2.29×10^{-4} and 0.60×10^{-4} g/cm^3, respectively. The rate of change of concentration in solution will be a function of both the rate of dissolution of the glass, and the rate of precipitation from solution. Assuming that solution supersaturation is the driving force for precipitation, and that the rate of precipitation is proportional to $(c - c_T)$, where c_T is the 'true' saturation concentration in solution, then

$$\frac{dc}{dt} = \frac{dQ'}{dt} \cdot \frac{SA}{V} - \alpha \cdot (c - c_T) \qquad (12)$$

where α is a constant of proportionality.

The 12 to 24 day data for Glass 871K was fitted by a third-order polynomial, constrained to have a slope of 1.48×10^{-4} g/cm^2.d, the slope observed at 24 d. Equation (12), with the aid of this polynomial, was used to perform a numerical integration between 12 and 24 days. Values were assigned to c_T, and values were found for α for which $c = 1.55 \times 10^{-4}$ g/cm^3 at 24 d. The result is shown as the lower curve in Figure 4. A similar approach was taken for Glass 871M, resulting in the upper curve in Figure 4.

The paths of the two curves in Figure 4 suggest that there is no prospect of them intersecting, to give common values of α and c_T. The results can be reconciled if a surface area-to-volume term is included in Equation (12) so that:

$$\frac{dc}{dt} = \frac{dQ'}{dt} \cdot \frac{SA}{V} - \alpha \cdot \frac{V}{SA} \cdot (c - c_T) \qquad (13)$$

when common values of $\alpha = 0.0103$ and $c_T = 5.13 \times 10^{-6}$ g/cm^3 can be found. The SA/V term in Equation (13) is not unreasonable if precipitation is by nucleation and precipitation from the bulk of solution.

SUMMARY

A model has been presented for the dissolution of a glass having an adherent insoluble surface layer. Equations derived from the model have been applied to data published by Macedo at al. for the dissolution of SRL TDS-131 glass under replenished solution conditions, and to the dissolution of Canadian reference glass 871 under static conditions. Good fits to both sets of data have been obtained. A complete description of the dissolution of Glass 871 has been obtained, using the current model, together with previous models that describe glass dissolution prior to the formation of a surface film. Dissolution behaviour during which supersaturation in solution is being reduced is consistent with an SA/V effect.

REFERENCES

1. A. Barkatt, W. Sousanpour, A.O. Barkatt, M.A. Boroomand, and P.B. Macedo, "Leach Behaviour of SRL TDS-131 Defense Waste Glass in Water in High/Low Flow Rates," Scientific Basis for Nuclear Waste Management, Vol. VII, p. 643, G.L. McVay, ed., North-Holland Publishing Company, New York, (1984).

2. P.B. Macedo, A. Barkatt, B.C. Gibson, and C.J. Montrose, "Long-Term Release Rates of Borosilicate Waste Forms," Nuclear Technology, 73, 199, (1986).

3. K.B. Harvey and C.A. Boase, "The Dissolution of a Glass Having an Adherent Insoluble Surface Layer," Advances in Ceramics, 20: Nuclear Waste Management II, p. 495, American Ceramic Society, Columbus, Ohio, (1986).

4. R.M. Wallace and G.G. Wicks, "Leaching Chemistry of Defense Borosilicate Glass," Scientific Basis for Nuclear Waste Management, Vol. VI, p. 23, D.G. Brookins, ed., Plenum, New York, (1983).

5. Final Report of the Defense High-Level Waste Leaching Mechanisms Program, PNL-5157, J.E. Mendel, Compiler, Pacific Laboratory, Richland, Washington, (1984).

6. G.G. Wicks, W.C. Mosley, P.E. Whitkop, and K.A. Saturday, "Durability of Simulated Waste Glass - Effects of Pressure and Formation of Surface Layers," J. Non-Crystalline Solids, 49, 413, (1982).

7. K.B. Harvey and C.A. Boase, "The Dissolution of a Simple Glass. Part 2: Behaviour in Closed Glass/Water Systems," Physics and Chemistry of Glasses 28, 11, (1987).

8. K.B. Harvey and C.A.B. Larocque, "Dissolution Model for A Glass Having an Adherent Insoluble Surface Layer," To Be Published.

9 Go. Okamoto, T. Okura and K. Goto, "Properties of Silica in Water," Geochemica et Cosmochimica Acta, 12, 123, (1957).

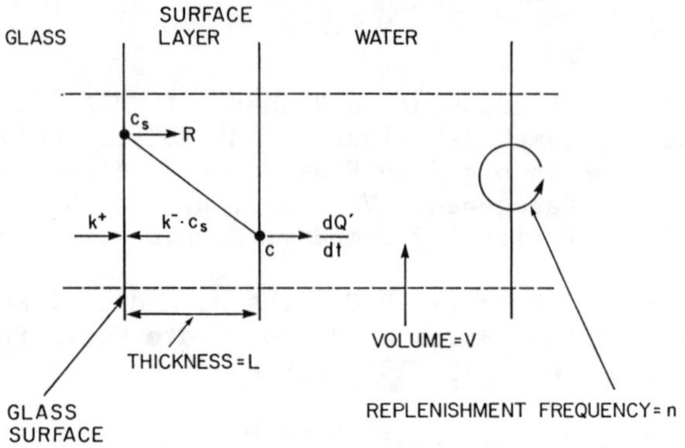

FIGURE 1: **Schematic Diagram for the Release of a Species Through a Surface Layer, Into a Replenished Volume of Water.**

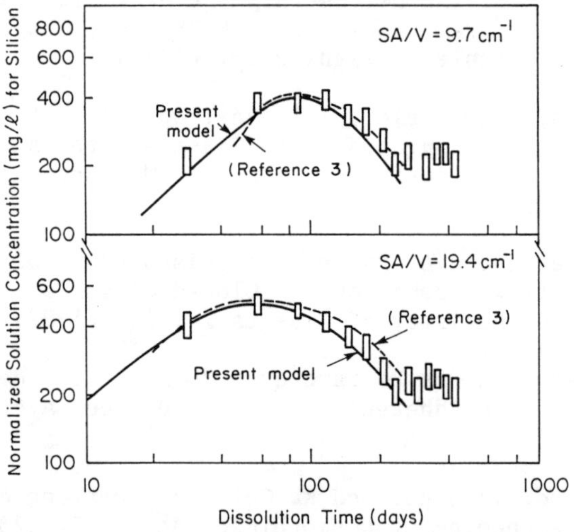

FIGURE 2: **Solution Concentration Data (after Macedo, et al., (2)), Showing Fits by Present Model (Full Line), Together With Fits by a Previous Model Which Used a Time-Dependent Diffusion Coefficient (Dashed Line).**

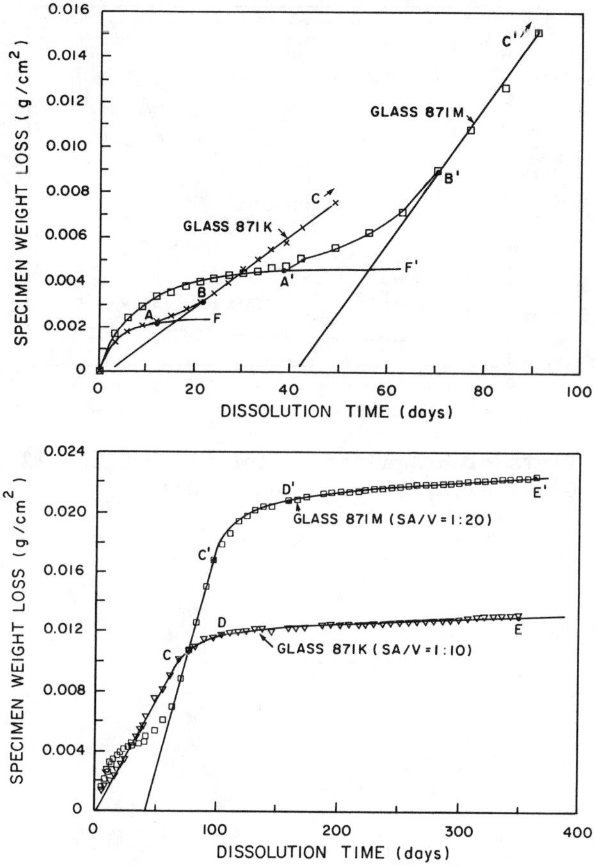

FIGURE 3: Weight Loss Per Unit Area Plotted as a Function of Time for Glasses, 871K and 871M. FIGURE 3a: Best Fit Curves to Data for First 100 days. FIGURE 3b: Best-Fit Curves to Long-Term Data.

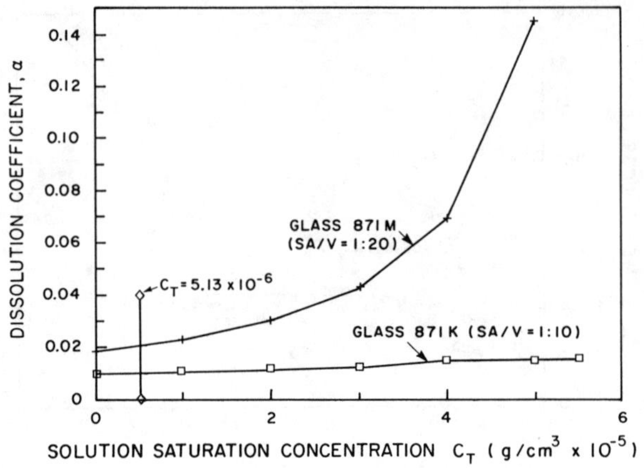

FIGURE 4: Plots of α Against c_T for Glass 871K, 12 to 24 day; Glass 871M, 39 to 70 days Data.

DISSOLUTION OF SPHENE GLASS-CERAMICS IN Ca-Na-Cl BRINES

P.J. Hayward, D.C. Doern and I.M. George
Atomic Energy of Canada Limited
Whiteshell Nuclear Research Establishment
Pinawa, Manitoba, ROE 1LO, Canada

INTRODUCTION

A major task of the Canadian Nuclear Fuel Waste Management Program is an assessment of the permanent disposal of nuclear fuel wastes by isolation of durable waste form materials in an engineered vault at a depth of 500 to 1000 m within a granitic pluton in the Canadian Shield. The candidate waste forms under study include used CANDU™ nuclear fuel, and also borosilicate and aluminosilicate glasses and sphene glass-ceramics for immobilizing high-level wastes (HLW) from possible future fuel recycling(1).

The results described here are intended to contribute to future development of a dissolution submodel for a sphene glass-ceramic waste form in contact with a Ca-Na-Cl brine typical of those found at depth in Shield-type granitic formations(2). In previous studies(3-5) with similar brines, chemical analyses of the leachants could not be used to follow the dissolution process because the dissolving ions were either major brine components (e.g., Ca, Na, Si, Sr) or formed species whose solubilities were below analytical detection limits (e.g., Ti, RE, U). To overcome this problem, brine-dissolution experiments were initiated on samples of aluminosilicate glass, sphene ceramics and sphene glass-ceramic doped separately with ^{22}Na ($t_{1/2}$ = 2.6 a) or ^{45}Ca ($t_{1/2}$ = 163 d), dissolution being measured by leachant analyses for β or γ activity(6). This report describes results from 496-d leaching experiments that constitute part of a larger test matrix, involving varying temperatures and solid-surface-area to leachant-volume (S/V) ratios; full results will be reported elsewhere.

SUMMARY OF PREVIOUS LEACHING STUDIES

Sphene glass-ceramics are formed by controlled crystallization during reheating to ~ 1050°C of precursor glasses from the Na_2O-Al_2O_3-CaO-TiO_2-SiO_2-X system, where X represents HLW and/or minor processing additives. They consist of 40-50 vol.% of micrometer-sized bulk-nucleated sphene ($CaTiSiO_5$) crystals dispersed within an aluminosilicate glass matrix. The waste ions are either contained within the sphene structure as solid-solution replacements for Ca and Ti or are incorporated in the glass phase(7,8). The glass-ceramics are significantly more durable than borosilicate glasses in many leachants at neutral or alkaline pH values, with leaching occurring predominantly in the glass phase(7).

Thermodynamic calculations(3,9) and experimental studies(3,4,10) have shown that sphene is a thermodynamically stable phase in Ca-rich, SiO_2-saturated solutions, such as the brines found deep in the Canadian Shield. The main dissolution reaction for sphene in aqueous media is

$$\underset{\text{sphene}}{CaTiSiO_5} + 2H^+ + H_2O = Ca^{2+} + \underset{\text{rutile}}{TiO_2} + H_4SiO_4 \quad .....(1)$$

although reactions with dissolved CO_2 and SO_4^{2-} are also possible(3).

Figure 1 shows the H^+-H_2O-CaO-TiO_2-SiO_2 activity diagram at 100°C, with the stability fields of sphene, rutile (TiO_2) and perovskite ($CaTiO_3$) being plotted as functions of the $[Ca^{2+}]/[H^+]^2$ ratio and of $[H_4SiO_4]$ (square brackets denote solution activities). The brine composition used in this study is also plotted at various pH values. At its unbuffered pH of 6.4 at 100°C, the brine plots just within the sphene stability field. However, a small reduction in solution pH or in $[H_4SiO_4]$ can move the brine's location to within the rutile stability field, when

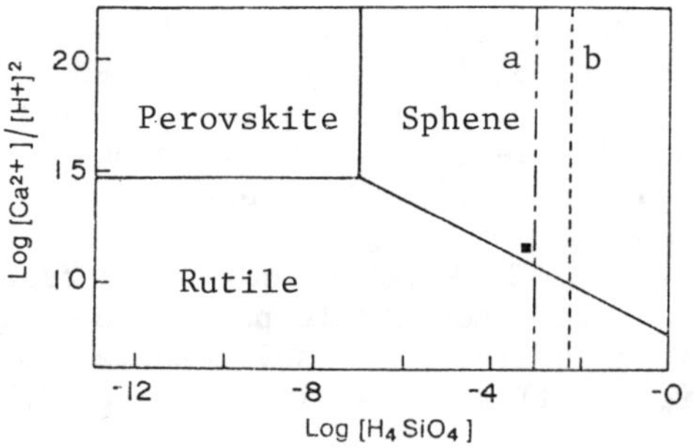

Figure 1. Activity diagram for the system H^+-H_2O-CaO-TiO_2-SiO_2 at 100°C. Dashed lines (a) and (b) represent quartz saturation and amorphous silica saturation, respectively. The position of unbuffered SCSSS brine at pH 6.4 is also shown.

sphene dissolution would commence according to equation 1, ceasing only when sufficient consumption of H^+ and production of Ca^{2+} and H_4SiO_4 occurred to restore equilibrium. Sphene equilibrium is particularly sensitive to pH changes, because of the $[H^+]^2$ term in the ordinate of Figure 1. The implications for sphene dissolution in a glass-ceramic are discussed later.

Nevertheless, in the absence of large pH or $[H_4SiO_4]$ changes, radionuclide release from sphene during leaching in Ca-Na-Cl groundwaters would occur mainly from surface ion exchange reactions rather than from bulk dissolution of sphene. For example, a rare earth (RE^{3+}) ion, together with a charge-compensating cation, can be substituted in the sphene structure, as in the formulation $(\underline{Ca}^{2+},RE^{3+},Na^+)TiSiO_5$. In this case, ion exchange could lead to the replacement of RE^{3+} and Na^+ by two Ca^{2+} ions from solution. Alternatively, replacement of Na^+ by Ca^{2+} and simultaneous protonation of an oxygen site is possible, leading to an idealized composition of $(Ca^{2+}_{1-x},RE^{3+}_x)TiSiO_{5-x}(OH)_x$. Similarly, substituted sphenes with charge compensation in the Ti^{4+} site, such as $(\underline{Ca}^{2+},RE^{3+})(\underline{Ti}^{4+},Al^{3+})SiO_5$, could undergo replacement of RE^{3+} by Ca^{2+} and simultaneous protonation of an O^{2-} site to form OH^-, giving a final formulation of $Ca^{2+}(\underline{Ti}^{4+},Al^{3+})SiO_4(\underline{O}^{2-},OH^-)$. Other similar ion exchange reactions leading to radionuclide release are also possible.

The development of a dissolution model for a composite material (e.g., a glass-ceramic or a polyphase ceramic) presents a significantly greater challenge than for a single-phase material (e.g., glass), because of factors such as synergism in leaching of the various phases, and the degree of phase interconnectivity. One purpose of the present study was to establish the importance of possible synergistic effects in the dissolution of the component glass and ceramic phases of the glass-ceramic.

EXPERIMENTAL PROCEDURE

The aluminosilicate glass composition used to represent the vitreous phase in a waste-loaded glass-ceramic (see Table 1) was based on previous transmission electron microscopic analyses(11) of the glass phase in a glass-ceramic containing 10 wt.% simulated HLW. Two 100-g batches of this glass were melted, separately doped with ^{22}Na (1.85×10^5 $Bq \cdot g^{-1}$) or ^{45}Ca (3.7×10^5 $Bq \cdot g^{-1}$), and then remelted, using previously described methods(6). In terms of overall Na_2O and CaO content, the radionuclide additions were insignificant.

The ^{45}Ca-doped sphene ceramic was formulated with a deliberate stoichiometric deficiency in CaO (see Table 1) to ensure that all Ca would enter the sphene structure. The ^{22}Na-doped sphene ceramic was similarly formulated, except that 10% CaO was replaced on a charge-balanced molar basis by Na_2O + La_2O_3

Mole% Oxide	Si-rich Glass	^{45}Ca-doped Ceramic	^{22}Na-doped Ceramic	Glass-Ceramic
Na_2O	7.4	–	0.76	6.4
Al_2O_3	9.3	–	–	4.9
CaO	5.8	30.0	27.77	16.0
TiO_2	2.1	35.0	35.36	14.3
SiO_2	71.9	35.0	35.36	55.1
Fe_2O_3	0.4	–	–	0.13
SrO	0.4	–	–	0.12
Y_2O_3	0.1	–	–	0.03
ZrO_2	0.6	–	–	0.43
UO_2	0.6	–	–	–
MoO_3	0.3	–	–	–
SnO_2	0.2	–	–	0.01
Cs_2O	0.3	–	–	0.16
BaO	0.3	–	–	0.16
Nd_2O_3	0.3	–	–	0.20
La_2O_3	–	–	0.76	0.12
CeO_2	–	–	–	1.60
Pr_2O_3	–	–	–	0.06
Cr_2O_3	–	–	–	0.04
NiO	–	–	–	0.22
Total	100.0	100.0	100.01	99.98

Table 1 Compositions of Glass, Sphene Ceramics and Glass-Ceramic

(see Table 1). This substitution level is within the previously determined[12] limits for charge-balanced replacement of 2 Ca by Na + La in sphene, and more accurately simulates the sphene phase in a waste-loaded glass-ceramic.

The ceramics were fabricated using the methods described earlier[4,6], with the same level of ^{22}Na- or ^{45}Ca-doping as for the aluminosilicate glass. Subsequent X-ray diffraction (XRD) analyses confirmed that sphene, together with traces of rutile (TiO_2) and cristobalite (SiO_2), were the only crystalline phases present in both ceramics. ASTM test C20[13], involving immersion of samples in boiling water and subsequent weighing, was used to establish that representative samples of each ceramic were impermeable to water. The existence of ~ 5 vol.% closed pores in these ceramics was subsequently deduced from scanning electron microscopy (SEM) observations.

The glass-ceramic base composition was similar to that studied previously[11], but with U being replaced by Ce and the noble metals Pd, Ru and Rh being omitted. Molybdenum was also omitted to avoid forming a minor powellite ($CaMoO_4$) phase[11], the presence of which could have complicated interpretation of the leaching results. The doped glass-ceramics were prepared by mixing precursor glass powders and nitrate solutions of ^{22}Na or ^{45}Ca, with the same doping levels as above. The mixtures were dried, remelted at 1400°C for 0.5 h, cast and annealed at 725°C. The glasses were then crystallized by reheating at $5°C \cdot min^{-1}$ to 1050°C and holding at 1050°C for 3 h. Subsequent XRD analyses and SEM examination confirmed that sphene was the only crystalline phase formed, occurring as bulk-nucleated, micrometer-sized crystals in a continuous aluminosilicate glass matrix.

The 100°C leach tests were performed in sealed PTFE vessels on 150-180 μm powdered samples, crushed, sieved and washed as described earlier(6). The leachant used was Standard Canadian Shield Saline Solution (SCSSS), with the following composition ($mg \cdot L^{-1}$): Ca^{2+}, 15 500; Na^+, 5047; K^+, 50; Mg^{2+}, 200; Sr^{2+}, 20; Cl^-, 34 310; SO_4^{2-}, 790; HCO_3^-, 10; NO_3^-, 50; and $SiO_{2(aq)}$, 50. An S/V ratio of 0.1 cm^{-1} was used in all tests, with a spherical-particle assumption being made to calculate solid surface areas. The tests with the powdered glasses and ceramics were performed in duplicate for 496 d, while those with the powdered glass-ceramic were performed in triplicate for 433 d.

In each test, 0.5-mL leachant samples were removed at regular time intervals, the leachant being replenished with fresh SCSSS and pH values being measured with colorfast indicator strips. Details of the liquid scintillation counting (LSC) measurements of β or γ activity in solution, and of initial ^{22}Na or ^{45}Ca activity levels in the unleached glass, ceramic and glass-ceramic samples have been reported earlier(6). The initial activity levels were used to convert β or γ activity levels in solution (after corrections for leachant replenishment and activity decay) to normalized(3) Na or Ca mass losses to solution from leaching.

RESULTS AND DISCUSSION

The leachant pH values in all experiments remained within the range 5.7-6.7, i.e., did not depart significantly from the unbuffered pH value of 6.4 for SCSSS at 100°C. Although some buffering by dissolved CO_2 would have occurred, neutral values of leachant pH were also maintained by precipitation of sepiolite from SCSSS, according to the equation

$$4Mg^{2+} + 8OH^- + 6H_4SiO_4 = \underset{\text{sepiolite}}{Mg_4Si_6O_{15}(OH)_2.6H_2O} + 9H_2O \quad (2)$$

Solution equilibrium calculations using the SOLMNEQ code(14) show that SCSSS is supersaturated with respect to sepiolite, and experiments have confirmed that minor temperature and pH perturbations can caused sepiolite precipitation from SCSSS[4]. Thus, a pH increase from reactions such as H^+-cation exchange during the initial stages of dissolution may have initiated sepiolite precipitation, which in turn would have reduced $[H_4SiO_4]$, causing accelerated glass dissolution, and possibly initiating sphene dissolution, as discussed earlier.

Figures 2(a) and 2(b) show, respectively, the normalized Na and Ca release rates for the various samples. Comparison of the corresponding plots for the glass, ceramic and glass-ceramic demonstrates that all the phases leach incongruently, with the Na release rates being at least an order of magnitude higher than those for Ca.

The apparent decrease in Na and Ca release rates with time are attributed to progressive elimination of surface ^{22}Na and ^{45}Ca ions available for exchange with H^+, Na^+, Ca^{2+}, etc. The decreases can not be due to solution saturation effects, since the leached Na or Ca will not change the total Na or Ca solution concentrations significantly. Also, no Ca- or Na-bearing precipitates were observed by SEM/EDX on the leached surfaces (see below). Dissolution rates that are controlled by solution saturation generally follow an equation of the form

$$C_t = C_s(1 - e^{-kt}) \qquad (3)$$

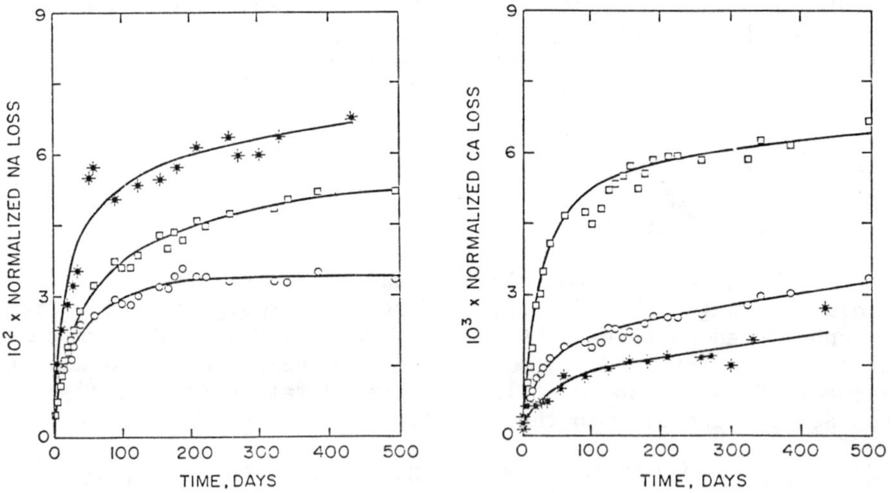

Figure 2. Mean normalized release values ($kg \cdot m^{-2}$) versus time at 100°C for glass (squares), ceramic (circles) and glass-ceramic (stars) powder samples in SCSSS: (a) Na, (b) Ca.

where C_t is the ionic concentration in solution after time t, C_s is the saturation concentration and k is a rate constant[15, 16]. The data in Figures 2(a) and 2(b), shown with hand-drawn curves, show initial release rates greater than would be predicted from equation 3, however, and the poor fits to this equation confirm that the dissolution processes are not simply governed by first-order solubility controls. Although the data show progressively decreasing Ca and Na release rates with time for each material, only the Na data from the ceramic samples had reached a maximum release value (i.e., zero release rate) when the experiments were stopped.

The Na release values from the ceramic phase are attributed mainly to ion-exchange processes, such as exchange of Na^+ in solution with $^{22}Na^+$ at the ceramic surface. Other ion-exchange reactions are also possible, such as Ca^{2+} (from solution) exchange for surface Na^+, with local charge compensation from OH^- replacement by O^{2-} at the solvated ceramic surface. A minor contribution to Na release may also have occurred from matrix dissolution in response to sepiolite precipitation, as discussed above; however, no evidence for matrix dissolution was observed in SEM examinations of leached surfaces (see below).

The Na losses from the aluminosilicate glass are attributed to matrix dissolution, initiated by Na^+/H^+ exchange at the surface[16] and/or by water diffusion into the glass[17]. Apparent losses will also arise from ion exchange between the ^{22}Na-doped surface and Na^+ in solution. The Na losses from the glass-ceramic are somewhat higher, but sufficiently close to those for the glass to suggest a similar release mechanism; this would be expected since the glass phase in the glass-ceramic contains most of the Na inventory[14].

Ca release from the glass phase is attributed mainly to matrix dissolution, initially involving exchange of cations, including Na^+ and Ca^{2+}, with H^+ or H_3O^+. The lower values for Ca release, compared with Na, reflect the relative ease of ion exchange with monovalent Na^+, rather than with divalent Ca^{2+}. Further contributions to Ca release from the glass phase arise from ion exchange between $^{45}Ca^{2+}$ at the glass surface and Ca^{2+} in solution, or from exchange between $^{45}Ca^{2+}$ (solid) and Na^+ (solution) with simultaneous protonation of an oxygen site, as discussed above.

Release of ^{45}Ca from sphene dissolution induced by sepiolite precipitation would be expected to obey a first-order equation similar in form to equation 3, i.e.,

$$Q_t = Q_{eq}(1 - e^{-k't}) \qquad (4)$$

where k′ is a rate constant, Q_t is the normalized Ca loss per unit surface area after time t, and Q_{eq} is the value at equilibrium, i.e., the value which, when added to the existing Ca in solution, restores the Ca concentration to that required at neutral pH for sphene equilibrium with SCSSS. Figure 3 shows fitted curves (equation 4) to the 100°C normalized release values for Ca from the ceramic and glass-ceramic, with respective values for k′ of 765 and 554 s^{-1} and for Q_{eq} of 2.28×10^{-3} and 3.32×10^{-3} $kg \cdot m^{-2}$. Despite the scatter in the data, there is a general trend for the experimental values to be greater than the predicted values at early times (< 100 d), converging with the curves at longer times. The plots imply that Ca release from the ceramic and the glass-ceramic cannot be simply explained by first-order sphene dissolution; a significant part of the apparent initial Ca release must be produced by surface ion exchange reactions, such as that between surface ^{45}Ca and leachant Ca^{2+}.

The Ca release curve for the glass-ceramic falls slightly below that for the ceramic, and significantly below that for the glass. Although the vitreous phase in the glass-ceramic contains some residual Ca, the majority (> 80 mol.%) of Ca resides in the sphene phase[7,11]. Thus, the comparable release rates for Ca from the glass-ceramic and ceramic suggest similar release mechanisms. Together with the Na release results, they support the contention of independent leaching of the component glass and ceramic phases in the glass-ceramic.

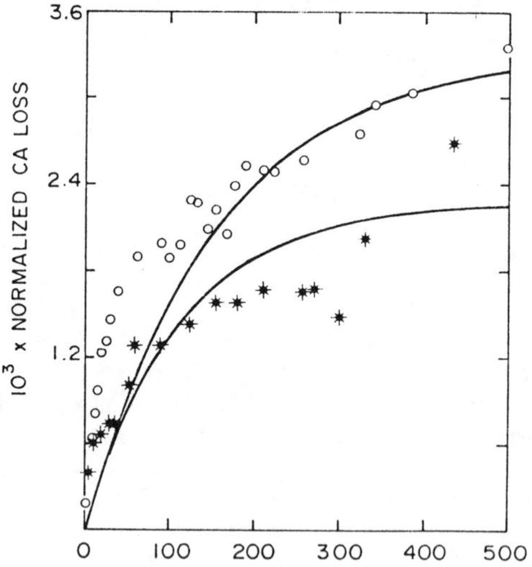

Figure 3. Fitted curves from equation (4) for the normalized Ca release values ($kg \cdot m^{-2}$) for the sphene ceramic (circles) and glass-ceramic (stars).

Leached layer thicknesses

The Na and Ca fractional losses and the equivalent leached layer thicknesses were calculated for the glass-ceramic, glass and ceramic powders from the data of Figures 2(a) and 2(b), using measured densities of 2.9, 2.4 and 3.5 $g \cdot cm^{-3}$, respectively, and assuming spherical particles of mean diameter 165 µm. After 496 d, the glass lost 79.1% of its Na inventory and 10.1% of its Ca inventory. These values correspond to leached layer thicknesses of ~ 33.5 µm and ~ 2.9 µm, respectively, and illustrate the incongruency of Na and Ca removal. The values also include contributions from exchanged ions, however, so that the net losses of Ca and Na are somewhat lower. Similarly, the ceramic sphene samples after 496-d leaching had lost 38.6% of the Na inventory and 3.7% of the Ca inventory, with corresponding leached layer thicknesses of ~ 12.4 µm and ~ 1.0 µm, respectively. As discussed above, these figures mainly represent surface Ca and Na ions exchanged with cations in solution and again demonstrate incongruency caused by the differing valence states of Na and Ca.

After 433-d leaching, the glass-ceramic had lost 85.2% of its Na content and 3.4% of its Ca content (although, as noted above, these values also include contributions from exchanged ions). Further interpretation of these results is difficult, however, because of the 2-phase nature of the glass-ceramic.

SEM observations

After termination of the leaching experiments, the samples were lightly rinsed with deionized water, air-dried at 60°C, sputter-coated with gold and then examined using SEM. The only visible alteration during leaching of the ceramic samples was the precipitation of a thin (< 1 µm) layer of sepiolite. However, the leached glass surfaces showed scalloped etch pits as well as extensive sepiolite precipitation. These observations confirmed that matrix dissolution had occurred in the glass phase, but not significantly in the sphene samples.

Figure 4(a), an unleached fracture surface of the glass-ceramic, shows the bulk nucleated microstructure of sphene crystals within a continuous glass matrix. Figure 4(b) shows the ^{22}Na-doped glass-ceramic surface after 433-d leaching. Although the ceramic phase is still visible, the glass phase has been preferentially leached and the surface is covered with a variable thickness of precipitated sepiolite. Preferential leaching of the glass phase had also occurred at the ^{45}Ca-doped glass-ceramic surface (Figure 4(c)). Only localized precipitates of sepiolite were found in this case, however, suggesting slight pH differences

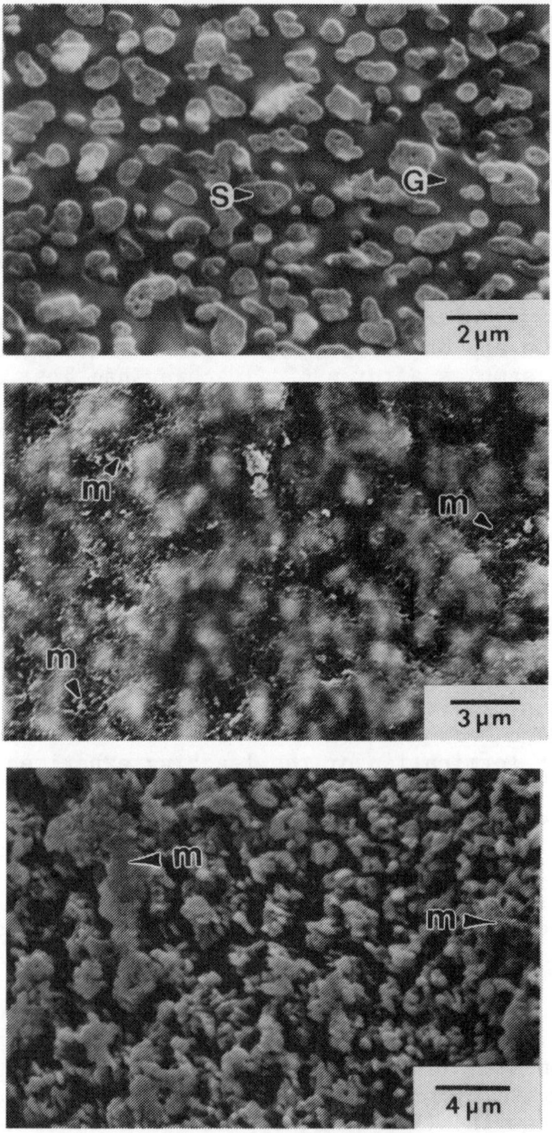

Figure 4. (a) unleached sphene glass-ceramic, showing discrete sphene crystals (S) within a continuous aluminosilicate glass matrix (G); (b) leached ^{22}Na-doped glass-ceramic, showing extensive precipitation of sepiolite (m); (c) leached ^{45}Ca-doped glass-ceramic, showing sporadic precipitation of sepiolite (m).

between the ^{45}Ca and ^{22}Na-doped experiments.

CONCLUSIONS

The results from 100°C leaching experiments in Ca-Na-Cl brine support the contention of independent leaching of the component glass and sphene phases in a waste-loaded sphene glass-ceramic. Radionuclide release from the glass-ceramic under static conditions would, thus, be governed by (i) the partitioning of individual radionuclides between the glass and sphene phases; (ii) the glass dissolution rate; (iii) ion exchange rates of radionuclides at the glass and sphene surfaces with species in solution, such as Ca^{2+}; and (iv) the solubilities of the leached radionuclide species. In the event of significant pH or $[H_4SiO_4]$ reductions from reactions, such as sepiolite precipitation, some release from sphene dissolution would also be expected.

ACKNOWLEDGEMENTS

We thank J.M.P. Ross for the LSC analyses, M. Duclos and J. Snider for XRD analyses, and K.B. Harvey, J.C. Tait, P. Taylor and E.R. Vance for useful manuscript reviews.

REFERENCES

1. W.T. Hancox, "Closing the Circle: Returning Used Uranium Fuel to the Geologic Environment", Proc. Canad. Nucl. Soc. Symp. on Uranium and Electricity, Saskatoon, SK, 1988 September 18-21.
2. S.K. Frape, P. Fritz and R.H. McNutt, "Water-Rock Interaction and Chemistry of Groundwaters from the Canadian Shield," Geochim. Cosmochim. Acta 48, 1617-27 (1984).
3. P.J. Hayward, F.E. Doern, E.V. Cecchetto and S.L. Mitchell, "Leaching Studies on Titanite, a Potential Host for Canadian Nuclear Reprocessing Wastes," Can. Mineral. 21, 611-23 (1983).
4. P.J. Hayward, D.G. Watson, A.K. McIlwain, I.M. George and S.L. Mitchell, "Leaching Studies of Sphene Ceramics Containing Substituted Radionuclides," Nucl. Chem. Waste Management 6, 71-80 (1986).
5. P.J. Hayward, W.H. Hocking, S.L. Mitchell and M.A.T. Stanchell, "Leaching Studies of Sphene-Based Glass Ceramics," Nucl. Chem. Waste Management 5, 27-37 (1984).
6. P.J. Hayward, J.C. Tait, I.M. George, A.A. Carmichael and J.M.P. Ross, "Progress in Development of a Source Term for Sphene Glass-Ceramic Dissolution Under Vault Conditions," pp.241-247 in Advances in Ceramics. Vol. 20, D.E. Clark, W.B. White and A.J. Machiels, Eds.,Amer. Ceram. Soc., Chicago, OH, 1987.

7. P.J. Hayward, "Glass-Ceramics", Chapter 7 in Radioactive Waste Forms for the Future, W. Lutze and R.C. Ewing, Eds.,North-Holland, 1988.
8. P.J. Hayward, "The Use of Glass-Ceramics for Immobilising High Level Wastes from Nuclear Fuel Recycling", Glass Techn. 29(4), 122-36, 1988.
9. H.W. Nesbitt, G.M. Bancroft, W.S. Fyfe, S.N. Karkhanis, A. Nishijima and S. Shim, "Thermodynamic Stability and Kinetics of Perovskite Dissolution," Nature 289, 358-62 (1981).
10. E.R. Vance, "Dissolution of Sphene and Sphene Glass-Ceramics in Aqueous Media," pp.231-9 in Advances in Ceramics. Vol. 20, D.E. Clark, W.B. White and A.J. Machiels, Eds., Amer. Ceram. Soc., Chicago, OH, 1987.
11. P.J. Hayward, E.R. Vance, C.D. Cann and I.M. George, "Influence of Heat Treatment and Redox Conditions on Crystallization of Titanosilicate Glass-Ceramics," pp. 347-54 in Scientific Basis for Nuclear Waste Management. Vol. IX. L.O. Werme, Ed., Elsevier, New York, 1986.
12. E.R. Vance and D.K. Agrawal, "Incorporation of Radionuclides in Crystalline Titanates," Nucl. Chem. Waste Management 3, 229-34 (1982).
13. American Society of Materials, Annual Book of ASTM Standards, 17, C2, 6-8 (1980).
14. Y.K. Kharaka and I. Barnes, "SOLMNEQ solution-mineral equilibrium computations", U.S. Geol. Surv. Comp. Contr., PB-215-899, 1973.
15. K.B. Harvey, "A Semiunified Approach to Leach Testing of Nuclear Wasteforms", Nucl. Chem. Waste Management 4, 201-5, 1983.
16. W.B. White, "Dissolution Mechanisms of Nuclear Waste Glasses: A Critical Review," pp. 431-42 in Advances in Ceramics. Vol. 20, D.E. Clark, W.B. White and A.J. Machiels, Eds., Amer. Ceram. Soc., Chicago, OH, 1987.
17. R.H. Doremus, "Glass Science", John Wiley and Sons, N.Y., 1973.

ANALYTICAL ELECTRON MICROSCOPY OF LEACHED NUCLEAR WASTE GLASSES

T. A. Abrajano, Jr. and J. K. Bates
Argonne National Laboratory, Argonne, IL 60439
J. P. Bradley
McCrone Associates Inc., Westmont, IL 60599

ABSTRACT

Preliminary results of an analytical electron microscopic study of alteration layers on SRL 131-based glasses are presented. The reacted surfaces consist of multiple reaction layers of systematically varying compositions and structures. The innermost layer adjacent to the fresh glass is depleted in mobile elements (e.g., alkali metals and B), but otherwise is compositionally similar to the fresh glass. Numerous crystalline and amorphous phases were observed in the alteration layers including clays (smectite- and serpentine-kaolinite types), manganese oxyhydroxides, iron oxides and hydroxides, and titanium oxide. Both the abundance and degree of ordering of the crystalline phases increased with distance from the fresh glass. Only the outermost secondary phases appear to have clearly precipitated from the leaching solution. Evidence indicates that most of the secondary phases present within the layers nucleated within the residual layers. These results, together with results of other surface and solution analyses, are examined in relation to the mechanisms of alteration layer formation. Implications of the study to the modeling of waste glass dissolution are also discussed.

INTRODUCTION

Much work has been devoted to elucidating the mechanism of nuclear waste glass dissolution in aqueous media. The amount of glass components present in solution is controlled by the processes occurring at the glass:solution interface region, most notable of which are sorption, diffusion, dissolution, and precipitation. Non-stoichiometric dissolution of nuclear waste glass could lead to preferential retention of certain glass components in a **residual** layer. Stoichiometric dissolution accompanied by incongruent precipitation could also lead to retention of insoluble elements in a

precipitate layer. In addition, formation of **protective** layers, primarily precipitated phases, that may limit interaction between the glass surface and the solution under certain circumstances has also been proposed.

Past interpretations of the primary mechanisms responsible for the retention of specific glass components in the solid have been largely based on solution data alone, and only rarely in conjunction with definitive surface analytical results. In the present communication, we present preliminary results of an analytical electron microscopic (AEM) study of reaction layers formed on leached Savannah River Laboratory (SRL) 131-based glasses. It should be noted that the 131 formulation is no longer the reference glass being considered for production in the Defense Waste Processing Facility. Our leaching experiments were performed several years ago when 131 was the defense waste reference glass. However, the present results still offer an excellent opportunity for defining micro-interactions during alteration of this extensively studied nuclear waste glass formulation. The results also should prove useful for understanding alteration layer formation in more durable glass formulations that were recently developed at SRL.

EXPERIMENTAL

MCC-1 type static leaching experiments (t = 3 - 546 days; T = 90°C, SA/V = 0.07 - 0.10 cm^{-1}; DIW) were performed using three different glasses: the base SRL 131 type glass (G1), the base glass modified by the addition of several minor waste elements (G2), and the base glass modified by the addition of waste elements and actinide elements (G3). Only the AEM results for the leached G2 glass will be considered in this communication. A complete description of the glass preparation and leaching procedures and detailed results of solution and complementary solid analyses (SEM, SIMS, XRD, IR, XPS) is given elsewhere[1].

Thin sections (<100 nm) of G2 glass leached for 28, 182, and 546 days were prepared using an ultramicrotome equipped with a diamond knife. In general, the ultramicrotomy procedure is similar to that employed for biological specimens[2,3]. Samples were mounted in a low-viscosity mounting media and, upon curing, the mount was trimmed until a working face was fashioned around the sample. Using a diamond knife, sections were then cut from the working face and floated (on water) away from the knife edge. In addition to ultramicrotomed samples, some sections of the fresh glass and of the layer:glass boundary were prepared using standard ion milling techniques. AEM images, selected area electron diffraction (SAED) patterns, electron microdiffraction patterns, parallel electron energy loss spectra, and quantitative chemical analyses were generated using a JEOL 2000-FX analytical electron microscope equipped with a Tracor-Northern TN 5500 X-ray energy-dispersive spectrometer. Quantitative chemical analyses were obtained on 50 nm ultramicrotomed sections using 200 KeV incident electrons. Thin film silicate, oxide, and phosphide minerals were used as standards[3].

RESULTS

Although the solution results and other surface analyses are described in detail elsewhere[1], we will briefly summarize the relevant aspects here. The measured room temperature final pH values for all experiments showed no systematic trends as a function of time. The initial pH of the DI water leachant is 6.3 and the final leachate pH values after 3 to 546 days range between 9 and 10. The calculated normalized leach rates of Na, Li, B, and Si showed a general decline with time. The highest elemental mass loss was noted for B, although the normalized releases of B, Li, and Na tended to overlap for the long duration periods (>200 days).

All experiments yielded reaction layers that were several microns thick. Examination of sample cross sections under the SEM revealed that after 3 days of reaction, only a convoluted single layer, about 3 μm thick, was observed (Fig. 1a). This layer faithfully follows the original surface morphology of the unreacted glass, and can be traced through original cracks on the glass that are at high angles to the surface. After 7 days of reaction, two distinct layers (total thickness of about 7 μm) were observed (Fig. 1b): (1) a thicker outer layer referred to as [1A] and (2) a thinner inner layer referred to as [2A] (with cracks perpendicular to the layering orientation). Backscattered electron images of the sample surface and cross section also revealed the presence of a thin (< 1 μm) veneer of secondary phases on the outermost surface (i.e., coating layer [1A]). After 91 days of reaction, a distinct additional layer (referred to as [2B]) was observed to form between the fresh glass and layer [2A] (Fig. 1c). In contrast to layer [2A], layer [2B] does not exhibit pronounced cracking and shows only slight density contrast with the fresh glass. Nevertheless, the boundary between this <u>new</u> innermost layer and the bulk glass is also distinct. After 91 days, the thin veneer of secondary phases has also evolved into a distinct outermost layer [1B]. X-ray diffraction analyses of layer [1B] suggest the dominant occurrence of birnessite ($Mn_7O_{13} \cdot 5H_2O$) and a smectite-type clay phase. Also identified in the XRD patterns are rutile (TiO_2) and barite ($BaSO_4$). The total layer thickness after 91 days is 11 to 13 μm.

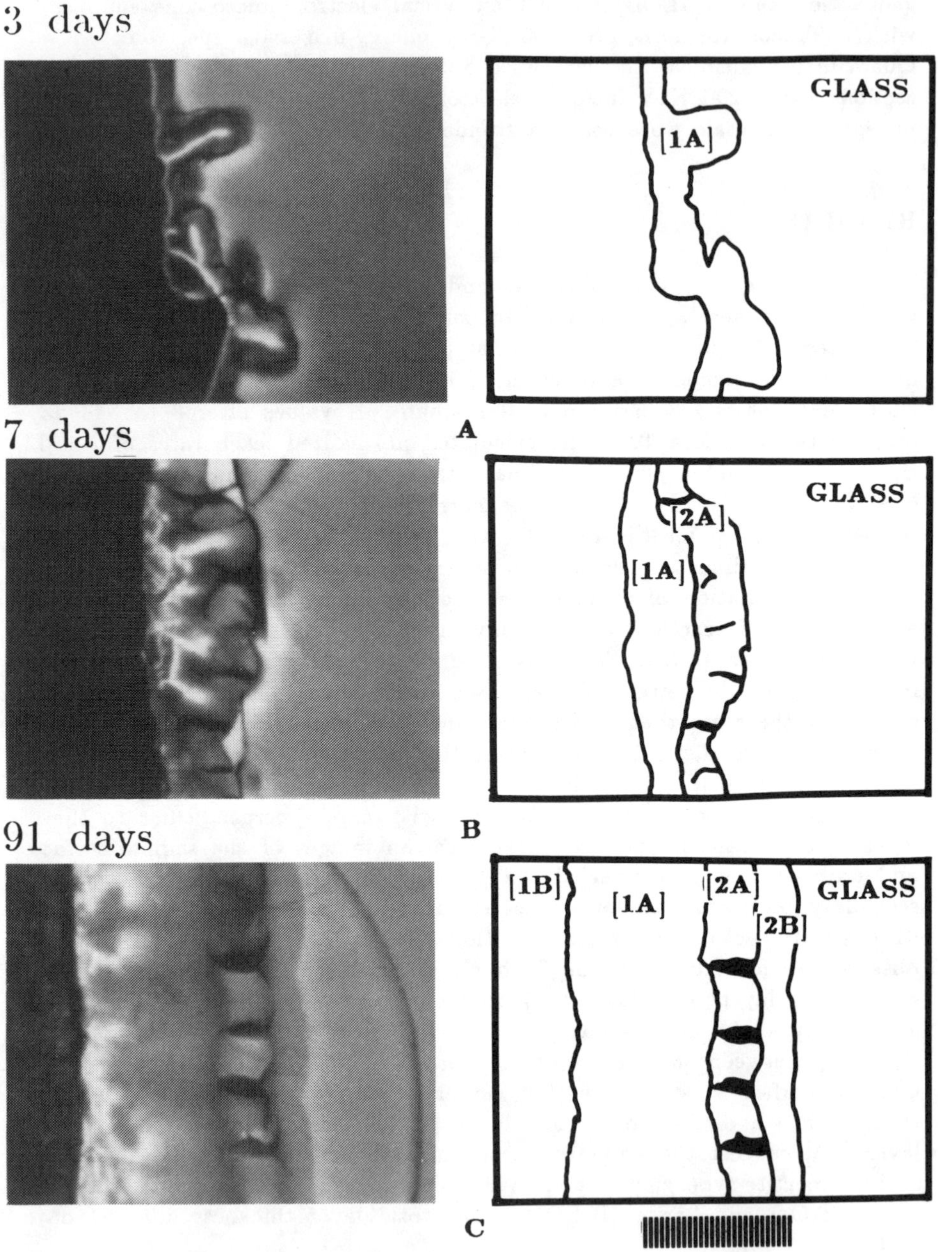

Figure 1. Reaction layer development in SRL 131. (a) 3 days. (b) 7 days. (c) 91 days. Bar scale at the bottom is 10 μm.

AEM images of G2 samples reacted for 28 days and 546 days are shown in Figures 2 and 3. SAED patterns are shown only for the sample reacted for 546 days (Fig. 3), although comparable patterns were collected from equivalent layers for the sample reacted for 28 days. Significant details of the structural characteristics of the reacted surface layers are clearly revealed in the AEM images. The unreacted bulk glass is relatively featureless (e.g., Fig. 2e). After 28 days of reaction, the two layers, [1A] and [2A], that were observable under the SEM are also distinguishable under the AEM. Layer [2A], which at this time is immediately adjacent to the unaltered glass, contains needle-like crystals that appear to decrease in abundance towards the fresh glass (Fig. 2d). Sharper SAED ring patterns were observed in the amorphous matrix of [2A] relative to the bulk glass indicating a higher degree of ordering in [2A] (see inset to Fig. 3g, below). In contrast to the distribution of the needle-like crystals in [2A], the crystalline phase observed in [1A] is more abundant and evenly distributed (Fig. 2c). The basal (001) spacing estimated from diffraction patterns of these crystals is consistent with the structure of smectites, the phase also suggested by the chemical composition. Inter-mixed with this smectite type phase and concentrated on the solution-side of [1A] are iron-rich lenses that appear to be largely amorphous (Fig. 2a). However, electron microdiffraction patterns were generated in some of these Fe-rich lenses and revealed the presence of magnetite (Fe_3O_4). The interface between [1A] and [1B] is delineated by plate-like crystals preferentially oriented at high angles to the interface (i.e., vertically-layered) (Fig. 2a and 2b). These crystals are depleted in Fe (compared to the needle-like smectites in [1A]) and the electron microdiffraction patterns, lattice images, and quantitative thin film analyses are suggestive of saponite $(Mg,Al)_3(Al,Si)_4O_{10}(OH)_2)$. Layer [2A] is depleted in B, alkali metals and Mn, and partly depleted in Si, Mg, and Ni relative to the fresh glass. Fresh glass values for Ca, Al, Ti, and Fe appeared to have been largely maintained in [2A]. Layer [1A] showed higher transition metal and alkaline earth contents but relatively lower Si content than layer [2A].

Extensive AEM data were collected for the G2 samples reacted for 182 and 546 days. Although the total layer thickness increased from about 20 to over 30 μm, no substantial changes in composition and structure occurred for the individual layers for reaction times between 182 and 546 days. Crystalline phases observed throughout the reaction layers became more ordered after 546 days compared to those observed after 28 and 182 days, enabling better indexing of diffraction patterns.

The general characteristics of [2A], [1A], and [1B] in the sample reacted for 546 days are largely similar to those for the glass reacted for 28 days. However, several differences exist. Large fiber-like crystals that are rich in Ni, Mg, Si, and Al were observed along the boundary

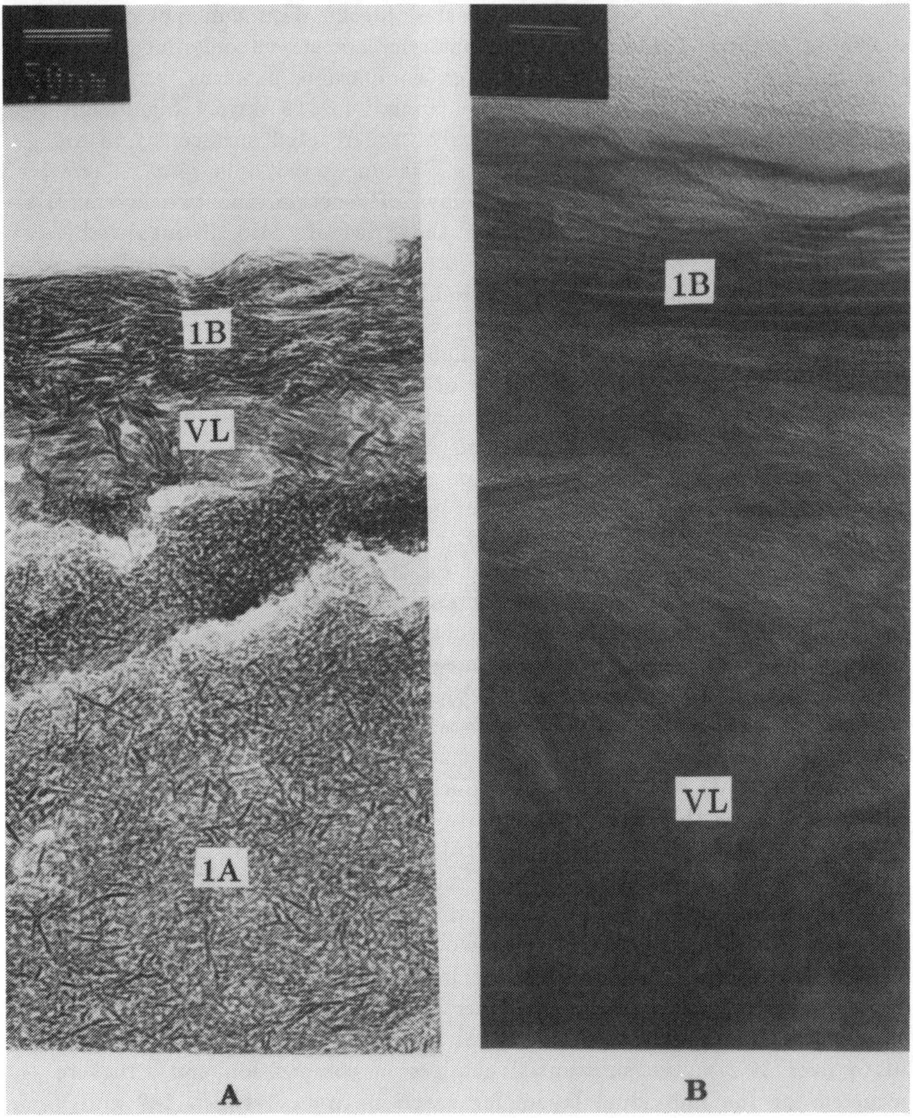

Figure 2. AEM images and SAED patterns for sample reacted for 28 days. (a) Low magnification image of layers [1A] and [1B]. (b) Higher magnification image of [1A]:[1B] interface showing layered birnessite and vertically-layered (VL) smectites. (c) Layer [1A]. (d) Layer [2A]. (e) Interface of [2A] and glass.

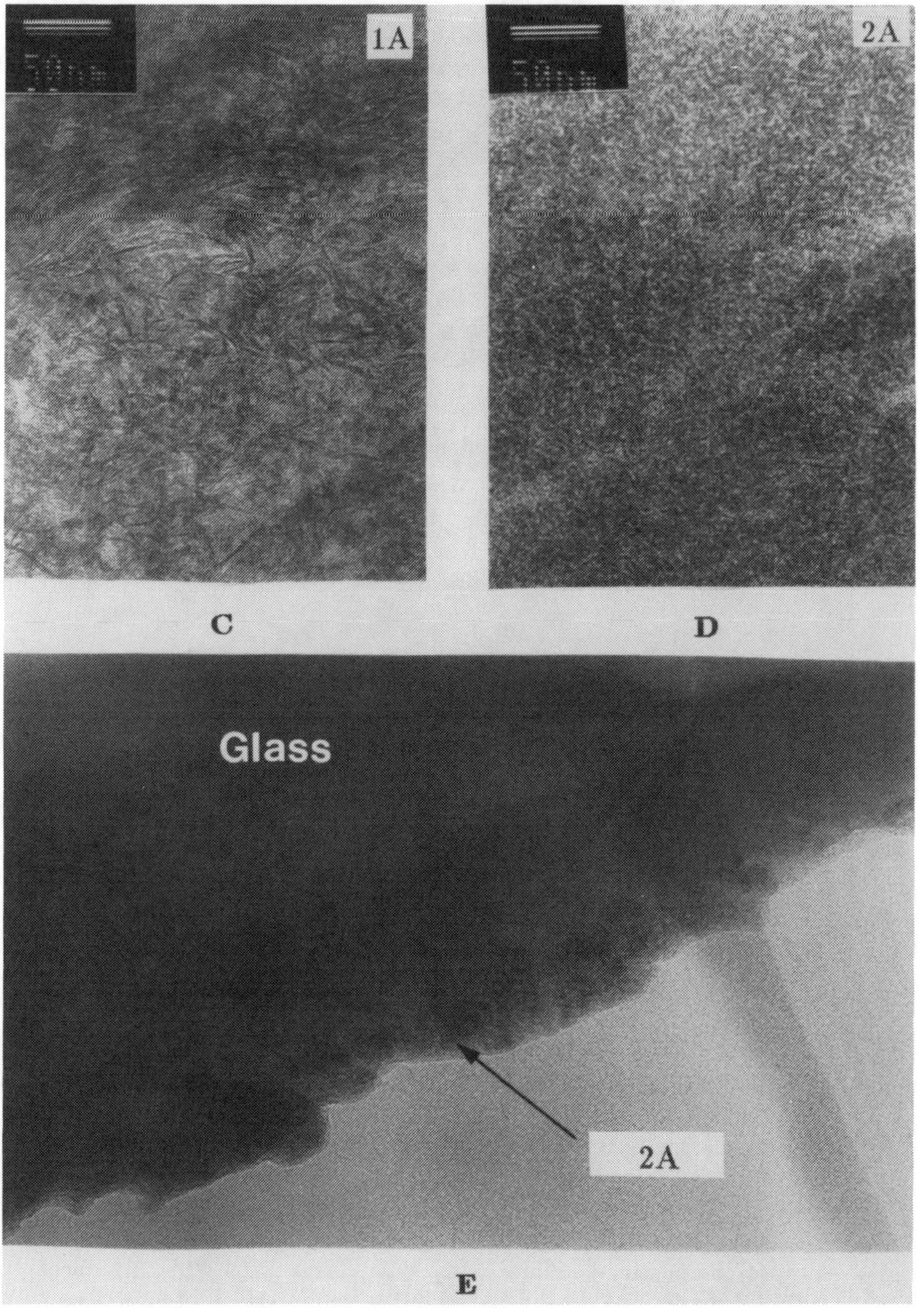

Figure 2 (Cont'd)

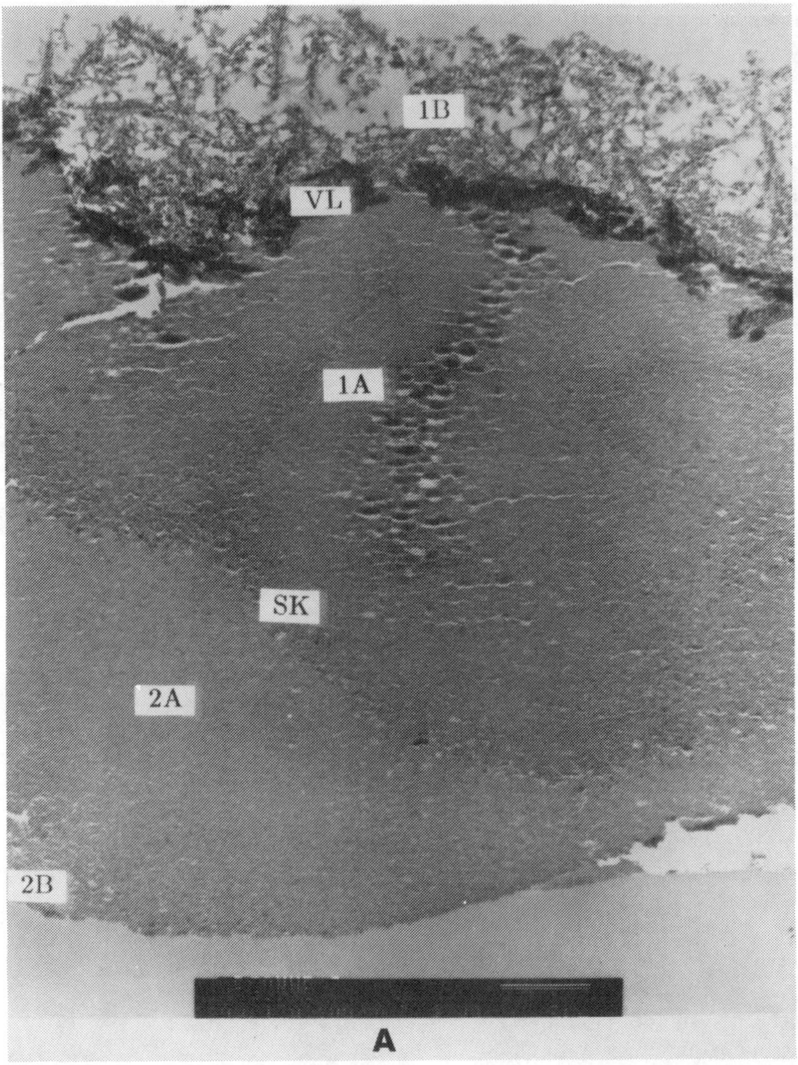

Figure 3. AEM images and SAED patterns for sample reacted for 546 days. (a) Low magnification image across all reaction layers. (b) Higher magnification image across [1A]:[1B] interface showing vertically-layered smectites and Fe-rich lenses. (c) Wedge (A) and disk (B) structure in birnessite in [1B]. (d) Vertically-layered (VL) smectites. (e) Layer [1A]. (f) Serpentine-kaolinite (SK) type clay at [1A]:[2A] interface. (g) Layer [2A]. (h) Layer [2B]. (i) Interface of [2B] and glass.

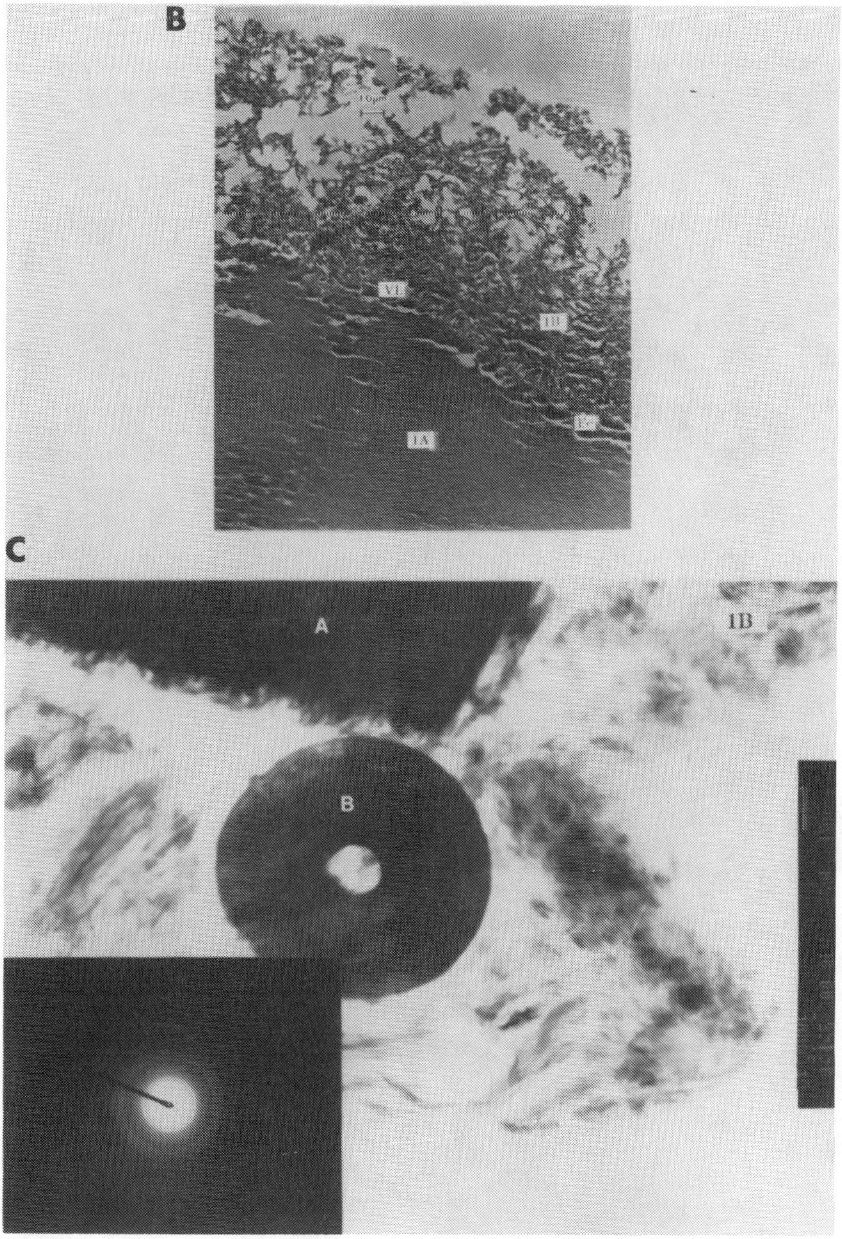

Figure 3 (Cont'd)

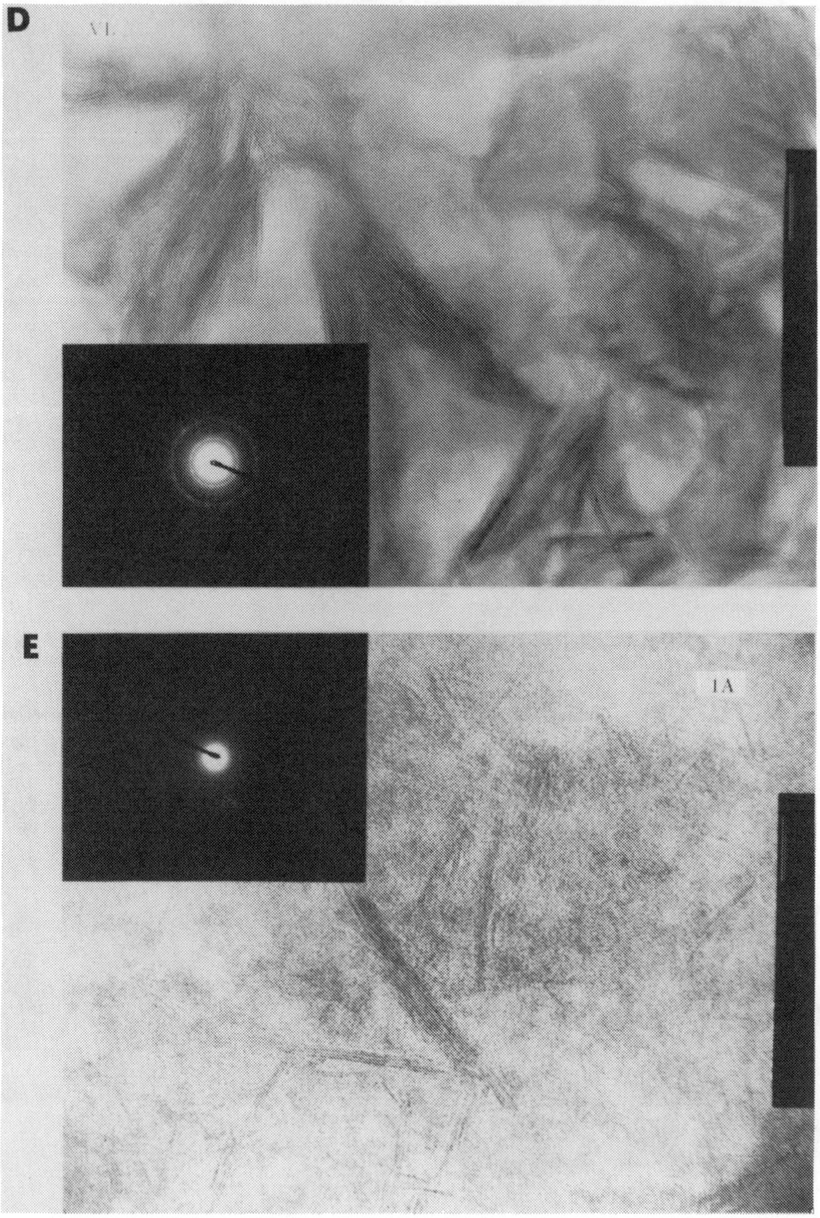

Figure 3 (Cont'd)

F

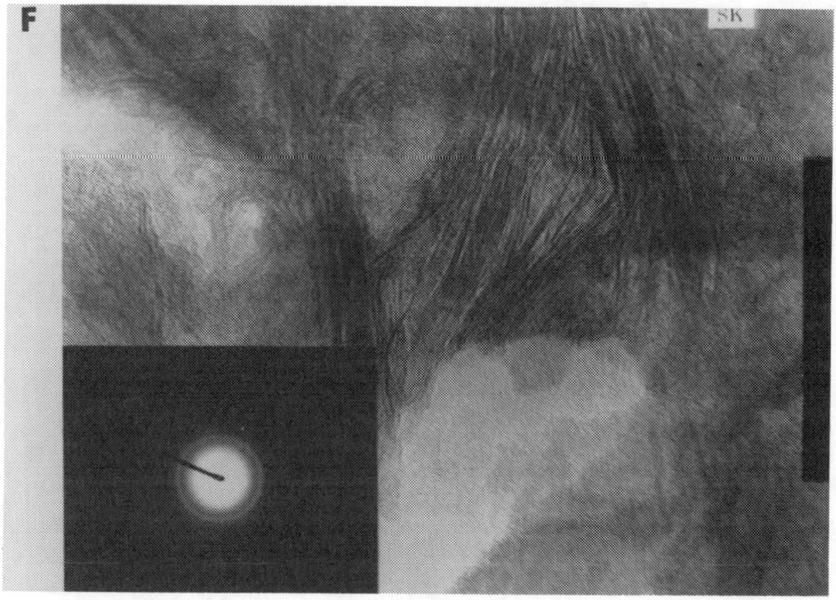

G

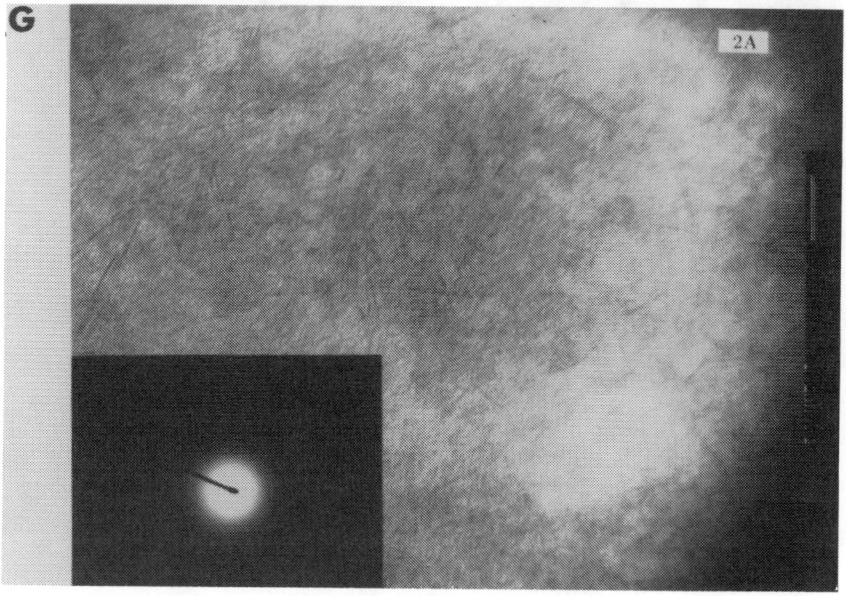

Figure 3 (Cont'd)

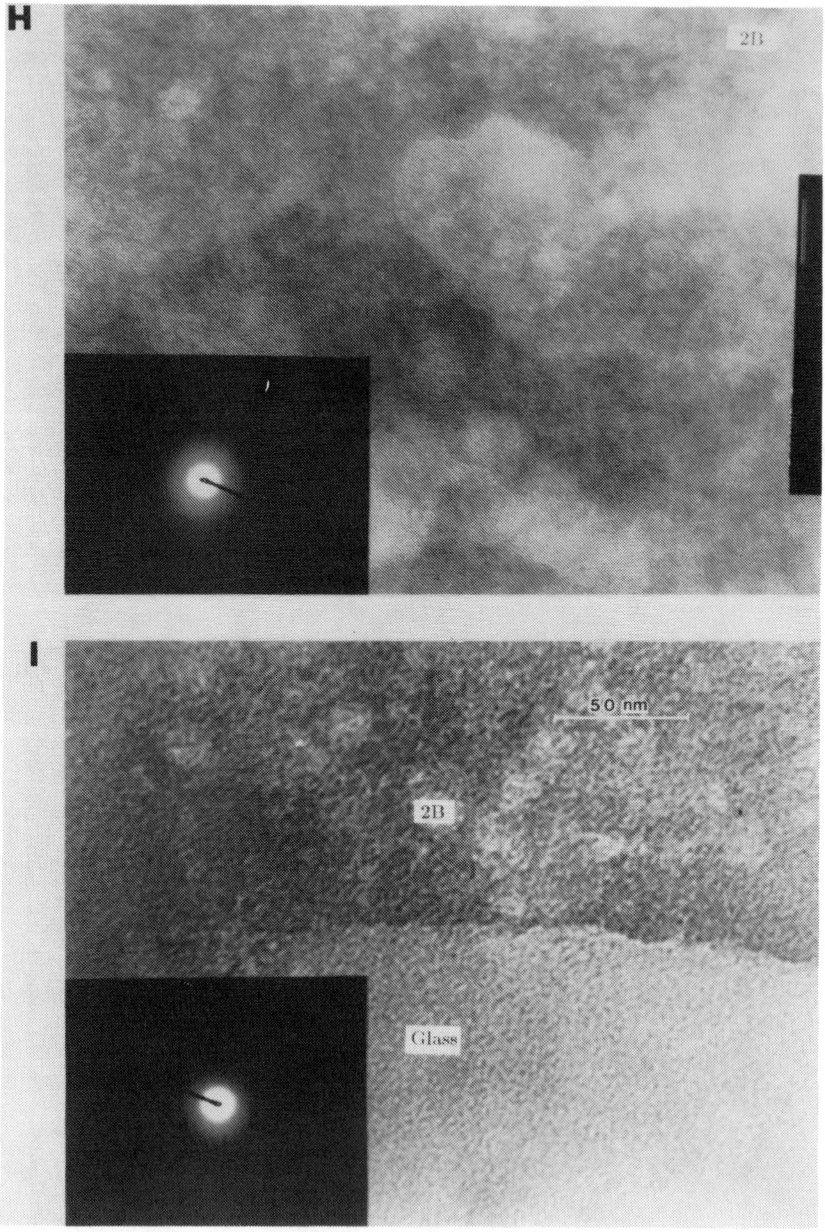

Figure 3 (Cont'd)

between [2A] and [1A] after 546 days (Fig. 3a and 3f). Both the basal (001) spacing and the chemical composition of this phase suggest a serpentine-kaolinite type clay, although a second unidentified phase may also be present. This serpentine-kaolinite clay "layer" was first observed in the sample reacted for 182 days. After 546 days, however, the presence of a smectite type phase in this serpentine-kaolinite clay "layer" is suggested by the increased basal (001) spacings (>1.0 nm) and lower cation/Si ratios in some of the crystal fibers. Compared to those observed in the layers formed after 28 days, the needle-like smectites present within [1A] and the vertically-layered smectites present between [1A] and [1B] are significantly larger in the sample reacted for 546 days (Fig. 3c and 3d). A more extensive distribution of the Fe-rich lenses occurs along the solution-side of [1A] after 546 days (Fig. 3a). Some of these Fe-rich lenses appear to preferentially occur symetrically on both sides of smectite lenses that are visually similar to the vertically-layered smectites noted above. This mode of occurrence is suggestive of preferential exclusion of Fe in the growing smectite crystals and consequent enrichment of Fe in the surrounding regions. As indicated by the XRD patterns, the dominant precipitate present in [1B] is birnessite. However, minor Mg-smectite type crystals (saponite) and amorphous precipitates are also present in [1B]. The compositions of layers [2A] and [1A] after 546 days of reaction do not appear to differ significantly from those measured after 28 days (except for the relative enrichment in Mg, Ni, Si, and Al due to serpentine-kaolinite clay formation along the boundary of [2A] and [1A] noted above).

A major difference between the layers present on G2 sample reacted for 28 and those reacted for 182 and 546 days is the presence of layer [2B] in the latter. It is evident from Fig. 1c that this layer is already present after 91 days. Indeed, most of the total layer increase between 91 and 546 days is due to the combined growth of layers [2A] and [2B]. AEM images and SAED patterns collected on [2B] suggest that this layer is totally amorphous. However, SAED rings observed on [2B] are generally sharper than those observed for the bulk glass, but less so than those observed on [2A]. The boundary between the glass and [2B] appears sharp. The boundary is very distinct due to the presence of a curious segregation texture on [2B] (Fig. 3h and 3i). This feature is similar to those previously described in leached borosilicate glasses[4,5]. No sharp boundary between [2A] and [2B] was observed under the AEM. However, this boundary between [2A] and [2B] is clearly observable during sample preparation for AEM analysis. The physical behavior (i.e., rigidity) of [2B] appears to be similar to that of the glass; consequently, preparation-induced fractures are preferentially present on [2B] and on the bulk glass (e.g., Fig. 3a). The portion of [2B] adjacent to the unaltered glass compositionally resembles the glass, although it is extensively depleted in B and the alkali metals. Similarly, the portion of [2B] adjacent to [2A] compositionally resembles [2A].

DISCUSSION

The data presented in the previous section permit a detailed description of surface layer evolution on SRL 131 glass in MCC-1 experiments over time. The earliest stage of the reaction progress is characterized by the maximum release rate of major glass components into solution. Solution results indicate that the release of various glass components is incongruent, with mobile elements (e.g., Li, Na, and B) being released faster than Si and other less mobile glass components. This incongruency persists to the longest reaction duration investigated (546 days), although the release rates of all components appear to decrease with time. Comparison of our solution results[1] with those presented by Strachan and others[6] and Wicks and others[7] suggests the persistence of the incongruent release behavior to reaction durations longer than those we have investigated. Our solution results for the major glass components and those of Strachan and others are identical to within the experimental uncertainties. Whether the incongruency noted in our experiments and those of others[6,7] resulted from non-stoichiometric dissolution or combined stoichiometric dissolution and secondary phase precipitation cannot be determined from the solution data alone. However, the structural and chemical characteristics of the reacted layers and their mode of development support a non-stoichiometric dissolution model.

Transmission electron images collected on [2A] and [2B] are strongly consistent with a **residual** origin of the layers. The SAED patterns collected on the unaltered glass, [2B], and [2A] (in that order) show a progressive increase in the degree of ordering of a largely amorphous structure. Features that may be attributed to secondary silicate or oxide-hydroxide precipitation[5,8,9] were not observed in [2B]. It is unclear whether extensive repolymerization of [2B] occurred during or after B and alkali metal mobilization. What is clear from both the SEM and AEM compositional analyses is that depletion of Si, as well as of Fe, Al, Ca, Sr, Ti, Zr and possibly the rare earth elements (REE), was not observed at the boundary between layer [2B] and the glass, i.e., where the B and the alkali metals were extensively depleted. Hence, dissolution was decisively non-stoichiometric. Note, however, that whatever mechanism enabled the release of B and alkali elements at the interface of [2B] and the glass, this process must have slowed down significantly with time. If there is no diffusional barrier to B transport through the reaction layers (also see below), it seems likely that the decrease in overall B release rate with time was due to the decrease (with time) of the activity of water at the interface of [2B] and the glass. The rate of hydrolysis of boron-oxygen bonds crucially depends on water activity at the reaction site. This temporal decrease in water activity at the [2B]:glass interface is consistent with the observation that the extent of silicon-oxygen bond hydrolysis was not sufficient for Si detachment and depletion to occur at the [2B]:glass interface for the long duration experiments (>91 days).

Because the release of Si requires hydrolysis of bridging oxygen bonds in the glass structure, it appears that significant network hydrolysis did occur within [2B]. However, the point at which network hydrolysis occurred to the extent necessary to detach some of the network-bound Si did not correspond to the interface of [2B] and the glass; instead, Si-depletion reactions occurred behind this interface. For the sample reacted for 546 days, the Si depletion front (i.e., where Si depletion was observed farthest into the fresh glass) was observed hundreds of nanometers behind the B and alkali metal depletion fronts (i.e., where B and alkali metal depletion was observed farthest into the fresh glass). For the sample reacted for 28 days, the locations of the Si depletion front and the B and alkali metal depletion fronts are spatially indistinguishable. In all cases, the maximum degree of network hydrolysis and Si release occurred on the solution side of [1A]. As for B, Si release rate decreased significantly with time and achieved steady-state values sooner than B[1]. As noted above, this decreased release could, at least in part, be attributed to the temporal decrease in water activity at the reaction site. SEM profiles and quantitative analyses across layer [2B] further suggest that Fe and other immobile elements (e.g., Ti, Zr, REE) were not depleted at the Si depletion front, again implying non-stoichiometric release.

Another important question concerns the ability of the reaction layers to physically inhibit further dissolution of glasses. Boron and the alkali elements were extremely depleted in all the alteration layers, indicating that these layers did not provide adequate protection against transport of these glass components. If B was transported across the reaction layers as H_3BO_3 (a relatively large molecule), the entire alteration layer must have had a porosity large enough for this transport to have occurred with ease. Indeed, these layers contain very significant amounts of water, largely as molecular water, as determined by IR spectroscopy[1]. Whether or not the same layer porosity can enable physical inhibition of the transport of other glass components (e.g., actinides or polymeric silicate units) remains to be determined.

Some insights as to the development of secondary phases can also be gained from the AEM and other surface analytical results. Because Fe and other immobile elements (see above) were not depleted from layers

[2A] or [2B], their enrichment in layer [1A] must have occurred as a consequence of the extreme depletion of other major matrix components (primarily Si) in the outermost portions of [1A]. These Fe-rich lenses are thus residues of extensive surface dissolution (etching) that occurred early in the reaction progress. Etching "scales" that are rich in Fe were indeed observed in samples reacted for short durations (<14 days). Later solid-state recrystallization of some silicate residues in [1A] resulted in further concentration of Fe in these lenses. The morphology of the vertically-layered smectites strongly suggests that they have precipitated from (or grew from) the aqueous solution. The same can be said for the birnessite in [1B]. In contrast, secondary phases present within [1A] (i.e., oxides, metal hydroxides, smectites, serpentine-kaolinite type clays) must have formed by nucleation and growth within the residual layer. In particular, the serpentine-kaolinite fibers found along the boundary of [2A] and [1A] appeared to have entrapped components (i.e., Si, Al, Mg, Ni) from the surrounding layers producing very localized anomalies in the compositional profiles[1].

The observations just presented must be considered in the formulation of reaction progress models for glass dissolution. It is clear that provisions for non-stoichiometric release of individual glass components must be incorporated in these models. One way to address this problem is to assume "stoichiometric" dissolution of a previously altered surface layer, i.e., a layer that has previously been depleted with respect to its mobile components and enriched with respect to its insoluble components. Indeed, such an approach has already been attempted with some promising results by Bourcier[10] in modeling the behavior of nuclear waste glasses in repository-relevant environments. Nevertheless, other observations presented in this paper deserve further consideration by modelers. Modeling calculations often assume that secondary mineral phases form or redissolve during leaching in response to changes in composition of the aqueous solution. Contrary to this assumption, formation of secondary phases within the residual layers (i.e., excluding [1B] and vertically-layered saponites) appears to reflect very localized equilibrium within microscopic regions of the layer. For example, the mineral phases observed, i.e., serpentine-kaolinite clay, smectites, metal oxyhydroxides, appear to reflect the changes in the cation/Si ratio in the residual layer where the secondary phases nucleated. Furthermore, the compositions of hydrosilicates and oxyhydroxides formed in different portions of the reaction layers were variable, again implying very localized equilibrium during crystallization. Indeed, the formation of magnetite in [1A] attests to the redox disequilibrium that could exist between the aqueous solution and the reacted layer immediately adjacent to it. Other workers have also suggested that pH gradients in reaction layers of other silicate glasses impacts on mineral stability within the alteration layer[9]. These observations suggest that it is possible that secondary phases, including those that could host important waste elements, may be stabilized at the glass:solution interface regardless of their degree of undersaturation in the bulk aqueous phase.

CONCLUSIONS

Analytical electron microscopy enabled detailed description of the multiple reaction layers formed on leached SRL 131-based glasses. It appears that much of the alteration layers present on these leached glasses are residues of non-stoichiometric dissolution. The sequential development of reaction layers of varying compositions and structures inward into the glass ([1A], [2A], and [2B] in that sequence) reflects continuous but decreasing flux of water to the fresh glass. After reaction for 546 days, B and the alkali metals showed the greatest depletion depths and the depletion fronts of these elements corresponded to the observed interface between the layer and the fresh glass. The Si depletion front was observed at lesser depths than this interface, whereas Fe and other immobile elements were not depleted in the residual layers. Secondary phases, predominantly birnessite and smectite-type clays, appeared to have precipitated from solution on the outermost surface. However, most of the secondary phases observed within the residual layers probably formed by crystallization within the residual layers. Both the type of secondary phases formed and their composition appear to reflect localized equilibration conditions modified by reaction with the bulk aqueous phase. These observations have important implications to modeling of waste glass dissolution.

ACKNOWLEDGMENTS

This research was supported by the Savannah River Laboratory. Very thoughtful reviews of an earlier version of this paper by B. Biwer, W. Ebert, and J. Mazer are sincerely appreciated. C. M. Jantzen and D. Stahl also provided helpful comments on the manuscript. R. Riel and D. Abrajano assisted in manuscript preparation.

REFERENCES

1. T. A. Abrajano and J. K. Bates, Argonne National Laboratory, unpublished information (1989).
2. N. Reid, "Ultramicrotomy," p. 213-353 in Practical Methods in Electron Microscopy, Edited by A. M. Glauert, Elsevier (1975).
3. J. P. Bradley, "Analysis of Chondritic Interplanetary Dust Thin Sections," Geochim. Cosmochim. Acta **52**, 889-900 (1988).
4. B. C. Bunker, G. W. Arnold, D. E. Day, and P. J. Bray, "The Effect of the Molecular Structure on Borosilicate Glass Leaching," J. Non-Crystalline Solids **87**, 226-253 (1986).
5. B. C. Bunker, D. R. Tallant, T. J. Headley, G. L. Turner, and R. J. Kirpatrick, "The Structure of Leached Borosilicate Glasses," Phys. Chem. Glass **29**, 106-120 (1988).
6. D. M. Strachan, L. R. Pederson, and R. O. Lokken, "Results from the Long-Term Interactions and Modelling of SRL-131 Glass with Aqueous Solutions," Pacific Northwest Laboratory Report PNL-5654 (1985).
7. G. G. Wicks, J. A. Stone, G. T. Chandler, and S. Williams, "Long-Term Leaching Behavior of Simulated Savannah River Plant Waste Glass," E.I. du Pont de Nemours & Co., Savannah River Laboratory DP-1728 (1986).
8. J. L. Crovisier, J. Honnorez, and J. P. Eberhart, "Dissolution of Basaltic Glass in Seawater: Mechanisms and Rate," Geochim. Cosmochim. Acta **51**, 2977-2990 (1987).
9. T. Murakami, R. C. Ewing, and B. C. Bunker, "Analytical Electron Microscopy of Leached Layers on Synthetic Basalt Glass," p. 737-748 in Scientific Basis for Nuclear Waste Management **XI**, Edited by M. J. Apted and R. E. Westerman, Materials Research Society (1988).
10. W. L. Bourcier, Lawrence Livermore National Laboratory, unpublished information (1989).

NATURE OF LEACHED LAYERS FORMED ON BOROSILICATE GLASSES DURING AQUEOUS CORROSION

L. Trotignon (1), J.-C. Petit (1), J.-C. Dran (2) and G. Della Mea (3,4)

(1) SESD/LECALT, CEN-FAR, BP 6, 92265 Fontenay aux Roses Cedex, France
(2) CSNSM/CNRS, 91405 Orsay, France
(3) Dipartimento di Ingegneria, Universita di Trento, Italy
(4) Unita CISM di Padova, 35131 Padova, Italy

ABSTRACT

Borosilicate glasses with compositions of increasing complexity were submitted to static leach tests in DI water at 90°C. Reaction progress was investigated by combining solution analyses with characterization of reacted surfaces with Scanning Electron Microscopy, Fourier Transform-Infrared Spectroscopy and MeV ion beam techniques (Rutherford Backscattering Spectrometry and Elastic Recoil Detection Analysis). Two types of layers were identified: for the less complex and less durable glasses, a thick one, constituted of hydrated silica compounds, is developed, in which hydrolyzable elements tend to accumulate; for more complex and durable glasses, the outer part of the leached layer is formed by a silica poor Fe/Zn/Zr oxy-hydroxide film, whereas the inner part is likely a hydrated residual silica network.

1. INTRODUCTION

The aqueous corrosion of nuclear waste glasses has been extensively studied [1], in order to infer meaningful prediction on their long-term durability. Dissolution of the glass network is accompanied by the formation of a hydrated layer, depleted in boron and alkalis and enriched in hydrolyzable elements such as Fe, Zr, lanthanides and actinides [2]. Due to the complex composition of these materials, the exact nature and properties of this layer are not well known. Several previous authors have attempted to relate the durability of relatively simple glasses to the presence of specific components, such as aluminum or zinc [3]. They have shown, in particular, that the corrosion rate is ruled by the accumulation of these elements which reportedly transform the altered layer into a diffusion barrier. However, the extrapolation of such dispersed data to the understanding of the behavior of the more complex nuclear glasses is not straightforward, because the relationship between composition and durability remains mainly empirical.

In order to overcome such a limitation, we have undertaken a study of a series of glasses with chemical composition of continuously increasing complexity up to that of the model R7T7 French nuclear glass (see table 1). Such a sampling helps identify the respective role of the different components, and in particular allows discontinuities in the leaching behavior to be attributed to specific elements. In addition, to fully investigate the corrosion reaction, we combined solution analyses with extensive characterization of the reacted solids (see section 2). This approach led us to specify whether the leached layer is simply a depleted residual glass or results from the precipitation of stoichiometric compounds, such as hydroxides or hydrosilicates, as already suggested [4].

2. EXPERIMENTAL TECHNIQUES

2.1. Investigated glasses

A series of 7 glasses has been fabricated by classical synthesis techniques. Their compositions (Table 1) satisfy three conditions: i. they are ranked by decreasing order of cation abundance in the R7T7 glass; ii. the molar ratios in each glass of the series are identical to those of the R7T7 glass; iii. the total amount of minor alkalis of R7T7 glass (Li and Cs) has been replaced in the simplified glasses by the equivalent amount of Na. Electron microprobe analyses have confirmed that the actual glass compositions do not significantly depart from the nominal ones. Furthermore, both the homogeneity and glassy structure have been checked.

2.2. Leaching experiments

Leach tests have been performed on centimeter-sized polished sections under static conditions, in deionized water at 90° C and with a ratio of surface area to solution volume of 20 m-1. Starting pH at 20° C was in equilibrium with the atmosphere (5.8). The run durations range from 4 h to 672 h.

2.3. Analytical techniques

Solution analyses relied on ion chromatography or atomic absorption spectrometry. Solid samples were characterized by scanning electron microscopy, by Fourier transform infrared spectrometry and two ion beam techniques for elemental depth profiling. Rutherford Backscattering Spectrometry (RBS), which is appropriate for intermediate and heavy elements, was performed with a 1.8 MeV He+ beam supplied by a Van de Graaff accelerator. Hydrogen profiling relied on Elastic Recoil Detection Analysis (ERDA) using a 2.2 MeV He+ ion beam. Details on the experimental set up for the last two techniques can be found in [5].

3. RESULTS AND DISCUSSION

The observed trend is an increase in durability with increasing complexity. Because of space limitation, we will only present the data concerning glasses V2, V3, V6 and R7T7, their leaching behavior being representative of the whole series.

3.1. Glass V2

For short leach times (< 40 h), dissolution is totally congruent : glass cations are leached out at a constant rate with the molar ratios of the bulk and the observed mass losses of the samples are compatible with Si, B and Na release in solution. For longer leach times, boron and sodium release slows down, their ratio keeping the stoichiometry of pristine glass, while silicon release is markedly damped. Simultaneously, a hydrated layer enriched in silicon builds up on the glass surface. The IR absorption spectrum suggests that this layer is constituted of a hydrosilicate. SEM observations show that it is zoned and cracked, with corrosion features on the outermost surface. The evolution of silicon in solution proves that equilibrium has not been reached, although the pH at 20° C of the leachates rapidly stabilizes around 9.65. Our observations agree with those of Bunker et al. [6] who have extensively studied a glass of similar composition.

3.2. Glass V3

From the very beginning (t < 4 h), dissolution is incongruent with a much lower release of silicon in comparison to sodium and boron than would be required by the glass stoichiometry; in contrast sodium and boron are released with mutual congruence. The pH of leachates at 20° C stabilizes at 9.1 after 60 h of leaching. One notes that the introduction of 3.9 % Al_2O_3 in the glass composition leads to a different leaching behavior, and in particular to a notably improved durability (by a factor of 6 to 10). The thickness of the superficial depleted zone, estimated through solution analysis, increases linearly with the square root of time. This suggests that its formation is ruled by diffusion with a diffusivity of ~$1.8 \times 10^{-17} m^2.s^{-1}$.

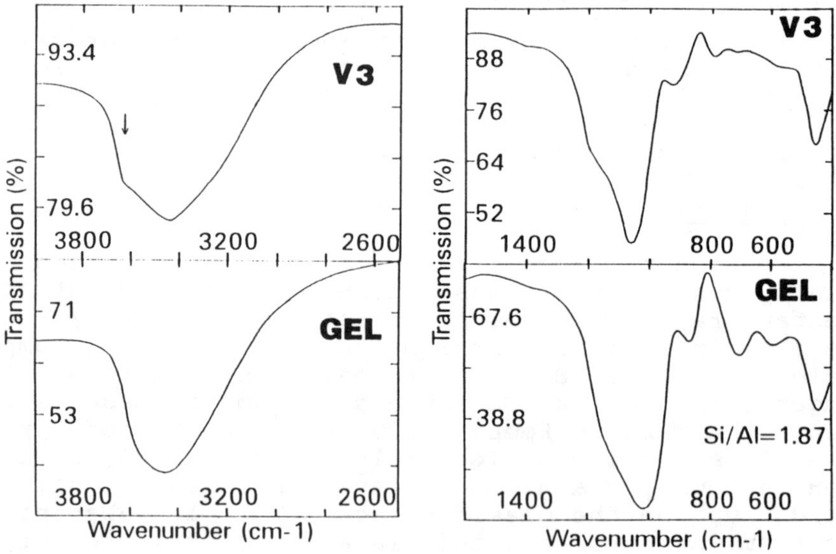

Figure 1 : IR-spectra of the hydrated layer of V3 (672 h) and of a synthetized aluminosilicate gel.

The transmission infrared spectrum of the external layer (Figure 1), obtained by embedding scratched material in a KBr pellet, exhibits characteristics similar to those of a synthetized hydroaluminosilicate gel (Si/Al = 1.87)(Figure 1)[7]. The differences between these two spectra in the 1200-400 cm-1 region could be explained by a very different Si/Al ratio (~5) for the leached layer. In addition, we attribute the shoulder at 3600 cm-1 in the water band of the V3 spectrum to a higher proportion of silanol groups [8]. The RBS spectrum after 199 h leaching essentially shows a marked sodium depletion and a slight enrichment in aluminum and/or silicon in the hydrated layer. The corresponding ERDA profile indicates an important hydration over a thickness larger than the investigation limit of 500 nm with a constant hydrogen concentration of 5.5×10^{21} H cm-3. All these data suggest that these corrosion products are formed in situ and are best described as a hydroaluminosilicate of definite composition.

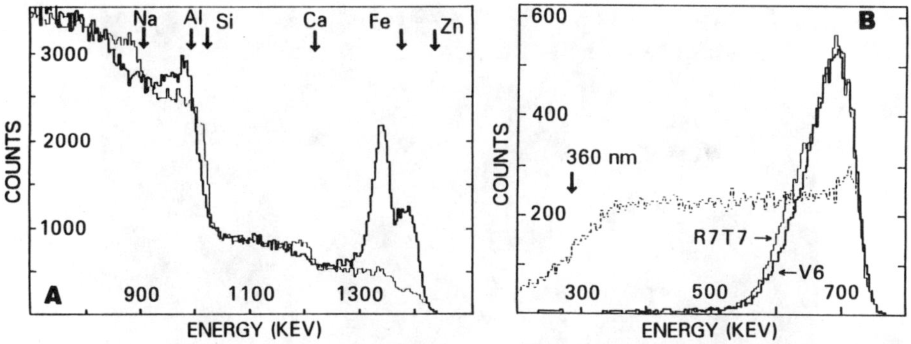

Figure 2: (A) RBS spectra of V6 before (light) and after (bold) leaching 185 h. (B) ERDA data on leached V6 (185 h) and R7T7 (193 h). The dashed line is a 5.5 10^{21} at.cm-3 H Si standard implanted on a 360 nm depth.

3.3. Glass V6

Table 2 shows that the amounts of released B and Na are mutually stoichiometric but are reduced by a factor ~4 in comparison to V5. The amount of released Si is reduced by a factor of 3 for the 185 h leach test and only 1.5 for the longest run. These data clearly indicate a marked improvement in glass durability, after a relatively small variation in composition, namely an addition of ~2% ZnO. The pH of leachates are low in comparison to V3 and V2 (7.8 at 20° C). The RBS spectrum of glass V6 corroded during 185 h (Figure 2A) exhibits a surface accumulation of Fe, Zn and to a lesser extend Al, in conjunction with a marked depletion of Na and Ca and a small shift on the Si edge. The computer simulation of the spectrum fits well with a heterogeneous coating, with mixed Fe,Zn hydroxides in the outer part and, in addition, some Si in the inner one. Its inferred thickness (~50 nm) is approximately half that of the hydrogen distribution deduced from the ERDA spectrum (Figure 2B). This reflects the hydration of the underlying glass. The corresponding maximum H concentration is about 10^{22} atoms cm-3, to be compared with the Fe and Zn concentrations of $2.4x10^{21}$ and $2x10^{21}$ atoms cm-3. These quantitative data support the hypothesis of a mixed Fe(III) and Zn hydroxide.

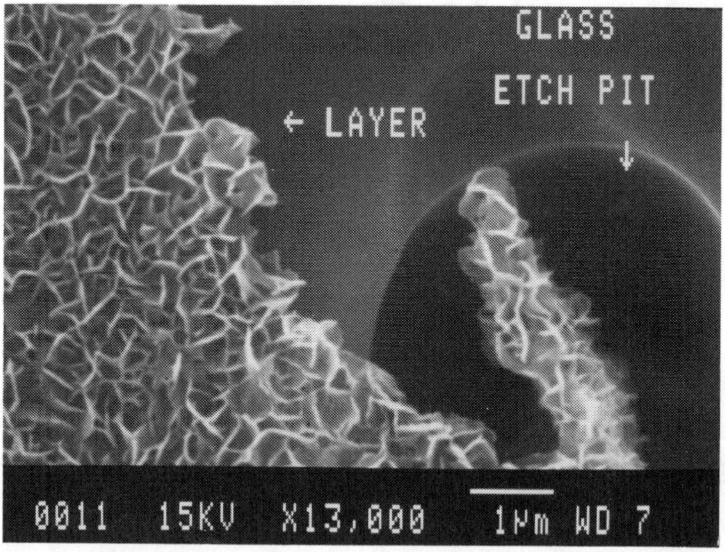

Figure 3: SEM micrograph of glass V6 leached 185 h.

The IR data on this glass were obtained by grazing incidence reflectance spectrometry because, due to its high durability, a sufficient amount of alteration products could not be recovered for classical absorption measurements. Unfortunately, these spectra are ambiguous and some characteristic bands (~1440 cm-1) could be attributed to carbonates. SEM micrographs of the reacted surface reveal the presence of a thin layer constituted of micro-lamellae, easily detached from the underlying glass (Figure 3). Qualitative EDX analyses exhibit the same trends as the RBS measurements, namely strong Fe and Zn enrichments and Si and Ca depletions.

3.4. Reference nuclear glass R7T7

The corrosion features of this glass have been extensively investigated [2]. We show here that this glass is the most durable of the entire series, the release rates of nearly all elements being lower than in V6 by at least a factor 4. This improvement can be mainly attributed to the adjunction of ZrO_2 in the glass. RBS and ERDA spectra (Figures 4 and 2B) as well as IR spectrometry show the same general features as V6 (see 3.3). Henceforth, we consider that these three techniques suggest that the altered layer is similarly formed of an hydroxide mixture. SEM micrographs are however quite different from those of V6, in displaying a very thin homogeneous coating with scattered micrometer-sized precipitates enriched in lanthanides.

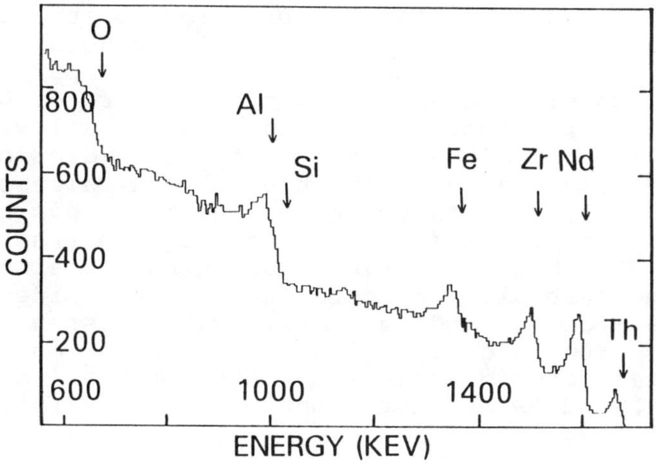

Figure 4: RBS spectrum of R7T7 leached 193 h.

We demonstrate in this paper that two types of

altered layers can form on leached glasses according to their composition. The simplest and less durable glasses (V2 to V4) form amorphous hydrosilicates, whereas the more complex and durable ones (V5 to R7T7) tend to form silica depleted layers likely constituted of hydroxides. In addition, the IR data on V6 suggest the presence of hydroxycarbonates in its altered layer, a quite likely compound for solubility reasons. It should be first noted that this varying nature of layers could also be explained by different evolutions of pH during leach tests. It should also be emphasized that on much longer tests the compositions of secondary phases are likely to evolve. In effect, Mouche and Vernaz [9] have shown in a two-years experiment on the R7T7 that the hydrated layer continuously enriches in silica up to about 70 % of its initial concentration in the pristine glass. This evolution is explained by the known property of hydroxides to very efficiently sorb dissolved silica [10]. In contrast, the estimated silica concentrations in the layers formed on V2 and V3 are constant. The observed variation in the glass durability clearly arises from the introduction of specific oxides in the composition.

Such an effect can be explained from a pure thermodynamic viewpoint or by considering the kinetics of glass dissolution. The variation of the intrinsic alterability of the glass in relation to its composition has been modelled by several authors (see in particular [11]). In this thermodynamic approach, the glass is considered as formed of a mixture of structural units (orthosilicates and oxides) allowing the computation of a mean free energy of hydration. The durability, as reflected by the silicon release, is in turn directly related to this quantity. This approach is not very conclusive when applied to our series of glasses, a better correlation being obtained when assuming that all structural units are oxides. However, a model of sodium borosilicate glasses established by Yun and Bray [12] and results on alkali-aluminosilicate glasses [13] suggest that even the simple V2 and V3 are not composed of oxides structural units.

The formation of a protective film has been invoked by many authors to explain changes in the kinetics of glass dissolution when introducing specific oxides in the glass composition (see for instance [14]). In addition, the retention and accumulation of minor or trace elements on the glass surface is directly linked to the formation of such a film [15]. The protective effect of leached layers is reflected by some of our results which show that the release of mobile elements (Na and B) is controlled by diffusion through the increasingly thick altered zone, whatever its exact nature. The respective contributions of thermodynamics (glass composition and structure) and kinetics (in particular leach test parameters: T, SA/V, flow rate, etc.) to glass durability has been discussed in detail by Jantzen [16] who investigated the leaching properties of over 155 natural and man-made glasses, including nuclear ones.

4. CONCLUSION

We have shown that hydrated layers formed on a series of glasses are constituted of either hydrosilicates for glasses of simple composition or hydroxides for more complex ones containing highly hydrolyzable elements. Moreover the introduction of these elements induces discontinuities in their chemical durability which markedly increases from V2 to R7T7. However, a precise link between glass composition, and hence its structure, and corrosion phenomenology (dissolution rate, properties of secondary phases) is still out of reach. Future investigations include experiments at buffered pHs as well as tests designed to discriminate between the thermodynamic effect of composition on glass durability and the protective effect of precipitates.

ACKNOWLEDGEMENTS

We would like to acknowledge the help of O. Michau for solution analyses and the skillful assistance of G. Egeni and V. Rudelo for operating the 2 MV accelerator of Legnaro (Italy).

REFERENCES

1) L.L. Hench, D.E. Clark and A.B. Harker,"Review - Nuclear Waste Solids", J. Mater. Sci., 21, 1457-1478 (1986).
2) J.L. Nogues,E.Y. Vernaz and N. Jacquet-Francillon, "Nuclear Glass Corrosion Mechanisms Applied to the

French LWR Reference Glass", Mater. Res. Soc. Symp. Proc., 44 89-98 (1985).
3) G. Calestani, A. Montenero, E. Ferraguti, G. Ingletto and M. Bettinelli, "Influence of Some Oxides on the Durability of a Borosilicate Glass", J. Non-Cryst. Solids 84 452-462 (1986).
4) J.-C. Petit, J.-C. Dran, L. Trotignon, J.-M. Casabonne, A. Paccagnella and G. Della Mea, "Mechanism of Heavy Element Retention in Hydrated Layers Formed on Leached Silicate Glasses", Mater. Res. Soc. Symp. Proc. (in press, 1989).
5) P. Mazzoldi and A. Carnera, "Structure Characterization of LiNbO3 Wave-Guides by Using Nuclear Techniques", Cryst. Latt. Def. and Amorph. Mat., 14 319-333 (1987).
6) B.C. Bunker, G.W. Arnold, D.E. Day and P.J. Bray, "The Effect of Molecular Structure on Borosilicate Glass Leaching", J. Non-Cryst. Solids 87 226-253 (1986).
7) R. Couty, F. Taulelle and H. Zanni-Theveneau, "Infrared and Al NMR.MAS spectra of aluminosilicate gels. Characteristic Vibrational Frequencies of AlO6 Octahedra and AlO4 Tetrahedra", C. R. Acad. Sc. Paris 304 (4) Serie II 165-167 (1987).
8) R.H. Doremus, "Infrared Spectroscopy of Surfaces of Glasses Containing Alkali Ions", J. Non-Cryst. Solids 41 145-149 (1980).
9) E. Mouche and E. Vernaz, "Aqueous Corrosion of the French LWR Solution Reference Glass : First Generation Model. Mater. Res. Soc. Symp. Proc., 112 703-712 (1988).
10) G.M Luciuk and P.M Huang, " Effect of Monosilicic Acid on Hydrolytic Reactions of Aluminum", Soil Sci. Soc. Am. Proc., 38 235-244 (1974).
11) C.M. Jantzen and M.J. Plodinec, "Thermodynamic Model of Natural, Medieval and Nuclear Waste Glass Durability", J. Non-Cryst. Solids, 67 207-223 (1984).
12) Y.H. Yun and P.J. Bray, "Nuclear Magnetic Resonance Studies of the Glasses in the System Na2O-B2O3-SiO2", J. Non-Cryst. Solids, 27 363-380 (1978).
13) P.I.K. Onorato, M.N. Alexander, C.W. Struck, G.W. Tasker and D.R. Uhlmann, "Bridging and Nonbridging Oxygen Atoms in Alkali-Aluminosilicate Glasses", J. Am. Ceram. Soc., 68 [6] C-148 C-150 (1985).
14) D.E. Clark and L.L. Hench, "Theory of Corrosion of Alkali-Borosilicate Glass", Mater. Res. Soc. Symp. Proc. 15 113-119 (1983).
15) G.G. Wicks, B.M. Robnett and W.D. Rankin,

"Chemical Durability of Glass Containing SRP Waste - Leachability Characteristics, Protective Layer Formation, and Repository System Interactions", Mater. Res. Soc. Symp. Proc. 11 15-24 (1982).
16) C. M. Jantzen, "Prediction of Glass Durability as a Function of Glass Composition and Test Conditions: Thermodynamics and Kinetics", to be published in Proc. Conf. Advances in the Fusion of Glass, Alfred University, Alfred, N-Y, June 14-17, 1988.

	SiO2	Na2O	B2O3	Al2O3	CaO	Fe2O3	ZnO	ZrO2
V2	64.6	18.1	17.0	-	-	-	-	-
V3	61.1	18.0	16.8	4.00	-	-	-	-
V4	57.9	17.3	15.1	3.70	5.6	-	-	-
V5	57.3	16.8	15.0	3.7	5.4	1.4	-	-
V6	56.4	15.8	(15)	3.5	5.2	1.4	2.3	-
V7	54.9	15.5	(16)	3.5	5.3	1.1	2.2	1.5
R7T7	52.5	11.0	14	3.4	5.0	1.3	2.1	1.5

Table 1 : Composition of investigated glasses in mole %. Values in parentheses are estimated. R7T7 also contains 4.6 % Li2O and 4.7 % other minor oxides.

	V2	V3	V6	R7T7
C(Si) :	8.654	0.983	0.762	0.182
C(B) :	24.51	2.137	0.485	0.111
C(Na) :	24.78	2.204	0.535	0.137
C(Li) :	--	--	--	0.046
pH(20 C):	9.66	9.16	7.85	7.09

Table 2 : Solution analysis for the 672 h run on V2, V3, V6 and R7T7 at 90°C and 20 m-1. Values are given in millimoles per kg water.

ALTERATION OF BASALTIC GLASSES IN WIPP IN-SITU CORROSION TESTS

M. J. Jercinovic, S. A. Kaser, and R. C. Ewing
Dept. of Geology, University of New Mexico
Albuquerque, New Mexico USA 87131

ABSTRACT

Surface precipitates on synthetic basaltic glass exposed to concentrated brine at 90°C during *in situ* repository corrosion tests are predominantly Mg-chloride and halite. This is in marked contrast to surface layers formed on basaltic glasses altered naturally in seawater at low temperature which are predominantly Fe-Al-silicates which replace the glass. Fundamental differences in surface layer composition between the synthetic basaltic glass and borosilicate waste form glass in the *in situ* repository tests has implications for the applicability of natural analogues to these conditions. If long term corrosion rates derived from the study of natural analogues are to be successfully applied to waste form glasses in repository conditions, the differences in the precipitation processes evident in the short term repository tests must be understood.

INTRODUCTION

The assessment of the long term corrosion behavior or borosilicate nuclear waste form glass relies upon the verification of predictive theoretical models[1,2,3,4]. Verification can be achieved by the application of natural analogues[5]. Basaltic glass (SiO_2 45-52 wt.%) has been shown to be a reasonable analogue to borosilicate waste form glass, as these glasses alter by similar processes at similar rates in short term laboratory experiments[6,7], and have similar hydration energies[8]. An international program for *in situ* testing of waste form glasses, called the Materials Interface Interactions Test (MIIT) was initiated in 1986[9]. The experiments, intended to approximate salt repository conditions, were conducted by subjecting various wasteforms and synthetic basaltic glass to brine solutions at 90°C in instrumented bore holes at the WIPP site[10]. The leachant composition used for the MIIT experiments was WIPP Brine A[10]. This study reports the results of 6 month, 1 year and 2 year corrosion tests for the synthetic basaltic glass and compares them to: 1) basaltic glass altered both naturally in seawater at low temperature (2-5°C, 10^5 to 10^6 years) and in laboratory corrosion tests (saturated NaCl-solution, 200°C, 30 days); 2) experimentally altered borosilicate glass (MIIT experiments); and 3) results predicted by reaction path modelling of seawater alteration of basaltic glass.

SAMPLES

The synthetic basaltic glass included in the Materials Interface Interaction Tests was made at the Savannah River Laboratory. The test samples included in this study are listed in Table 1.

Table 1. Synthetic basaltic glass samples.

Sample	Exposure Time
SRL BAS-8	6 months
SRL BAS-17	1 year
SRL BAS-21	2 years

ANALYTICAL TECHNIQUES

The mineralogical and geochemical characterization of the altered glasses was completed using: Optical Microscopy (OM), Scanning Electron Microscopy (SEM), Electron Microprobe Analysis (EMPA), and Analytical Electron Microscopy (AEM).

Samples were prepared for scanning electron microscopy (Hitachi S-450 Scanning Electron Microscope) by mounting centimeter-sized chips onto aluminum mounts using household cement. The samples were coated with carbon and gold-palladium using vacuum evaporation.

Thin sections used for EMPA were prepared using kerosene as a lubricant and organic solvents as cleaners in order to preserve water soluble materials. Electron microprobe analysis of glasses and alteration products is complicated by the effects of electron beam irradiation, which can result in dehydration and alkali migration in the analyzed region of the specimen[11]. Therefore, some precautions are necessary. The techniques outlined by Jercinovic and Keil[11] were generally followed (e.g., lowering beam current and increasing beam diameter in order to minimize current density and, therefore, heating rate). For glass, a 40 μm diameter, 20 nA beam was used. For analysis of surface layers, a 2 μm diameter, 2 nA beam was used. Elements to be included in the analysis of glass and alteration products were determined by qualitative EDS analyses in both the SEM and electron microprobe. Glass standards were used for most elements. Mineral standards were used for elements where either glasses were unavailable or concentrations were inappropriate (Marjalati olivine for Mg, doped diopside for Mn, WIPP halite for Cl, celestite for S). Analyses were done using a fully automated, 5 wavelength-dispersive (WDS) spectrometer JEOL 733 Superprobe equipped with a Tracor Northern TN-2000 x-ray analysis system and Li-drifted, Si-solid state energy dispersive system (EDS). Microprobe automation and implementation of Bence-Albee matrix corrections were done via the Sandia Task8 pseudo-operating system[12].

Two sample preparation techniques were used for AEM: 1) Ultra-thin sections were prepared by ultramicrotoming resin impregnated sample chips; 2) a layer stripping technique, as outlined by Vernaz and Godon[13], was done prior to ultramicrotomy. A Reichert-Jung Ultracut-E Microtome was used. AEM was done using a JEOL 2000FX electron microscope equipped with a Tracor Northern TN-

5500 solid state energy dispersive x-ray analysis system (EDS). A Gatan low-background, double-tilt cryogenic stage (LN_2, -150°C) was used in order to minimize electron irradiation effects. Standard bright field imaging and electron microdiffraction techniques were used.

RESULTS

6 Month MIIT

Surface layers exist on all glass surfaces exposed to the brine solution. Where surface layers have been removed, scanning electron microscopy shows the underlying glass surface to be smooth (Figure 1c), lacking the dissolution features seen on naturally altered basaltic glass (Figure 1a,b). The surface layers are predominantly Mg-chloride (or, perhaps, Mg-oxychloride[14,15]), forming rosettes and continuous matts of clay-like, crenulated sheets (Figure 1d). Pre-existing minerals, since dissolved, have left "casts" in the Mg-chloride layer (Figure 2a). The casts are equant to prismatic and appear to have resulted from crystallization of one or more euhedral phases within the layer. SEM and EMPA analyses of the glass (including a sample of pre-test glass) show no indication of phase separation within the glass which may account for the casts if they were a result of dissolution during replacement of the glass. The total layer thickness is 5-10 μm.

The surface layer was not preserved during specimen preparation for analytical electron microscopy. Some layers were preserved in the thin section used for microprobe analysis. Layer EMPA analyses, however, could not be determined because the preserved material is thinner than the beam diameter.

1 Year MIIT

Similar to the 6 month MIIT results, the glass does not appear to be extensively dissolution pitted (Figures 2b through 2d). The surface layers are predominantly Mg-chloride with halite crystals interspersed throughout. The total layer thickness ranges from 10 to 20 μm.

A discontinuous Cl-rich layer, up to 0.2 um thick, exists between the glass and the Mg-chloride layer (Figure 2c and 2d). High magnification reveals a fibrous texture for this layer (Figure 2d). Compositional determination on this layer is difficult due to the generation of x-rays from the underlying glass. The only compositionally distinguishing feature of EDS patterns obtained from this layer is the relatively high Cl content compared to glass.

AEM investigation reveals a mottled texture in the preserved surface layer (Figure 3). There is no evidence of crystallinity in the region of the layer observed in AEM. EDS analyses indicate that the observed material is predominantly Mg oxide or oxy-hydroxide with a variable silicate component (Table 2).

EMPA of surface layers generally agrees with AEM results (Table 2) but not with qualitative compositions obtained by SEM (primarily Mg-chloride). It is likely that the material preserved in thin sections for EMPA is not representative of the overall layer as seen in SEM, or this material has been altered during preparation.

Table 8. Summary of AEM-EDS and EMPA-WDS analyses (wt.%) of surface layers formed on synthetic basaltic glass (SRL)* and compositions of SRL basaltic glass and HMI borosilicate glass for comparison. Surface layer analyses are normalized to 100% total oxides.

	SRL Glass**	HMI Glass**	SRL 1-Year Layer AEM	SRL 1-Year Layer EMPA	SRL 2-Year Layer AEM	SRL 2-Year Layer EMPA
SiO_2	51.34	52.14	36.5	26.4	31.8	32.2
TiO_2		4.54	1.0	1.0		3.4
Al_2O_3	17.68	3.62	6.8	7.9	6.7	10.8
FeO***	14.23	1.54	10.4	2.4	4.6	7.9
MnO_2	0.33	0.76				
MgO	3.84	2.05	40.0	50.3	50.7	30.5
CaO	6.08	4.54		0.2	3.3	1.3
Na_2O	4.42	9.15		0.2		2.8
K_2O				0.4		1.0
P_2O_5		0.01		0.5		2.2
S			1.1	0.3	1.3	0.2
Cl			1.1	10.3	4.9	7.6
NiO		0.43	3.0			

*Does not include Mg-chloride.
**Compositions of glasses given by Wicks et al.[16].
***All Fe reported as FeO.

2 Year MIIT

As is the case with 6 month and 1 year MIIT results, SEM reveals the surface layers formed on the glass exposed to brine for 2 years to be predominantly composed of Mg-chloride (Figure 4). Halite crystals, mostly less than 10 μm across, are common within, and on top of, the Mg-chloride layer. The total layer thickness is 5 to 20 μm. There is no evidence for extensive dissolution of the glass surface beneath the Mg-chloride layer.

AEM observations indicate an amorphous character for the portion of the layer preserved for observation. There is no evidence of crystallinity in the layer. EDS analyses are similar to those of the layer formed after the 1 year MIIT experiment, with most of the preserved layer being composed of Mg-Al-silicate (Table 2).

EMPA of the surface layers reveals none of the Mg-chloride indicated as the dominant phase by SEM. The preserved material which could be analyzed is predominantly Mg-Silicate (Table 2). Halite crystals were preserved in the thin section and WDS analyses yielded good results.

DISCUSSION

COMPARISON TO NATURAL ALTERATION AND LABORATORY EXPERIMENTS

There is no direct evidence, such as dissolution pitting of the glass, that glass/solution interactions were significant in determining the nature of the precipitates in the *in situ* tests for synthetic basaltic glass. The presence of silicate material in preserved surface layers analyzed by both EMPA and AEM may indicate that some glass network dissolution has taken place during the MIIT experiments, however, this dissolution is not as pervasive as is generally seen in materials that have been naturally altered in seawater (see the comparison in Figure 1). Although the products analyzed by EMPA and AEM are apparently not representative of the bulk of the precipitate which results from the test (SEM indicates Mg-chloride, EMPA and AEM analyses are of silicate material), they do represent the residue which remains after sample preparation, and may, therefore, be important in determining the total system chemistry. Natural alteration of basaltic glass in seawater results in replacement of the glass by palagonite, a clay-like Fe-Al-silicate gel, and cementation by zeolites, clay minerals, and calcite at high reaction progress[17]. In contrast, the MIIT produced (concentrated brine solution, 90°C) surface layers on synthetic basaltic glass primarily consist of brine controlled chloride precipitates with little evidence for glass network dissolution/replacement reactions (palagonitization).

Short term laboratory corrosion tests (saturated NaCl solution, 200°C, 30 days) result in surface layers formed on basaltic glass which more closely resemble natural palagonites[6,7]. The experimentally formed surface layers are primarily smectite clay-like (Fe-Al phyllosilicate) precipitates with abundant analcime crystals ($NaAlSi_2O_6.H_2O$). Phyllosilicates are also formed as the major alteration phase in experimental alteration of basaltic glass in seawater at 50°C[18].

Glass dissolution and subsequent reprecipitation of silicate material is the principal mechanism which is observed in alteration of basaltic glasses in nature and in most experiments. The apparent scarcity of similar materials in the MIIT altered basaltic glass is enigmatic.

COMPARISON TO BOROSILICATE GLASS

Although there is some evidence that silica-bearing phases were precipitated in the 1 year and 2 year MIIT experiments on basaltic glass, borosilicate glass exposed in the same tests have surface layers dominated by Mg-silicate[19]. Mg-chloride is present, but not dominant, in the 6 month MIIT experiment on borosilicate glass. Halite is seen in the borosilicate glass surface precipitates. Clearly, there is some difference in the alteration processes affecting the basaltic and borosilicate glasses in the MIIT experiments which results in this fundamental difference in surface layer compositions. If glass reaction is not significant, it is not apparent why silicate precipitates would be found in the case of formation of surface layers on borosilicate glass and not on basaltic glass. This disparity can only be explained by differences in corrosion phenomena (mechanisms or rates) between basaltic and borosilicate glass. Note that, in the case of SRL basaltic glass and HMI borosilicate glass (Table 2), the silica contents of the glasses are similar. Thus, if the respective degrees of polymerization are comparable between the two glass types, network dissolution should proceed to similar degrees in the same environment[2]. Corrosion kinetics vary primarily as a result of silica activity in solution and temperature[2]. The applicability of basaltic glasses as analogues to borosilicate waste form glass, at least for the *in situ* environment of the MIIT experiment, may be in question if the compositional difference between surface precipitates on the two glass types indicates a corresponding difference in corrosion mechanism.

MODELLING COMPARISONS AND IMPLICATIONS

Glass network dissolution will proceed as long as water has access to the glass surface[2]. At a defined temperature and surface area to solution volume ratio, the dissolution rate will decrease (final rate) in comparison to the initial (forward) rate, as the silica concentration in solution reaches saturation relative to the glass surface[2]. Therefore, even in concentrated brine solutions, some glass components should be released into solution and be incorporated into precipitates. This, without Si-bearing phases precipitating, solution silica concentrations would rise (in a closed system), and glass surface silica saturation would be attained more rapidly than in similar natural systems where aluminosilicate materials are precipitated through most of the alteration history. Calculated reaction products in low temperature, closed-seawater systems are dominated by the precipitation of Fe-rich silicates (as clay minerals), Al-hydroxide, and Fe-oxy-hydroxides through most of the sequence, with the appearance of zeolites (alkaline aluminosilicates) at high reaction progress[20]. Magnesium in such systems is readily incorporated into clay minerals. Modelling predictions agree reasonably well with some naturally altered samples[20] but differ considerably from the phases seen in the MIIT experiments. In natural systems of high reaction progress, net Si-mass loss is low (less than 10% relative to the glass reacted) due to precipitation of clay minerals and zeolites[17]. Therefore, most of the Si released may participate in back reactions and the forward rate may dominate. Solutions supersaturated with respect to $MgCl_2$ and NaCl, as in the MIIT experiments, will still be able to react with glass as long as water has access to the glass surface. The type of rapid reprecipitation reactions producing the clay-like gels which replace the glass as dissolution proceeds in natural systems is not seen in the MIIT reacted basaltic glasses. This would indicate that either the brine precipitates have sealed the glass surface from further reaction (impermeable surface layer) or that constituents of the glass are incorporated into the chlorides precipitating at the surface. As Mg- and Na-chloride precipitate, the residual solution which remains in the near-glass environment to react with the glass will have lower brine concentrations and silicate precipitation is expected. This becomes progressively more likely as dissolution penetrates the glass below the initial chloride layers. This is not clearly evident from SEM observation, but the existence of thin silicate layers in residual material observed in AEM may be indicative of this possibility. If the Si, Al, and Fe in the surface layers observed by AEM is from an external source, such as impurities in the salt, then glass reaction cannot be inferred from the experimental results[21].

Predicting long term corrosion behavior of wasteform glass requires knowledge of reaction rates over the long term. As the rate is theoretically higher when the solution silica concentration is below the glass surface silica saturation value compared to silica-saturated conditions[2], both rates must be known. Short term rates can be measured in the laboratory but long term (silica-saturated conditions?) may only be obtained from natural analogues. Besides the obvious difficulties in comparing reaction products between seawater at low temperature (0 to 5°C) and concentrated brine at 90°C, there are also difficulties in estimating reaction rates applicable to the repository environment from natural alteration systems. There are no natural systems involving basaltic glass which approximate the MIIT WIPP brine conditions. Therefore, pertinent dissolution rate information can only be extracted from natural analogues if the long term corrosion mechanisms are truly comparable between basaltic and borosilicate glass and the short term rate differences, as possibly indicated by the MIIT results, can be disregarded or considered in the predictions.

CONCLUSIONS

Surface layers formed on synthetic basaltic glass during 6 month, 1 year, and 2 year repository corrosion tests are predominantly Mg-chloride and halite. There is evidence that Si, Al, and Fe are included in the precipitates as impurities or discrete phases. This may imply that glass corrosion has taken place despite the lack of dissolution pitting. These surface layers do not compositionally or mineralogically resemble surface layers formed on basaltic glasses altered either naturally in seawater or experimentally in saturated NaCl solutions. Borosilicate glasses subjected to the same repository corrosion tests develop surface layers primarily composed of Mg-silicate. This striking difference between surface layers formed on basaltic and borosilicate glasses points out a possible difficulty in the use of natural analogues in the verification of predictive geochemical models applied to the long term corrosion of waste form glasses in a brine repository.

ACKNOWLEDGEMENTS

Samples of the Savannah River synthetic basaltic glass were supplied by George Wicks. Thin sections for microprobe analysis were prepared by Tom Servilla. Analyses were completed in the Electron Microbeam Analysis Facility at the Department of Geology of the University of New Mexico, supported, in part, by NSF, NASA, DOE-BES and the state of New Mexico.

REFERENCES

[1]A. Barkatt, P.B. Macedo, B.C. Gibson, and C.J. Montrose "Modelling of Waste Form Performance and System Release"; in Scientific Basis for Nuclear Waste Management VIII. Materials Research Society, Pittsburgh, PA. 3-14 (1985).

[2]B. Grambow "A General Rate Equation for Nuclear Waste Glass Corrosion"; in Scientific Basis for Nuclear Waste Management, Vol. IX. Materials Research Society, Pittsburgh, PA. 263-272 (1985).

[3]K.B. Harvey and C.D. Litke "Model for Leaching Behavior of Aluminosilicate Glasses Developed as Matrices for Immobilizing High Level Wastes," *J. Amer. Cer. Soc.* **67** 553-556 (1984).

[4]C. Pescatore and A.J. Machiels "Mechanistic Approach to Modelling of Nuclear Waste Form Leaching"; in Advances in Ceramics VIII, Nuclear Waste Management. Amer. Cer. Soc., Columbus, OH. 509-518 (1983).

[5]R.C. Ewing and M.J. Jercinovic "Natural Analogues: Their Application to the Prediction of the Long-Term Behavior of Nuclear Waste Forms"; in Scientific Basis for Nuclear Waste Management, Vol. X. Materials Research Society, Pittsburgh, PA. 67-83 (1987).

[6]G. Malow, W.L. Lutze, and R.C. Ewing "Alteration Effects and Leach Rates of Basaltic Glasses: Implications for the Long-Term Stability of Nuclear Waste Form Borosilicate Glasses," *J. Non-Cryst. Solids*, **67** 305-322 (1984).

[7]W. Lutze, G. Malow, R.C. Ewing, M.J. Jercinovic, and K. Keil "Alteration of Basalt Glasses: Implications for Modelling the Long-Term Stability of Nuclear Waste Glasses," *Nature (London)*, **314** 252-255 (1985).

[8]C.M. Jantzen and M.J. Plodinec "Thermodynamic Model of Natural, Medieval and Nuclear Waste Glass Durability," *J. Non-Cryst. Solids*, **67** 207-223 (1984).

[9]G.G. Wicks and M.A. Molecke "WIPP/SRL International In Situ Testing Program- MIIT"; in Proceedings of the Workshop on Testing of High Level Waste Forms Under Repository Conditions, to be published as EUR 12 017 EN in 1989, The Commission of The European Communities.

[10]M.A. Molecke "Technical Operations and Data Collection Details of the In Situ WIPP Materials Interface Interactions Test"; in Proceedings of the Workshop on Testing of High Level Waste Forms Under Repository Conditions, to be published as EUR 12 017 EN in 1989, The Commission of The European Communities.

[11]M.J. Jercinovic and K. Keil "Electron Microprobe Analysis of Basaltic Glasses and Associated Alteration Products"; in Microbeam Analysis-1988. San Francisco Press, Inc., San Francisco. 495-497 (1988).

[12]W.F. Chambers "SANDIA TASK8. A Subroutined Electron Microprobe Automation System," Sandia Laboratories Research Report SAND85-2037. (1985).

[13]E. Vernaz and N. Godon "Examination of Ultrathin Cross Sections From R7T7 Glass Samples After 6 Months and 1 Year of Alteration in the WIPP"; in Proceedings of the Workshop on Testing of High Level Waste Forms Under Repository Conditions, to be published as EUR 12 017 EN in 1989, The Commission of The European Communities.

[14]C.M. Jantzen, D.R. Clarke, P.E.D. Morgan, and A.B. Harker "Leaching of Polyphase Nuclear Waste Ceramics: Microstructural and Phase Characterization," *J. Amer. Cer. Soc.*, **65** 292-300 (1982).

[15]A.B. Harker, J.F. Flintoff, and D.G. Howitt "Dissolution Behavior of HLW Borosilicate Glass in Brine Environments"; in Proceedings of the Workshop on Testing of High Level Waste Forms Under Repository Conditions, to be published as EUR 12 017 EN in 1989, The Commission of The European Communities.

[16]G.G. Wicks, W.G. Ramsey, and K.A. Molen "Correlation of MIIT Glass and Waste Glass compositions"; in Proceedings of the Workshop on Testing of High Level Waste Forms Under Repository Conditions, to be published as EUR 12 017 EN in 1989, The Commission of The European Communities.

[17]M.J. Jercinovic and R.C. Ewing "Basaltic Glasses from Iceland and the Deep Sea: Natural Analogues to Borosilicate Nuclear Waste-Form Glass," JSS Project Technical Report 88-01, SKB, Stockholm. (1987).

[18]J.L. Crovisier, J.H. Thomassin, T. Juteau, J.P. Eberhart, J.C. Touray, and P. Baillif "Experimental Seawater-Basaltic Glass Interaction at 50°C: Study of Early Developed Phases by Electron Microscopy and X-Ray Photoelectron Spectrometry," *Geochim. Cosmochim. Acta*, **47** 377-387 (1983).

[19]M.J. Jercinovic, S. Kaser, and R.C. Ewing "Observations of Surface Layers Formed on Basaltic and Borosilicate Glass: 6 Month and 1 Year MIIT Experiments"; in Proceedings of the Workshop on Testing of High Level Waste Forms Under Repository Conditions, to be published as EUR 12 017 EN in 1989, The Commission of The European Communities.

[20]B. Grambow, M.J. Jercinovic, R.C. Ewing, and C.D. Byers "Weathered Basalt Glass: A Natural Analogue for the Effects of Reaction Progress on Nuclear Waste Glass Alteration"; in Scientific Basis for Nuclear Waste Management, Vol. IX. Materials Research Society, Pittsburgh, PA. 263-72 (1986).

[21]M.A. Molecke, N.R. Sorensen, and J.L. Krumhansl "Summary of WIPP Materials Interface Interactions Test Data on Metals Interactions and Leachate Brine Analyses"; in Proceedings of the Workshop on Testing of High Level Waste Forms Under Repository Conditions, to be published as EUR 12 017 EN in 1989, The Commission of The European Communities.

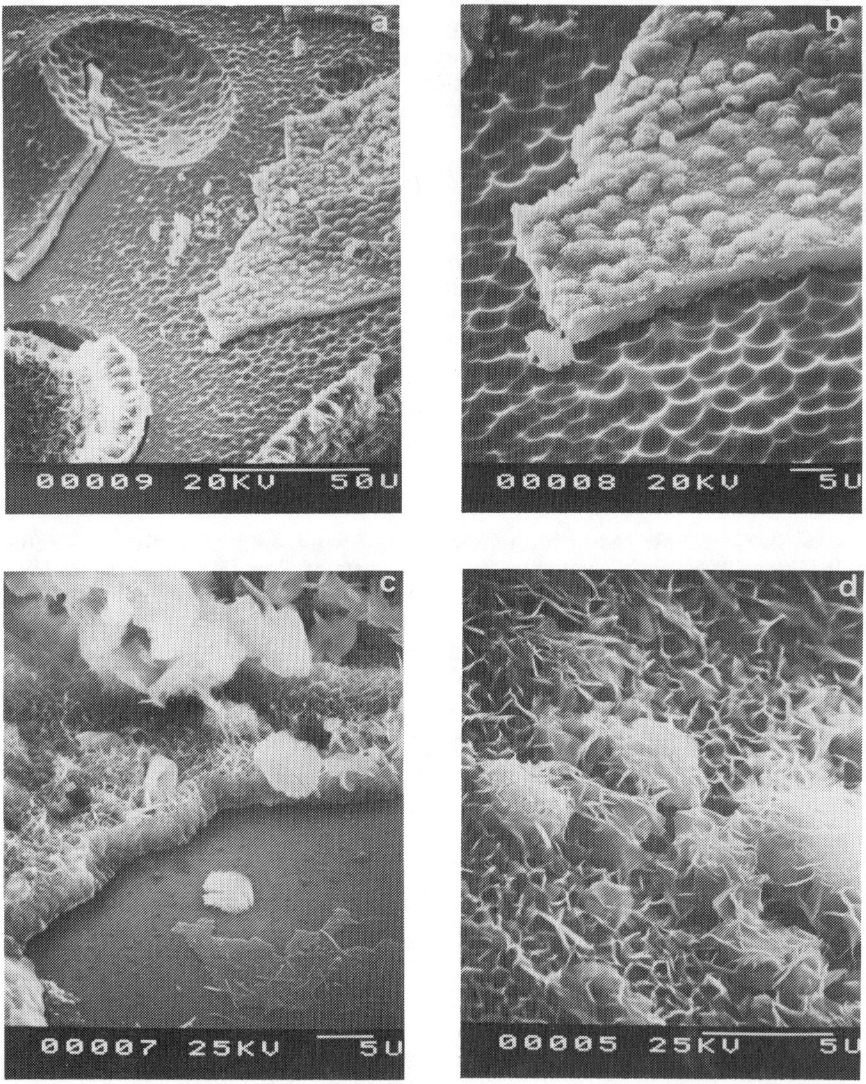

Figure 1. SEM photomicrographs of natural and synthetic basaltic glasses. A) A deep sea dredge sample (USNM 113432-24) showing pitted glass surface and overlying silicate (palagonite) layer. b) Higher magnification image of a). c) SRL BAS-8 (6 month MIIT sample). Note the relative lack of dissolution features on the glass surface. d) SRL BAS-8 Mg-chloride surface layer. This figure is from Jercinovic et al.[19].

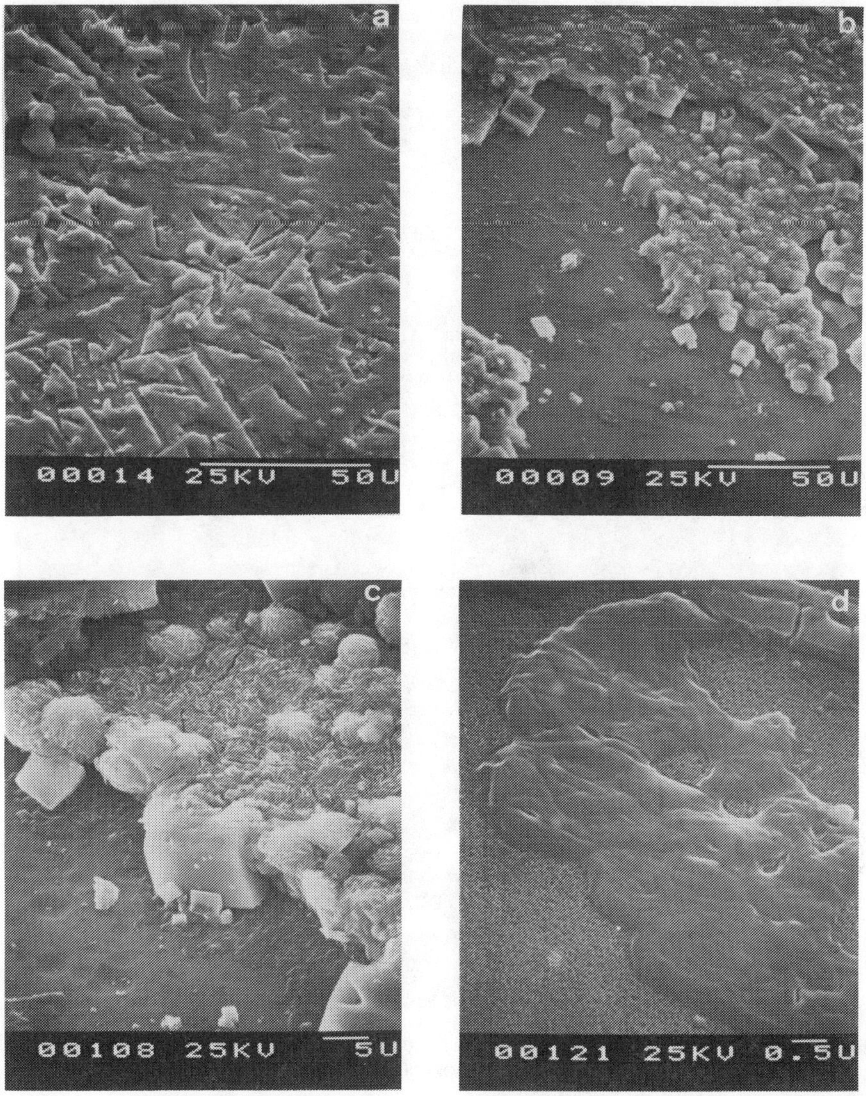

Figure 2. SEM photomicrographs of MIIT experimental results on synthetic basaltic glass. a) SRL BAS-8 "casts" of pre-existing crystals in Mg-chloride layer. b) SRL BAS-17 (1 year MIIT) layers of Mg-chloride overlying glass. c) Higher magnification image of layer on glass in b). Note halite crystals. d) High magnification image of glass/layer interface region in b) and c). The thin deposit on the surface is Cl-rich. This figure is from Jercinovic et al.[19].

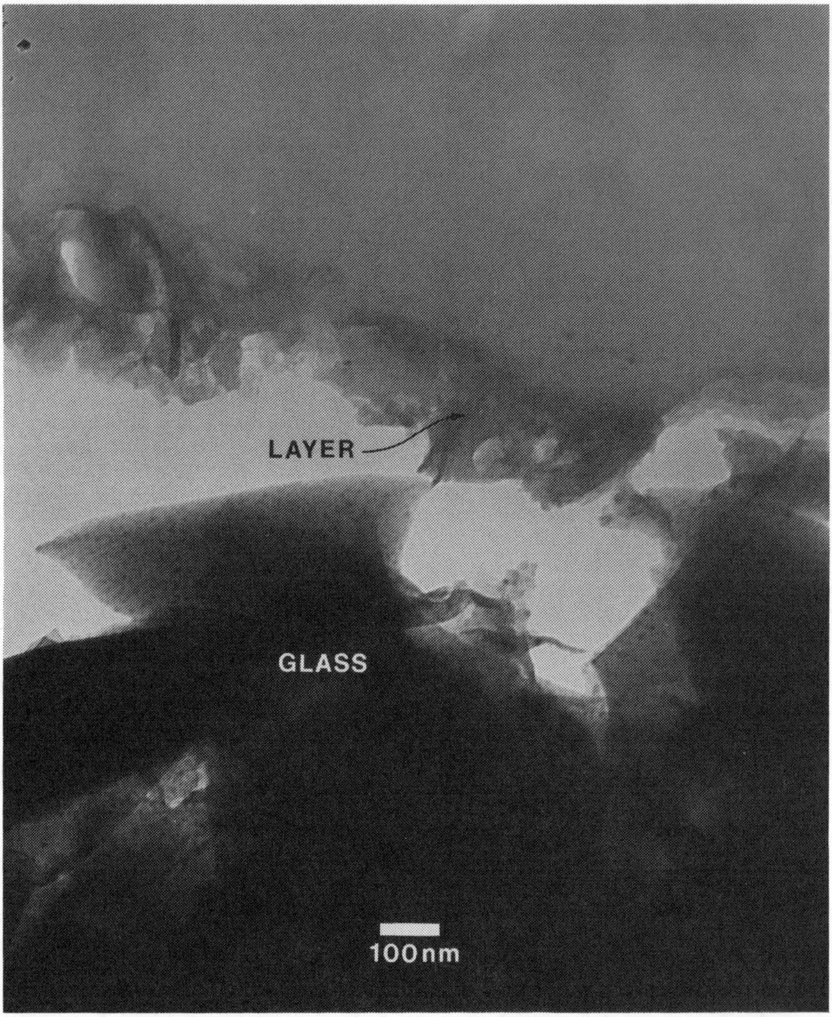

Figure 3. AEM micrograph of surface layer on synthetic basaltic glass SRL BAS-17 after 1 year exposure to brine solution. This figure is from Jercinovic et al.[19].

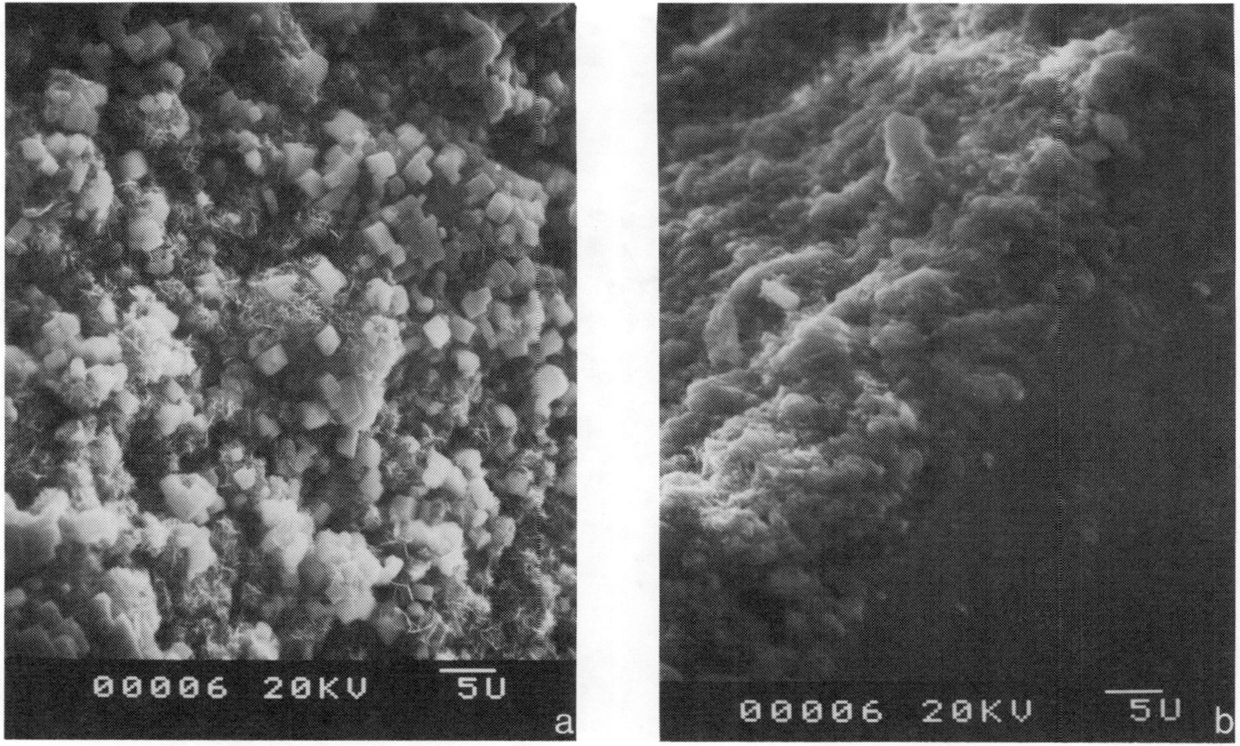

Figure 4. SEM photomicrographs of surface layers formed on synthetic basaltic glass SRL BAS-21 (2 year MIIT). a) Mg-chloride and halite. b) Surface layers overlying fresh glass (lower right).

Section IV. Systems and Field Testing

WIPP/SRL IN-SITU TESTS; COMPOSITIONAL CORRELATIONS OF MIIT WASTE GLASSES

W. G. Ramsey
Department of Ceramic Engineering
Clemson University
Clemson, SC 29634

G. G. Wicks
Westinghouse Savannah River Company
Savannah River Laboratory
Aiken, SC 29808

ABSTRACT

The Materials Interface Interactions Test (MIIT) is the only *in-situ* program involving the burial of simulated high-level waste forms operating in the United States.[1-5] Fifteen glass and waste form compositions and their proposed package materials, supplied by 7 countries, are interred in salt at the Waste Isolation Pilot Plant (WIPP) in Carlsbad, New Mexico. A joint effort between Sandia National Laboratories and Savannah River Laboratory, MIIT is the largest international cooperative *in-situ* venture yet undertaken. The objective of the current study is to document the waste form compositions used in the MIIT program and then to examine compositional correlations based on structural considerations, bonding energies, and surface layer formation. These correlations show important similarities between the many different waste glass compositions studied world wide and suggest that these glasses would be expected to perform well and in a similar manner. The compositional correlations define a 'commonality' between all MIIT waste forms and assist in furthering our understanding of the performance of high-level waste glass.[6,7]

INTRODUCTION

A major objective of the Savannah River Laboratory (SRL) during the past decade has been to understand the behavior of high-level waste forms on a geologic time scale. In order to evaluate waste forms and the effects of various canister and overpack materials, as well as proposed geologic environments on those waste forms, a multi-phase experimental program has been in progress for about a decade. Laboratory scale experiments initiated the investigation by developing testing procedures, comparing overpack and canister materials, and optimizing key properties such as composition of the waste form system. Next, repository relevant tests were used to study the behavior of waste glasses in specific geologic settings, under laboratory controlled conditions. This included evaluating interactions between waste form, support material, and geologic components.[8,9] The final phase of this effort is in-situ testing. Non-radioactive waste forms and package components are placed in actual geologic settings. While field tests do not contain the controls of the laboratory, they yield the most realistic, although most complex data, which cannot always be fully reproduced in the laboratory. By correlating the results of in-situ tests to laboratory studies, more accurate models of glass behavior can be formulated.[8,9]

SUMMARY

- MIIT waste form systems, representing formulations of 7 nations, can be correlated by relating individual components to their common glass structural role.
- This correlation suggests that the leaching behavior and associated mechanisms of Savannah River (SR) waste glass may be similar to other waste glasses, such as those in actual production overseas, as well as to a natural glass analogue (basalt) which has been stable for millions of years.

MIIT BACKGROUND

The Materials Interface Interactions Test (MIIT) is the largest in-situ experiment, in terms of scale and international participation, yet undertaken. MIIT is also the first in-situ test involving burial of simulated nuclear waste glasses to be conducted in the United States. MIIT is a joint venture between

Savannah River Laboratory (SRL) and Sandia National Laboratories (SNL/WIPP), and involves burial of 15 waste form systems supplied by 7 nations. In addition to the waste forms, there are also eleven different potential canister or overpack materials and several proposed backfills buried in the salt geology at WIPP.[1-3]

The MIIT program includes almost 1000 waste form specimens, 300 metal specimens and more than 500 salt and backfill geologic samples. MIIT consists of 50 boreholes located in Room J of the WIPP mine, 650 meters below the earth's surface, containing seven assembly configurations of test specimens. These assemblies consist of various combinations of waste forms, overpack materials, container metals, backfills and either WIPP 'Brine A' or salt, in contact with one another.

Glass, metal, and geologic samples are fabricated into 'pineapple slice' configurations of varying diameter, allowing multiple interfaces for each specimen. In all, almost 2000 relevant interactions characterize the MIIT program. The tests are accelerated by heating the assemblies to 90°C. The experiment is designed to assess waste form and package component behavior via solution and surface studies of leached components. The test is monitored at intervals of 6 months, 1 year, 2 years, and 5 years.

MIIT OBJECTIVES

The overall objectives of the WIPP/SRL MIIT in-situ program are as follows:[5]

A) To assess and compare the relative performance or behavior of Savannah River simulated waste glass compositions with the performance of other U.S. and foreign waste glasses, under the same anticipated "real-world" repository-relevant test conditions.
B) To contribute to mechanistic leaching studies aimed at understanding the behavior of waste glass(es) by coupling leachant solution analyses with surface-interactions studies of glass waste forms and other waste package components. This information can help in development of source terms and performance assessments that can be used to describe long-term

waste package materials behavior in salt.
C) To provide a standardized, in-situ data base on waste glass leaching including relevant interactions with metal canister or overpack components, salt, and brine. This in-situ data can then be correlated with other in-situ data and the large body of existing laboratory data. The data, results, and test experience are to be freely shared by all participants and interested parties.
D) To provide waste form data that will ultimately be useful in the technical data base to help demonstrate compliance with emerging U.S. federal regulations.

This paper focuses, in part, on the first objective as stated above. In this effort, the MIIT waste form compositions were grouped according to the structural role of components, relative oxide ratios and surface layer elements. The mean bonding energy was also used to assist in correlation of MIIT compositions.

MIIT WASTE FORMS

As indicated earlier, there are almost 1000 waste form samples comprising 15 systems derived from 7 countries. The MIIT waste forms, relevant information, and their U. S. and international affiliations are as follows:[5]

- United States Samples

1) Savannah River Laboratory, SRL 165/28% 'black frit' waste glass (reference DWPF composition).
2) Savannah River Laboratory, SRL 165/28%, chemically doped and 60Co irradiated waste glass.
3) Savannah River Laboratory, SRL 131/35% TDS waste glass.
4) Savannah River Laboratory, SRL 165/28%, prepared by SRL for Pacific Northwest Laboratory (PNL) studies.
5) Pacific Northwest Laboratory (PNL) 76-68 high-level waste glass (prepared at SRL).
6) Rockwell Hanford Operations, HWVP-HW39 DHLW glass.
7) Materials Characterization Center, ARM-1 standard reference glass.
8) Catholic University, CU developmental borosilicate waste glass.
9) Basalt, natural glass analogue (prepared at SRL).

- International Samples

10) Canada: AECL-GC-NACTS 33, sphene glass ceramic,

supplied by Atomic Energy of Canada, Ltd., Whiteshell Laboratory, Pinawa, Manitoba.
11) Canada: AECL-AS-981, aluminosilicate glass, supplied by Atomic Energy of Canada, Ltd., Whiteshell Laboratory, Pinawa, Manitoba.
12) France: SON 68-18-17-L1-C2-A2-Z2 (JSSA/AVH; R7T7 glass), as used in the Japanese, Swedish, and Swiss programs, supplied by Centre D'Etudes Nucleaires ge la Vallee du Rhone (CEN/VALRHO).
13) West Germany: SM 513LV11, Pamela glass, supplied by Hahn-Meitner Institut, Berlin.
14) Belgium: FLK TRUW; WG124 transuranic waste glass, supplied by Studiecentrum Voor Kernenergie/Centre D'Energie Nucleaire (SCK/CEN).
15) Belgium: AVB/SAN 602519 L3 C2 HLW glass, supplied by (SCK/CEN).
16) Japan: JAERI-Z-20/JW-A, supplied by Japan Atomic Energy Research Institute (JAERI).
17) Great Britain: BNFL WG, from British Nuclear Fuels Ltd. (BNFL).

CORRELATION TECHNIQUES AND RESULTS

The MIIT glasses represent diverse compositions with some systems having as many as 30 different components. In earlier studies, compositional correlations were found using ternary representations of other waste glass systems.[10,11] In the present study, the various components in the MIIT glasses were cataloged and grouped in a similar, though more detailed, manner. The correlation model is based on structural considerations, bonding strengths, and surface layer formation.[10,11] The glasses have been plotted on ternary and bar diagrams to examine the compositional correlations.

- Structural Roles

It is common to classify constituents as performing one of three roles in amorphous structures: network former, modifier, or intermediate.[12,13] Network formers readily assume vitreous structure and are the backbone of the random glass network. They are normally small, highly charged cations, covalently bonded to larger, less highly charged anions. Zachariasen's rules, which apply to most glasses, allow a cation to bond with 3 or more anions, limiting the anion to 2 or fewer cations. Silicon is a classic network former and the principal component in all of the MIIT glass compositions.[12,13]

In order to reduce the high melting temperatures and viscosities of silicate glasses, modifiers are introduced to the system. Alkali and alkaline earth oxides tend to act as modifying agents, breaking the network Si-O bonds. Many metals also act as modifiers in their lower valence states.[12,13]

Intermediates can have several effects on the nature of glass, eg. Ti^{+4}, can increase the durability and melting temperature, while Fe^{+3} may act as a flux (these effects are dependent on the amounts of constituents that are present). Several metals have multiple structural roles depending on their charge and coordination. Zr is either a network former or intermediate while Pb is both an intermediate and a modifier.[12]

The MIIT waste form compositions have been assembled based on the three basic structural roles of their constituent components. This allows comparison by summing the intermediates, modifiers, and network formers and diagraming their sums with ternary plots (Figures 1 and 2). Figure 2 shows the overall similarity of the high-level waste glasses: SRL 165/28%, Catholic U., PNL 76-68, RHO HWVP HW39, SON 68-18, HMI PAMELA, BNFL WG, AVB SON, and AECL-AS. These glasses all fall within a narrow structural role/composition range.The low silica compositions, JAERI Z-20 and SRL 131/35%, are high-level waste glasses not within the tight cluster, but still within the general area. The natural analogue, basalt, low-level TRUW WG 124, and glass ceramic AECL GC are also separate from the central grouping.

- Mean Bonding Strength

As stated previously, the structural role of an ion is normally related to its charge and coordination. A model proposed by Sun, defines the strength of a single bond between a cation and oxygen (M-O) as the dissociation energy divided by the coordination number of the cation. The single bond strength of a cation (MO_x basis) is 90 Kcal/mole or higher, for a network former, between 60 and 90 Kcal/mole for an intermediate, and 60 Kcal/mole or lower for a modificr.[12] Figurc 3 is a ternary plot of the MIIT compositions (MOx basis) and Figure 4 depicts the corresponding mean bonding strengths of the glasses. TRUW WG 124, basalt, and JAERI Z-20 (the only high-level borosilicate composition outside of the central grouping) are still removed from the majority of waste forms, as is AECL AS. Moreover, it is now possible to differentiate between the remaining borosilicates. It is interesting to note that the foreign borosilicate compositions have a slightly higher mean bonding energy (78-79 Kcal/mole) and cluster separately than the U. S. compositions (74-76 Kcal/mole). MCC ARM-1, the reference borosilicate, has similarities to both groups.[12]

It is likely the mean bonding energy value given for these compositions will be more relevant when comparing the durability of glasses of a similar type, for example, high-level borosilicate waste glasses. Likewise, the various surface layers that form during the leaching of glass may be highly indicative of the relative durability of the borosilicate compositions.

- Surface Layer Formation

Surface layer effects have been studied for several years. It is believed that these layers relate directly to the overall leachability of the glass. Several types of leached layers have been classified; the most important of which are the transition metal and alkaline earth enriched layers. The SRL 131/35%, SRL 165/28%, CATHOLIC U., and PNL 76-68 compositions each contain over twice the amount of transition metals versus alkaline earths, RHO HWVP HW39, MCC ARM-1, and SON 68-18 contain roughly equivalent amounts, while the other glasses have more alkaline earths, proportionately. This may be reflected in the chemistry of the surface layers that form in the respective glasses.[14-17]

Only one of the high-level waste glasses in the MIIT study is not a borosilicate, AECL-AS. Since the amount of SiO2 is fairly consistent among all MIIT systems, it is possible that classification of the borosilicate compositions based on the B2O3 content may be useful. Figure 5 shows the total percentage of oxides accounted for by surface layer and SiO2-R2O-B2O3 classifications in addition to the graphical relations.

An important effect on the performance of MIIT glasses could be the formation of surface layers which may affect important properties of the product. These layers could be a function of both initial structure of the waste glass and bonding energy considerations of glass components, as well as leaching test conditions. Consequently, comparisons and correlations are most valid when made within a class of waste forms, such as the borosilicate compositions.

CONCLUSIONS

A close correlation was observed among MIIT waste forms from 7 countries by relating components in waste form systems to structural roles in a glass matrix. These data suggest that there are important similarities between the many different systems studied world-wide, which help to tie together the behavior and subsequent performance of these many sytems.

FUTURE WORK

The MIIT tests, now more than 2 years old, will run for 5 years, with analyses continuing beyond that time. The data base evolved will be available to all interested parties. Analytical services provided by each of the participating nations help to achieve a cost effective, maximum benefit from the program. The analytical tools used include Wide-Angle X-ray Diffraction (WAXD), Scanning Electron Microscopy (SEM), Fourier Transform Infrared Reflection Spectroscopy (FTIRRS), Rutherford Back Scattering (RBS), Inductively Coupled Plasma Spectroscopy (ICP), Atomic Absorption, (AA), Electron Microprobe analysis (EMP), Transmission Electron Microscopy (TEM), and Secondary Ion-Mass Spectroscopy, (SIMS). Each tool provides information about species of interest in, on, and through leached glass surfaces. MIIT waste form data and analyses after 6 months, 1 year, and to a limited extent, 2 years of burial in WIPP were summarized at an international workshop in Cadarache, France during October 1988 and will be published in the proceedings.[18]

A main objective of the MIIT effort is not only to assess the effects of WIPP salt geology on high-level waste glass, but to also correlate the behavior of all participating compositions. The comparison of SRL waste glass in the various geologies (salt, limestone, clay, and granite) is a major consideration as well. Once these objectives are met, information can be extrapolated from waste glass systems worldwide, including those in actual production, and related to SR compositions. Furthermore, the performance of our systems can be defined in a wide range of geologies and for a larger variety of potential storage conditions. Utilization of the in-situ testing data in conjunction with concurrent laboratory studies will result in the comprehensive understanding of leaching behavior of high-level waste glass, useful in supporting the safe and effective disposal of potentially hazardous waste in a repository environment.

REFERENCES

1 G. G. Wicks, "WIPP/SRL In Situ and Laboratory Testing Programs-Part I: MIIT Overview, Nonradioactive Waste Glass Studies," USDOE Report DP-1706, Savannah River Laboratory, E. I. du Pont de Nemours & Co., Aiken, SC (1985).

2 G. G. Wicks, M. E. Weinle, and M. A. Molecke, "WIPP/SRL In-Situ Tests-Part II: Pictorial History of MIIT and Final MIIT Matrices, Assemblies, and Sample Listings," USDOE Report DP-1733, Savannah River Laboratory, E. I. du Pont de Nemours & Co., Aiken, SC (1987).

3 M. A. Molecke and G. G. Wicks, "Test Plan: WIPP Materials Interface Interactions Test [MIIT]," Sandia National Laboratories, Albuquerque, NM (1986).

4 G. G. Wicks and M. A. Molecke, "WIPP/SRL In-Situ Testing Program," Nuclear Waste Management II, Advances in Ceramics, American Ceramic Society, D. E. Clark, W. B. White, and A. J. Machiels, Eds., Vol. 20, pp. 657-667 (1986).

5 G. G. Wicks and M. A. Molecke, "WIPP/SRL In-Situ Testing-MIIT Update," Waste Management '88, Vol. II, pp. 383-392 (1988).

6 G. G. Wicks, W. D. Rankin, K. A. Molen, and W. G. Ramsey, "Compositional Correlations of MIIT Waste Glasses," to be presented at the Workshop on Testing of HLW Forms Under Repository Conditions, Cadarache France, October 1988, to be published in proceedings.

7 W. G. Ramsey and G. G. Wicks, "WIPP/SRL In-Situ Tests-Part III: Compositional Correlations of MIIT

Waste Glasses," USDOE Report DP-1769, Savannah River Laboratory, E. I. du Pont de Nemours & Co., Aiken, SC (1988).
8 G. G. Wicks, "Nuclear Waste Glasses," in Treatise on Materials Science and Technology, Glass IV, M. Tomozawa and R. H. Doremus, Eds., Academic Press, Vol. 26, pp. 57-118 (1985).
9 G. G. Wicks, "Nuclear Waste Vitrification - The Geology Connection," J. Non-Crystalline Solids, Vol. 84, pp. 241-250 (1986).
10 A. A. Hench and L. L. Hench, "Computer Analysis of Compositional Effects in Nuclear Waste Glass Leaching," Nuclear and Chemical Waste Management, Vol. 4, pp. 231-238 (1983).
11 G. G. Wicks, W. D. Rankin, and S. L. Gore, "International Waste Glass Study - Composition and Leachability Correlations," in Scientific Basis for Nuclear Waste Management VIII, Materials Research Society, C. M. Jantzen, J. A. Stone, and R. C. Ewing, Eds., Vol. 44, pp. 171-177 (1985).
12 W. D. Kingery, H. K. Bowen, and D. R. Uhlmann, Introduction To Ceramics, John Wiley and Sons, pp. 95-100 (1976).
13 G. G. Wicks, "Structure of Glasses," Encyclopedia on Materials Science and Engineering, M. B. Bever, Ed., Vol. 3, pp. 2020-2025, Pergamon Press Ltd. (1986).
14 G. G. Wicks, W. C. Mosely, P. G. Whitkop, and K. A. Saturday, "Durability of Simulated Waste Glass-Effects of Pressure and Formation of Surface Layers," Journal of Non-Crystalline Solids, Vol. 49, pp. 413-428 (1982).
15 D. E. Clark, B. F. Zhu, R. S. Robinson, and G. G. Wicks, "Preliminary Report on a Glass Burial Experiment in Granite," Nuclear Waste Management I, Advances in Ceramics, G. G. Wicks and W. A. Ross, Eds., Vol. 8, pp. 324-336 (1984).
16 B. F. Zhu, D. E. Clark, A. R. Lodding, and G. G. Wicks, "Two Year Leaching Behavior of Three SRL Nuclear Waste Glasses in Granite," Nuclear Waste Management II, Advances in Ceramics, D. E. Clark, W. B. White, and A. J. Machiels, Eds., Vol. 20, pp. 591-599 (1986).
17 A. R. Lodding, E. V. Engstrom, D. E. Clark, L. O. Werme, and G. G. Wicks, "SIMS Analysis of Leached Layers Formed on SRL Glasses During Burial," Nuclear Waste Management II, Advances in Ceramics, D. E. Clark, W. B. White, and A. J. Machiels, Eds., Vol. 20, pp. 567-581 (1986).
18 Proceedings of the Workshop on Testing of High

Level Waste Forms Under Repository Conditions, October 17-21, 1988, Commission of the European Communities (CEC), EUR 12 017 EN, R. Simon, C. Sombret and G. Wicks, Eds., to be published (1989).

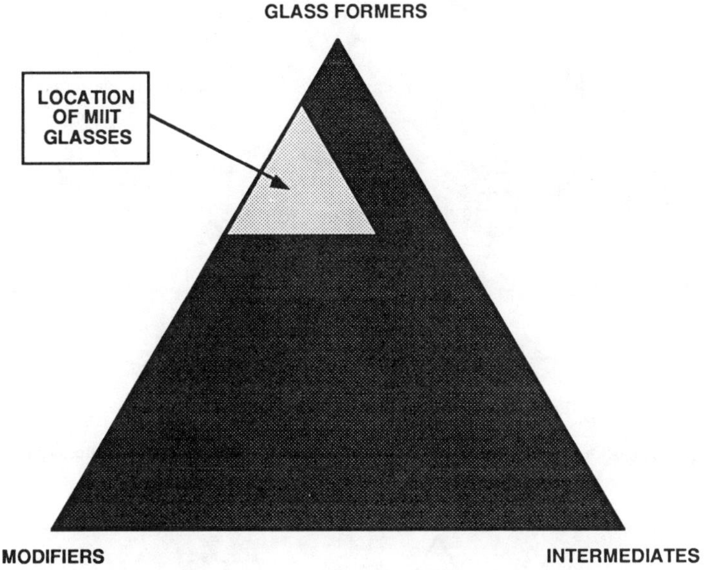

Figure 1. Ternary diagram depicting the compositional region of the MIIT glasses with respect to structural role.

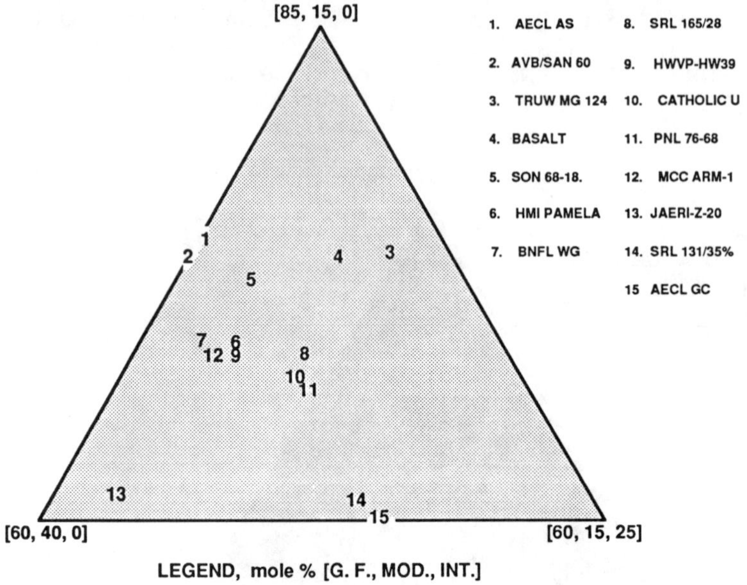

Figure 2. Expanded ternary diagram detailing the MIIT waste glass compositions with respect to stuctural role.

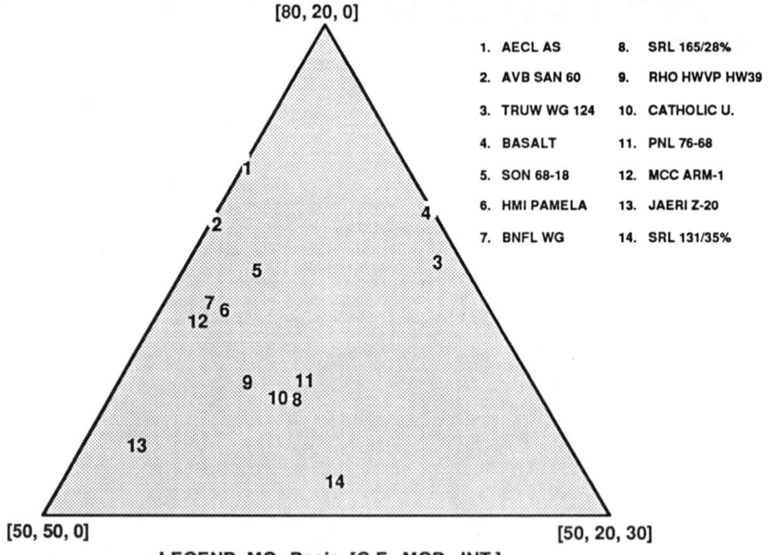

Figure 3. Expanded ternary diagram detailing the MIIT waste glass compositions on an MOx basis..

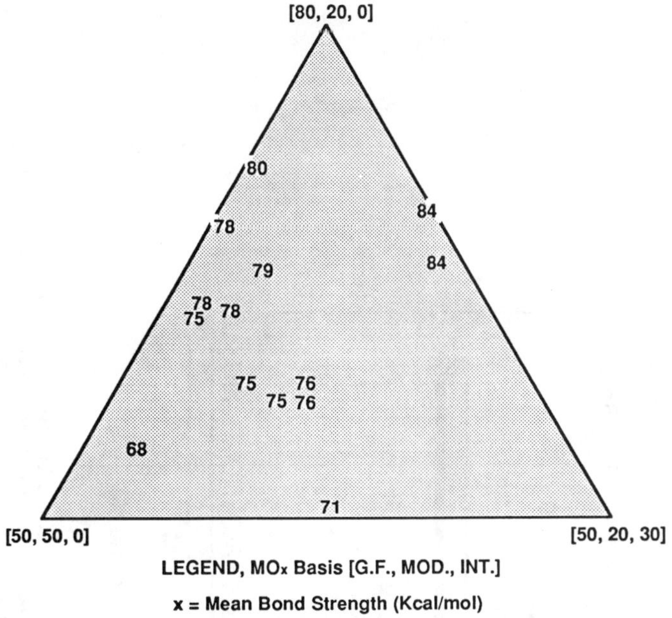

Figure 4. Expanded ternary diagram illustrating the mean bonding strengths of the glasses depicted in Figure 3.

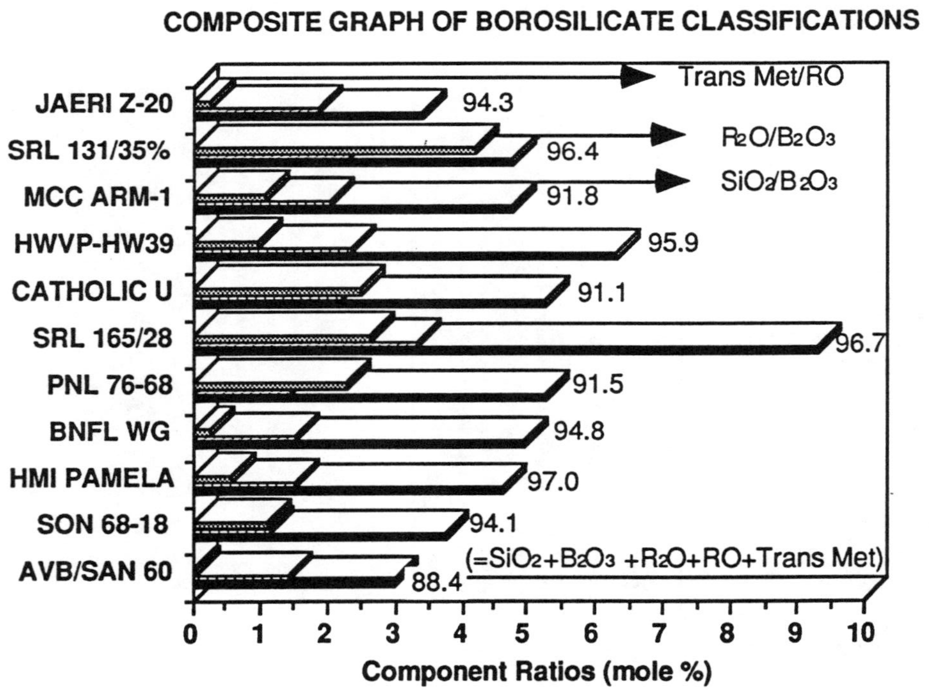

Figure 5. Is a stacked bar graph which shows the total percentage of oxides accounted for by surface layer and SiO_2-R_2O-B_2O_3 classifications. The top level of each stack represents the transition metal and alkaline earth enriched layers, the middle level represents SiO_2 with respect to B_2O_3, and the bottom level of the stack represents R_2O with respect to B_2O_3.

WIPP/SRL IN-SITU TESTS: MIIT PROGRAM-SURFACE STUDIES OF SRL WASTE GLASSES

J. A. Tacca
Department of Ceramic Engineering
Clemson University
Clemson, SC 29634

G. G. Wicks
Westinghouse Savannah River Company
Savannah River Laboratory
Aiken, SC 29808

ABSTRACT

The first in-situ tests involving burial of simulated, non-radioactive waste glasses and package components in the United States was begun in 1986. The effort involves almost 2000 specimens from seven nations buried in the salt site at the Waste Isolation Pilot Plant (WIPP) in Carlsbad, New Mexico. Among these samples are non-radioactive Savannah River Laboratory (SRL) waste glasses which have been buried for 6-mo., 1-yr. and 2-yrs., as part of a 5-yr. study. Scanning electron microscopy (SEM), energy dispersion x-rays (EDX), and wide angle x-ray diffraction (WAXD) were used to assess the behavior of these glasses and will be presented.

INTRODUCTION

Approximately one-half of the 2000 waste form and package component samples buried in the bedded salt site at WIPP are simulated waste glasses, comprising 15 different systems supplied by seven countries. This program, called the Materials Interface Interactions Tests (MIIT), is the largest, most cooperative effort of this type underway[1-6]. Among the primary interactions studied are glass/salt, glass/glass, and glass/metal interactions. The glass, fabricated into 'pineapple slices', is polished on one side and stacked on heater rods, often with other package components, into one of seven types of assembly configurations. There are a total of 50 assemblies that are inserted into boreholes in WIPP. Brine is added to the boreholes and the tests are conducted at 90°C. Aliquots of solution are withdrawn and analyzed at various time intervals and samples exhumed at 6 months, 1 year, 2 years, and 5 years. As of this date, only the 5 year samples remain in test.

The objective of the current study is to characterize and assess glass/salt and glass/glass interfaces, involving interactions of SRL Y (165/TDS) waste glass with brine. These studies were performed as a function of sample orientation, interfaces of interest, and leaching time. To accomplish these tasks, scanning electron microscopy (SEM), energy dispersion x-ray (EDX), and wide angle x-ray diffraction (WAXD) were used. These data will ultimately be coupled with other studies using different analytical tools to achieve as complete a picture as possible of the performance of SRL waste glass in WIPP[7].

SUMMARY

- Two distinct regions on surfaces of leached SRL waste glass were identified: an outermost *precipitated layer* and an *interaction zone* underneath.
- The *precipitated layer*, approximately 1-3 microns after two years in test, was generally enriched in Mg, Cl, Ca, Na, S and Si. The *interaction zone* is also about 1-3 microns after two years in test and was characterized by major glass and waste components, Si, Na, Al and Fe along with major constituents from the brine, Mg and Cl.
- The total reacted depth (both layers) after 6 months was 1-2 microns, while after 1 and 2 years the depth was 3-6 microns. These data suggest that long term leaching tests of 1 year or greater are most appropriate to assess field leaching behavior of waste glass systems.
- Glass/salt interactions (4-6 microns) appear to be slightly more extensive than glass/glass interactions (2-4 microns) and orientation of the glass, up versus down, appears to have little affect on glass leaching

behavior.

• Layers formed on surfaces of leached waste glasses in WIPP are different than leached layers formed in tests with other geologic media.

EXPERIMENTAL PROCEDURES

Five SRL Y waste glass samples were studied, all from MIIT assembly #7, involving glass-brine interactions. The samples were designated Y001, Y004, Y007, Y008, and Y009. The actual orientation and time intervals in which the samples were buried in WIPP is depicted schematically in Figure 1[2]. After the samples were exhumed from the ground, the pineapple slices were cut into sections for various analyses, as shown in Figure 2. For SRL Y glasses, Fourier transform infrared reflection spectroscopy (FTIRRS) is being performed by Prof. Clark at University of Florida, secondary ion mass spectroscopy (SIMS) analysis by Prof. Lodding at Chalmers University in Sweden, brine analyses by Prof. Macedo of Catholic University of America and SEM, EDX and WAXD at SRL. Additional analyses of SRL glasses and salt precipitates are being performed by Prof. Rodney Ewing of University of New Mexico, Dr. Alan Harker of Rockwell International and Dr. Etienne Vernaz of the CEA in France. Figure 3 summarizes the variety of analytical tools used in the MIIT program for characterizing constituents of interest in and throughout leached glass surfaces[8].

• SEM/EDX (scanning electron microscope/energy dispersion x-rays):

- *Sample Preparation:* Initial sample preparation involved mounting samples in epoxy and grinding and polishing surfaces to a one micron finish. This is a technique often used to study cross-sections of leached glasses. Several different epoxies were considered but an inherent difficulty was noted. The mounting technique adversely affected the relatively friable outer precipitated salt layer. When the sample was mounted in epoxy, the precipitated layer often lifted away from the glass surface as the epoxy dried and shrunk. In addition, the water used in the grinding and polishing operations dissolved salt-enriched phases on the glass surfaces.

A simpler procedure was used. Post-test glasses were

cut into desired sizes by scribing one side (opposite the side to be observed) and breaking them. The samples were then carbon coated to minimize charging effects and finally, directly inserted in the SEM and characterized. This provided an easy, reproducible method to prepare samples and resulted in a minimum of surface damage.

- *Analysis:* A three part approach was used to study leached layers and precipitated phases of SRL Y waste glasses on the SEM. This included (a) characterizing the morphology of surfaces and cross-sections using SEM photographs (b) determining major elements present using EDX elemental profiling and (c) defining elemental distributions by EDX elemental mapping.

• WAXD (wide angle x-ray diffraction):

- *Sample Preparation:* For this analysis, a small piece of glass was cut from a pineapple' slice and directly mounted into a wide angle x-ray diffractometer.

- *Analysis:* Scans were performed using CuK_a radiation. The major elements present found with EDX were used in a computer search of observed WAXD peaks to determine possible phases. The phases found by the computer were then checked manually.

RESULTS

• Surface:

On the surfaces of each of the waste glass samples, an inhomogeneous precipitated layer was noted and characterized. In Figures 4A and 4B, discontinuous layers formed on SRL Y glasses after 6 months and 2 years of burial are shown. In Figure 4C, corresponding elemental mapping is given for the precipitated phases formed after six months, showing the inhomogeneous distribution of Si, Mg, and Cl . This outermost layer was generally enriched in Mg, Cl, Ca, Na, and S and consisted primarily of salt phases. Most of these components are from the brine or surrounding salt while silicon may also be derived from the glass. Data from WAXD [Figure 5] in combination with EDX indicate that major precipitated phases include sodium chloride (NaCl) and magnesium chloride ($MgCl_2$). Other possible phases, consistent with the observed spectrum, include potassium magnesium silicate ($K_2Mg_5Si_{12}O_{30}$), calcium sulfate ($CaSO_4$), calcium aluminum silicate hydrate ($CaAl_2Si_2O_8 \cdot 4H_2O$), and potassium calcium silicate ($K_4Ca(SiO_3)_3$). Glass/salt interfaces generally had larger amounts of these salt deposits on surfaces than glass/glass interfaces.

• Cross-Section:
Two distinct regions on the surface of leached SRL waste glass were observed and denoted as the precipitated layer (outermost) and interaction zone (under the precipitated layer).

Elemental analysis of the outermost, precipitated layer in the cross-section was similar to that observed from the surface study described above. One additional feature noted was the presence of Si, which was most enriched adjacent to the glass surface and sometimes, but not always seen in the outermost region of the precipitated layer [Figures 4C and 7].

Analyses of the interaction zone showed different morphological and chemical differences compared to the precipitated layer [Figure 7]. There were no apparent phases, such as salt precipitates seen, but instead, a usual glass morphology was noted. Principal elements present included major components of the glass and waste such as Si, Na, Al and Fe along with major elements of the brine, Mg and Cl. It was interesting to note that as one moved into the bulk glass, the amounts of Mg and Cl decreased continuously until they were not detectable with the SEM. The distance from the glass surface to where the Mg and Cl were no longer detected was taken to be the overall depth of the interaction zone and the beginning of the unaltered or pristine bulk glass.

The two observed layers were about the same thickness and increased with time [Figures 6 and 7]. For the glass/salt interactions, after six months (Y001), each of these layers was less than 1 micron, for a total thickness of 1-2 microns. After one year (Y004), the precipitated layer and interaction zone were 1-2 and 2-3 microns, respectively, for a combined thickness of 3-5 microns. After two years (Y007 and Y009), each investigated layer was 2-3 microns, for a combined thickness of 4-6 microns. Measurements were taken randomly in a radial direction on the pineapple slice.

It was also noted that the total interaction depths for the glass/salt interfaces (4-6 microns for Y007 and Y009) were slightly more extensive than the glass/glass interface (2-4 microns for Y008). The effects of orientation was seen to be insignificant. For Y007 (orientation up) and Y009 (orientation down), each had precipitated layers of of 2-3 microns and interaction zones also of 2-3 microns, respectively. A schematic representation of the leached layers formed on SRL Y waste glass and measured layer thicknesses are summarized in Figure 6.

CONCLUSIONS

Interactions observed between brine and SRL Y waste glass produced leached layers different than observed in other in-situ tests involving other geologic media[9-11]. Among the important constituents characterizing glass leaching, leached layers, and precipitates formed, are Mg and Cl from the brine and salt surroundings. The formation of the dual layers is attributed to precipitation of salt phases onto the glass surfaces and to subsequent interaction of brine and surrounding salt with the glass underneath.

Time had a significant effect on the leached layers produced. From analysis of leached cross-sections, the total interaction region increased nonlinearly with time. Samples left in the ground for 6 months possessed the shallowest interaction depth while the 1 year sample had a total interaction depth roughly similar to the 2 year samples. In addition, the morphology of 1 and 2 year glass/salt samples appeared to be more developed and better defined than the 6 month glass/salt sample. This time dependence suggests that long term leaching tests (~1 year or greater) are most appropriate for studying leaching behavior of waste glass systems in environments such as WIPP.

Based on surface and cross-section data, interactions for glass/salt interfaces appeared more extensive than for glass/glass interfaces. In addition, sample orientation did not produce any significant effect on the leaching behavior of the glasses.

Even after two years of leaching, based on surface studies and complimentary solution analysis[7], the leaching of SRL Y glass is relatively slow and improves with time.

REFERENCES

1 G. G. Wicks, "WIPP/SRL In-Situ and Laboratory Testing Programs-Part I: MIIT Overview, Nonradioactive Waste Glass Studies", USDOE Report DP-1706, Savannah River Laboratory, E. I. du Pont de Nemours & Co., Aiken, SC (1985).

2 G. G. Wicks, M. E. Weinle, and M. A. Molecke, "WIPP/SRL In-Situ Tests-Part II: Pictorial History of MIIT and Final MIIT Matrices, Assemblies, and Sample Listings", USDOE Report DP1733, Savannah River Laboratory, E. I. du Pont de Nemours & Co., Aiken, SC (1987).

3 W. G. Ramsey and G. G. Wicks, "WIPP/SRL In-Situ Tests-Part III: Compositional Correlations of MIIT Waste Glasses", USDOE Report DP1769, Savannah River Laboratory, E. I. du Pont de Nemours & Co., Aiken, SC (1988).

4 M. A. Molecke and G. G. Wicks, "Test Plan: WIPP Materials Interface Interactions Test [MIIT]", Sandia National Laboratories, Albuquerque, NM (1986).

5 G. G. Wicks and M. A. Molecke, "WIPP/SRL In-Situ Testing Program", Nuclear Waste Management II, Advances in Ceramics, American Ceramic Society, D. E. Clark, W. B. White, and A. J. Machiels, eds., Vol. 20, pp. 657-667 (1986).

6 G. G. Wicks and M. A. Molecke, "WIPP/SRL In-Situ Testing-MIIT Update," Waste Management '88, Vol. II, pp 383-392 (1988).

7 Proceedings of the Workshop on Testing of High Level Waste Forms Under Repository Conditions, October 17-21, 1988, Commission of the European Communities (CEC), EUR 12 017 EN, R. Simon, C. Sombret and G. Wicks, eds., to be published (1989).

8 G. G. Wicks, "Nuclear Waste Glasses," in Treatise on Materials Science and Technology, Glass IV, M. Tomozawa and R. H. Doremus, eds., Academic Press, Vol. 26, pp. 57-118 (1985).

9 G. G. Wicks, "Nuclear Waste Vitrification - The Geology Connection," J. Non-Crystalline Solids, Vol. 84, pp. 241-250 (1986).

10 G. G. Wicks, W. C. Mosely, P. G. Whitkop, and K. A. Saturday. "Durability of Simulated Waste Glass-Effects of Pressure and Formation of Surface Layers", J.

Non-Crystalline Solids, Vol. 49, pp. 413-428 (1982).
11 A. R. Lodding, E. V. Engstrom, D. E. Clark, L. O. Werme, and G. G. Wicks, "Two Year Leaching Behavior of Three SRL Nuclear Waste Glasses in Granite," Nuclear Waste Management II, Advances in Ceramics, D. E. Clark, W. B. White, and A. J. Machiels, eds., Vol. 20 (1986).

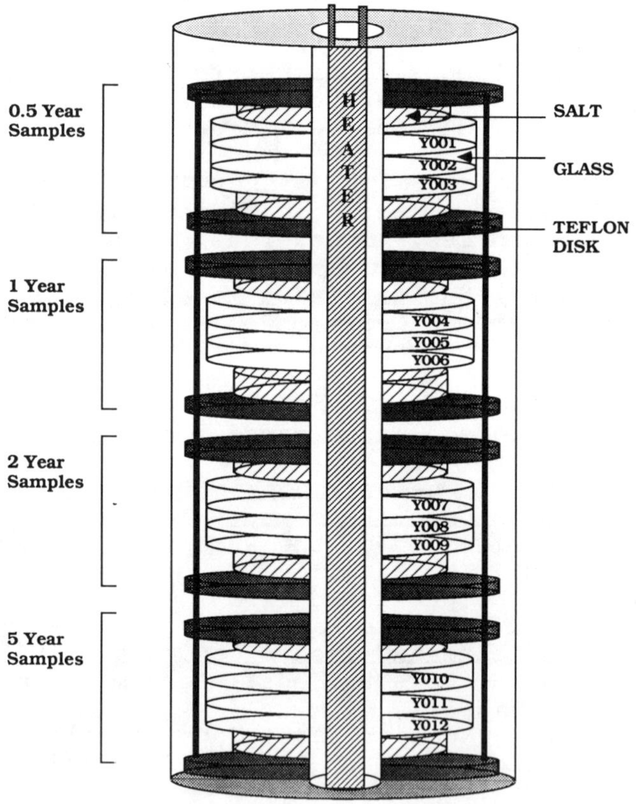

Figure 1

WIPP MIIT Sample Assembly #7

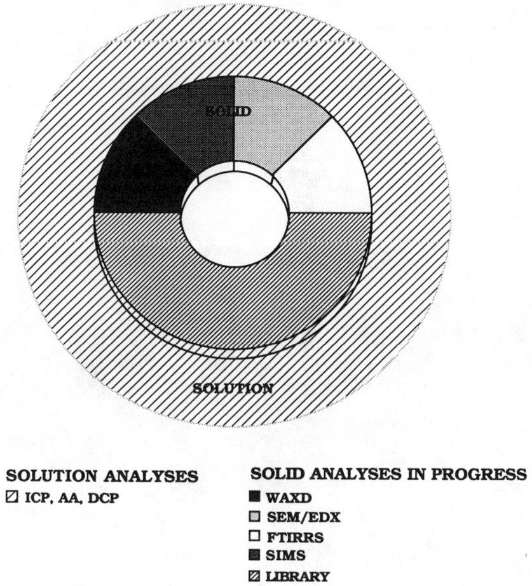

Figure 2

SRL Y Glass 'Pineapple Slice'

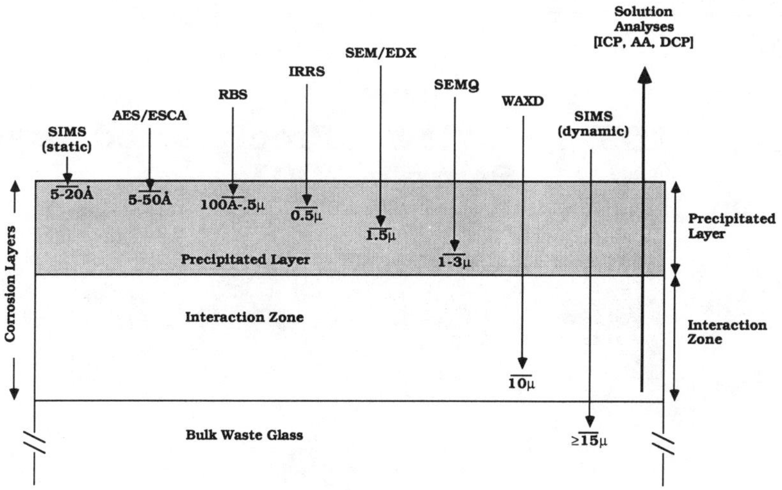

Figure 3

Surface Analyses of SRL Y Leached Glass Surface Layers

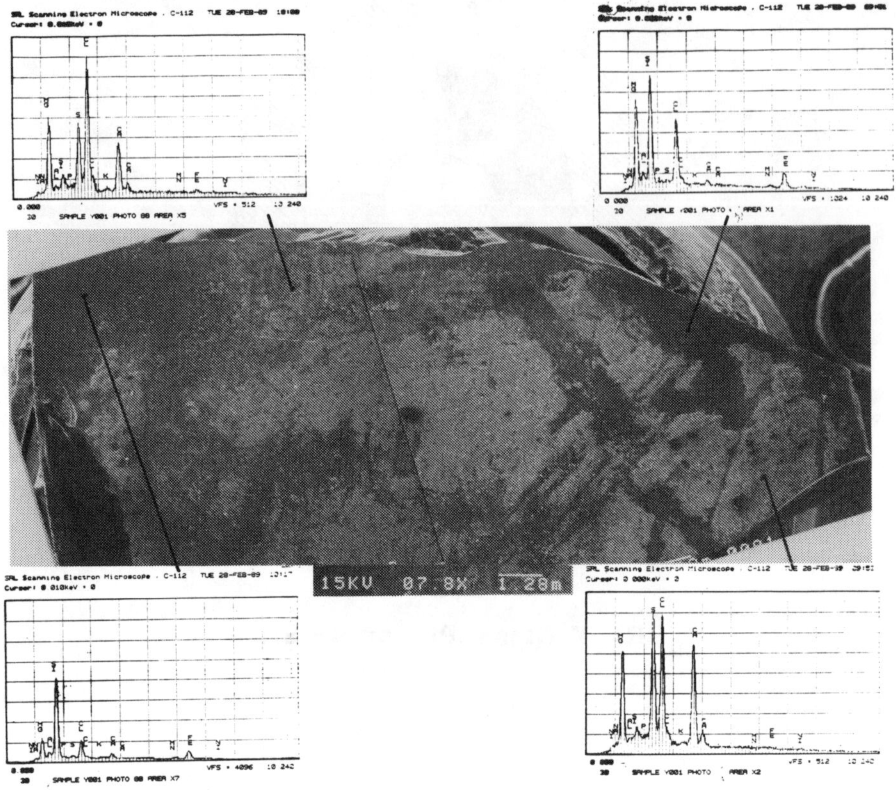

Figure 4A

SEM/EDX Top View - Precipitated Layer Sample Y001

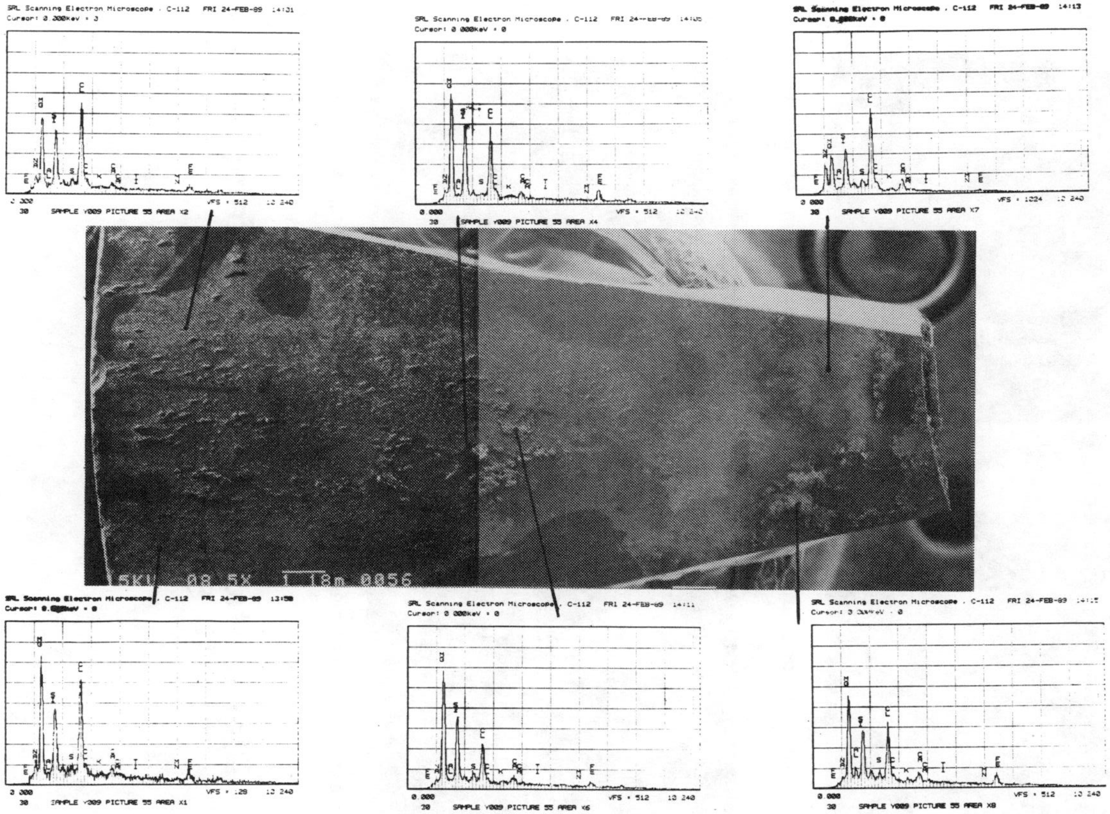

Figure 4B

SEM/EDX Top View - Precipitated Layer
Sample Y009

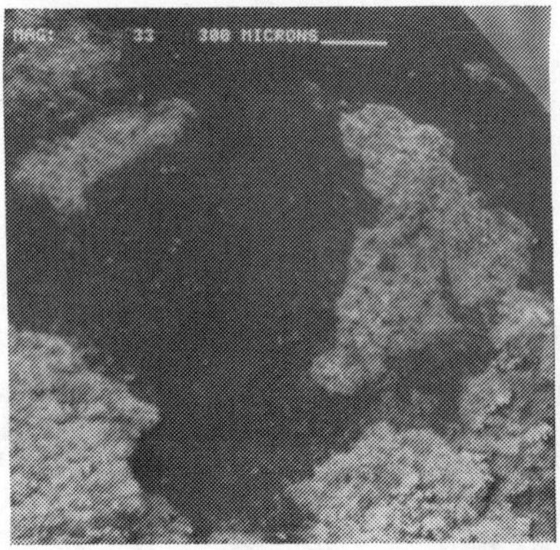

1. Picture of area under analysis

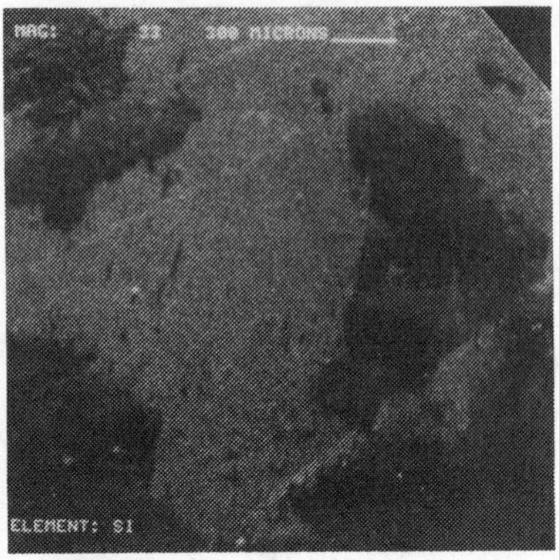

2. Element mapping of silicon

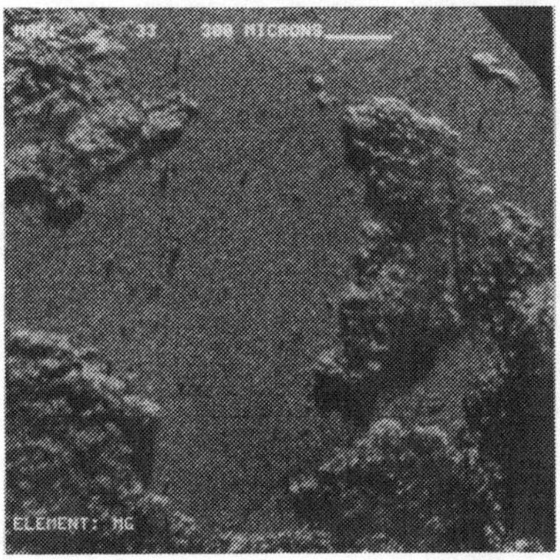

3. Element mapping of magnesium

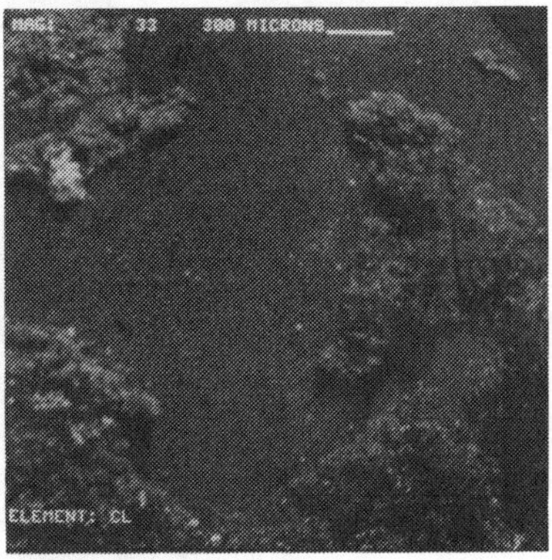

4. Element mapping of chlorine

Figure 4C

Elemental Mapping of Surface Sample Y001

Figure 5

WAXD of Leached SRL Y Glass Surface

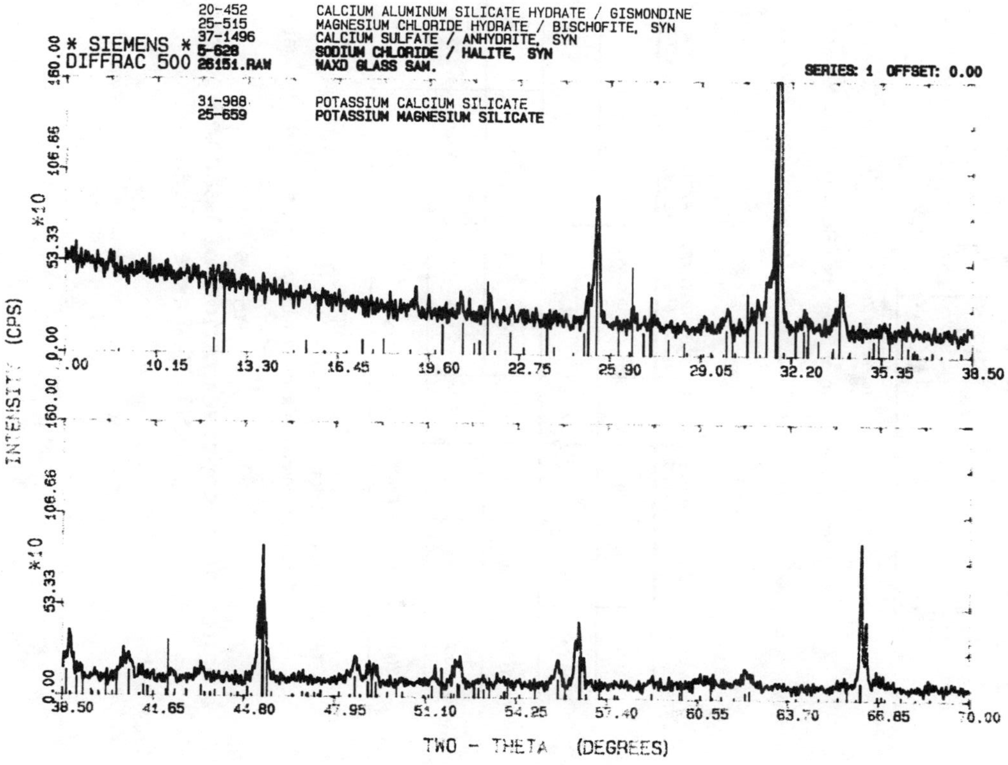

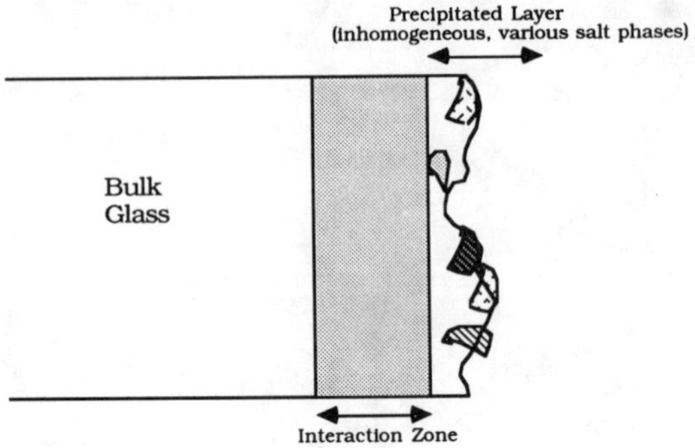

A. Schematic Representation of 2 Leached Regions of SRL Y Waste Glass

SAMPLE	TIME	ORIEN-TATION	INTER-FACE	PRECIPITATED LAYER (μm)	INTERACTION ZONE (μm)	TOTAL DEPTH OF INTERACTION (μm)
Y001	6 mos.	Up	G/S	<1 microns	<1 microns	1-2 microns
Y004	1 yr.	Up	G/S	1-2	2-3	3-5
Y007	2 yrs.	Up	G/S	2-3	2-3	4-6
Y008	2 yrs.	Up	G/G	1-2	1-2	2-4
Y009	2 yrs.	Down	G/S	2-3	2-3	4-6

B. Observed Region Thicknesses
(G/S denotes glass/salt interface, G/G denotes glass/glass interface)

Figure 6
Schematic Representation of Leached Regions of SRL Y Waste Glass

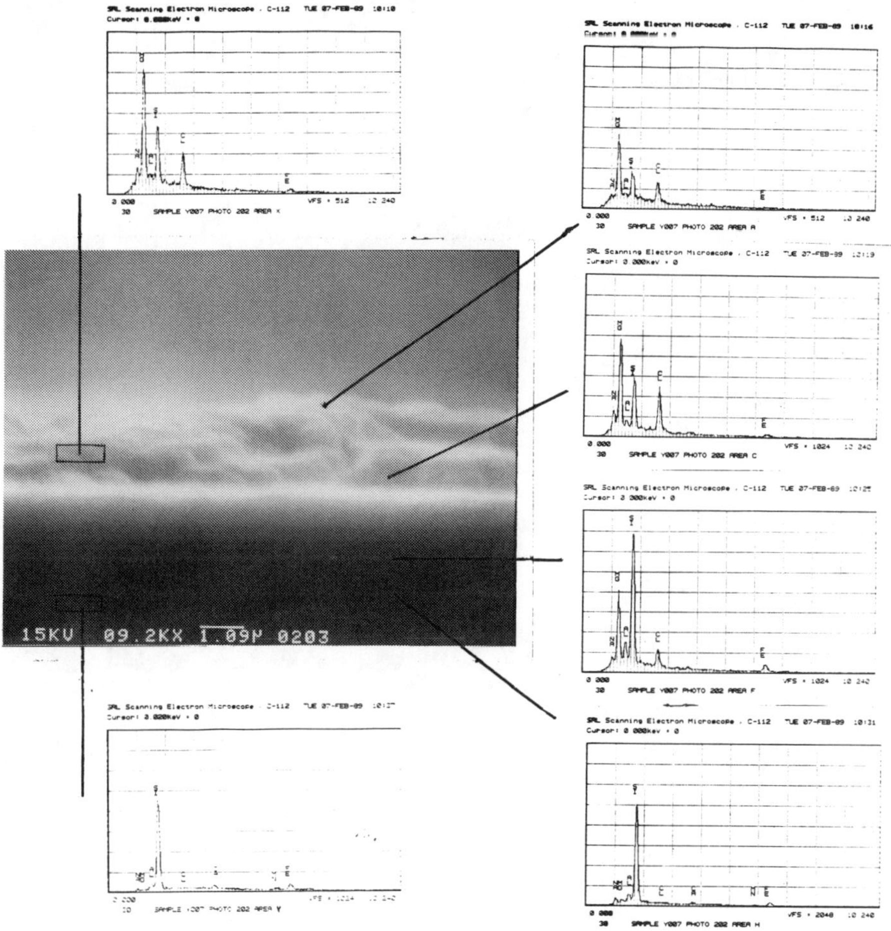

Figure 7

SEM/EDX Cross Section - Precipitated Layer Sample Y007

ANALYSIS OF BRINE LEACHATES FROM MATERIALS INTERFACE INTERACTIONS TESTS
2. Leaching of Lithium and Zirconium from Nuclear Waste Glass

Marek Brandys, Meiling Gong+, Richard E. Sassoon, Aaron Barkatt and Pedro B. Macedo

Vitreous State Laboratory
The Catholic University of America
Washington, DC 20064

ABSTRACT

Analysis of brine leachate samples from borehole #49, which contains SRL-Y glass, and from reference borehole #53, which contains no glass, was carried out for Li and Zr using the Inductively Coupled Plasma-Mass Spectrometry (ICP-MS) technique. Determination of the isotope ratio of Li-6 to Li-7 led to inconclusive results regarding the leaching of lithium from the nuclear waste glass. Accurate analysis of Zr, however, led to an upper limit for the leach rate based on this element from the glass of less than 1.6×10^{-4} gm^{-2} d^{-1}. This result together with surface data obtained elsewhere using the Secondary Ion Mass Spectrometry (SIMS) technique are discussed in order to clarify the uncertainty of the estimate of the position of the original glass-solution interface and to reach the conclusion that all of the suggested precipitated layer, the α-layer, is deposited. A leach rate for Li of 0.36μm yr^{-1} could then be estimated based on the SIMS data.

INTRODUCTION

Analysis of brine leachates from the Materials Interface Interactions Test (MIIT) program is very important in understanding the leaching process in the boreholes. In

+ In partial fulfillment of Ph.D requirement

this article we wish to describe in detail the leaching of a particular element, namely zirconium, from a nuclear waste glass, SRL-Y, into a brine leachant and to relate its behavior to the results of studies made on the surface of the glass at the same time period carried out by Lodding and co-workers[1] using the Secondary Ion Mass Spectrometry (SIMS) technique. From these measurements we wish to estimate the position of the original glass solution interface and the leach rate of Li from the glass.

EXPERIMENTAL

The Inductively Coupled Plasma - Mass Spectrometry (ICP-MS) technique described in the previous article of this series[2] was used to analyze the leachates. Samples were taken from MIIT boreholes #'s 49 and 53 for a period of 12 months beginning 12 months after the initial test turn-on, when 6 of the 12 SRL-Y glass slices had already been removed from borehole #49. In borehole #49 the "pineapple-shaped" glass slices were separated by small spacers which allow for a very accurate determination of the surface area of the glass and this consequently leads to a more accurate calculation of leach rates. Borehole #53 is a reference borehole in which no glass assembly is present.

RESULTS AND DISCUSSION

The isotope ratio method was attempted in order to determine the leaching of lithium from the glass. Since the ICP-MS technique can analyze for individual isotopes, the source of lithium may be determined if the isotopic ratio of Li-7 to Li-6 is different in the glass than in the leachant. As leaching proceeds the isotope ratio of lithium in the leachate will tend towards the value in the glass. This technique was discussed in some detail in a previous communication[3]. In this investigation, Li-7 to Li-6 isotope ratios were determined for all the samples from the sampling period beginning 12 months after initial test turn-on from boreholes #'s 49, and 53. The Li-7 to Li-6 ratio was found to fluctuate in the range 14.2 to 14.8 for borehole #53 while in borehole #49 which contained the SRL-Y glass, a slight decrease from 15.5 at day 65 to 14.8 at day 275 of the sampling period could be observed. The Li-7 to Li-6 ratio for SRL-Y glass was determined by dissolving the glass in an HF/HNO_3 matrix to be 14.6. Interpretation of this data is made complicated since again it appears that other processes such as ion exchange with the rock salt walls of the borehole or leakage of groundwater or brine into or out of the borehole, are also involved in the lithium concentration profile as is evident from the behavior of the reference borehole.

Very accurate analysis of the zirconium leached from the glass samples was also performed on the ICP-MS instrument and the results are presented in Table I.

Table I. Analysis of Zr by ICP-MS

Borehole/Sample	Dilution[a]	Response in Area Counts Per Second (ACPS)	
		Zr-90	Zr-91
1% HNO_3	1	28 ± 3.6	14 ± 0.6
Brine A	100	24 ± 6.0	14 ± 0.9
MIIT #53-day 245	100	32 ±10.0	16 ± 2.7
MIIT #53-day 275	100	22 ± 1.9	13 ± 1.3
MIIT #53-day 275 + 200 ppt Zr	100	409 ± 24	99.0 ± 5.2
MIIT #49-day 245	100	19 ± 2.0	13 ± 2.1
MIIT #49-day 275	100	20 ± 2.1	12 ± 1.0

[a]Dilution factor required for analysis by ICP-MS

It may immediately be noted on comparing the data from borehole #49, which contained SRL-Y glass, to analogous data from borehole #53, that negligible leaching of zirconium from the glass took place over the entire period studied. From the data of borehole #53 at day 275 with and without 200 parts per trillion (ppt) of standard zirconium added it may be calculated that 1.93 ± 0.12 area counts per second corresponds to 1 ppt Zr based on measurements of Zr-90. An upper limit of the leach rate may also be determined based on 3 standard deviations of a blank sample matrix, namely Brine A. This corresponds to a concentration of 9.3 ppt in the sample solution which after correcting for the dilution factor of 100, would yield a concentration of 0.93 parts per billion in the original brine. Based on the leach rate calculations given in the previous article[2] an upper limit of the leach rate of 1.6 x 10^{-4} g m^{-2} d^{-1} corresponding to a layer of glass of about 0.02 μm depth leached per year. This depth of glass is calculated assuming that all the leaching takes place just over this thickness and not within deeper layers of the glass.

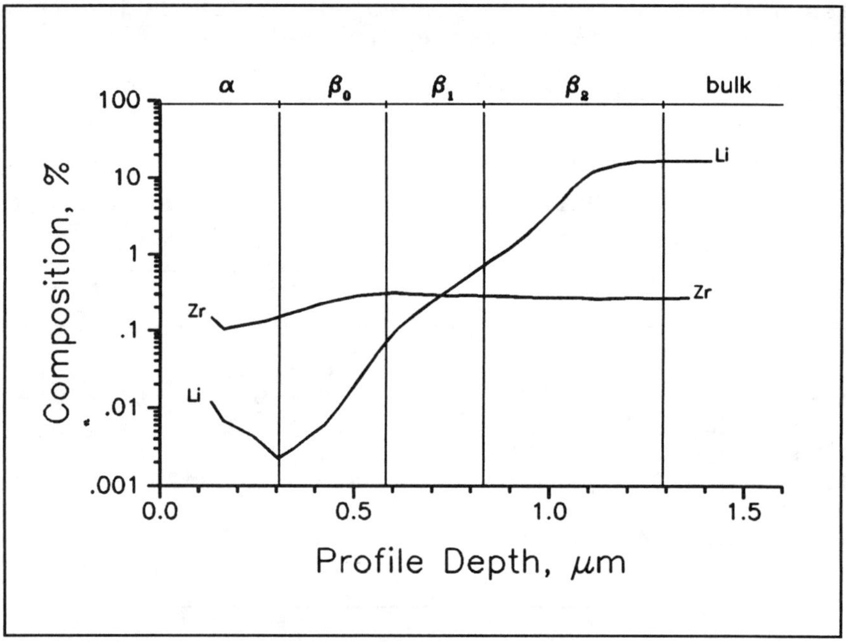

Figure 1. Depth profiles of Li and Zr.

According to the work of Lodding et al.[1] the surface of the glass may be described as in Figure 1 in which the exterior consists of the α-layer, which is assumed to be deposited on the glass during the test, and 3 β layers which represent layers within the surface of the glass where concentration changes have occurred. These 3 layers are labelled β_0, β_1, and β_2, and are termed the interaction or substitution layer, the gradient layer and the diffusion layer respectively. It is clearly seen that the lithium begins to be depleted already in the most interior of these layers, namely the β_2 layer, while an element such as Zr is only leached out strongly in the β_0 layer.

The data obtained in this work, together with the surface studies data, now provide a clearer picture of the situation at the glass-solution interface. This may be represented as shown in Figure 2 in which the percentage zirconium content of the glass normalized to constant

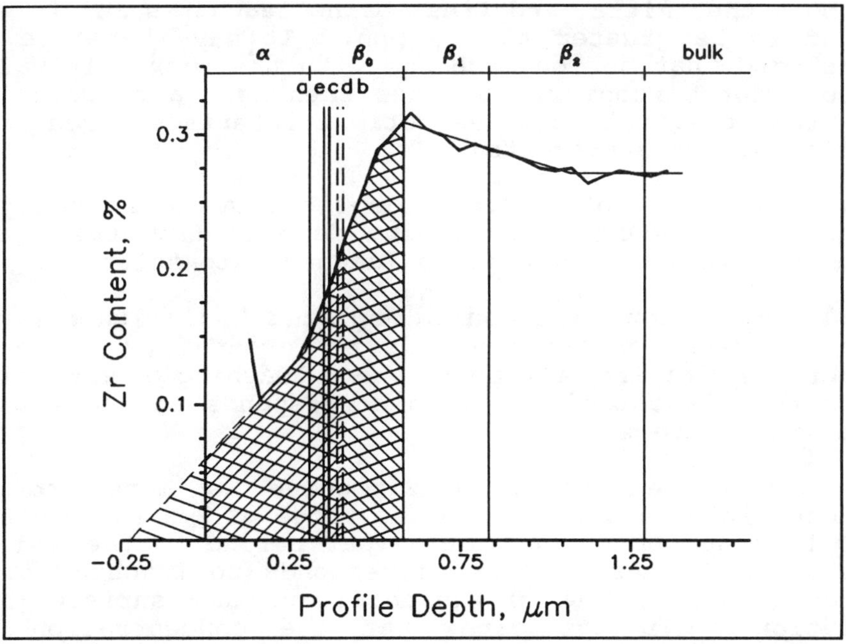

Figure 2. Depth profile of Zr normalized to constant Al content. See Text for meaning of lines and areas.

aluminum content throughout the glass is plotted against depth from the surface found at the end of the 12 month test.

Line "a" of Figure 2 represents the α - β_0 interface obtained from the SIMS data and according to the estimate of Lodding et al.[1] this corresponds to the original surface of the glass. We may also make an estimate of the position of the original glass surface from both the SIMS data shown in Figures 1 and 2 and from our leachate analysis data.

One must first consider what is the ultimate fate of zirconium which leached from the glass. Some of the zirconium may be deposited on to the surface of the glass while our data shows that there is less than 1 ppb of zirconium in the brine leachate at the end of the 12 month test. We also checked the solubility of zirconium in both the initial and final brine leachates and it was found to be greater than 1 ppm. It may therefore be concluded that no ion exchange with the rock salt walls occurs for zirconium. One can then make a calculation of the original glass-solution interface from the following two statements:

(a) the amount of zirconium depleted in the leach layer equals the amount of zirconium deposited on the glass plus the amount of zirconium leached into solution.

(b) the amount of zirconium depleted in the leach layer and deposited on the glass is represented by the area under the plot from the point where leaching of zirconium from the glass begins. This corresponds to the shaded area in Figure 2.

Now the concentration of zirconium estimated to be leached into solution is in the range of 2 x (0 to 0.93) = 0 to 1.86 ppb over the two year period to the end of the 12 month test. This corresponds to leaching of a range of 0 - 0.04 μm of the total original surface into solution. In the case that the concentration of zirconium leached into solution equals 0 ppb, then the area under the graph corresponding to deposited zirconium equals the area under the graph corresponding to depleted zirconium and an estimate of the original glass solution interface is given by line "b". If 0.04 μm of the total original surface was leached into solution the estimate of the original glass-solution boundary is shifted to the left of the figure and is given by line "c".

There is also uncertainty in the position of the outer edge of the α - layer since the samples used for SIMS analysis were found to have some of the α - layer scraped away. In the calculation of the positions of lines "b" and "c" above, the outer edge of the α - layer was assumed to be 0 μm on the depth axis given with the SIMS data. A more valid assumption may involve extrapolation of the plot to 0% zirconium content and to use this point as the outer edge of the α - layer. On repeating the calculations of the positions of lines "b" and "c" based on this assumption we obtained lines "d" and "e" respectively. A summary of the positions of the various estimates of the original glass-solution interface is given in Table II.

Table II. Estimates of Original Glass-Solution Interface

Line	Depth[a] (μm)	Comments
a	0.31	α- βo interface of ref. 1
b c	0.40 0.36	Range based on (b) no Zr leached into solution, (c) 1.86 ppb Zr leached into solution with outer edge of α-layer at 0 μm.
d e	0.39 0.35	Range of values as for lines b and c with outer edge of α-layer at 0% Zr in glass.

[a]Based on scale given in Fig. 2

Reasonable agreement is observed between the positions of the original glass-solution interface determined by the SIMS method and those made on using our zirconium data, particularly on consideration of the uncertainties in both methods. Based on our best estimate of the position of the original glass-solution interface, a leach depth within the glass for lithium, which is one of the most leachable elements in the glass, may be determined from the plot given in Figure 3. Line AB corresponds to the position of line "e" in Figure 2. The area of the rectangle ABCD equals the total shaded area and thus yields a leach depth, L, equal to 0.36μm for leaching of Li from the glass over the 2 year period.

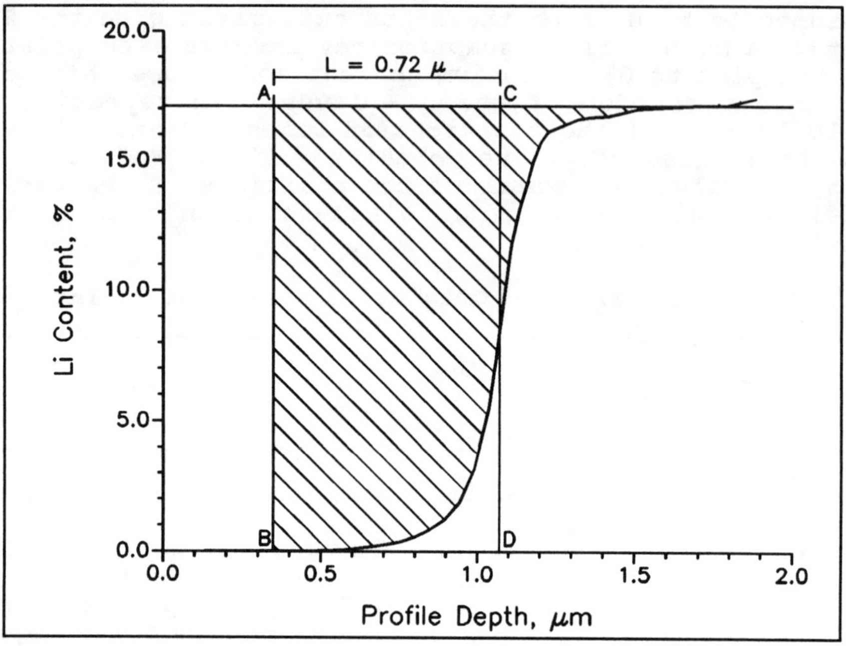

Figure 3. SIMS profile of Li in SRL-Y glass

CONCLUSIONS

It is clear from our solution analysis data that very little leaching of the glass occurs into solution. A leach rate of less than 0.02 μm yr^{-1} was estimated based on the data for zirconium. The NRC set a requirement for the annual fractional loss from a nuclear waste material of 10^{-5} yr^{-1}. Based on a typical cannister size and assuming leaching from all sides of the cannister this requirement is equivalent to a leach rate of about 1μm yr^{-1}. It is clear therefore that the leach rate of Zr from SRL-Y glass is considerably below this requirement even after accounting for possible internal fracturing within the nuclear waste glass. Detailed analysis of this data together with data from surface studies suggests that all the α - layer is certainly depleted and it has also been possible to make an estimate of the position of the original glass-solution interface. From this point it may be estimated that lithium leaches from the glass at a rate of 0.36μm yr^{-1} a value which is also considerably below the NRC requirement. Moreover, this value of 0.36μm can be considered as a maximum value of the uniform annual leaching rate for any element present within the glass matrix.

REFERENCES

1. A. R. Lodding et al; Proceedings of the 91st Annual Meeting of the American Ceramic Society, April 1989, Indianapolis, Indiana, U.S. To be published in this volume.

2. R. E. Sassoon, M. Gong, M. Brandys, M. Adel-Hadadi, A. Barkatt, and P. B. Macedo; Analysis of Brine Leachates from MIIT 1. Leaching of Nuclear Waste Glass Doped with Chemical Tracers. Proceedings of the 91st Annual Meeting of the American Ceramic Society, April 1989, Indianapolis, Indiana, U.S. To be published in this volume.

3. R. E. Sassoon, M. Gong, M. Adel-Hadadi, M. Brandys, A. Barkatt and P. B. Macedo; Monitoring of MIIT Glass Solution Interfaces by Brine Analysis. Proceedings of Workshop on Testing of High Level Waste Forms under Repository Conditions, October 1988, Cadarache, France. To be published as EUR 12 017 EN in 1989.

A MARKER STUDY OF SURFACE LAYER FORMATION AND PARTIAL DISSOLUTION

B.K. Zoitos, R.L. Schulz and D.E. Clark

University of Florida
Department of Materials Science and Engineering
Gainesville, FL 32611

A.R. Lodding

Chalmers University of Technology
Physics Department
41296 Gothenburg, Sweden

Phosphorus markers, implanted to a shallow depth in simulated nuclear waste glasses, have been used to directly measure the degree of total congruent dissolution which occurs during leaching. Two-day leach tests of 165/TDS and ABS 39 simulated nuclear waste glasses in deionized water (modified MCC-1) and in granite rock cups using Stripa water were conducted, followed by secondary ion mass spectrometry (SIMS) profiling to determine the position of the markers with respect to the leached surface. SIMS data indicate that within the leached interface, both glasses underwent a similar amount of total congruent dissolution (between 0.1 and 0.2µm); however, 165/TDS leached to an average depth of 0.36µm--an amount 4-12 times less than ABS 39. This work has demonstrated the efficacy of marker use to examine dissolution mechanisms and will serve as a basis for future, longer term leaching studies.

Surface analyses are used extensively in the study of simulated nuclear waste glasses leached in burial environments. However, they are limited by their inability to measure portions of the glass surface which may have undergone total congruent dissolution and are therefore, no longer available for analysis. Experiments were conducted with the objective of extending the capabilities of these techniques to allow measurement of the amount of glass completely lost to solution during leaching. This was accomplished by leaching glasses on which the original surface had been marked, using aluminum masking[1,2] and ion implantation[3] as shown schematically in Figure 1. After being leached, glasses were examined using SIMS and profilometry to assess the position of the markers and aluminum coating with respect to the leached surface.

Review of Glass Leaching Mechanisms

For the purpose of interpreting subsequent data, it may be informative to review the ways in which glasses leach and the effect these reactions have on the position of the leached glass surface. Glasses may react with aqueous solutions by three fundamental mechanisms: selective dissolution, congruent dissolution, and precipitation. Selective dissolution occurs when non-structural elements such as lithium and sodium exchange with hydrogen or other ions present in solution. This particular process is also known as ion exchange. Additionally, non-major structural elements such as boron may undergo selective leaching. These processes result in a SIMS elemental profile characterized by a region depleted with respect to the bulk glass in the selectively leached elements. ABS 39 displays this type of behavior[4].

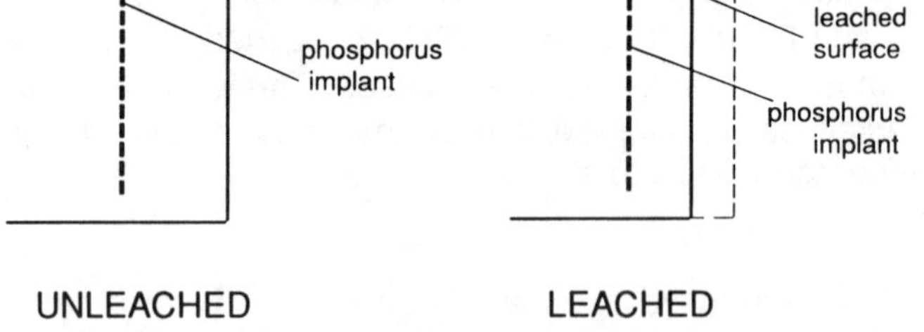

Fig. 1. Schematic of coated, implanted specimen: (a) prior to leaching; (b) after leaching.

If leaching is purely by selective dissolution, the sample should not undergo any dimensional changes, and the leached surface will coincide with the original surface.

Congruent dissolution occurs when the Si-O-Si network is attacked. This results in glass species being released into solution in proportion to their concentration in the glass. For purposes of this work, it is useful to distinguish between total congruent dissolution, in which all glass in a given volume dissolves, and partial congruent dissolution, where surface roughening and microporosity result from incomplete dissolution of the glass. Elemental profiles of congruently leached glasses show a depletion zone in which elemental concentrations decrease from their bulk values to a minimum at the leached surface. This behavior is observed in SIMS profiles of leached 165/TDS glasses. Congruent dissolution results in glass being lost from the surface and causes the surface of the leached specimen to recede below that of the original glass.

Finally, either of the above processes may be coupled with precipitation which occurs when the solution concentration of a given leached species exceeds solubility limits with respect to some solid phase[5]. This results in an elemental profile which displays multiple zones controlled by elemental solubilities. Precipitation acts to extend the leached surface beyond its initial position. It should be noted that these three phenomena may (and frequently do) occur in combination with one another[6].

Experimental

The glasses 165/TDS and ABS 39 were chosen for this study for a number of reasons. These two compositions represent the extremes in leached layer thicknesses measured in the Stripa burial study[7]. In addition, 165/TDS is believed to leach mainly by a congruent dissolution mechanism, while ABS 39 leaches primarily by a selective mechanism.

Glasses were melted, cast into 1cm x 1cm x 10cm bars and annealed. Individual specimens were cut into 0.2cm segments from the centers of the bars and polished to a 0.25μm surface. Samples were then wiped with an ethanol soaked cloth, cleaned ultrasonically three times in absolute ethanol, and dried for one hour at 110°C.

A thin coating of aluminum (0.1μm) was applied to a small portion of the surface of each sample, using electron beam evaporation. Specimens were implanted with 150 keV phosphorus ions. Two specimens of each glass received doses of 5×10^{16} ions cm^{-2} ("high" dose) and the remaining two received doses of 5×10^{15} ions cm^{-2} ("low" dose). Implanted profiles of 165/TDS and ABS 39 are shown in Figures 3(a) and 4(a).

The samples were leached for two days at 90°C at an SA/V of $0.15 cm^{-1}$. One set of samples was leached in deionized water (modified MCC-1). The remainder were leached in granite rock cups using Stripa groundwater. Such rock cup studies produce surfaces similar in composition to those formed under burial conditions. However, because rock cup tests are usually conducted at a lower SA/V than burial studies, they generally result in thicker leached layers[8,9]. The sample matrix is shown in Figure 2.

After leaching, specimens were removed from the leachate, rinsed in absolute ethanol for 5 seconds and allowed to air dry. Aluminum/glass interfaces were examined using profilometry, and implanted regions were examined using SIMS.

Results and Discussion

Upon removing the samples from the leaching containers, it was observed that most of the aluminum on the coated region had dissolved. Attempts to measure a step between the coated and uncoated region were unproductive due to the excessive roughness of the partially dissolved aluminum surface. Fortunately, SIMS analysis to determine the position of the implanted markers was more successful.

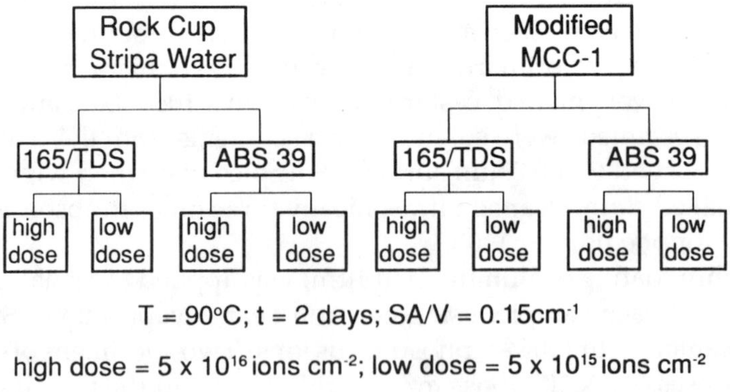

Fig. 2. Sample matrix.

Examples of SIMS profiles for the leached, high-dose implanted specimens are shown in Figures 3 and 4. Results from the low dose specimens were very similar, but the trends are more clearly visible in the profiles of the high dose samples. Between three and five SIMS profiles were recorded for each specimen. In all cases the glass reaction front had proceeded past the marker; however, the markers survived and are clearly visible, although the outer portion of each marker appears to have been removed through total congruent dissolution.

All profiles in Figures 3 and 4 are normalized to aluminum. This is done on the assumption that aluminum is conserved within the leached layer and allows interpretation of the profiles as atomic concentration per unit volume[10].

The SIMS profile of 165/TDS leached in the rock cup (Figure 3(b)), shows the centroid of the phosphorus implant very near to the surface, indicating that a surface layer of approximately 0.16μm has been lost in total congruent dissolution. The profile further shows a layer which has undergone partial congruent leaching (only very slightly compensated by hydrogen and potassium influx) to a depth of approximately 0.2μm. The total reacted depth is thus 0.36μm. The average values for the mean of four profiles were found to be 0.14μm congruently dissolved, 0.22μm partially congruently dissolved and 0.36μm total affected depth.

As shown in Figure3(c), when leached in deionized water, 165/TDS has similarly lost approximately 0.19μm in total dissolution. The profile, down to 0.16μm depth, shows more pronounced near-congruent dissolution tendency than in the Stripa water--not only lithium and sodium, but also boron, calcium and magnesium are depleted to a similar degree. The averages of three profiles are 0.18μm total congruent dissolution, 0.19μm partial congruent dissolution and 0.37μm combined affected depth.

The SIMS profile of the ABS-39 glass leached in the rock cup, shown in Figure 4(b), indicates a reacted depth of 1.5μm, of which 0.15μm has undergone total congruent dissolution. The 1.35μm remaining leached layer shows the characteristic plateau indicative of selective leaching--primarily loss of boron and sodium. This loss is to a large extent compensated by influx of hydrogen, calcium, magnesium and potasssium from the Stripa water. An average of five profiles yields 0.15μm totally dissolved layer, 1.45μm depleted layer and 1.6μm affected depth.

In deionized water, ABS-39 is leached considerably deeper, as shown in Figure 4(c). The affected layer is seen to be 4.5μm, of which 0.2μm is totally dissolved. The remaining 4.3μm layer displays typical selective leaching, with boron and sodium strongly depleted in the plateau region, compen-

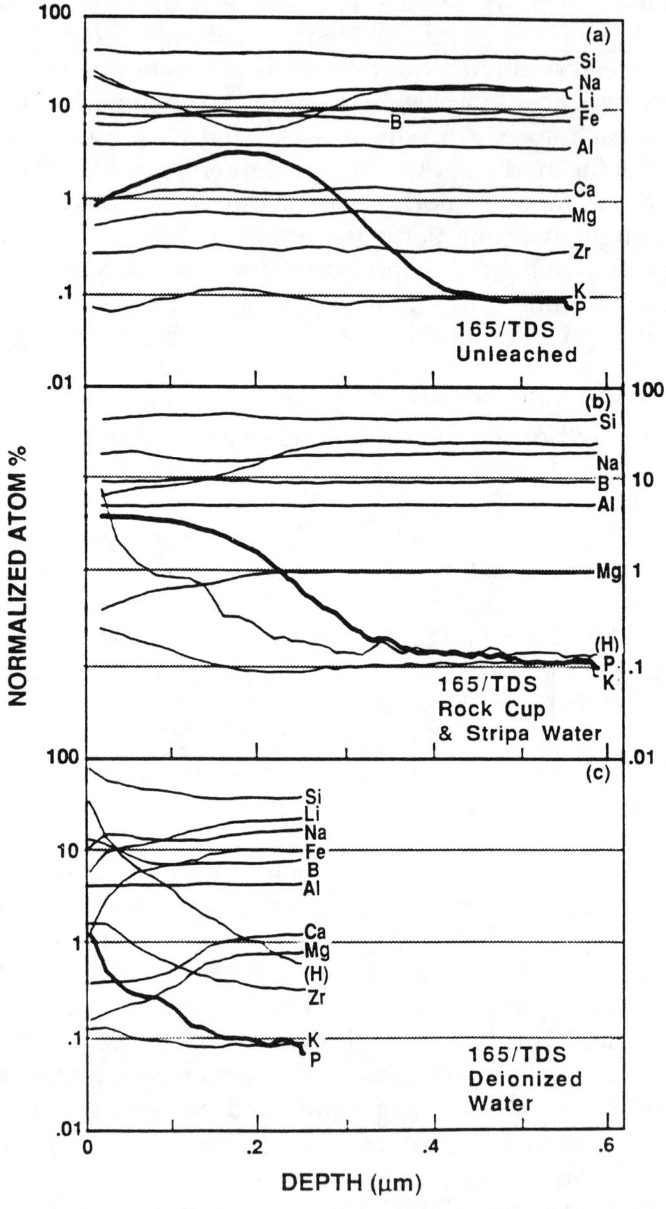

Fig. 3. SIMS profiles of phosphorus-implanted 165/TDS: (a) unleached; (b) leached two days in granite rock cup with Stripa water; (c) leached two days in deionized water.

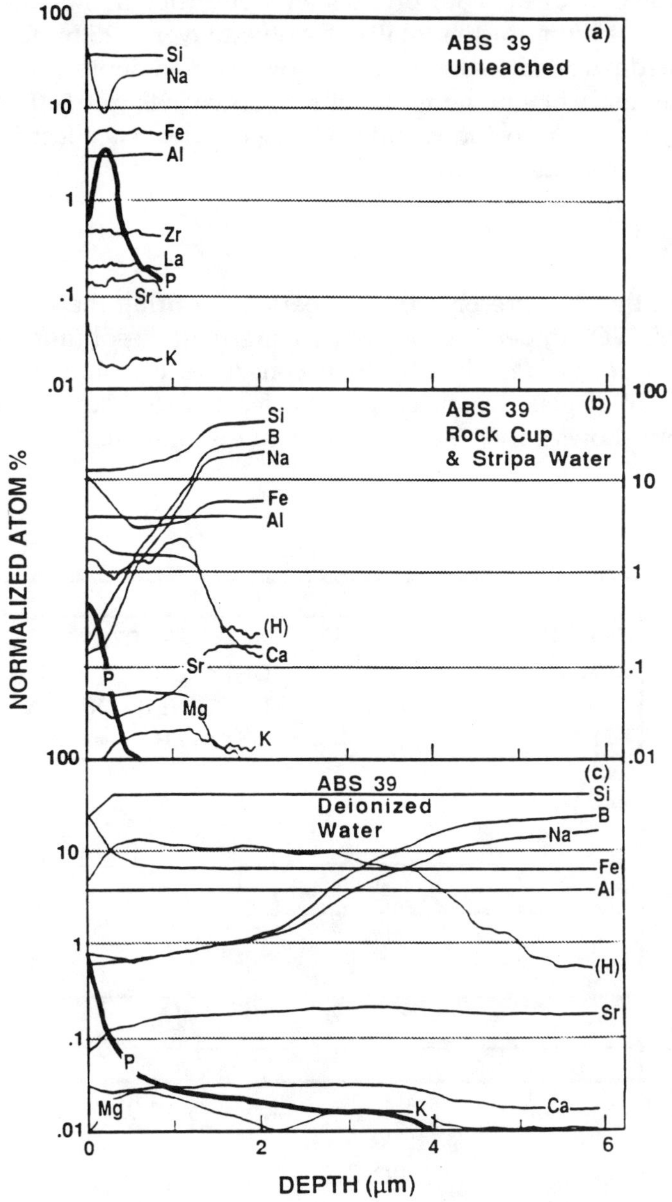

Fig. 4. SIMS profiles of phosphorus-implanted ABS 39: (a) unleached; (b) leached two days in granite rock cup with Stripa water; (c) leached two days in deionized water.

sated only by hydrogen. Four profiles were recorded in this case, yielding an average of 0.17μm for the totally dissolved layer, 4.25μm for the selectively leached part and 4.4μm for the whole affected depth.

A summary of the average results is shown in Figure 5. The scatter of the observed marker positions and leaching depth at each test surface was less than 15% from the mean.

Conclusions

From the data presented, it appears that during the early stages of leaching 165/TDS undergoes primarily congruent dissolution, with up to 40% of the leached layer lost by total congruent dissolution. This is in contrast to ABS 39 glass which leaches to a depth up to 12 times greater than 165/TDS but shows only 5% total congruent dissolution.

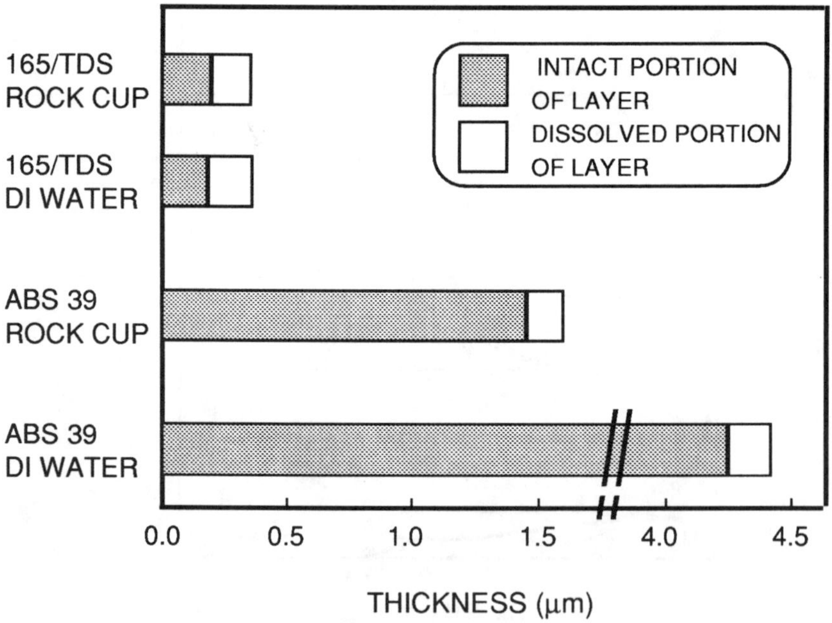

Fig. 5. Summary of two-day test results. Shaded bar indicates thickness of leached layer remaining intact; open bar indicates thickness of layer which has undergone total congruent dissolution.

This work has demonstrated that implanted markers combined with SIMS profiling may be used to directly examine leaching mechanisms in glass, including congruent dissolution. Leaching depths encountered in this work (less than 4.5μm) are within the range of those found in burial studies, suggesting that implanted specimens may be of use in such studies. Although dissolution of the aluminum coating made step height measurements impossible in this work, the ability to check results by a second technique is highly desirable. It is likely that overcoating the aluminum with gold will act to protect the aluminum, thus preserving the underlying surface. This will be attempted in future work.

Finally, caution must be used in trying to apply this short term data to the interpretation of longer term burial or laboratory results. Solution analyses from the Materials Interface Interaction Testing (MIIT) program indicate 165/TDS (SRL Y) undergoes a much smaller degree of total congruent dissolution in burial than was observed in this study.[11,12] This may be due to SA/V effects. Clearly, more extensive studies will be necessary before any general conclusions can be reached.

Acknowledgments

This work was supported by the Westinghouse Savannah River Co. under subcontract Nos. AX-175234 and AX-843931. The authors are indebted to Mr. E.U. Engstrom for performing the SIMS analyses and to Dr. G. Holmen for expert ion implantation.

References

[1]J.C. Dran, M. Maurette and J.C. Petit, "Radioactive Waste Storage Materials: Their Alpha-recoil Aging," *Science*, **209**, 1518 (1980).

[2]J.C. Dran, M. Maurette, J.C. Petit and B. Vassent, "Radiation Damage Effects on the Leach Resistance of Glasses and Minerals: Implications for Nuclear Waste Storage"; p. 449 in *The Scientific Basis for Nuclear Waste Management*, Vol. III. Edited by J.G. Moore. Plenum Press, New York, 1981.

[3]R.H. Doremus, Y. Mehrotro and T. Wassick, "Nuclear Reaction Analysis of Glass Surfaces: The Study of the Reaction Between Water and Glass"; p. 549 in *Advances in Materials Characterization*. Edited by D.R. Rossington, R.A. Condrate and R.L. Snyder. Plenum Press, New York, 1983.

[4]D.E. Clark, A.R. Lodding, H. Odelius and L.O. Werme, "Elemental Analysis of Swedish Nuclear Waste Glasses: Leachability vs. Composition," *Mat. Sci. Eng.*, 91, 240 (1987).

[5]B. Grambow, "The Role of Metal Ion Solubility in Leaching of Nuclear Waste Glasses": p. 93 in *The Scientific Basis for Nuclear Waste Management*, Vol. V. Edited by W. Lutze. Elsevier, New York, 1982.

[6]A.R. Lodding, E.U. Engstrom, D.E. Clark and G.G. Wicks, "Quantitative Concentration Profiling and Element Balance in SRL Glass After Two Years in the WIPP Repository"; in these proceedings.

[7]B.F. Zhu, "Nuclear Waste Glass Leaching in a Simulated Granite Repository," Doctoral Dissertation, University of Florida, Gainesville, FL (1987).

[8]B.K. Zoitos and D.E. Clark, "The Role of Surface Layers in the Leaching Behavior of Glasses"; p. 169 in *Materials Stability and Environmental Degradation.* Edited by Aa. Barkatt, E. D. Verink, Jr. and L. R. Smith. Elsevier, New York, 1988.

[9]B. K. Zoitos, D. E. Clark, A. R. Lodding and G. G. Wicks, " Correlation of Laboratory and Stripa Field Leaching Studies"; p. 145 in *The Scientific Basis for Nuclear Waste Management,* Vol. XII. Edited by W. Lutze and R. C. Ewing. Elsevier, New York, 1989.

[10]A.R. Lodding, E.U. Engstrom, D.E. Clark, L.O. Werme and G.G. Wicks, "SIMS Analysis of Leached Layers Formed on SRL Glasses During Burial"; p. 567 in *Advances in Ceramics,* Vol. 20. Edited by D.E. Clark, W.B. White and A.J. Machiels. The American Ceramic Society, Westerville, OH, 1986.

[11]M. Brandys, M. Gong, R.E. Sassoon, Aa. Barkatt and P.B. Macedo, "Leaching of SRL Y Nuclear Waste Glass in MIIT by Analysis of Brine Solutions"; in these proceedings.

[12]R.E. Sassoon, M. Gong, M. Brandys, M. Adel-Hadadi, Aa. Barkatt and P.B. Macedo, "Leaching of Nuclear Waste glasses Doped with Chemical Tracers in MIIT by Analysis of Brine Solutions"; in these proceedings.

ANALYSIS OF BRINE LEACHATES FROM MATERIALS INTERFACE INTERACTIONS TESTS
1. Leaching of Nuclear Waste Glass Doped with Chemical Tracers

Richard E. Sassoon, Meiling Gong+, Marek Brandys, Mohammed Adel-Hadadi, Aaron Barkatt and Pedro B. Macedo

Vitreous State Laboratory
The Catholic University of America
Washington, D.C. 20064

ABSTRACT

The Inductively Coupled Plasma-Mass Spectrometry (ICP-MS) technique was found to be very useful for analyzing brine leachates for the presence of chemical dopants from nuclear waste glasses. The standard addition method was used to determine the concentrations of La, Eu, Ce, Nd, Y, Pr, Zr, Th and U in samples from leach tests performed over a 12 month period beginning 12 months after initial test turn-on. By utilizing the low detection limit of the ICP-MS instrument for these elements values of the upper limits of leach rates from Aged SRL-Y, FR-JSSA and CUA-WV glasses were determined to be lower than $8x10^{-4}$ $gm^{-2}d^{-1}$. This is equivalent to leaching of less than 0.1 μm of total glass surface per year.

INTRODUCTION

As part of our participation in the Materials Interface Interactions Test (MIIT) program[1], our laboratory has been engaged in analyzing leachate samples from some of the Waste Isolation Pilot Plant (WIPP) emplacements in the Materials Interface Interactions Test (MIIT) leach tests in which 12 "pineapple-like" slices of one

+ In partial fulfillment of Ph.D requirement

or various glasses were immersed in boreholes containing WIPP Brine A at 90°C. These assemblies, which contain glasses only, provide a unique opportunity to follow an in situ test by combining data from surface studies with data from solution analysis. After periods of 6, 12 and 24 months, 3 slices were removed and fresh leachant was added. Samples of the leachates were collected at various times during these periods and were analyzed for the leached out elements.

Analyses of brine samples taken from borehole MIIT #8 at the WIPP site have been carried out earlier[2]. These samples were from the approximate half-year periods following the removal of the first and second set of 3 glass slices. Using the Direct Current Plasma (DCP), Atomic Absorption (AA) and Inductively Coupled Plasma-Mass Spectrometry (ICP-MS) techniques, by which we can determine the concentrations of individual cationic elements in solution, we followed the behavior of many of the major and minor components of the leachates. A Dionex ion exchange method was used to determine the concentrations of anions in the leachates.

The concentration profiles of the major components in the brines, namely K, Na, Mg and Cl were all found to increase steadily with the time with the exception of Na which decreased. This behavior could be accounted for simply by ion exchange with the rock salt walls and evaporation of the brine during the test period under investigation. These two processes appear to dominate the concentration behavior of all except perhaps some of the very minor components in the brine. A steady increase is also observed for Li during a sample test period beginning 12 months after the initial test turn-on. A study of the isotope ratios of lithium-6 to lithium-7 from the glass, leachant and leachate only yielded an upper limit for the leach rate of the glass based on the detection limit of the ICP-MS instrument. (Also see next article in this series)[3].

In this article we shall describe our more recent efforts to obtain leach rates for various glass samples which contain appropriate chemical tracer elements. These elements were present in the glass samples at relatively high content while at a low concentration in Brine A. This allows for easier detection using the

ICP-MS technique. Table I lists the boreholes together with the glass slices present and the elements that were analyzed.

Table I. Chemical Tracers in Nuclear Waste Glasses

MIIT Borehole #	Glass	Element	Weight % Composition of Oxide in Glass
11	Aged SRL-Y	La	0.40
		Eu	0.40
20	FR-JSSA	Ce	0.93
		Nd	1.59
		Y	0.20
		Pr	0.44
		Zr	2.65
		Th	0.33
		U	≈0.5
39	CUA-WV	Zr	3.1

EXPERIMENTAL

In this study the Inductively Coupled Plasma Mass Spectrometry (ICP-MS) technique was the sole method used to analyze for the concentrations of the chemical tracers which may have leached out into the brine. A schematic description of the apparatus is shown in Figure 1. The sample to be analyzed was diluted by weight by a factor of about 100 in 1% ultrapure doubly distilled HNO_3 in order to reduce the total solid content of the brine to below 0.5% and the sample solution was introduced into the instrument using a peristaltic pump. It passes through a nebulizer in the spray chamber and carried into the core of a very high temperature plasma (several thousand degrees Kelvin) sustained by a radio frequency field delivered from a coil surrounding the torch. Energy is transferred from the plasma to the sample, dissociating, atomizing and ionizing it in turn. The ionic plasma produced is then extracted into a quadruple mass spectrometer through sampling and skimmer cones which allows for a drop in pressure from atmospheric pressure to around 10^{-7} torr. The individual isotopes of the sample are separated according to their charge/mass ratio and then detected by an electron multiplier

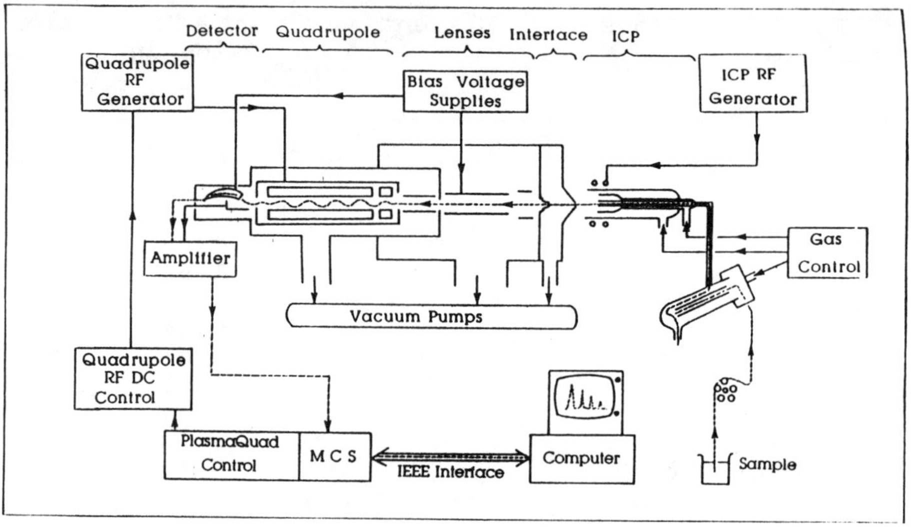

Figure 1. A schematic description of the ICP-MS experimental system

detector. A Tracor-Northern multichannel analyzer connected to an IBM PC/XT computer is then used to analyze the signals.

The integrated signal in units of area counts per second obtained from the instrument for a particular isotope is proportional to the concentration of that isotope in the original sample solution. Interferences do sometimes occur when isotopes of different elements or molecular ions possess the same charge/mass ratio although there is for almost every element one individual isotope which may be analyzed without substantial interferences from other species. In fact, the total mass spectrum consists of only several hundred lines corresponding to the roughly one hundred elements of the periodic table. This should be compared to the tens of thousands of lines seen in a regular ICP-emission spectrum which does not therefore allow for the high resolution obtained with ICP-MS. Indeed a detection limit of around 10 parts per trillion can be obtained for many elements using ICP-MS compared to a concentration about 20 times higher using the regular ICP technique.

In this study the standard addition method was used to determine the concentration of a particular element in the sample and also to check the linearity of the reading of the instrument with concentration. In this method the signal for the diluted sample solution was measured on the ICP-MS as well as the signals for the same solution spiked with three different known concentrations of a standard of the element being analyzed. A plot of the integrated signal versus the added concentration of a standard is shown in Figure 2 for a typical element determination.

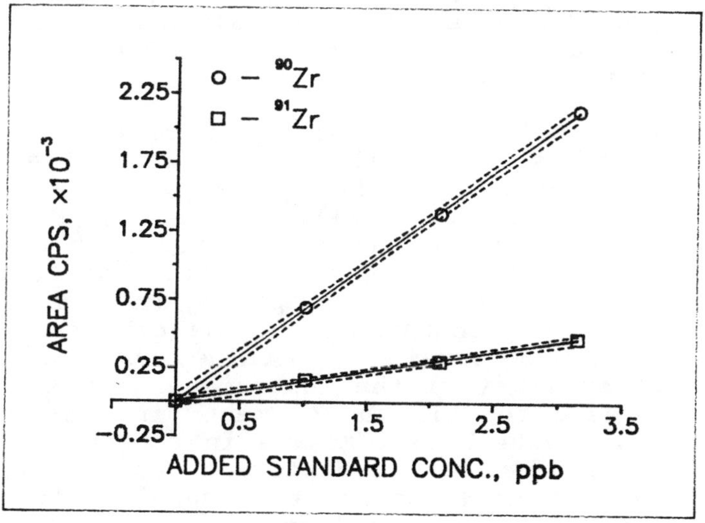

Figure 2. A typical standard addition plot for the isotopes ^{90}Zr and ^{91}Zr. The dashed lines represent the 95% confidence limits for the plots.

Since the signal can be related to the concentration of added standard by the expression

$$S^i = A^i\,(Ci_{sample} + Ci_{standard}) \qquad (1)$$

Where S = Signal
A = Instrument Coefficient
C = Concentration
and i = Element analyzed

the concentration of the element may be determined by dividing the intercept by the gradient for a straight line plot such as that shown in Figure 2.

RESULTS AND DISCUSSION

The concentrations of the elements in the brine leachates as determined by the standard addition method and multiplying up by the dilution factor from the tests begun 12 months after initial turn-on are shown in Figures 3(a), 3(b) and 3(c) for the 3 MIIT borehole assemblies under investigation.

It may immediately be noted that the concentrations of the analyzed elements in the samples determined using the standard addition method appear to be scattered around zero parts per billion. Noting that the detection limit of the instrument for most of these elements is about 10-50 ppt and a dilution factor of about 100 was necessary in preparing the samples for analysis, any concentrations in the original sample below about 5 ppb correspond to values obtained within the detection limit of the instrument. This appears to be the case for all the elements investigated in this study. Indeed no increase in concentration of elements with time can be discerned from these plots indicating that no leaching of these elements may be detected from our investigation.

It is, however, possible to determine a value for the upper limits of the leach rates of these elements from their respective glasses from the standard deviations of the concentration values determined by the standard addition method. Assuming that the upper limits of concentrations to be equal to 3 standard deviations of the concentration values themselves, upper limits of

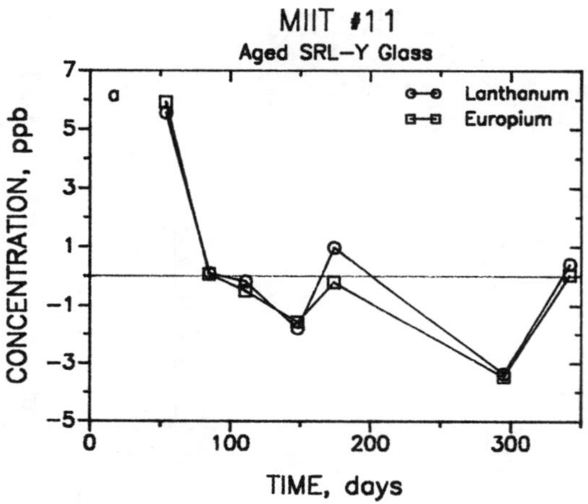

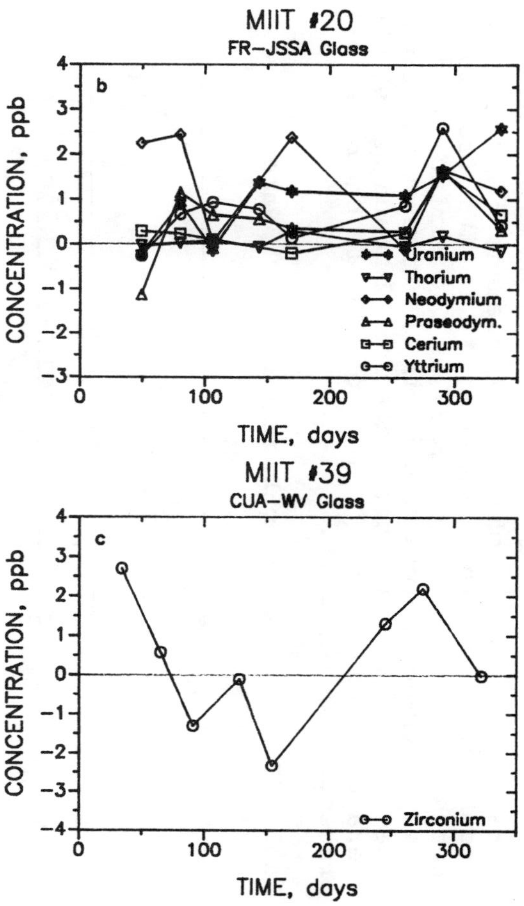

Figure 3. Plots of concentration in the brine leachate calculated according to the standard addition method versus time for 12 month tests for (a) La and Eu from MIIT #11, (b) Y, Ce, Pr, Nd, Th and U from MIIT #20, and (c) Zr from MIIT #39.

the leach rates may be calculated from the expression.

$$LR_i = C_i/\{f_i(SA/V)t\} \quad (2)$$

where LR_i is the leach rate of element i, C_i is its concentration in the leachate, f_i is its fractional composition in the glass, SA/V is the surface area of glass/volume of solution and t is the time of sampling. Based on the concentration data for 12 month tests, the averages of the upper limits of the leach rates are presented in Table II.

Table II. Average Upper Limits of Leach Rates [a]

Element	$10^{-4} LR_i$ ($gm^{-2}d^{-1}$)	LR_i[b] ($\mu m\ yr^{-1}$)
MIIT #11 - Aged SRL-Y glass		
La	3.5	0.042
Eu	3.0	0.036
MIIT #20 - FR - JSSA glass		
Y	7.4	0.090
Ce	1.2	0.014
Pr	3.1	0.037
Nd	1.2	0.015
Th	0.9	0.011
U	2.1	0.026
MIIT #39 - CUA - WV glass		
Zr	0.9	0.011

[a] Averages calculated for samples from period 12 months through 24 months after initial turn-on of test.

[b] Leach rates calculated assuming a density of glass of 3 g cm^{-3}. See discussion in following paper for significance of these numbers.

The values of the upper limits of these leach rates all fall below around 8×10^{-4} $gm^{-2}d^{-1}$ which is equivalent to about 0.1 μm of surface leached per year. Some numbers even fall nearly an order of magnitude lower than this value. This data is very much in line with the data obtained for accurate analysis of zirconium leached from SRL-Y glass in borehole #49[3] and represents a more than one hundred-fold decrease in the value of the upper limit of leach rate from SRL-Y glass obtained earlier using the ICP-MS and other techniques on samples from 6 month tests in borehole #8[2]. A detailed analysis of the significance of this leach data together with SIMS data regarding the surface of SRL-Y glass is given in the accompanying article[3].

CONCLUSIONS

This investigation has demonstrated that very low levels of detection of leached out elements from nuclear waste glasses may be achieved even in brine solutions using the ICP-MS technique. With a careful choice of glasses containing appropriate chemical tracers, upper limits of leach rates of less than 8×10^{-4} $g\ m^{-2}\ d^{-1}$ were determined for these elements. The leach rates of these glasses, regarding the more leachable elements such as B and Li, are somewhat higher and are discussed for SRL-Y glass in the accompanying article.

REFERENCES

1. G. G. Wicks, M. E. Weinle and M. A. Molecke; WIPP/SRL in Situ Tests - Part II: Pictorial History of MIIT and Final MIIT Matrices, Assemblies and Sample Listings. Report DP-1733 from E.I. du Pont de Nemours and Co., Savannah River Laboratory, Aiken, SC 29808. January 1987.

2. R. E. Sassoon, M. Gong, M. Adel-Hadadi, M. Brandys, A. Barkatt and P. B. Macedo; Monitoring of MIIT Glass Solution Interactions by Brine Analysis. Proceedings of Workshop on Testing of High Level Waste Forms under Repository Conditions, October 1988, Cadarache, France. To be published as EUR 12 017 EN in 1989.

3. M. Brandys, M. Gong, R. E. Sassoon, A. Barkatt and P. B. Macedo; Analysis of Brine Leachates from MIIT 2. Leaching of Lithium and Zirconium from Nuclear Waste Glass. Proceedings of the 91st Annual Meeting of the American Ceramic Society, April 1989, Indianapolis, Indiana, U.S. To be published in this volume.

QUANTITATIVE CONCENTRATION PROFILING AND ELEMENT BALANCE IN SRL GLASS AFTER TWO YEARS IN WIPP

Alexander R. Lodding and E. Urban Engström
Chalmers University of Technology, S-41296 Gothenburg, Sweden

David E. Clark
University of Florida, Gainesville, FL 32611

George G. Wicks
Savannah River Laboratory, Aiken, SC 29808

ABSTRACT

Elemental in-depth profiling has been performed on specimens of the SRL-Y (165/TDS) nuclear waste glass after 24 months' leaching in the WIPP in-situ test. The concentrations of 17 elements were determined quantitatively by means of secondary ion mass spectrometry (SIMS). The technique permitted the calculation of the loss or influx of each element, in atoms/cm^2, in the leached glass layers.

Between the unchanged bulk glass and the specimen surface, three distinct depletion zones and an outer precipitation layer could be identified. The outermost layer, of variable thickness, dominated by Mg, B, Si and H in the approximate proportions 4:1:2:3, is permeable to leachant but may to some extent act as a barrier to further dissolution of the glass network. In the interacted layers of the glass, main network formers are only moderately depleted, B somewhat less than Si; the relative resistance of B is in contradistinction to other leaching environments. Other cations are found to be leached differentially, from nearly complete elimination of the alkalis to near-conservation of Al, Zr, Fe and Ni. The alkalis and boron appear, to a high extent, replaced by influx mainly of Mg and H from the surrounding salt. The thickness of the interacted glass layers after 2 years is found to be, on an average, 1.0 μm at glass-glass interfaces, 1.2 μm at glass-salt, and 0.5 μm at glass-brine interfaces.

INTRODUCTION

The assessment of long-term performance and reliability of high level radioactive waste forms has in recent years required realistic burial testing, closely simulating the interactions of the waste form with the actual geological environment of the repository. The most extensive investigations of this kind, the Materials Interface Interaction Tests (MIIT /1,2/), are performed on material stored at 650 m depth in a salt formation in New Mexico, the Waste Isolation Pilot Plant (WIPP). So far, the effects of ½ year, 1 year and 2 year burials have been made available for evaluation /3-5/; 5 year specimens will be brought to the surface in 1991.

Among the wide range of techniques for characterizing the leached waste form within the MIIT program, secondary ion mass spectrometry (SIMS) has been particularly adapted /6-8/ for quantitative elemental profiling in glasses, combining high microanalytical sensitivity with high topographic resolution. The main objectives of SIMS studies within the SRL-WIPP test are the following:

A) The measurement of leaching depths in glass at different interfaces; identification of different interaction layers.
B) The quantitative determination of leaching losses in all cations; substitutions and microporosities; total element balance with no recourse to leachant analysis.

The present paper gives an account of 2 years' interactions in the WIPP salt environment of a waste glass developed at the Savannah River Plant (Y-glass; SRL 165-TDS /8b/). This high performance glass had earlier been studied in a long-term burial test in a granite environment /8/ and is also discussed in several parallel contributions to this volume /4, 9-11/.

EXPERIMENTAL

The cation composition of the SRL-Y glass is listed in Table I. The specimens were sectors of "pineapple slices", as described in

Table I. Cation Composition of SRL-Y (165-TDS) Glass, in Atom-% of the Sum of All Cations

Si	B	Li	Na	Cs	Mg	Ca	Sr	Al	Zr	Mn	Fe	Ni
41.9	8.6	14.5	15.9	0.7	0.7	1.4	0.4	4.4	0.7	1.4	7.2	0.7

other documentation on the WIPP test (see, e.g., ref./10/). The test surfaces before leaching were polished to <1 μm finish. In the WIPP leaching assembly the heaters, maintaining about 90°C, were accommodated inside the inner circular rim of the slices. The major part of the test surface, from the inner rim to about 2 mm from the outer rim, is in contact either with another glass surface (GG-interfaces), or with original dry salt (GS), or (due to spacers between consecutive specimens) with liquid brine (GL-interfaces). The 2 mm zone just inside the periphery (the "lip" portion of the slice) is in contact with the fluid only.

After leaching, separation and removal from the WIPP assembly, the specimens were subjected to ultrasonic bath in water at ca 30°C, which removed nearly all remnants of salt and clay deposits on the surfaces. Nevertheless, more or less shiny or cloudy patches could still be distinguished, probably because the access of the leaching fluid to the test surface had varied with the contact conditions. E.g., the areas just inside the outer "lip" and those close to the inner rim (i.e. next to the heaters) occasionally showed visibly stronger corrosion than the central parts of the contact surfaces. The GL surfaces, however, exposed to liquid brine only, showed much lesser variation both in deposit thickness and in actual reacted depth. The presently reported SIMS profiles were all recorded about half-way between the inner rim and the outer "lip".

The details of SIMS instrumentation /12/ and of the actual profiling procedure as developed for glasses have been described in earlier publications /5-8/. A few essential features are summarized in Table II. The number of elements routinely profiled in the SRL-Y glass is 17, of which H, Cl, K and Ba are not intentionally included in the making of the glass. The elemental sensitivity factors are based both on nominal composition and (in leached layers) on cross-calibration with analytical SEM-EDX. While the primary data are obtained as ion counts versus sputtering depth, the computer expresses the results as element concentrations (in % of the sum of all cations) versus depth. This is effected by the employment of relative sensitivity factors and of measured sputtering rates. To obtain the different sputter speeds in the successive leaching layers, several profiles to different depths have to be made at each studied area, and each sputter crater is measured by means of a Talysurf profilometer.

Table II. Experimental Parameters of the SIMS Routine for Profiling in Glasses

Instrument: Cameca IMS-3F.

Extra software features: Automatic adjustment of specimen potential to compensate for surface charge buildup. Program to convert primary data (counts vs sputter time) to concentration profiles (atom-% vs depth) and to element loss profiles (residual cation levels vs depth).

Primary ions: O^-; ca 17 keV.
Beam current: 0.05 - 0.15 µm. Diameter: 20 - 50 µm.
Bombarded (rastered) area: ca 100 - 250 µm diam.
Analyzed area: 7 x 7 µm square.

Secondary ions: Positive; energy acceptance window 75 to 125 eV.

Detection limits: ca 0.1 - 10 ppm.

Quantitation: By elemental sensitivity factors.

Depth resolution: ca 5 - 100 nm.

Speed of profiling: ca 0.5 - 10 µm/hour.

Duration of a 24 element cycle: ca 1.5 min.

RESULTS AND DISCUSSION

a) Shape of leaching profiles; qualitative tendencies

Figure 1 shows a typical set of element concentration profiles in SRL-Y glass after leaching in the SALT environment of WIPP. The selected example is a glass-salt (GS) interface, but similar tendencies apply at the GG or GL interfaces. The concentrations are here expressed in atom-% of the total sum of cations (H excepted). The elements in parentheses, H and Cl, are not included in the sum, i.e., are counted in excess of 100%.

In an initial stage, the 17 elements in Fig.1 may be roughly divided according to the following patterns of leaching behavior: i/ Strongly depleted: Li, Na, Cs, Sr, Ba; Mn. ii/ Moderately depleted: Si, B; Ca. iii/ Slightly depleted or preserved: Al, Zr, Fe, Ni. iv/ Penetrated from fluid: Mg, H, Cl; K; locally Ni, Fe.

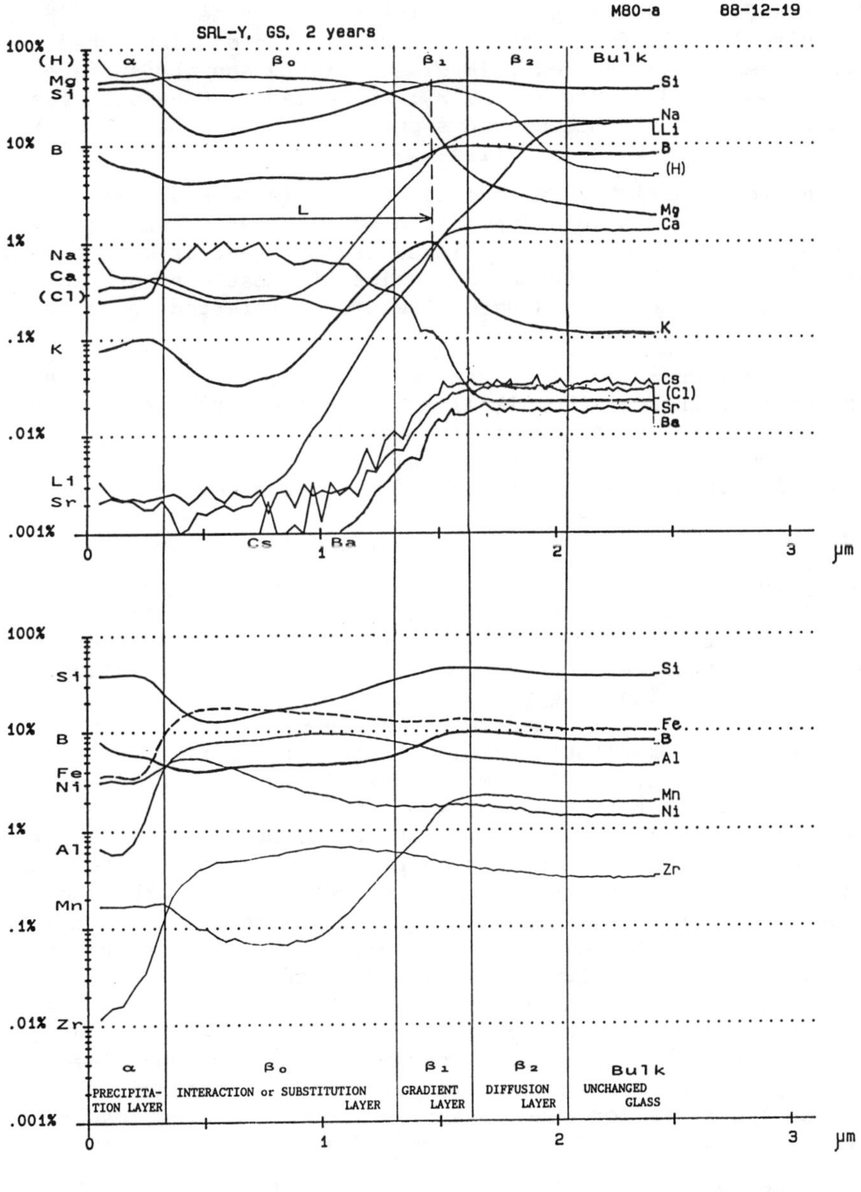

Figure 1. SIMS Concentration Profiles of Elements in Reacted Layers on SRL-Y Glass Leached for 24 Months in WIPP-MIIT; % of Cation Sum Plotted vs. Depth. Above: Si, B, Mono- and Divalent Elements. Below: Si, B, Trivalent and Transition Elements.

This classification applies to the interaction zones denoted β in the figure. In all, four distinct layers may be distinguished in Fig.1 between the unchanged bulk glass (right) and the surface (left), designated α, β_0, β_1 and β_2. These layers have been identified by SIMS in all measurements on SRL glass leached in WIPP (in ref./5/ the β_0 zone is called α_2).

Starting from the right, in zone β_2 only Li is seen significantly depleted, and this loss appears to be compensated chiefly by hydrogen. The thickness of this layer is, in order of magnitude, reasonably compatible with the diffusion of these elements in bulk glass, and the zone is here, therefore, referred to as the "diffusion layer".

The layer denoted β_1, called "gradient layer", shows significant depletion gradients of the alkali and alkaline earth metal with the exception of Mg. Potassium arrives from the outside brine as well as from the interior, and actually shows a distinct peak in the β_1 zone. This typical behavior of K is observed in nearly all glass specimens after leaching in WIPP. The K peak may be considered to mark the actual reaction front of glass corrosion.

The main depletion zone β_0, showing clear impoverishment not only of the modifier elements but also of the main network formers Si and B, may be referred to as the "interaction" or "substitution layer". The latter term corresponds to the very strong influx of Mg and H from the leaching fluid, which appears nearly quantitatively to replace the leached-out bulk elements (see section c below). Another substitution element, however entering in a much lower concentration, is Cl. Some influx of Na, K and Ca from the brine probably inhibits these elements from being depleted as completely as Li. The observed enrichment of Ni is almost certainly due to the presence of corrosion products from the system components, such as the heaters and the thermocouple sheaths. No such enrichment had been observed in profiles recorded after $\frac{1}{2}$ year leaching, and the Ni enrichment after 1 year was comparatively slight. Another argument for the outside origin of the Ni enhancement is that particularly high amounts of Ni (and also Fe, Mn, Cr) were found in the deposits and surface layers on the outer "lip" of the pineapple slice specimens. The observed rise in the Fe and Mn profiles in the left part of β_0 is probably due to the same effect. As for the apparent enrichment of the leach-inert elements Al and Zr, it will be discussed in section b below.

The surface zone α is characterized by i/ a steep concentration drop of the normally leach-resistant elements Al, Zr and Fe, and ii/ a significant rise in the Si and B levels. The depletion of the three elements, found nearly inert in the selectively leached zones of all hitherto studied glass systems /4-6, 8/, suggests that the α zone is not a leached-out part of the original glass. Further, while the steep rise of the B concentration could possibly be associated with the percentage presence of this element in the brine /13/, the concentration of Si in the salt is relatively low, and so the high level of Si in the α zone is most likely due to partial re-precipitation of leached-out material. The zone is therefore here referred to as the "precipitation layer". It should be noted that zone α is quite different from the precipitated salt deposits which are always observed on specimens freshly taken up from WIPP; here such deposits have been removed by ultrasonic cleaning. The dominant cation in zone α is Mg, and the concentrations of Na and Cl are here usually only in the 0.1 - 1% range. However, patchwise on a number of roughly cleaned specimens, SIMS profiling may reveal an additional layer outside α, with a rise in Cl and the alkalis, and with a gradual decrease in Si. Such tendencies have also been observed by SEM on specimens as-received fresh out of the repository /10/ and probably indicate a transition layer between α and the salt deposit.

The example in Fig.1 (and 2a, 2b below) is one of 11 sets of profiles recorded within the present 2-year leaching study. Table III lists the collected data on the thicknesses of the observed reacted layers. Further discussion of the leaching depths at different interfaces is given in section c below.

Glasses may react to leaching by three basic mechanisms: congruent dissolution, selective depletion, and precipitation. Partial contributions of the three modes may compete. Zone α in the present study appears to be an example of the last-mentioned case. The depletion zones β_{012} show features of selective leaching behavior (differential element depletion) similar to those observed in earlier experiments on various glasses /8, 9, 14/; the SRL-Y glass here exhibits more typical selective leaching than in granite burial /8/, where more evidence of partial congruent leaching was observed.

Congruent leaching behavior, in which either the glass totally dissolves, or the different components are partially released into solution in similar proportions, is not implied in the present profiles. Conceivably a transition zone between α and β may have dissolved congruently. In that case, amounts even of the most "inert" components would have entered the leachant. For Al and Fe, which are appreciably present in the brine (each in ca 0.1 atom-% of total cations /13/), this is difficult to detect. However, Zr is not found in the geology of WIPP (the Salado formation). Leachant analyses after experiments under WIPP conditions /15/ have recently revealed some release of Zr from SRL-Y glass. However, if interpreted as congruent leaching, the results correspond to a dissolved zone less than ca 0.03 μm thick in 2 years

Table III. Thickness (μm) of Different Layers in Specimens of SRL-Y Glass, as Determined by SIMS Profiling after 2 Years Leaching at Different Interfaces in WIPP.

Specimen	Profile number	α	β_0	β_1	β_2	L
07-Y008-GG-2/U	77b	0.45	1.05	0.35	0.45	1.20
	77d	0.1	0.65	0.50	0.55	0.80
	77e	0.2	0.7	0.55	0.55	0.95
Mean, GG		0.25	0.80	0.47	0.52	0.98
07-Y009-GS-2/U	76a	0.5	1.15	0.35	0.65	1.30
	76b	0.45	1.00	0.30	0.55	1.15
07-Y009-GS-2/D	78a	0.25	1.2	0.20	0.55	1.35
	80a	0.30	1.00	0.20	0.50	1.10
Mean, GS		0.37	1.09	0.26	0.56	1.22
49-Y021-GL-2/D	81b	0.40	0.55	0.20	0.60	0.70
	81c	0.25	0.30	0.25	0.50	0.40
	81d	0.35	0.30	0.10	0.45	0.35
	81g	0.25	0.30	0.30	0.65	0.50
Mean, GL		0.31	0.36	0.21	0.55	0.49

exposure. It may therefore be reasonably concluded that total congruent leaching, if any, may be neglected in the present calculation of element loss. To gain more certainty on this point, WIPP leaching studies on SRL-Y glass with ion-implanted inert depth markers are under preparation. Such studies, conducted along lines described in a recent pilot report /9/, may quantitatively determine the extent of congruent leaching, without the need of leachant analysis.

b) Quantitative element depletion-substitution profiles

For more detailed discussion, the profiles of the main elements are shown in Fig.2a on a larger vertical scale than in Fig.1. A few relatively "inert" elements actually appear enriched in zones β_0 and β_1. As discussed in earlier publications /5, 6, 8/, the concentration rise is a rigorous fact, but only in relation to the material remaining in the depletion layers after leaching. In order to evaluate the actual (absolute) amount of material removed by leaching, the assumption can be made (arbitrary, but empirical-

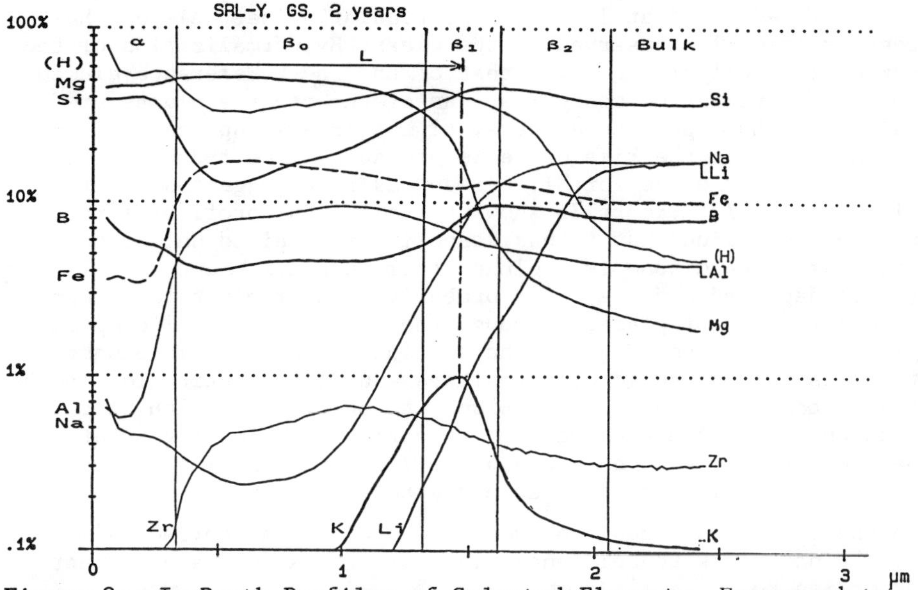

Figure 2a. In-Depth Profiles of Selected Elements, Expressed in Atom-% as Relating to Sum of all Cations Remaining after Leaching.

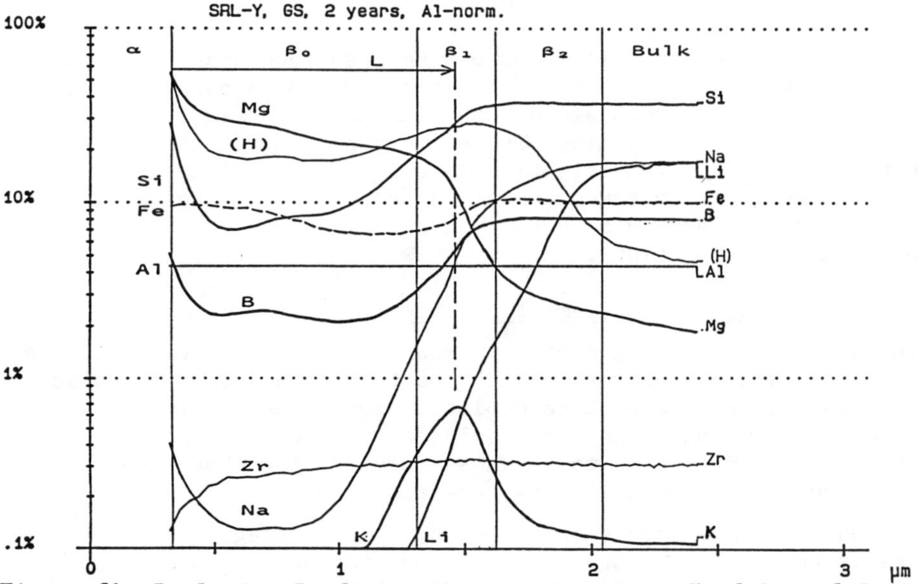

Figure 2b. Depletion Profiles, Expressed in Atom-% of Sum of Cations in Unleached Glass; under Assumption that Al is Preserved Undepleted in the Reacted Layers.

ly justified) that at least one particularly inert element has been practically conserved in the glass. By normalization of the curves in Fig.1 or Fig.2a to that of the inert reference element, the actual remaining concentrations of the other elements are obtained. In this particular case, Al and Zr are applicable as references (while the rise of Fe is probably in part due to influx, as pointed out above for Ni, Cr, and as also suggested in other WIPP work /16/). Figure 2b shows the elemental profiles corrected on the assumption that Al has essentially remained unleached in the glass. The α zone is excluded from such treatment, as Al and Zr are depleted and the layer probably is a precipitate rather than a residue of glass. Although the actual elemental composition of α is given (Fig.1), the absolute balance is more uncertain, as the knowledge of porosities and SIMS quantitation in the precipitate is insufficient. Also, the part of the high Si plateau that may have been removed in ultrasonic cleaning is at present uncertain; even on a given specimen, the profiled thickness of α could exhibit considerable scatter.

In the depletion zones, however, the profiles as corrected in Fig.2b now allow the elemental loss or influx for each element, according to

$$\Delta n = \frac{1}{100} n_{tot} \int_{A}^{A+L} (c_o - c)\, dl \qquad (cm^{-2}) \quad , \qquad (1)$$

where Δn is the loss, in atoms per unit surface area, of a particular element, n_{tot} (cm^{-3}) is the total cation density of unleached glass, c the percentage of the element as in Fig.2b, and c_o its percentage before leaching. The A coordinate marks the boundary between α and β_0. The effective leaching depth L (see Figs. 1-2) is here arbitrarily taken as the in-depth distance between A and the potassium peak in zone β_1 (neglecting only a very minor part of Li and Na depletion).

c) Reacted depths and element balance

The effective leaching depths L (mean values from Table III) are listed in Table IV together with the corresponding total cation mass loss in the respective depletion zones, as summed after separately calculating Δn (via Eq.1) for each cation and multiplying with its mass (mg/atom). These results, together with those of the $\frac{1}{2}$ year and 1 year studies, are also shown in Fig.3.

The GS interfaces are seen to leach, on an average, by about 20% faster than the GG contacts. More qualitatively such a difference has also been observed in recent SEM studies /10/. It may possibly be attributed to facilitated access of leachant to the GS surface (i.e., decreasing SA/V ratio) when the original dry salt is gra-

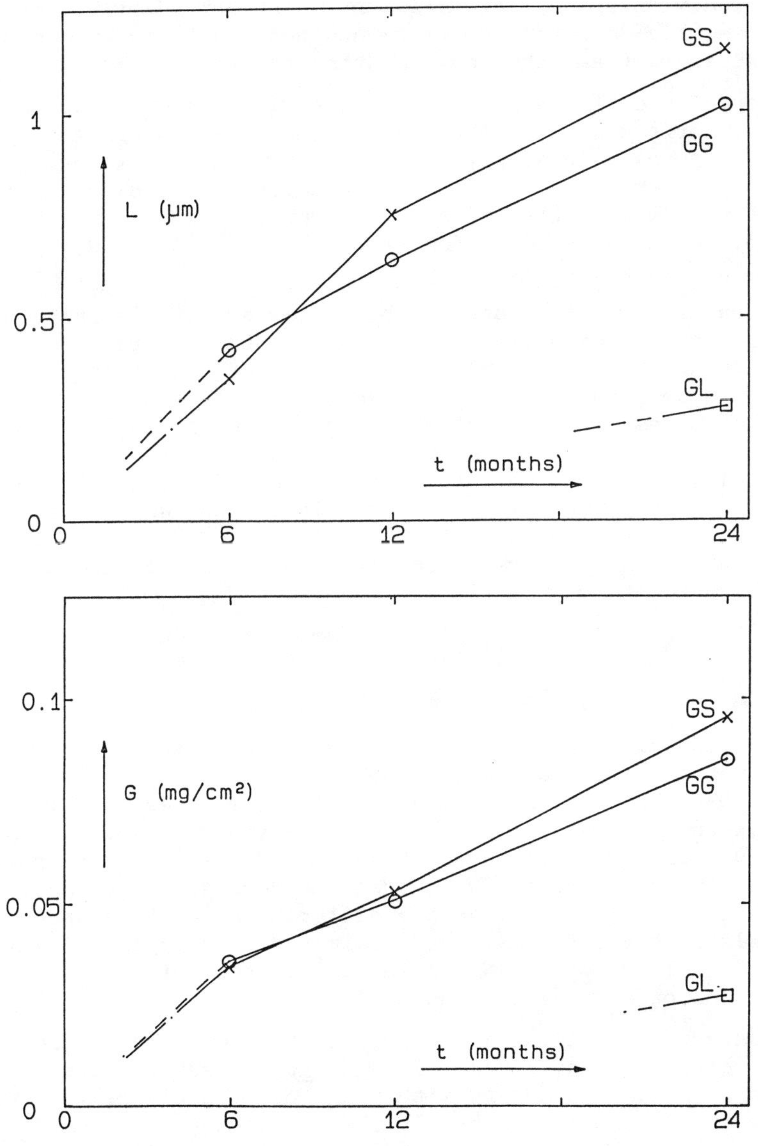

Figure 3. Time Dependence of Effective Leaching Depth and of Total Cation Mass Loss for SRL-Y Glass Leached at Different Interfaces in the WIPP Test.

dually infused with brine. This argument is supported by the observation (see Fig.3) that the difference between GS and GG does not apply at short (less than ca 6 months) exposure.

The time dependence of leaching is in line with results reported for the GG interface earlier /5/. The rate of thickening of the reacted layer appears to slowly decrease with time. The slowing down is less obvious as regards the weight loss, for which a nearly constant rate seems attained after 6 months. However, the error margin involved in the calculation of G does not allow a definite conclusion on this point.

Somewhat surprising is the considerable resistance to leaching of the glass in contact with liquid brine only (GL): the reacted depth is only about half of that at the GG interface. This is in disagreement with expectation, if based only on the SA/V ratio. However, the GL test was conducted in a different borehole of the WIPP assembly (nr 49; the GG and GS specimens were in nr 07), and the possibility of some temperature or geologic difference cannot be ruled out. The GL specimens also exhibited much less contamination from Ni, Cr and Mn than did those from the 07 borehole. This implies that the heaters and other structural details in borehole 49 were relatively little corroded.

Table IV also lists the mean depletion factors, defined as

$$DF = \int_{A}^{A+L} (c_o/c)\,dl \quad , \qquad (2)$$

measured for the different elements. These factors (again averages from three sets of profiles for GG and four sets each for GS and GL) are based on the assumption that Al remains essentially unleached in the β zones. Although the three types of interfaces yield somewhat different figures, these differences are marginal. Values of DF in excess of ca 20 in practice correspond to nearly complete leach-out. The considerable differences in DF between different elements are typical for a mainly selective leaching mechanism. The <1 listings for Ni correspond to the influx of presumed corrosion products in borehole 07.

A unique feature of leaching in the WIPP site, reported earlier /5, 7, 16/, is the strong influx of magnesium, together with hydrogen apparently interchanging with the atoms (mainly alkali) released from the β zones. In Table IV the interchange balance is illustrated. Although the quantitation for H is relatively uncertain, it is obvious that the combined Mg and H influx approximately corresponds to the nearly total release of Li, Na, Cs, Ca and Sr (i.e., ca 32 atom-% of cations, see Table I). The balance also suggests that the leached-out boron (amounting to about 4-6 % of cations) may be substituted by Mg and H, while about a half of the

Table IV. Two-Year WIPP Test; Leaching at Different Interfaces. GG: Glass-Glass. GS: Glass-Salt. GL: Glass-Liquid Brine.

Interface		GG	GS	GL
Mean leaching depth, L (μm)		1.0	1.15	0.50
Cation loss, G (mg/cm^2)		0.085	0.095	0.045
Mean depletion factors, DF, in β layers	Si	3.8	3.1	3.0
	B	2.5	2.4	1.6
	Al	1	1	1
	Fe	1.1	1.3	1.2
	Li	500	600	1100
	Na	80	45	40
	Cs	150	110	70
	Ca	12	8	5
	Sr	15	15	10
	Ba	16	15	12
	Zr	1.1	2.8	1.1
	Mn	20	45	10
	Ni	(<1)	(<1)	1.4
	Mg	(≅0.07)	(≅0.06)	(≅0.05)
β-layers				
Atom-% retained cations		28	31	32
" Mg influx		22	22	27
" H influx, approx.		(14)	(17)	(15)
" "microvoids", approx.		(36)	(30)	(26)
Composition of α layer (atom-% of cation sum; H not included in 100%)	Mg	52	49	56
	Si	20	32	20
	B	11	10	18
	(H)	(55)	(55)	(25)

vacated cation sites, corresponding mainly to released Si (i.e., about 28-31 %), remains as microporosities.

Another typical feature observed only in the WIPP leaching environment /5/ is the very moderate depletion of boron. In laboratory or burial experiments under other geological conditions /8, 9, 14/, boron is nearly completely removed from the reacted zone, B responding to leaching in about as high extent as Li or Na. Here, on the other hand, it is noted (see Fig.2b and Table IV) that the depletion factor of B is relatively low, in fact only some 2/3 of that of Si, the main network former. The fact that in the brine environment the B network remains largely intact (with high substitution of Mg) is likely to have positive consequences for the durability of the glass.

By comparison, in SRL-Y glass leached for 2 years in granite environment (Stripa mine /8/), the reacted layer was found to be nearly emptied of B as well as Li and Na, altogether ca 40% of all cations. The only sizable influx was H, to the amount of ca 15%. The share of microporosities in the depletion zone could be calculated to be 50% or more.

The depth of the reacted zone (at GG interface) after the 2 year Stripa test was, according to SIMS profiling, only ca 0.5 µm, i.e. only about half of that of the WIPP test. However, recent studies /9/ have yielded evidence that SRL-Y glass in Stripa environment may have more of a tendency to dissolve in the total congruent mode. Extrapolated to 2 years' exposure, the experiments suggest that the total reacted depth may be about 1 µm, i.e. similar as in the WIPP results. As, moreover, the cation release and porosity formation in Stripa is higher than in WIPP, the leaching resistance of SRL-Y in brine can hardly be considered as poorer than in granite repository. In both of these very different geologies, the waste glass appears to perform well.

The effective element balance is probably further shifted in favor of the WIPP test by the occurence of the precipitation layer. The α zone, although evidently permeable to the leaching fluid, is likely to act as a dissolution barrier for the main network forming elements.

The chemical nature of the material in α is yet uncertain. Probably several phases are present /10/, but it is obvious from Figures 1-2 and from Table IV that the dominant components are Mg, Si, B and H. The element composition seems to be roughly similar for all three types of interface, although on the GL surfaces the precipitate appears to contain more B and Mg, and less H. The averaged proportions for the α layers in GG, GS and GL are Mg : B : Si : H ≅ 20 : 5 : 9 : 17. Magnesium is the dominant cation also in the brine, ca 50% of the cation sum /13/. The brine concentrations of B and Si are, respectively, of the order of 1% and $3x10^{-4}$% /13/, which may reasonably imply that at least the precipitated silicon is practically all of glass origin. As seen in Fig.1, the dominant cation of the brine, chlorine, is represented in the α zone only to about 1%. It seems likely that the material in the precipitate is mainly of an oxide-hydroxide type.

CONCLUSIONS

Following features are typical for the leaching behavior of SRL-Y glass in the WIPP-MIIT environment:

a) The thickness of the reacted layer at glass-glass and glass-salt interfaces after 2 years exposure is of the order of 1 μm. Its rate of growth slowly decreases with time.

b) Unique for the WIPP environment: the glasses exhibit a very strong influx of magnesium in the leached layers.

c) In distinction to tests in other environments: the leaching rate of boron is relatively low (less than that of silicon); partial preservation of both networks in the depletion layers.

d) The following reacted layers, counted from surface to bulk, are identified:
   - α, "precipitation" layer. Dominated by Mg, Si, B and H; permeable to leachant.
   - β_0, "interaction" or "substitution" layer. All cations depleted, in various degrees, by substitutions of Mg and H or by formation of microporosities.
   - β_1, "gradient" layer at the reaction front. Lattice formers only slightly depleted. End of influx of Mg.
   - β_2, "diffusion" layer. Practically only Li significantly depleted, by substitution of H.

e) Quantitative depletion factors for all elements, as determined by SIMS, are in line with a mainly "selective" leaching mechanism, with partial re-precipitation on glass surface. No clear evidence of a "congruent" leaching mode is noted.

REFERENCES

1. G.G. Wicks and M.A. Molecke, "WIPP/SRL In-Situ Testing Program". Nuclear Waste Management II, Adv. in Ceramics, Amer. Cer.Soc. (D.E. Clark, W.B. White and A.J. Machiels, eds.), 20, 657-667 (1986).

2. G.G. Wicks and M.A. Molecke, "WIPP In-Situ Testing - MIIT Up-Date". Proc. of Waste Management '88, Tucson Ariz., Vol II, 383-392 (1988).

3. G.G. Wicks and M.A. Molecke, "WIPP/SRL International In-Situ Testing Program - MIIT". Workshop on Testing of High Level Waste Forms under Repository Conditions; Cadarache, France; Eur 12 017 EN (1989).

4. M. Brandys, M. Gong, R.E. Sassoon, Aa. Barkatt and P.B. Macedo, "Leaching of SRL-Y Nuclear Waste Glass in MIIT by Analysis of Brine Solutions". This Volume.

5. A. Lodding, E.U. Engström and H. Odelius, "Elemental Profiling by SIMS of Leached Layers in Repository Tested SRL Waste Glass". Workshop on Testing of High Level Waste Forms under Repository Conditions; Cadarache, France; EUR 12 017 EN (1989).

6. A. Lodding, H. Odelius, D.E. Clark and L.O. Werme, "Element Profiling by SIMS of Surface Layers in Glasses". Mikrochim. Acta Suppl.11, 145-161 (1985).

7. A. Lodding, E.U. Engström and H. Odelius, "Quantitative SIMS, Particularly of Insulating Materials". Fortschrittliche REM-Analyse in der Hochtechnologie; Deutscher Verb. für Materialprüfung, Berlin (V. Thien, ed.), 13, 351-369 (1988).

8. A. Lodding, E.U. Engström, D.E. Clark, L.O. Werme and G.G. Wicks, "SIMS Analysis of Leached Layers Formed on SRL Glasses during Burial". Nuclear Waste Management II, Adv. in Ceramics, Amer.Cer.Soc. (D.E. Clark, W.B. White and A.J. Machiels, eds.), 20, 567-581 (1986).

8a. P.D. Soper, D.D. Walker, M.J. Plodinec, G.J. Roberts and L.F. Lightner, "Optimization of Glass Composition for Vitrification of Nuclear Waste at Savannah River Plant". Am.Cer.Soc. Bulletin 62-9, 1013-1018 (1983).

9. B.K. Zoitos, R.L. Schulz, D.E. Clark and A. Lodding, "A Marker Study of Surface Formation and Partial Dissolution". This Volume.

10. J.A. Tacca and G.G. Wicks, "Leaching of SRL Waste Glass in the MIIT Program; Surface Studies". This Volume.

11. W.G. Ramsey and G.G. Wicks, "Leaching of SRP Waste Glass in Brine; Laboratory Support of MIIT Program". This Volume.

12. A. Lodding, "Secondary Ion Mass Spectrometry". Chapter 4, Inorganic Mass Spectrometry (F. Adams, R. Gijbels and R. van Grieken, eds.); Chem. Analysis Series 95, John Wiley & Sons, New York (1988).

13. M.A. Molecke, N.R. Sorensen and J.L. Krumhansl, "Summary of WIPP Materials Interactions Test Data on Metals Interactions and Leachate Brine Analyses". Workshop on Testing of High Level Waste Forms under Repository Conditions; Cadarache, France; EUR 12 017 EN (1989).

14. D.E. Clark, A. Lodding, H. Odelius and L.O. Werme, "Elemental Analysis of Swdish Nuclear Waste Glasses; Leachability versus Composition". Materials Sci. & Eng. 91, 241-256 (1987).

15. R.E. Sassoon, M. Gong, M. Brandys, M. Adel-Hadadi and P.B. Macedo, "Leaching of Nuclear Waste Glasses Doped with Chemical Tracers in MIIT by Analysis of Brine Solutions". This Volume.

16. J.C. Tait, P.J. Hayward, W.H. Hocking, J. Betteridge and D. Doern, "Analysis of Aluminosilicate Glass Samples from the In-Situ WIPP Salt Burial Test". This Volume.

WIPP/SRL IN-SITU TESTS; LABORATORY SUPPORT EXPERIMENTS

W. G. Ramsey
Department of Ceramic Engineering
Clemson University
Clemson, SC 29634

G. G. Wicks
Westinghouse Savannah River Company
Savannah River Laboratory
Aiken, SC 29808

ABSTRACT

Simulated Savannah River waste glass was leached in brine and deionized water, with and without metals present, at 90°C, in support of the WIPP/SRL in-situ testing program. The behavior of Savannah River Laboratory [SRL] 165/TDS waste glass was assessed by combining solution analysis with bulk and surface studies, and correlating laboratory results with field data. The glass, whose performance was affected by salt phases formed, performed well. The metals 304L stainless steel, Ticode 12, and lead had no significant effect on glass leaching, while A216 adversely affected the leaching of the waste glass system.

INTRODUCTION

The WIPP/SRL Laboratory Leaching Study (SRL Study) was conducted to help assess the behavior of simulated Savannah River waste glass in a controlled salt environment. These tests were performed in support of the Materials Interface Interactions Tests (MIIT), the largest in-situ testing program involving burial of simulated high-level waste forms currently in progress.[1-5] The laboratory study was performed under controlled conditions using a modified MCC-1 leaching procedure.[6] This study is intended to simulate the salt environment of the Waste Isolation Pilot Plant (WIPP) and consisted of 48 500cc Teflon leaching vessels, each containing either three SRL 165/TDS simulated waste glass 'pineapple slice' segments or three MCC ARM-1 (a reference borosilicate composition) specimens. The glasses were leached in one of three environments: deionized water, brine with and without additional salt, or brine with additional salt and a metal sample [Figures 1 and 2]. Data were obtained from aliquots of leachates taken at 14, 30, 90, and 180 days after beginning the test. Glass and metal samples were evaluated after the 30, 90, and 180 days of testing.

SUMMARY

- SRL 165/TDS waste glass exhibited good leaching behavior. Leached glass layers, under the salt deposits, were 1-2 microns thick after 3 mo. of leaching in brine and enriched in chlorine.
- The presence of metals 304L stainless steel, Ticode 12, and lead had no significant affect on glass leaching, while A216 increased leaching of the SRL 165/TDS waste glass system.
- Laboratory leaching data from the SRL Study were similar to MIIT field data on the same glass (SRL 165/TDS), when the larger SA/V ratios in the field were taken into account.
- Outermost salt deposits, indexed primarily as halite (NaCl), formed on all glasses leached in brine.

OBJECTIVES

The major objectives of the SRL Study were to:
- Assess the leaching behavior of SRL 165/TDS simulated waste glass in a controlled laboratory environment
- Assess the effects of potential canister and overpack materials, i.e., 304L stainless steel, TiCode 12, lead, and A216 carbon steel, on leaching behavior of SRL 165/TDS waste glass.
- Correlate laboratory leaching data to data currently available in MIIT in-situ field tests conducted at WIPP.

EXPERIMENTAL PROCEDURES

- <u>Sample Preparation</u>

The glass compositions used in the WIPP/SRL Laboratory Leaching Study (SRL 165/TDS and MCC ARM-1) were the same as those used in the MIIT program.[1] MCC ARM-1 samples were prepared from blocks supplied by the Materials Characterization Center and analyzed by the National Bureau of Standards. These are materials that will be used as standards in future studies. SRL waste glass samples were prepared (batching, casting, grinding, and polishing) at Savannah River with 165/TDS simulated waste glass. The samples were fabricated in the shape of 'pineapple slices' similar to those used in the MIIT test and then fractured into quarters or sixths for the leaching study.

- <u>Sample Matrix</u>

Figure 1 is a schematic representation of the WIPP/SRL Laboratory Leaching Study (SRL Study). The effort consists of three groups of experiments. Group I contains glass samples leached in deionized water, which allows this leaching study to be compared with previous SRL durability studies and standardized MCC-1 leaching test data. The effects of salt and brine leachant on waste glass and the ARM standard was the focus of Group II. By comparing Groups I and II, it was possible to assess surface and leach rate variations between the two leachants. The glass-metal-brine interactions (Group III) section was the largest and most significant portion of the program. In Figure 2, a schematic representation of a leaching vessel used in the SRL Study is given. Surface studies and solution analyses of those samples were used to determine the effects of proposed canister and overpack materials on leaching behavior of SRL 165/TDS simulated waste glass. The systematic testing effort of the SRL Study was designed to allow a direct progression from 'idealized' laboratory conditions (MCC-1 test; Group I), to a salt environment (Group II), to a salt environment with support materials (Group III), to actual in-situ field data (MIIT).[7] All tests were conducted at 90°C and involved low and high ratios of surface area of sample to volume of leachant (10 and 50m-1).

- <u>Sample Analyses</u>

Brine leachates derived from the SRL Study were analyzed by Inductively Coupled Plasma Spectroscopy (ICP). Salt deposits were characterized by Wide Angle X-ray Diffraction (WAXD) and leached glass samples investigated by Scanning Electron Microscopy (SEM) and Energy Dispersive X-rays (EDX).

ANALYSIS AND DISCUSSION

Brine leachates were diluted and analyzed by ICP. Analyses focussed on lithium, one of the few elements that is present in the glass and not present in large amounts in the brine or surrounding salt. Therefore, the lithium found in solution, corrected for background levels, was attributed primarily to that released from the glass. More quantitative techniques, using lithium isotope separations[8] and monitoring of other species[9], have recently been developed by Macedo et al.

Salt deposits were found on all samples of SRL 165/TDS waste glass and ARM-1, when exposed to simulated WIPP brine. They appeared inhomogeneous and non-uniformly deposited on glass surfaces. The WAXD pattern for the deposited salt, shown in Figure 3, indicated that the main phase formed was halite (NaCl).

Leaching of the ARM-1 standard glass did not produce any characteristic leached surface layers. In contrast, leached layers were observed after leaching SRL 165/TDS waste glass for 3 months at 90°C, in WIPP A brine [Figure 4]. A leached layer was produced that was about 1-2 microns in size. The composition of this layer, determined by EDX, shows that it is characterized by the major components of the glass along with chlorine from the brine.

In Figure 5, the normalized elemental mass losses (NL) for SRL 165/TDS waste glass, based on lithium extraction, are shown after leaching in deionized water and brine. From these data, it is clear that deionized water is a much more aggressive leachant than the brine. Since many 'standardized' laboratory leach tests use water as leachant, this provides a 'worst case' scenario concerning the leaching of the waste glass system and actual performance in a geologic setting may be significantly better.

The effects of four metals, 304L stainless steel, Ticode 12, lead, and A216 mild steel, on the leaching behavior of SRL 165/TDS, is shown in Figure 6. Waste glass leached in the presence of 304L stainless steel performed similarly to glass leached in the presence of Ticode 12, and both glasses leached similar to glass leached in the absence of these same metals. Based on release of lithium, the leach rates were about 0.025 g/m^2-day after 90 days. Visual examination of the 304L metal samples showed a small amount of surface corrosion while no observable corrosion was detected on the Ticode 12 metal samples.

Leachates removed from vessels containing A216 carbon steel showed a similar solution pH (6.0) as those containing Ticode 12, 304L, or no metal, yet had a 'rust red' color. The leachates also exhibited a higher concentration of lithium. (Based on release of lithium, the leach rate was about 0.125 g/m^2-day, five times higher than the leach rate of glass in contact with the other metals.) The adverse effects of carbon steel on leaching of SRL glass have been observed before[10] and attributed to formation of iron silicate precipitates which can form in solution, as postulated elsewhere[11]. This results in a depletion or undersaturation of silicon in solution and can cause a larger driving force for increased dissolution. The effect of lead on the durability of the waste glass was not significant and may have actually been slightly beneficial. The brine leachates that contained lead had an equilibrated solution pH of approximately 7.5 and leachates showed consistently low lithium levels (based on release of lithium, the leach rate was about .02 g/m^2-day). Type 304L stainless steel, Ticode 12, and A216 metal systems have been proposed as potential canister or overpack materials.

COMPARISION WITH MIIT

One of the more important parameters affecting glass leaching is the ratio of glass surface area to leachant volume [SA/V]. The SA/V ratios of the laboratory SRL Study were 10 and 50 m^{-1}. The value of 10 m^{-1} is a standardized laboratory leaching test ratio which is also roughly equivalent to the starting value in the MIIT field experiments. The value of 50 m^{-1} was used to approximate a larger SA/V that could be achieved in the lab, but still using solid samples. In the the actual field experiments, SA/V rapidly increased due to evaporation and migration of brine from the boreholes so the final SA/V most likely ranged from 10 m^{-1} to orders of magnitude larger. This change is realistic and to be expected in a salt repository environment and cannot be easily imitated in a laboratory using solid samples. To compare the laboratory and field data, the changes in SA/V were taken into account by plotting concentrations of lithium found in leachates in the SRL Study and the MIIT program as a function of the scaling parameter, SA/V * time. This scaling parameter has been used successfully on other studies involving SRL waste glass[12]. An empirical fit was made with adjustment of SA/V. As shown in Figure 7, for three sets of data the adjustment in SA/V was 100x and for one set of data, 25x. These corrections, which are empirically derived, are due to the higher SA/V conditions occurring in the field, due to loss and migration of brine leachant.

In Figures 7A and 7B, the lithium concentrations resulting from the leaching of SRL 165/TDS in brine in the presence of 304L stainless steel and Ticode 12 are shown and found to be similar. A good correlation between laboratory and field data is achieved by multiplying the lithium concentrations observed in the laboratory tests by a factor of 100. In Figure 7C, a similar correlation is observed for leaching of the glass in the presence of lead and also correlated with field data using the same factor. In addition, these curves appear to be leveling off at SA/V * time of about 1000.

The 165/TDS waste glass leached in the presence of A216 also showed a correlation between laboratory and field data, this time with a 25x correction factor (Figure 7D). Unlike the interactions described above, waste glass leached in the presence of A216 carbon steel did not exhibit a leveling off of lithium measured in solution. Instead, the concentration continued to increase, thus indicating more significant corrosion of the waste glass. Future studies will focus on improved brine analyses using additional elements as indicators of leaching behavior.

DISCUSSION

SRL 165/TDS waste glass leached in the laboratory exhibited similar leaching characteristics compared to the same glass leached at WIPP in the MIIT in-situ testing program. Based on both surface and solution analyses, the waste glass system performed well in both environments and exhibited low leaching rates.

Salt deposits observed on laboratory leached glass surfaces were indexed primarily as halite. Under these deposits, glass leached layers were enriched in chlorine and 1-2 microns thick. In the MIIT field tests, similar deposits and leached glass layers were noted.[13] In the field tests, however, the deposited layers also contained a series of minor salt phases in addition to halite and the leached glass layers were enriched in both magnesium as well as chlorine.[13]

Laboratory leaching data also agreed with field data when the higher SA/V ratio experienced in the field was taken into account. The use of the SA/V*t as a scaling parameter also provides a means for extrapolation of leaching data to longer time periods.

The effects of metals on the leaching behavior of the SRL waste glass system were assessed in the laboratory and found to be consistent with behavior observed in the field.[14] Type 304L stainless steel (the canister material for the Defense Waste Process Facility (DWPF)) and Ticode 12 (possible overpack material) had no significant effect on waste glass leaching in the salt/brine environment. The only significant effect observed was due to A216 carbon steel, which increased leaching of the SRL waste glass system and therefore, would be undesirable in this environment.

REFERENCES

1 G. G. Wicks, "WIPP/SRL In Situ and Laboratory Testing Programs-Part I: MIIT Overview, Nonradioactive Waste Glass Studies," USDOE Report DP-1706, Savannah River Laboratory, E. I. du Pont de Nemours & Co., Aiken, SC (1985).

2 G. G. Wicks, M. E. Weinle, and M. A. Molecke, "WIPP/SRL In-Situ Tests-Part II: Pictorial History of MIIT and Final MIIT Matrices, Assemblies, and Sample Listings," USDOE Report DP1733, Savannah River Laboratory, E. I. du Pont de Nemours & Co., Aiken, SC (1987).

3 M. A. Molecke and G. G. Wicks, "Test Plan: WIPP Materials Interface Interactions Test [MIIT]," Sandia National Laboratories, Albuquerque, NM (1986).

4 G. G. Wicks and M. A. Molecke, "WIPP/SRL In-Situ Testing Program," Nuclear Waste Management II, Advances in Ceramics, American Ceramic Society, D. E. Clark, W.

B. White, and A. J. Machiels, Eds., Vol. 20, pp. 657-667 (1986).
5 G. G. Wicks and M. A. Molecke, "WIPP/SRL In-Situ Testing-MIIT Update," Waste Management '88, Vol. II, pp. 383-392 (1988).
6 U. S. Department of Energy, Report DOE/TIC-1400 "Nuclear Wste Materials Handbook," MCC-1P Static Leach Test Method," Materials Characterization Center, Pacific Northwest Laboratory, Richland, WA (1982).
7 G. G. Wicks, "Nuclear Waste Vitrification - The Geology Connection," J. Non-Crystalline Solids, Vol. 84, pp. 241-250 (1986).
8 R. E. Sasoon, M. Gong, M. Adel-Hadadi, M. Brandys, A. Barkatt, and P. B. Macedo, "Monitoring of MIIT Glass Solution Interactions by Brine Analysis," presented at the Workshop on Testing of HLW Forms Under Repository Conditions, Cadarache, France, October 1988, to be published in proceedings (1989).
9 M. Brandys, M. Gong, R. E. Sasoon, A. Barkatt, and P. B. Macedo, "Leaching of SRL-Y Waste Glass in MIIT by Analysis of Brine Solutions", presented at the 4th International Symposium on Ceramics in Nuclear Waste Management, Indianapolis, IN, April 23-27, 1989, to be published in proceedings (1989).
10 D. B. Burns, B. H. Upton, and G. G. Wicks, "Interactions of SRP Waste Glass with Potential Canisters and Overpack Metals," J. of Non-Cryst. Solids, Vol. 84, pp. 258-267 (1986).
11 G. L. McVay and C. Q. Buckwalter, PNL-SA-10474, Pacific Northwest Laboratory, Richland, WA (1982).
12 G. T. Chandler, G. G. Wicks, and R. M. Wallace, "Effects of SA/V and Saturation on the Chemical Durability of SRP Waste Glass," Nuclear Waste Management II, Advances in Ceramics, American Ceramic Society, D. E. Clark, W. B. White, and A. J. Machiels, Eds., Vol. 20, pp. 455-463 (1986).
13 J. A. Tacca and G. G. Wicks, "WIPP/SRL In-Situ Tests; MIIT Program-Surface Studies of SRL Waste Glasses," presented at the 4th International Symposium on Ceramics in Nuclear Waste Management, Indianapolis, IN, April 23-27, 1989, to be published in proceedings (1989).
14 Proceedings of the Workshop on Testing of High Level Waste Forms Under Repository Conditions, October 17-21, 1988, Commission of the European Communities (CEC), EUR 12 017 EN, R. Simon, C. Sombret, and G. Wicks, Eds., to be published (1989).

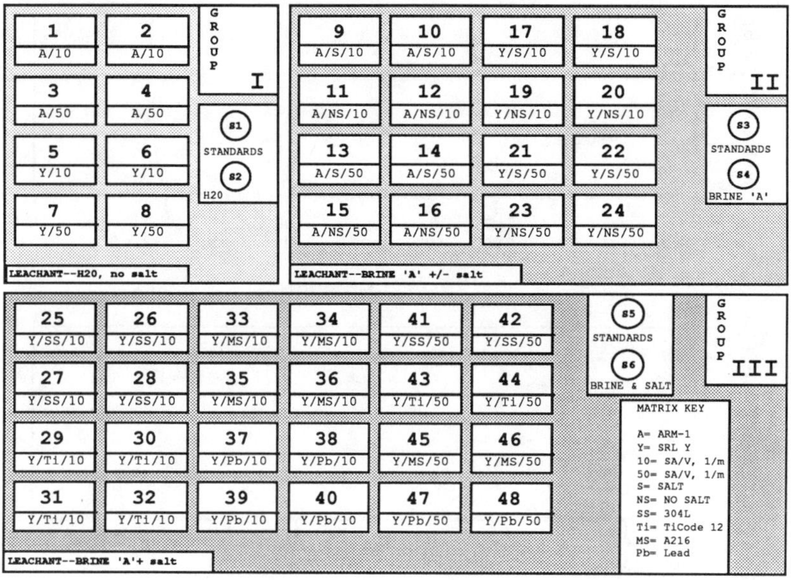

Figure 1. Schematic of the Leaching Study Test Matrix. Each Rectangle Represents One Leaching Vessel.

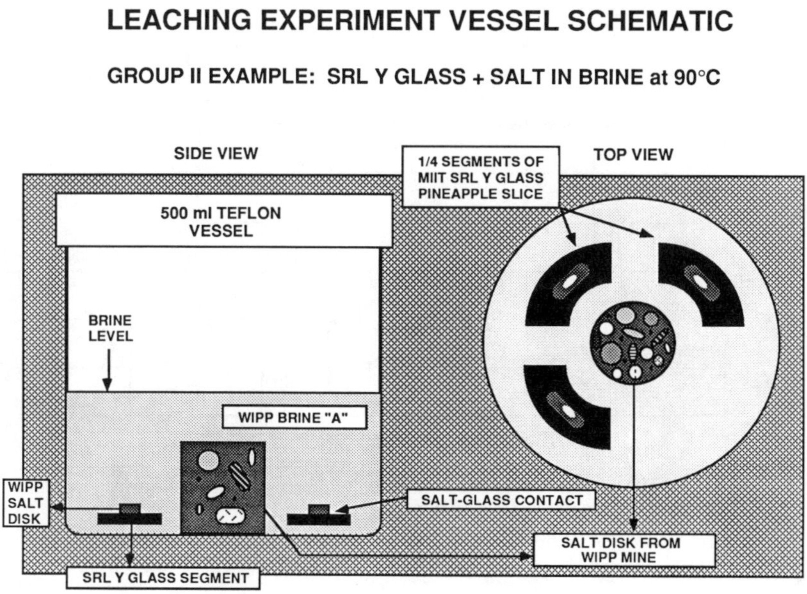

Figure 2. Schematic Representation of Leaching Study Test Vessel.

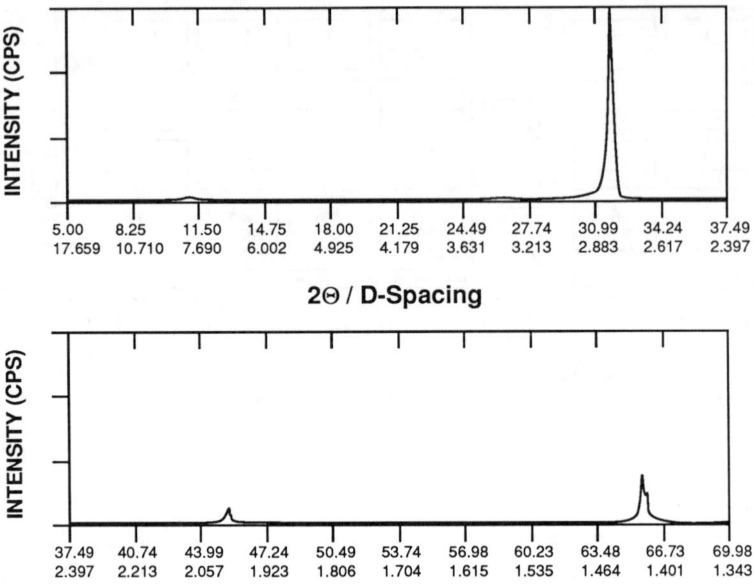

Figure 3. X-ray Diffraction Pattern (Computer Simulation of Observed Data) of SRL 165/TDS Simulated Waste Glass and Salt Deposit After Leaching in WIPP Brine A.

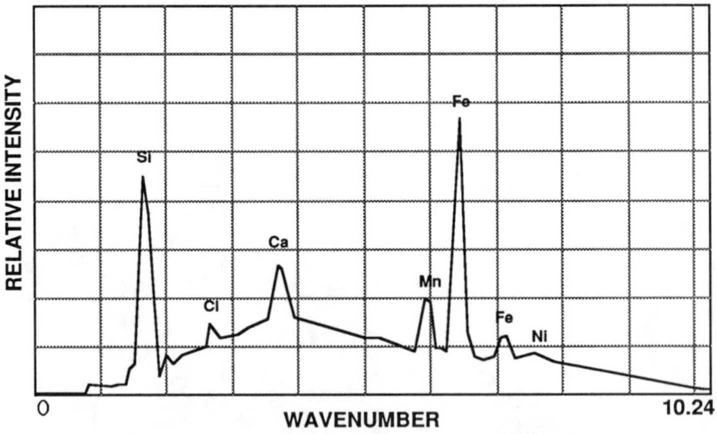

Figure 4. SEM/EDX Analysis (Computer Simulation of Observed Data) of Surface Layer of SRL 165/ TDS Waste Glass (Brine Leached).

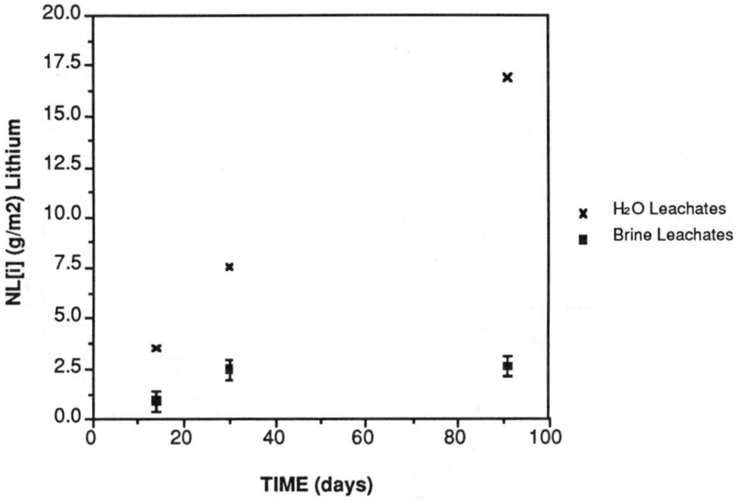

Figure 5. Normalized Elemental Mass Losses, Based on Lithium Release, of SRL 165/TDS Waste Glass After Leaching in Deionized Water and Brine.

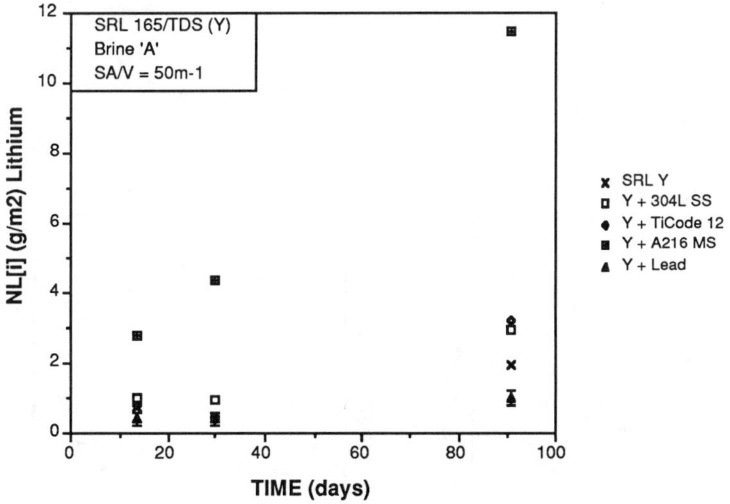

Figure 6. Normalized Elemental Losses, Based on Lithium Release of SRL 165/TDS Waste Glass Leached in the Presence of Metals.

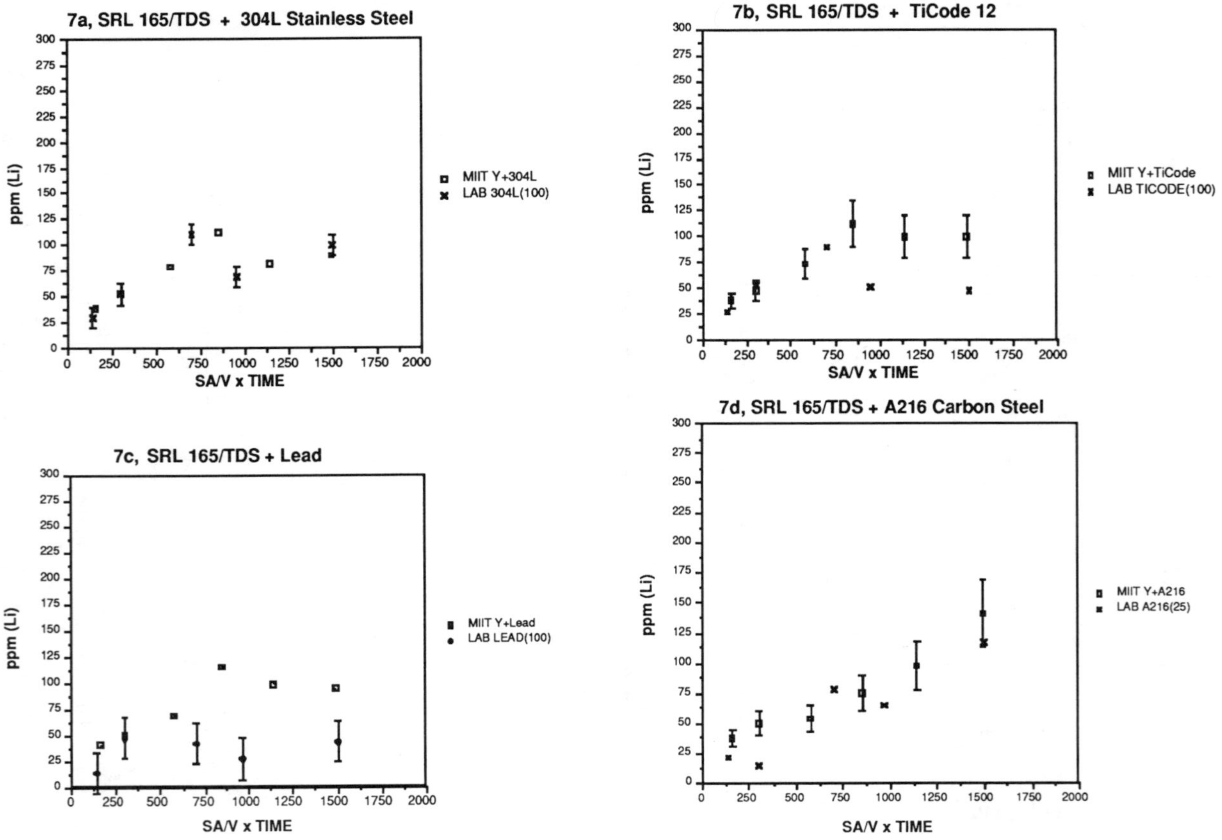

Figure 7. Effects of Metals on the Leaching Behavior of SRL 165/TDS Waste Glass and Correlation of Data in Laboratory and Field Studies.

PREDICTION OF NEAR-FIELD FLUID COMPOSITION IN HIGH-LEVEL NUCLEAR WASTE REPOSITORIES: DATA FROM GEOTHERMAL AND EXPERIMENTAL ANALOGS

D. E. Grandstaff, G. C. Ulmer, G. H. Kacandes
Temple University, Department of Geology,
Philadelphia, PA 19122 USA

ABSTRACT

Fluids from active geothermal fields and hydrothermal experiments have been used to predict the composition of groundwaters passing through a high-level nuclear waste repository. Predictions from both analogs are similar over most of the temperature range. This suggests that possible deficiencies in the experiments (short duration) and geothermal fields (composition differences from proposed repositories) do not invalidate their use as analogs. Differences in fluid composition parameters found at high temperatures (>250°C) appear largely due to magmatic gas inputs into the geothermal fluids.

INTRODUCTION

Performance assessments for a high-level nuclear waste (HLW) repository require prediction of the composition of heated groundwaters passing through the repository. Fluid composition parameters (for example, pH, oxidation state, and content of specific complex-forming species) influence rates of container corrosion and waste-form dissolution, oxidation states of some redox-sensitive elements (U, Np, Tc, S, Se, Fe, Mn), and mobility of radionuclides. The fluid composition also influences the nature of alteration minerals formed, affecting adsorption and incorporation of radionuclides and retardation of their movement.

Two of the most promising analogs for predicting fluid compositions in a HLW repository are active geothermal systems and laboratory-based hydrothermal experiments. Active geothermal systems are widespread and well studied and should offer, within their range of host environments, an analog to the artificial geothermal system which may be generated in a HLW repository. Hydrothermal experiments can and have been designed to closely reproduce repository conditions (e.g., pressure, temperature, rock and solution composition). Results of such experiments have proven successful in modeling natural counterparts of other systems. Thus, both of these analogs should be of some forecasting value to nuclear waste isolation. The purpose of the present study was to identify possible agreements between and possible shortcomings in each of these analogs and assess their predictive value.

DATA SOURCES AND DATA REDUCTION

Published fluid data (164 analyses) have been collected from several continental geothermal fields (Table 1) and from results of hydrothermal experiments (67 solution analyses) from a number of investigators (Table 2). High temperature solution parameters were calculated from published low temperature data using the program HIPH/FOLLY[1]. Where necessary, values of association constants have been pressure-corrected following the method of Bjerrum[2]. Gas, liquid, and deep steam fractions for data from wet-steam wells were calculated assuming instantaneous boiling in the well (single stage flashing at the wellhead), using the sampling pressure, reservoir temperature, and measured enthalpy (for high enthalpy wells), unless literature interpretations accompanying the data suggested that other models were required.

Concentrations of uncharged dissolved species (H_4SiO_4, H_2CO_3, HCl°, etc.) may be directly compared in various fluids; however, the concentrations of ionic species are usually affected by fluid ionic strength, therefore, comparisons for these species usually involve cation/cation, cation/hydrogen, or anion/hydroxyl ion activity ratios[1,3] (Na/K, $Na/Ca^{0.5}$, Na/H, Ca/H^2, etc.).

TABLE 1. **GEOLOGICAL CHARACTERISTICS OF GEOTHERMAL FIELDS WHOSE SOLUTION DATA ARE CONSIDERED IN THIS PAPER.**

FIELD	ROCK TYPE	SOURCE WATERS	TECTONICS	REF
Iceland	basaltic	meteoric, glacial,and seawater low to high salinity	Mid-Atlantic Ridge	3,8,9,13 14,16,17
New Zealand (Taupo Volc. Zone)	rhyolite, tuff and sediment	meteoric $\pm$ seawater	transform zone	5,6,7
Kenya (Olkaria)	rhyolite, trachyte basalt, sediments	low salinity	African rift valley	18
Japan	andesite	intermediate salinity	Subduction	7, 19
Mexico (Cerro Prieto)	granite and sediments	high salinity	East Pacific Rise	7, 20
Idaho (various)	granite and sediments	low salinity	Hot-spot volcanism	21, 22
Guatamala	andesite	low salinity	Subduction	23
Hawaii (HGA-P)	basaltic	seawater $\pm$ meteoric	Hot-spot	24
Philippines	andesite	high salinity	Subduction	7, 25

PROPERTIES OF GEOTHERMAL FLUIDS

Figure 1 shows data for some ion activity ratios and neutral species concentrations from several geothermal fields (Table 1). These geothermal fields are set in several very different host environments (e.g. Iceland vs. New Zealand). Rock types range from basalt to granite. Source water compositions vary from near meteoric to near seawater salinities and the fluid flow is influenced by diverse hydrologic factors. Despite these and other major differences, inspection of data in Figure 1 reveals that most variation in individual parameters is temperature dependent; there appears to be little dependence on other parameters such as rock or water composition, hydrology, or pressure. Such concordance has been observed by a number of workers. Some temperature dependences are so distinctive that Fournier and Truesdell[4] and others[5,6,7,8] have previously used certain cation ratios (for example Na/K and $Na/Ca^{0.5}$) and dissolved silica concentrations (Figures 1A-C) as geothermometers for a variety of geothermal fluids. From an analysis of Icelandic geothermal fluids, Arnórsson *et al.*[3,8] have shown that values of cation/proton ion activity ratios, neutral species concentrations, and dissolved gas fugacities are also, for the most part, simply functions of aquifer temperature (Figs. 1D-J). The similarity in fluid composition parameters in these varied geothermal environments probably results from control by approach to equilibrium with secondary phases coupled with similarity in secondary phase assemblages found in most geothermal regions[10].

Redox state and pH are two very important factors influencing repository performance. The oxygen fugacity (redox) state of geothermal fluids (as calculated from SO_4^{2-}/H_2S) is shown in Figure 1I. Geothermal fluids plot near magnetite-hematite (MH) or

TABLE 2. **WATER/WHOLE ROCK EXPERIMENTS COMPARED IN THIS PAPER**

REACTED ROCK	STARTING SOLUTION	T(°C)	MR*	REF
Qtz tholeiite-Columbia Plateau (RUE)	SGW* (GR-3) I = 0.017	100, 200, 300	5-50	11
Qtz tholeiite-Columbia Plateau (RCE)	SGW (GR-4) I = 0.015	100, 200, 300	10-140	26, 27
Olivine tholeiite-Krafla, Iceland	SGW I = 0.003 $\pm$ CO_2	100, 200, 300	10	1, 28
Olivine tholeiite-Krafla, Iceland	SGW (GR-4)	300	10	28
Basalt, Clear Lake, California	DDW*, 0.1% and 2% NaCl	200, 300	3	29
Glassy rhyolite, McDermitt, CA	DDW, 2% NaCl	200, 300	3	29
Andesite, Hakone,Japan	0.1% NaCl	200,300	10, 50	29
Basalt, New Zealand	DDW	150-350	2	30, 31
Andesite, New Zealand	DDW	150-350	2	31
Dacite, New Zealand	DDW	150-350	2	31
Rhyolite pumice, New Zealand	DDW	150-350	1	31
Monzogranite Ossian,Scotland	DDW	100	10	32
Carnmenellis Granite, UK	Meteoric water (I=0.002)	80-250	10	33
Sedimentary and metamorphic Rocks	DDW	200, 300	1-2	34
Well-cuttings (Nesjavellir, Iceland)	SGW (KRAGW) I = 0.003	300	10	1

* SGW = simulated groundwater, for compositions of specific groundwaters see individual references
DDW = distilled deionized water, I = ionic strength. MR = Mass Ratio: the initial water/rock mass ratio.

between (MH) and quartz-fayalite-magnetite (FMQ) buffer values[9]. The redox values may be controlled by equilibrium with mineral assemblages containing magnetite, hematite, epidote, chlorite, and other iron-bearing phases. Figure 1L shows calculated pH values for fluids from various geothermal fields. Superimposed on these data is the line for neutral pH as a function of temperature. The pH values of most reservoir fluids are somewhat basic at low temperatures and near neutral at higher temperatures. Ellis and Mahon[5] have suggested that the pH of geothermal fluids may be described, as a first approximation, as if resulting from equilibrium between quartz, potassium feldspar, and muscovite. Superimposed on data in Figure 1L is the pH predicted for equilibrium between sodium feldspar-potassium feldspar, muscovite, and quartz (line KMQ) as a function of temperature for $m_{Na} + m_K = 0.01$, near the average for Icelandic geothermal data (I) displayed in this figure. The pH values for Icelandic geothermal fluids (and other low salinity geothermal fluids) do form a rough envelope about the calculated KMQ line. The value of pH is also a function of ionic strength. Therefore, the pH values for some other geothermal fields, for example New Zealand (Z) and Cerro Prieto (M), are somewhat more acidic because of their higher salinities. This illustrates that part of the pH scatter in Figure 1L is due to variations in fluid ionic strength.

The absence of major *systematic* differences in fluid parameters between these geothermal fields, despite differences in rock type and ionic strength of geothermal solutions, tends to support the conclusion that variations in rock composition, mineralogy, geothermal fluid origin and composition, overburden pressure, permeability and residence time do not greatly perturb the water geochemistry development at a given temperature[3]. These trends allow prediction of groundwater chemistries in geothermal fields. This suggests that differences in rock or fluid composition do not necessarily invalidate use of natural analogs in repository predictions. This is the type of predictability that is required for proper design and performance evaluation for a HLW repository and its components.

Although there are no systematic differences, examination of Figure 1 (and 2) reveals that fluid parameter data do scatter, forming bands having widths from less than one to more than three log units at a given temperature. The scatter may be due, in part, to analytical errors, but also arises from fluid equilibration with minerals of different compositions and structural states[3] and different degrees of fluid/mineral equilibration. This scatter imposes limits on the precision of predictions which can be made from data obtained from geological analogs and experiments. That the isothermal scatter is generally less than 2 log units despite the major differences in host environments (Table 1) still places important restrictions on predicted repository fluid compositions.

It it tempting to use these data directly for performance assessments; however, these geothermal analogs differ somewhat from proposed repositories in rock and fluid composition and host environment, for example the geothermal fluids are magmatically driven and contain important components (acidic magmatic gases, e.g. CO_2 and H_2S) that are not expected in a mined repository. The effects of these compositional differences can be evaluated by carefully designed experiments.

OBSERVATIONS FROM EXPERIMENTS - COMPARISONS WITH GEOTHERMAL FLUIDS

Water/whole rock studies have been conducted by several investigators. Rock types reacted have included basalt, granite, rhyolite, volcanic glass, and sedimentary rocks.

Initial fluids reacted range from distilled water to solutions approximating compositions of specific low-salinity groundwaters. Reactants and conditions employed by various investigators examined in this paper are summarized in Table 2. Long-term stabilized solution compositions are used in the following comparisons. Two types of comparisons can be made: comparisons between experiments and comparisons of experiments with geothermal analogs. Symbols designating data from various experiments and geothermal fields are given in the caption to Figure 2.

Possible repository parameters such as temperature, pressure, and rock and mineral composition may be precisely reproduced in these experiments. However, even the longest experiments last only a few thousands of hours whereas a repository must retain radionuclides for many thousands of years. This time-scale difference is one factor which has lead some to question the validity of experiments as a predictive tool.

High-temperature solution parameters calculated from these experiments are remarkably similar (Figure 2). For example, for data at 300°C, the temperature at which many of the experiments have been conducted, there are no major systematic differences between data of various investigators; it appears that differences in rock composition and mineralogy or starting solution composition as well as initial water:rock mass ratio do not have a major *systematic* effect on the nature of the resulting solutions. This behavior thus closely parallels observations from geothermal fields.

Comparisons of results from experiments with data from geothermal fields reveal many similarities[11,12]. For example, the pH values for fluids from both experiments and analogs are basic to neutral, redox values are near magnetite-hematite (MH), and values of most cation/cation and cation/proton ratio overlap. Over most of the temperature interval and for most fluid parameters it appears that no significant offsets between the data can be attributed to the short-term nature of the experiments or variability of geothermal environments.

However, the comparisons do reveal some differences. For example, Figure 2A shows comparisons for Na/K ion activity ratio, The Na/K ratio in geothermal fluids shows marked dependence on aquifer temperature (Figure 1A and [•] symbols in 2A); however, although values of ion ratios are similar, the long-term Na/K free ion activity ratios from experiments show little or no such temperature dependence. The lack of concordance in the two data sets may result from slow exchange rates and the short-term nature of the experiments. The Na/K geothermometer equilibrates slowly in geothermal fields and the experiments may not have had sufficient time to produce valid Na/K values and geothermometer temperatures, particularly at lower temperatures. In contrast, comparison is better for the $Na/Ca^{0.5}$ ion activity ratio (Figure 2B), probably because this ratio is established by more rapid ion-exchange between clay minerals. Dissolved silica concentrations in most experiments are slightly higher than in geothermal fields (2C). Silica concentrations in most closed-system experiments are near cristobalite saturation values whereas in higher temperature geothermal fluids values tend toward quartz saturation. This difference may be due to sluggish quartz growth kinetics and the presence of rapidly dissolving mesostasis. Thus this difference is due to formation of a metastable mineral in the experiments.

Cation/proton activity ratios (Figures 2D-H), CO_2 fugacity (Figure 2J), total dissolved carbonate (ΣCO_2) values (Figure 2K), and pH (Figure 2L) are similar at lower temperatures (<250°C) but differ somewhat, particularly at higher temperatures (>250°C). The difference is shown most clearly in comparisons of cation/proton activity ratios for experiments at 300°C and geothermal field data between 270° and 325°C. Figure 3 shows an example for Ca/H^2 *vs.* Na/H activity ratios. Although the $Na/Ca^{0.5}$ ratio is similar in both the experiments and geothermal fields (both sets of data lie along a diagonal line having 1:2 slope), the data do not appreciably overlap. The difference in cation/proton activity ratios is due largely to a difference in pH between geothermal fluids and experiments (Figure 2L) rather than differences in cation concentrations. The calculated high-temperature pH of geothermal fluids is usually between 5 and 7, depending on ionic strength. However, the pH from the 300°C experiments is generally near 7 to 8 (Figure 2L) a difference of one to two pH units. Similar offsets may also be seen in potassium- and magnesium-proton ion activity ratios[1]. Although the data do scatter somewhat, in both cases the cation activity ratios (e.g. Ca/Mg and Na/K) produced in experiments are very similar to those in the geothermal fields (along the same diagonal lines having 1:1 slopes), and the offset is produced primarily by a difference in the hydrogen ion activity. The nature of these offsets place certain constraints on the mechanisms which produce them.

POSSIBLE MECHANISMS PRODUCING THE OBSERVED OFFSET AND pH DIFFERENCES: IMPLICATIONS FOR REPOSITORY ASSESSMENTS

A number of possible reasons might be cited to explain the difference in pH, among these are: (1) formation of metastable (i.e. saponite rather than chlorite) or poorly crystalline phases and lack of equilibration in the experiments due to their short-term nature, and (2) input of magmatic acids (CO_2, H_2S, HCl, etc.) into high-temperature geothermal fields (natural open systems vs. closed system experiments).

Formation of metastable minerals in experiments is often cited as a reason for the differences observed in Figure 3. Although metastable minerals or mineral assemblages are found in geothermal fields they are very common in solid reaction products from whole-rock experiments. For example, smectite, cristobalite and sanidine are often found in reaction products from 300°C experiments whereas chlorite, quartz, and microcline are the stable equivalents most often found in geothermal fields at this temperature. Furthermore, minerals common in geothermal fields, such as epidote or prehnite, are absent from experiments. If the solution composition is controlled by equilibrium with secondary phases then substitution of metastable phases or omission of minerals would alter the resulting fluid composition. To test whether metastability has an appreciable effect an experiment (NG-7) has been conducted in which well-cuttings (chlorite, quartz, amphibole, feldspar, calcite with minor basalt) from 1800 m depth of Nesjavellir geothermal field (southern Iceland) well NG-7 (this corresponds with the 300°C isotherm[13]) were reacted with simulated low-salinity Icelandic groundwater in Dickson autoclaves at 300°C and 30 MPa[1]. Although the stable solution parameters achieved in that experiment (some results shown by [□] in Figure 4) are closer to those observed in the Nesjavellir geothermal field fluids (region of geothermal fluid composition values shown by results given by stippled area in Figure 4), the experiment did not entirely reproduce the natural values. The pH and Ca/H^2 and Mg/H^2 ratios (Figure 2, 3) were slightly higher and fCO_2 lower than observed values[1,13,14]. Thus substitution of a stable phase assemblage did not seem entirely able to reproduce the fluid composition from this field, at least within the duration of the experiment (1100 hrs).

Further, it seems unlikely that offsets in composition parameter data are due solely to the short-term nature of the experiments, because at higher temperatures reactions would be much more rapid and equilibrium most likely to be achieved during experiments. Thus, if the short-term nature of experiments were the factor causing the offset it would be most reasonable that the offset would be greatest at low temperature and less at high temperatures. This pattern is not observed except in the Na/K trends.

Although dissolved carbonate (ΣCO_2) is an abundant constituent in high-temperature geothermal fluids it is present only in much lower concentrations in low-temperature geothermal waters or groundwaters feeding geothermal zones (Fig 1K). This suggests that CO_2 is added, from either magmatic or metamorphic sources, to cold groundwaters as they enter high-temperature regions. A continuous flux of CO_2 is necessary to support emanations of this gas found in most geothermal settings[15]. This type of open-system process has not been modeled in most of the closed-system autoclave experiments.

Introduction of CO_2 or other acidic gases would inevitably lower the pH of the fluids. Indeed, in extreme cases, addition of sulfuric or hydrochloric acid vapors, due to seawater interactions or H_2S oxidation in near-surface environments, can produce pH's near 1. To simulate magmatic gas additions a high concentration of CO_2 (*ca.* 0.1 m total CO_2) was injected into the autoclave during one experiment[1] conducted at 300°C and 30 MPa (KRA-5) to reproduce the highest ΣCO_2 concentrations found in high-temperture geothermal fluids. The long-term (1100 hrs) high-temperature pH value achieved in that experiment was 6.1 and resulting Na/H and Ca/H^2 cation/proton ratios are shown by (✕) in Figure 4. The values of pH and fluid parameters achieved during that experiment do appear to reproduce those found in some geothermal fluids. Therefore, it appears that open-system behavior, addition of acidic gases to magmatically-driven geothermal fluids, might account for much of the discrepancy in the high-temperature data. However, such additions of CO_2 or other acidic gases should be absent from engineered nuclear waste repository environments. Therefore, in this respect, fluids from high-temperature geothermal fields may not be strict analogs for a nuclear waste repository. In this case the pH and other fluid properties from autoclave experiments may be more directly comparable to those expected in nuclear waste repositories. Finally, the temperature range in which the differences occur (> 250°C) is generally higher than that expected in HLW repositories.

CONCLUSIONS

Although data from geothermal fields do scatter the properties of geothermal fluids appear predictable and solely a function of temperature. Differences in rock type, fluid composition, fluid residence time and other parameters do not appear to generate any major systematic differences between fluids from various geothermal fields. This appears to be paralleled in laboratory experimental data. Fluid compositions produced in experiments are quite similar to those of geothermal analogs over most of the temperature range examined. This suggests that the possible deficiencies in experiments (short time period) and analogs (differences from proposed repositories) are not critical and do not invalidate their predictive use over most of the temperature range. Differences in fluids which are found at high temperatures (> 250°C) appear primarily to result from inputs of acidic magmatic gas (e.g. CO_2 and H_2S) into the geothermal fluids. Because such inputs should be absent from nuclear waste repositories, the high temperature geothermal fluid parameters must be used with discretion as analogs for a nuclear waste repository. High-temperature geothermal fluids having low total CO_2 may prove to be better analogs than those chosen in this study.

Data from both analogs and experimental fields suggest that repository fluids will have redox values near or below magnetite-hematite (MH) values (neglecting radiolysis effects) and pH values will be neutral to basic. Solutions should be saturated with respect to quartz (at higher temperatures) or chalcedony (at lower temperature) and should have ionic ratios (e.g. Na/K and $Na/Ca^{0.5}$) within the range predicted by geothermal analogs. The scatter in fluid parameters from geothermal fields and experiments imposes limits on the precision of predictions which can be made. However, the data still place important restrictions on predicted repository fluid compositions. Given anion concentrations the absolute cation concentrations expected in a nuclear waste repository, as a result of water-rock interactions, may be calculated from empirical temperature functions developed from geothermal data. Such predictions have important implications for design of repository components and modeling of repository performance. The sincere criticisms about the short duration of experiments can be answered: in moving from thousands of hours to thousands of years, no major geochemical surprises await repository engineers.

REFERENCES

1. G.H. Kacandes and D.E. Grandstaff, "Differences between geothermal and experimentally derived fluids," *Geochim. Cosmochim. Acta*, **53**, 343-358 (1989).

2. R.M. Fuoss and C.A. Krauss, "Properties of electrolyte solutions. III. The dissociation constant," *J. Amer. Chem. Soc.*, **55**, 1019-1028 (1933).

3. S. Arnórsson, E. Gunnlaugsson, and H. Svavarsson, "The chemistry of geothermal waters in Iceland. II, Mineral equilibria and independent variables controlling water compositions," *Geochim. Cosmochim. Acta*, **47**, 547-566 (1983a).

4. R.O. Fournier and A.H. Truesdell, "An empirical Na-K-Ca geothermometer for natural waters," *Geochim. Cosmochim. Acta* **37**, 1255-1275 (1973).

5. A. J. Ellis and W.A.J. Mahon, Chemistry and Geothermal Systems. Academic Press, New York, 1977.

6. A.J. Ellis, "Explored Geothermal Systems"; pp. 632-683 in Geochemistry of Hydrothermal Ore Deposits. Edited by H.L. Barnes. 2d ed. Wiley, New York, 1979.

7. R.W. Henley, A.H. Truesdell and P.B. Barton, Fluid-mineral Equilibria in Hydrothermal Systems. Reviews in Economic Geology, Vol. 1, Society of Economic Geologists, El Paso, TX, 1984.

8. S. Arnórsson, E. Gunnlaugsson, and H. Svavarsson, "The chemistry of geothermal waters in Iceland, III. Chemical geothermometry and geothermal investigations," *Geochim. Cosmochim. Acta*, **47**, 567-577 (1983b).

9. S.R. Gislasson and H.P. Eugster, "Oxygen fugacities in a basalt-meteoritic water geothermal system: Krafla, N.E. Iceland", *Abst. with Prog. Geol. Soc. Am.*, **16**, 520 (1984).

10. P.R.L. Browne, "Hydrothermal alteration in active geothermal fields," *Ann. Rev. Earth Planet. Sci.*, **6**, 229-250 (1978).

11. E.L. Moore, G.C. Ulmer and D.E. Grandstaff, "Hydrothermal interaction of Columbia basalt from the Umtanum flow (Washington, U.S.A) with its coexisting groundwater," *Chem. Geol.*, **49**, 53-71 (1985).

12. G.C. Ulmer, G.H. Kacandes, and D.E. Grandstaff, "Icelandic geothermal fields as an analog for nuclear waste disposal in basalt"; pp. 755-764 in Nuclear Waste Management II, Advances in Ceramics, Vol. 20, 1986.

13. H. Hreggvidsdottir, "The greenschist to amphibolite facies transition in the Nesjavellir hydrothermal system, Southwest Iceland. M.S. Dissertation, Stanford

University, Stanford Ca., 1987.

14. H.Franzson and H. Sigvaldason, "Nesjavellir, Hola NG-7: Jardlög, ummyndun, maelingar og vatnsaedar." Orkustofnun Report OS-85124/JHD-18, Reykjavik, 80 p., 1985.

15. W.A.J. Mahon, G.D. McDowell and J.B. Finlayson, "Carbon dioxide: Its role in geothermal systems," *New Zealand J. Sci.*, **23**, 133-148 (1980).

16. S. Arnórsson, and E. Gunnlaugsson, "New gas geothermometers for geothermal exploration - Calibration and application," *Geochim. Cosmochim. Acta*, **49**, 1307-1325 (1985).

17. B. Steingrimmson, H. Franzson, G. Sverrisdottir, H. Sigvaldason, O. Sigurdsson, and E. Gunnalugsson, "Nesjavellir, Hola NJ-12: Borun, rannsoknir og vinnslueiginleikar." Report OS-86026, Orkustofnun, Reykjavik, Iceland, 147 p., 1986.

18. Z. W. Muna, "Chemistry of well discharges in the Olkaria geothermal field, Kenya." U.N. University Geothermal Training Programme, Iceland. Report 1982-8, 38p., 1982.

19. A. Koga, Geochemistry of the waters discharged from drillholes in the Otake and Hatchobaru Areas," *Geothermics, Special Issue Vol. 2*, 1422-1425 (1970).

20. N.L. Nehring, and F. D'Amore, "Gas chemistry and thermometry of the Cerro Prieto, Mexico, geothermal field," *Geothermics*, **13** 75-89 (1984).

21. M. Nathenson, N.L. Nehring, E.G. Crosthwaite, R.S. Harmon, C. Janick, and J. Borthwick, "Chemical and light-stable isotope characteristics of waters from the raft river geothermal area amd environs, Cassia County, Idaho, Box Elder County, Utah," *Geothermics.*, **11**, 215-237 (1982).

22. S.G. Spencer, B.F. Russell, and J.F. Sullivan, "Potential use of geothermal resources in the snake river basin: an environmental overview," U.S. Department of Energy (DOE) Report EGG-2001. Vol. 1. EG&G, Idaho Falls, Idaho, 1979.

23. R.O. Fournier, and B.B. Hanshaw, "Geochemical evaluation of the geothermal resources in the San Marcos region, Guatemala," *Applied Geochem.*, **1**, 189-197 (1986).

24. D. M. Thomas, personal communication, 1987.

25. A.G. Reyes, "Acid hydrothermal occurrences in Philippine Geothermal areas"; pp.461-465 in Fourth International Symposium on Water-Rock Interaction, 1986.

26. R. Korn, G.C. Ulmer, and D.E. Grandstaff, "Hydrothermal reactions of Columbia Plateau basalts from the Umtanum and Cohassett Flows (Washington) with simulated Hanford Groundwaters"; pp 333-336 in Fifth International Symposium on Water-Rock Interactions, 1986.

27. M.A. Gardiner, G.H. Kacandes, G.C. Ulmer, and D.E. Grandstaff, "Application and results of a reinjection technique for hydrothermal experiments in Dickson rocking autoclaves," pp 383-392 in Scientific Basis for Nuclear Waste Management, Mat. Res. Soc. Symp. Proc. Vol 112, Edited by M. Apted and R. E. Westerman, 1988.

28. G. H. Kacandes, G.C. Ulmer, and D.E. Grandstaff, "Icelandic geothermal fields as an analog for nuclear waste isolation in Columbia Plateau (Washington, U.S.A.) basalt"; pp. 302-305 in Fifth International Symposium on Water-Rock Interactions, 1986.

29. F.W. Dickson and J.M. Potter, "Rock-Brine Interactions". EPRI AP-2258. Electric Power Research Institute, Palo Alto, Ca, 1982.

30. A.J. Ellis, and W.A.J. Mahon, "Natural hydrothermal systems and experimental hot water-rock interaction I," *Geochim. Cosmochim. Acta*, **28,** 1323-1357 (1964).

31. A.J. Ellis and W.A.J. Mahon, "Natural hydrothermal systems and experimental hot water-rock interactions II," *Geochim. Cosmochim. Acta*, **31,** 519-531 (1967).

32. D. Savage, "Granite-water interactions at 100°C, 50 MPa: an experimental study". *Chem. Geol.*, **54**, 81-95 (1986).

33. D. Savage, M.R. Cave, A.E. Mildowski, and I. George, "Hydrothermal alteration of granite by meteoric fluid: an example from the Carnmenellis Granite, United Kingdom". *Contrib. Mineral. Petrol.*, **96**, 391-405 (1987).

34. W.A.J.Mahon, "Natural hydrothermal systems and the reactions of hot water with sedimentary rocks," *New Zealand J. Sci.*, **10**, 206-221 (1967).

35. H.C. Helgeson, J.M. Delany, H.W. Nesbitt and D.K. Bird, "Summary and critique of the thermodynamic properties of rock-forming minerals," *Amer. J. Sci.*, **278-A**, 1-229 (1978).

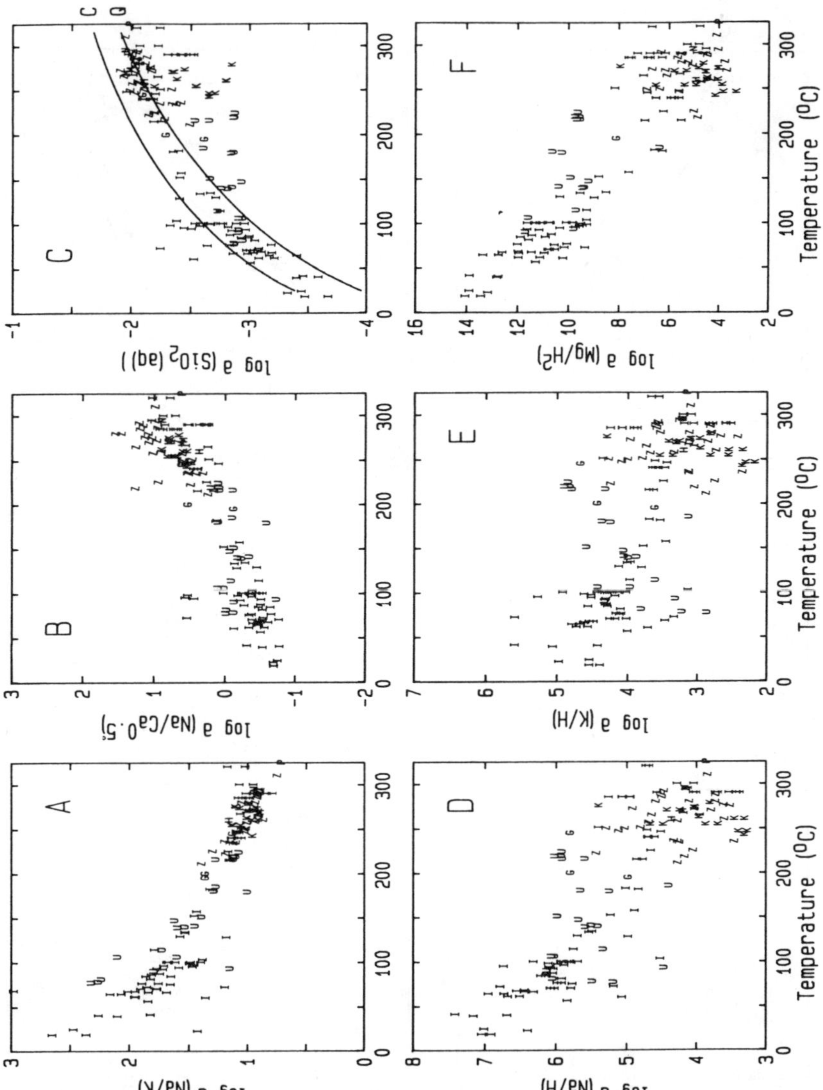

Figure 1. Density plot showing variations in values of selected solution parameters (free cation activity ratios, neutral species activities, gas fugacities, and pH) as a function of temperature for several geothermal areas: (G) Guatemala, (H) Hawaii, U.S.A, (I) Iceland, (K) Olkaria, Kenya; (M) Cerro Prieto, Mexico; (N) Japan, (P) Philippines, (U) Idaho and Yellowstone, United States; (Z) New Zealand. High temperature speciation calculated using computer program HIPH/FOLLY. Sources of data and some characteristics of these regions are presented in Table 1. The following equilibrium or trend lines are also

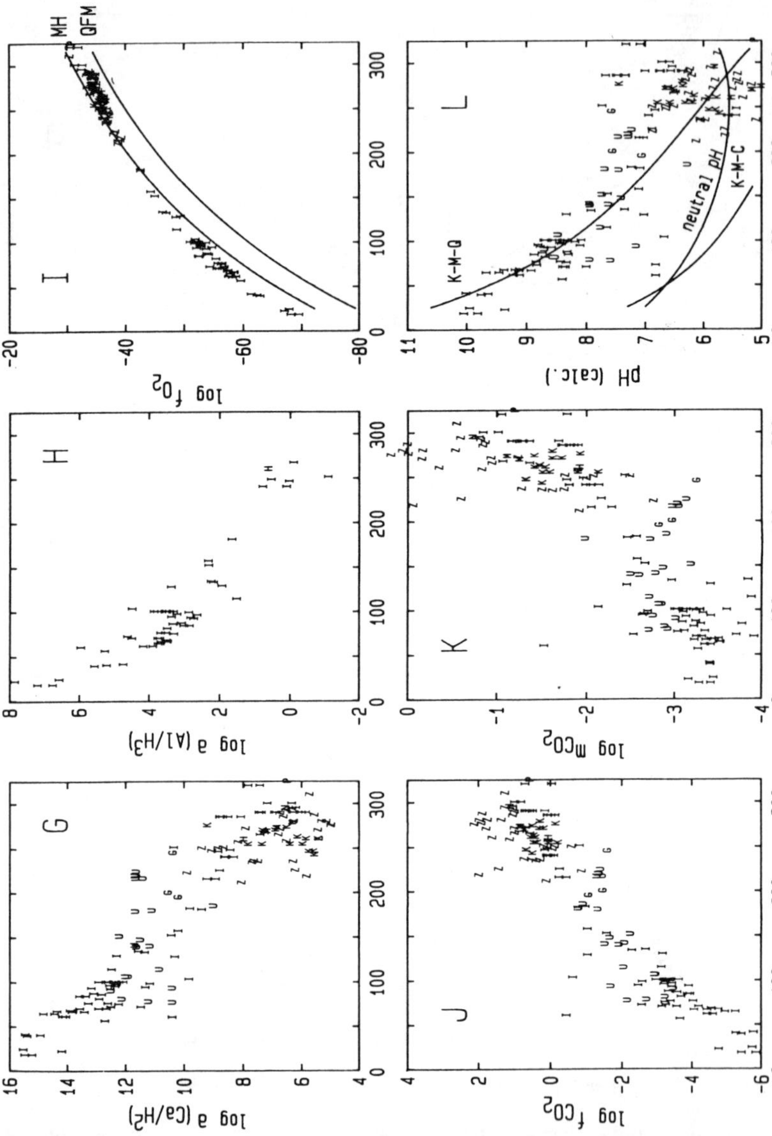

shown in Figure 1: (C) cristobalite saturation, (Q) quartz saturation, (neutral pH), (MH) magnetite-hematite and (FMQ) fayalite-magnetite-quartz buffer redox values as a function of temperature. (K-M-Q) pH value predicted from equilibrium with sodium and potassium feldspar-muscovite-quartz equilibrium at (m_{Na} + m_K) = 0.01, the approx average of Icelandic geothermal fluids presented in the figure. Thermochemical data are from Helgeson *et al.*[35] and the Na/K ratio for the (K-M-Q) calculation is from Arnórsson *et al.*[3].

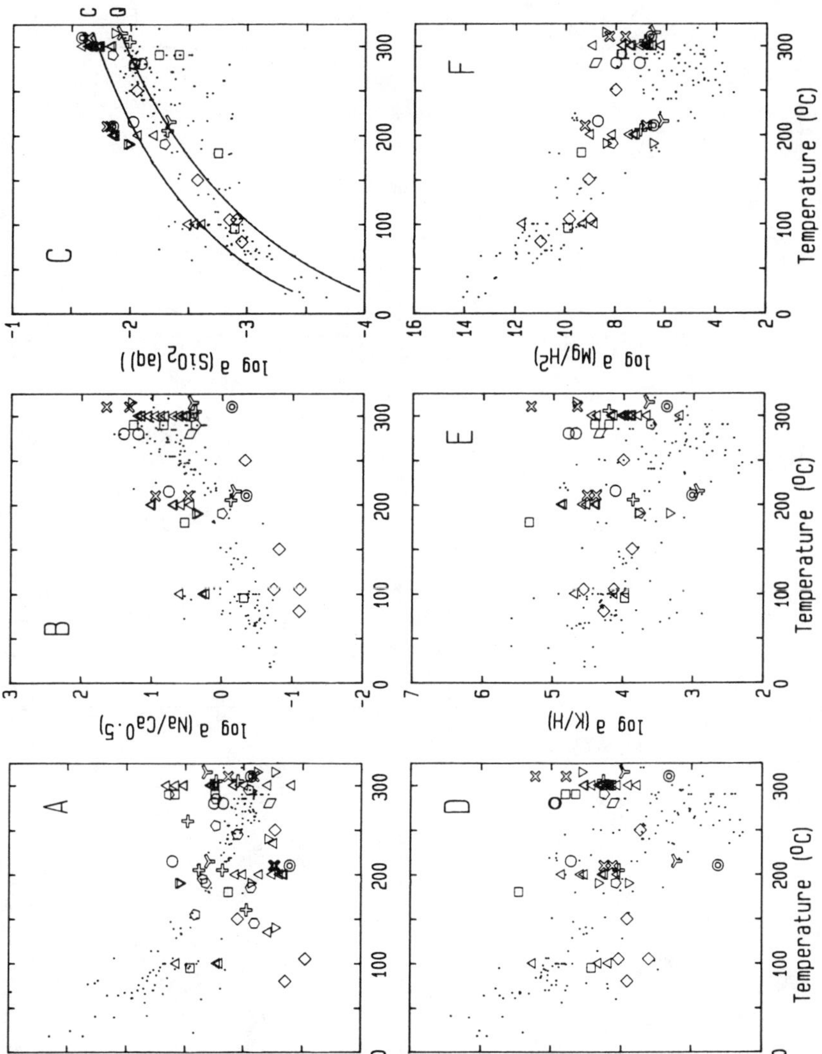

Figure 2. Comparison of selected solution parameters obtained from geothermal fluids (•) and various rock-water experiments: (△) quartz normative basalt, (□) olivine tholeiite basalt, (○) basalt, (✧) glassy rhyolite, (▽) andesite, (⬡) dacite, (◇) rhyolite pumice, (✚) greywacke, (Y) shale, (◎) siltstone, (⬠) mudstone, and (▱) schist. Sources of data and experiment parameters are given in

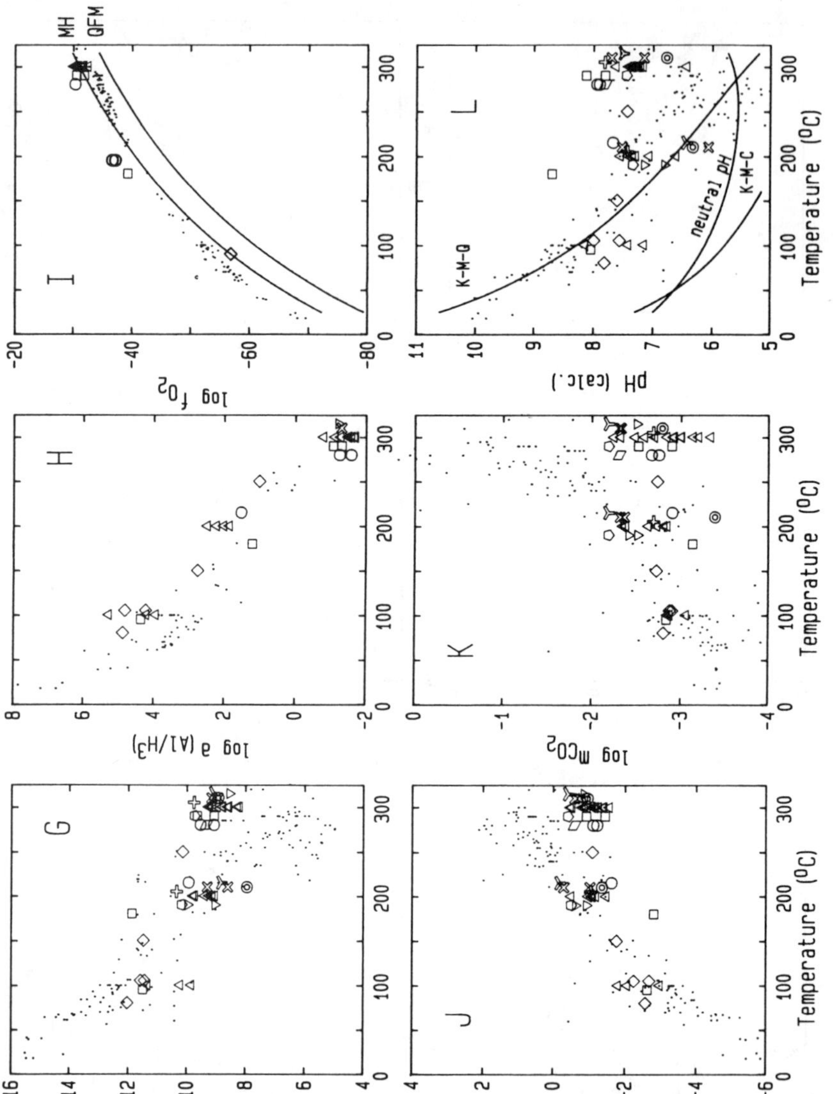

Table 1. Experiments have been conducted isothermally at 80°, 100°, 150°, 200°, 250°, and 300°C. Some symbols sets have been offset slightly (to lower or higher temperatures) to avoid overlap. Trend lines as in Figure 1.

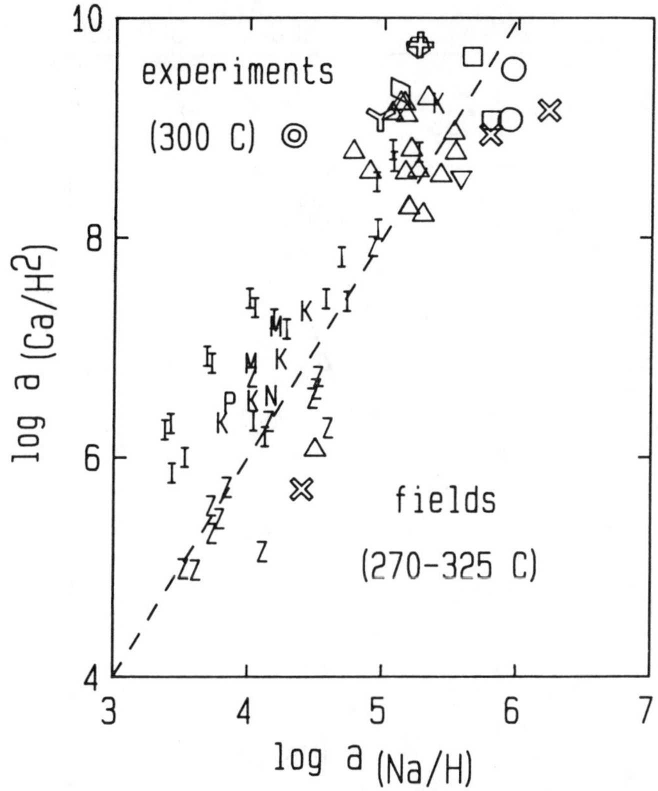

Figure 3. Comparison of Ca/H^2 and Na/H ion activity ratios calculated from various 300°C rock-water experiments and geothermal fluids having reservoir temperatures between 270° and 320°C. Data in Figure 1 show little systematic temperature-dependent trends in values of solution parameters in this temperature region. Symbols as in Figures 1 and 2. The diagonal line has a slope consistent with a constant $Na/\sqrt{Ca}$ ion activity ratio. Ca/Mg and Na/K show similar behavior[1]. The comparison shows that values of the cation/proton parameters from experiments and geothermal fields do not appreciably overlap, largely as a result of the 1 to 2 pH unit difference in experimental and geothermal fluids. However, cation/cation ion activity ratios are similar for both data sets.

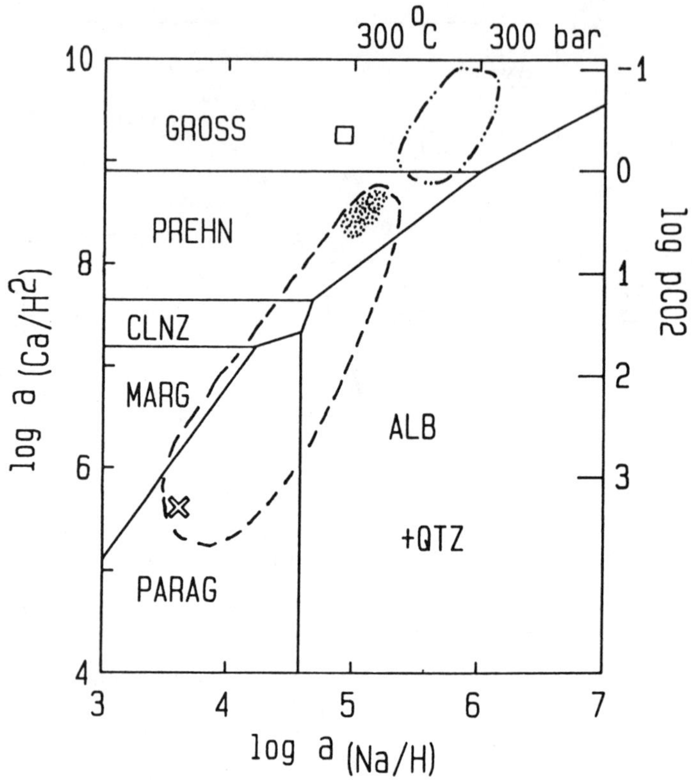

Figure 4. Cation/proton activity ratios produced during experiments NG-7 and KRA-5 (both at 300°C and 30 MPa with an initial water:rock mass ratio of 10) and comparisons with previous experimental results (region within dot-dash lines), with fluids from Nesjavellir Geothermal Field (stippled), and other geothermal fluids having reservoir temperatures between 270° and 325°C (region within dashed lines). Stable results from experiment NG-7 (containing a mineral assemblage from Nesjavellir, Iceland stable near 300°C) are indicated by (□). Results of addition of 0.1 m CO_2 (KRA-5) are indicated by (✕). Equilibrium boundaries for end-member mineral phases, shown for reference, were calculated for quartz saturation and 300°C and 30 MPa from thermochemical data of Helgeson *et al.*[35]. The solid near-vertical phase boundary in Figure 4C is for 7Å-clinochlore, the dashed boundary for 14Å-clinochlore. Mineral abbreviations: gross (grossularite), prehn (prehnite), clnz (clinozoisite), marg (margarite), parag (paragonite), alb (albite), 14A-cl (14Å clinochlore), 7A-cl (7Å clinochlore), pyroph (pyrophyllite), +qtz (quartz saturation).

ANALYSIS OF ALUMINOSILICATE GLASS SAMPLES FROM THE MIIT IN SITU WIPP SALT BURIAL TEST

J.C. Tait, P.J. Hayward, W.H. Hocking,
J. Betteridge, and D.C. Doern,
Atomic Energy of Canada Limited,
Whiteshell Nuclear Research Establishment,
Pinawa, Man. ROE 1LO
Canada

INTRODUCTION

In 1983, the Savannah River Laboratory, in cooperation with the Sandia National Laboratories, undertook to design and coordinate a series of in situ field tests on fuel recycle waste glasses to be conducted at the Waste Isolation Pilot Plant (WIPP) site near Carlsbad, New Mexico. The Materials Interface Interactions Tests (MIIT), which are a part of the overall test matrix, represent an international cooperative effort to test the durability of simulated high-level waste glasses and candidate container materials supplied by the United States, France, West Germany, Belgium, Canada, Japan and the United Kingdom. The samples were buried during July 1986 in boreholes in mined rooms at the 650-m level in the Salado salt formation at the WIPP site. The samples were in contact with clay and salt specimens, representing the likely environment in a future waste disposal vault sited in a salt deposit. Overviews of the MIIT program and details of the in situ burial test configurations have been previously described in the literature [1-4].

One of the MIIT tests, designated MIIT-MI (multiple interactions), is intended to investigate the durability of different glass compositions in the presence of various waste package components in both initially dry salt-boreholes and in boreholes containing a saturated brine solution, representative of the Salado formation. Another test, designated MIIT-SS (solution/surface interactions), involves corrosion of glass/glass, glass/salt or glass/metal/salt interfaces in boreholes saturated with brine. The solutions in these boreholes were analyzed as a function of time, to provide data to correlate with studies of the the glass surfaces.

Since all other participant countries had submitted borosilicate glass waste forms, it was decided that the contribution from the Canadian Nuclear Fuel Waste Management Program would be two alternative waste-form materials. Thus, an aluminosilicate glass based on the Na_2O-CaO-Al_2O_3-SiO_2 system and a sphene glass-ceramic based on the Na_2O-Al_2O_3-CaO-TiO_2-SiO_2 system were chosen as the Canadian waste-form materials, both having already been extensively characterized in laboratory leaching studies.

A wide variety of characterization tests have already been conducted on the test samples at other laboratories (surface examination, solution analysis, etc.) [5,6]. This paper describes only surface analytical results for aluminosilicate glass samples retrieved after burial at 90°C for six months in the initially dry borehole, and after six months or one year in the saturated brine borehole. The analyses were performed using secondary ion mass spectrometry (SIMS), scanning electron microscopy (SEM) coupled with energy dispersive X-ray analysis, and X-ray diffraction.

DESCRIPTION OF THE MIIT TESTS AND TEST ASSEMBLIES

The aluminosilicate glass, with the composition shown in Table 1, was prepared at the Whiteshell Nuclear Research Establishment by melting well-mixed reagent-grade oxides, nitrates and carbonates in a platinum crucible at 1400°C, and pouring the melt into a 50-mm-diameter graphite mold containing a central 19-mm cylindrical graphite core. The resulting glass body was annealed at 650°C, removed from the mold and cut into ~ 2-mm-thick slices, with one cut surface being ground and polished to a 6-μm finish. These "pineapple slices" were then sent to the Savannah River Laboratory for incorporation in the test assemblies.

The test assemblies consisted of a series of glass and salt "pineapple slice" discs mounted so that the surfaces of the various components were in contact with one another. The configurations of the assemblies used here in the MIIT-MI Part I-B and MIIT-SS Part II-A tests are shown in Figure 1; a thermocouple-controlled electric heater element was located in the central core of each assembly to maintain the temperature at 90°C. In each MIIT-SS assembly, four identical test segments (sub-assemblies) were separated by Teflon discs and the entire assembly was held together by Teflon support rods. Each assembly was placed in a borehole, 1.2 or 2.1 m deep and 83 mm in diameter. In the MIIT-SS test, the borehole had been previously filled and equilibrated with a simulated brine representative of the fluid phase of the Salado formation. The "Brine A" pH ~ 6.5, was of composition (mol/L) Na^+, 1.83; Mg^{2+}, 1.44; K^+, 0.02; Cl^-, 5.35; SO_4^{2-}, 0.04; BO_4^{3-}, 0.02;

HCO_3^-, 0.01; Br^-, 0.01. Both series of tests will run until July 1991 and involve removal of individual sub-assemblies for analysis after specified time intervals. The remainder of each assembly was replaced in its borehole for further testing after removal of each test segment.

The MIIT-SS samples discussed in this report (aluminosilicate glass/salt/brine, assembly #22) are those removed after six months (samples CAN-AS-13, 14, 15) and one year of corrosion (samples CAN-AS-16, 17, 18). Also discussed are samples from the MIIT-MI dry-salt tests (aluminosilicate glass/salt, assembly #46) after six months corrosion (sample CAN-AS-51).

Table 1. Aluminosilicate Glass Composition (wt%)

Na_2O	8.24	CeO_2	1.74	SrO	0.08	SnO_2	0.01
CaO	11.72	La_2O_3	0.24	BaO	0.15	Fe_2O_3	0.13
Al_2O_3	20.48	Pr_2O_3	0.12	Y_2O_3	0.05	Cr_2O_3	0.04
TiO_2	4.76	Nd_2O_3	0.14	MoO_3	0.69	NiO	0.11
SiO_2	50.22	Cs_2O	0.29	ZrO_2	0.33	Pd	0.45

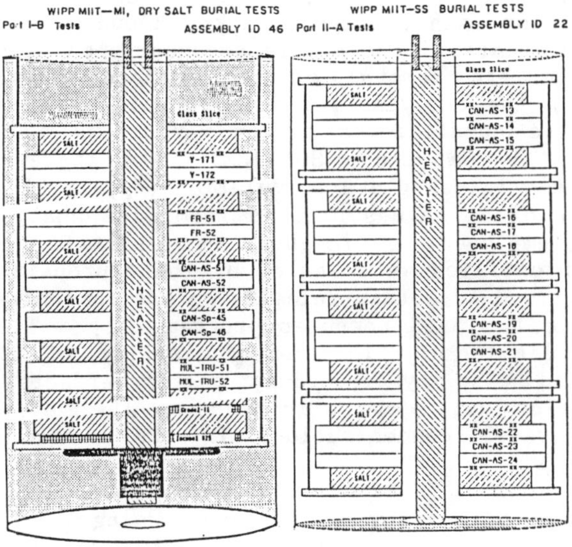

Figure 1. Glass "pineapple slice" configuration for the MIIT-MI (dry-salt; borehole 46) and MIIT-SS (salt brine; borehole 22) tests showing location of CAN-AS-51 and CAN-AS-13 to 18 samples.

ANALYTICAL PROCEDURES

The surface alteration layers on the various glass slices were composition depth-profiled using a SIMS II analyzer mounted on a SAM 590A scanning Auger microscope (Physical Electronics Division of Perkin-Elmer Corporation). Secondary ion emission was excited by 4.5-keV argon-ion bombardment, using ~ 330 nA of Ar^+ current focussed into a ~ 200-μm-diameter beam. The Ar^+ beam was rastered over an area of ~ 2 mm^2 on each sample, with an average ion current density of 18.0 μA/cm^2. Prior calibration showed the mean sputter etching rate was ~ 0.003 μm/min. Charge neutralization was provided by 2-keV electron bombardment using ~ 2.5 μA of electron current in a large-diameter beam directed at the centre of the sputter crater. The analyses were performed under standardized conditions to maintain consistent element sensitivities. Signal gating (70%) was employed throughout to minimize crater-edge effects [7].

For the cation species, the raw SIMS data were converted into composition depth profiles using the empirical relative sensitivity coefficient (S_x) method [8-11]. A new suite of S_x values was calculated for each profile from the known glass composition and from measurements of the stable secondary-ion signal intensities after complete removal of the surface alteration layer [9,10]. During the one-year period over which the analyses were performed, these coefficients showed variations about the mean between ± 25% (Al, Ca) and ± 80% (Cs). For those elements not included in the original glass formulation (Mg, B, Li, K, and Ni), a constant set of S_x values derived from SIMS analysis of glass standards was employed. Throughout each profile, the measured SIMS signal intensities were scaled by the appropriate S_x values and then normalized to obtain relative cation concentrations (atom percent), using the MASS computer program [11]. The analytical accuracy was limited by the reproducibility of the analysis conditions, and also by secondary matrix effects, since the ionization yields will depend in part on the variable cation, chlorine and water concentrations in the surface alteration layers.

Scanning electron microscopy was performed using an ISI-DS 130 SEM equipped with a Tracor Northern TN 2000 EDX analyzer. Samples were gold-coated to avoid charging. To identify the phases present on the more heavily altered glass surfaces, XRD measurements were made using CuKα radiation and a Rigaku diffractometer equipped with a graphite monochromator.

RESULTS

Dry-Salt Burial Tests: 6 Months

After 6 months in the 90°C dry-salt burial test, the CAN-AS-51 sample showed evidence of leaching in the presence of an extremely corrosive fluid phase likely produced by brine migration into the borehole. Brine pH values as low as 0.93 have been reported for fluid inclusions within the Carlsbad salt dome, where brine has migrated towards a heat source and the MgCl components have reacted with aqueous carbonate to form magnesite ($MgCO_3$) and HCl vapor. (Note, however, that in the saturated brine experiments, solution pH values were typically in the range 4.4 - 7. [4].)

Scanning electron microscopy of the upper surface of CAN-AS-51, which had initially been in contact with dry salt, showed it to be encrusted with an alteration layer of variable thickness (see Figure 2a) containing Si, Al, Mg, and Ca (in decreasing order of EDX peak intensities), together with traces of Cl and Ti. Less dense

Figure 2. SEM micrographs of aluminosilicate glass corroded in a dry-salt borehole; (a) alteration layer on the glass/salt interface; (b) alteration layer near the outer edge of the glass/glass interface; (c) glass/glass interface beneath corrosion layer.

regions of this layer showed a morphology typical of smectite clay minerals, with a generalized formula of $(\frac{1}{2}Ca,Na)_{0.7}(Al,Mg,Fe)_4$-$[(Si,Al)_8O_{20}](OH)_4$, suggesting formation of a gel layer on the glass surface from contact with the leachant, followed by crystallization of this layer to form smectite.

SIMS depth profiles of this surface revealed an alteration layer whose composition varied in the range (cation atom %): Si (20-40), Al (23-25), and Mg (25-35), together with minor amounts of Ca (1.5), Ti (0.5-1), Na (1.5-6), and B (1.5-5) as well as O and Cl anions. The thickness of this layer was found to decrease from ~ 0.36 μm near the outer edge to ~ 0.05 μm near the centre. A SIMS depth profile from the centre region is shown in Figure 3.

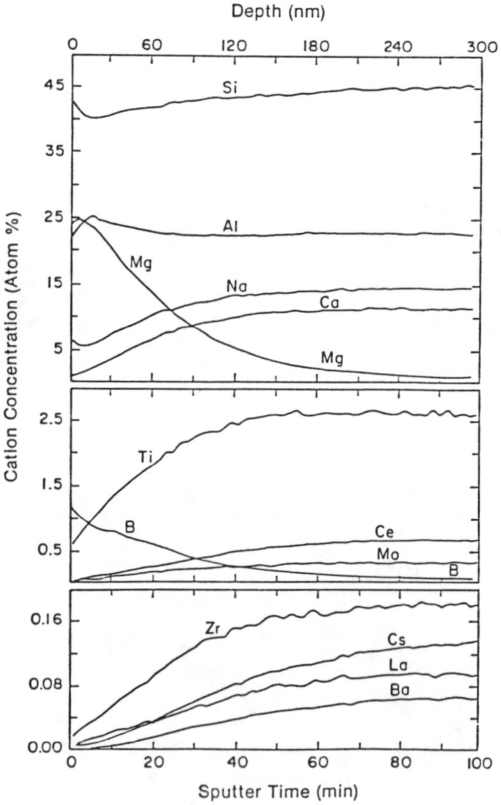

Figure 3. SIMS depth profile of the aluminosilicate glass/salt interface corroded for 0.5 year in the dry-salt borehole.

The lower surface of this sample, in contact with an identical glass slice, showed clear glass near the centre and localized patches of heavy corrosion along the inner and outer edges, indicating limited penetration of fluid between the mating glass surfaces (Figure 2b). Among the alteration products, the acicular crystals (labelled X) gave strong EDX peaks for Mg, Cl, Al, Si and minor peaks for Ti and Ca, while the anhedral phase (labelled Y) gave strong EDX peaks for Si, Al, Mg and Cl and lesser peaks for Ca and Ti. Some or all of the Si and Al peak intensities will derive from the underlying glass, and further identification of these phases was not attempted. In regions where the altered layer had become detached, the underlying heavily pitted glass surface was visible (Figure 2c).

Depth profiling by SIMS confirmed that the clear glass mentioned above was essentially undisturbed glass, with only a thin ($\leq$ 0.015 μm) altered surface film, slightly enriched in Si, Ti, Mo, and Zr, slightly depleted in Al, Na, Ca, Cs, and Ba, and contaminated with trace levels of Mg and B. Analysis by SIMS of the heavily corroded patches was difficult because of charging artifacts, which may have arisen from spalling of the thick surface alteration layer. This layer was well in excess of 1 μm thick (the greatest depth reached in the SIMS depth profiles). It appeared to contain not only all the elements present in the underlying glass, in approximately the same relative proportions, but also significant amounts of Mg (~ 30 cation atom %) and Cl and low levels of B.

Salt/Glass/Brine Test: 6 Months and 1 Year

After 6 months, the glass surfaces in direct contact with salt coupons from assembly 22 (upper surface CAN-AS-13 and lower surface CAN-AS-15) showed an alteration layer of variable thickness that had undergone localized crystallization to form a smectite-like phase (Figures 4a, b) that gave Si, Al, Mg, and Ca peaks in the EDX spectrum. Analysis by SIMS of this layer showed the presence of Si, Mg, B, and Al, with minor amounts of Ca, Ti, and Na, as well as O and Cl anions. Other constituents of the bulk glass (Zr, Mo, Cs, Ba, Y, La, Ce, Pr, and Nd) were present in this layer at only trace levels, i.e., concentrations were more than 10-fold lower than in the bulk glass.

Examination by SEM of the surfaces of six-month test samples that had been in contact with other glass surfaces (lower surface CAN-AS-13, upper surface CAN-AS-15, both surfaces of CAN-AS-14) revealed the presence in each case of a fairly uniform alteration layer (Figure 4c), indicating that complete brine penetration between the contacting glass slices had occurred. The EDX spectra for these layers showed Si, Mg, Al, and Ca in order of decreasing

Figure 4. SEM micrographs of aluminosilicate glass corroded for 0.5 year in a brine borehole; (a) glass/salt interface (CAN-AS-13); (b) glass/salt interface (CAN-AS-15); (c) glass/glass interface (CAN-AS-14).

peak intensities. The alteration layers were shown by SIMS depth profiling to have a similar composition but were consistently thicker than those formed at the glass/salt contact surfaces. Representative SIMS depth profiles from glass/salt and glass/glass contact surfaces are shown in Figures 5a and 5b, respectively. Mean thicknesses of the alteration layers are reported in Table 2. The glass/alteration-layer interface was arbitrarily defined as the midpoint of the transition zone between stable concentrations of relatively immobile elements (e.g., Zr, Ti, and La) in the altered layer and in the glass bulk. Figures 5a and 5b show that most elements present in the glass achieved their bulk concentrations within ~ 0.1 μm of this interface; however, Cs, Na, Ba, K, and Li showed evidence of leaching to greater depths (~ 0.3 μm into the glass bulk). The variable composition of the alteration layers is summarized in Table 3.

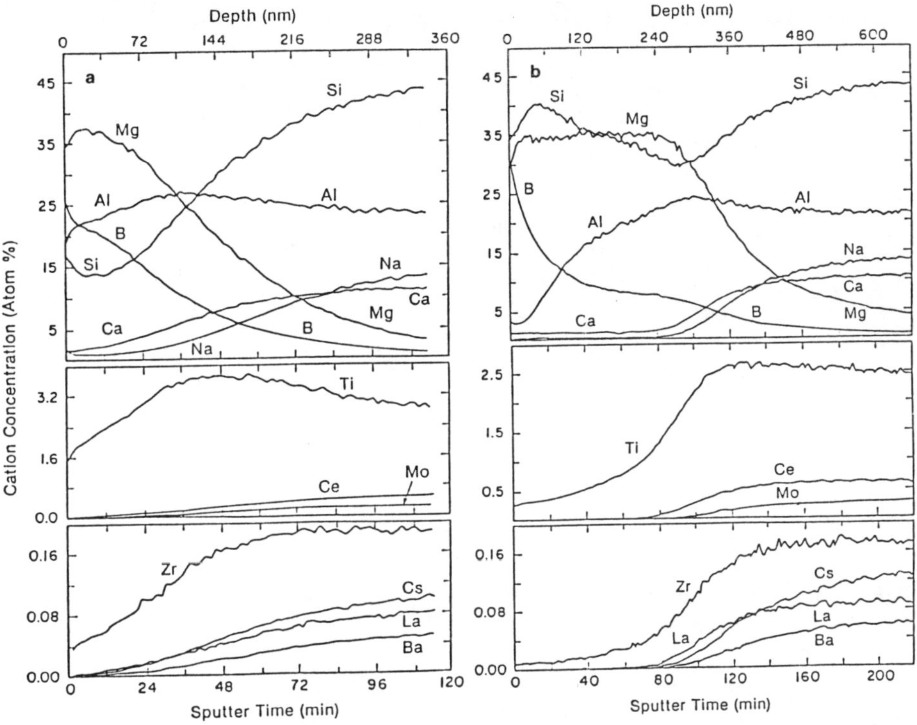

Figure 5. SIMS depth profile of (a) the aluminosilicate glass/salt interface and (b) the aluminosilicate glass/glass interface corroded for 0.5 year in the brine borehole.

Glass/salt contact surfaces from the 1-year test (CAN-AS-16 upper surface, CAN-AS-18 lower surface) showed more extensive corrosion than the 6-month samples (see Figure 6a), with further development of smectite together with localized growth of an Mg, Si phase, probably talc ($Mg_3Si_4O_{10}(OH)_2$). The glass/glass contact surfaces from the 1-year test (CAN-AS-16 lower surface, CAN-AS-18 upper surface and CAN-AS-17 both surfaces) also showed further growth of the alteration layer (Figure 6b), with conversion of the gel-like alteration products to smectite (Figure 6c) in many regions. The EDX spectra for the gel layers from the 6-month and 1-year samples were virtually identical to the smectite spectra, indicating that the smectite had formed by crystallization of the gel layer. XRD of localized white patches on the surfaces indicated additional precipitation of gypsum ($CaSO_4 \cdot 2H_2O$), carnallite ($KMg(Cl)_3 \cdot 6H_2O$) and halite (NaCl), all mineral components of the Salado formation [4].

Table 2. Thickness of Alteration Layer on Aluminosilicate Glass Exposed to Brine, determined by SIMS Depth Profiling

Time/ Year	Glass Slice	Surface	Contact	Alteration-Layer Thickness* (μm)
0.5	AS13	TOP	GS	0.16
0.5	AS13	BOTTOM	GG	0.34
0.5	AS14	TOP	GG	0.34
0.5	AS14	BOTTOM	GG	0.33
0.5	AS15	TOP	GG	0.33
0.5	AS15	BOTTOM	GS	0.11
1	AS16	TOP	GS	0.54
1	AS16	BOTTOM	GG	0.25
1	AS17	TOP	GG	0.36
1	AS17	BOTTOM	GG	0.36
1	AS18	TOP	GG	0.48
1	AS18	BOTTOM	GS	0.45

* Estimated as midpoint in transition from layer to glass composition on SIMS depth profiles - corresponds approximately to maxima in Al and Ti concentrations.

Table 3. SIMS Analysis of Alteration Layer on Glass Exposed to Brine for 0.5 year

Cation	Surface	Typical[a]	Range[b]	Trend
Si	GG/GS	30	15-45	Variable through layer with no regular pattern between layers
Mg	GG/GS	35	25-45	Uniform through layer except for slight dip at surface
B	GG/GS	10	45-5	Decreasing steadily from peak value (25-45) at outer surface
Al	GG GS	10 22	2-25 20-25	Increasing steadily from minimum at surface to shallow maximum near layer/ bulk interface
Ca	GG/GS	1.5	1-2	Uniform through layer
Na	GG/GS	0.5	0.2-1	Uniform through layer
Ti	GG GS	0.3 1.5	0.2-0.5 1-2	Uniform through most of layer rising to a maximum (3-4) near layer/ bulk interface

[a] Typical average concentration in alteration layer (cation atom%)
[b] Range in concentration across (between) alteration layer(s)

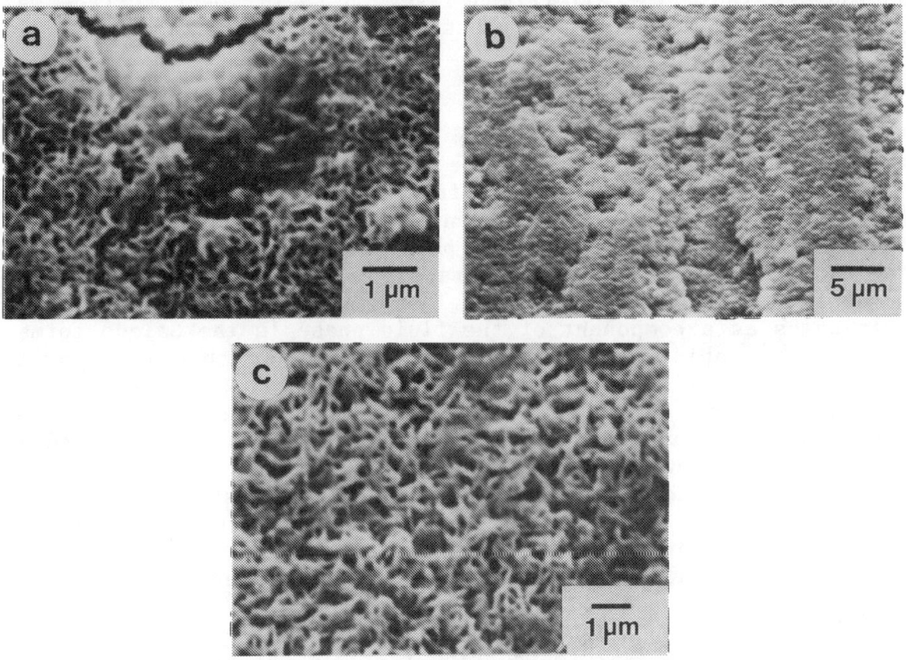

Figure 6. SEM micrographs of aluminosilicate glass corroded for 1 Year in the brine borehole; (a) glass/salt interface (CAN-AS-16); (b) glass/glass interface showing gel layer; (c) glass/glass interface showing smectite layer formation.

The SIMS depth profiles of the 1-year samples exposed to brine also showed progressive evolution of the alteration layer during the second 1/2 year. Growth of the glass/salt alteration layer to a comparable extent to that on the glass/glass surfaces was the most significant change (Table 2). The composition gradients within the alteration layers were generally similar to those observed for the 1/2-year exposure. Surface enhancement of B for the 1-year samples was less extreme (~ 20 cation atom %, as opposed to $\leq$ 45 cation atom % for the 6-month samples). However, the layer on one glass/glass surface (AS-18 TOP) showed high B concentrations (30-40 cation atom %) throughout. Appreciable concentrations of Cr and Ni (~ 0.5-5 cation atom %) were also found in the 1-year alteration layers (both were below ~ 0.05 cation atom % in the alteration layers from the 6-month samples). Variability in the thicknesses of the various alteration layers may have resulted from differing degrees of brine penetration between the mating surfaces. Thus, the alteration layer on AS-16 lower surface was less developed than in other glass/glass contact surfaces.

DISCUSSION

A number of general observations can be made from a preliminary assessment of these results. The enrichment of the glass alteration layers in boron in samples from the dry borehole tests indicate that these boreholes had been infiltrated with a fluid phase. The probable source of the fluid phase is from fluid inclusions in the salt deposit that have migrated to the heat source in the borehole and accumulated at the glass surfaces. The boron likely originates as a component of the fluid phase in the Salado formation. The rapid decrease in alteration layer thickness moving inward from the sample edges, decreasing from 0.36 to 0.15 μm within 3 mm from the outer edge, suggests limited penetration of the fluid phase and indicates that most corrosion has occurred at the outer surface. The fact that corrosion thicknesses were least (~ 0.05 μm) at the centre of the "pineapple slices" suggests there is little fluid convection between the glass surfaces; however, corrosion was evident at the edge of the central core, indicating fluid migration along the heater rod/glass surface interface.

The observed boron enrichment in the alteration layers of the 0.5- and 1-year samples (saturated brine test) is due to the boron component of the saturated brine. In these experiments, the glass/salt alteration layer showed a more rapid growth between 0.5 and 1 year than during the first 0.5 year of the test. The apparent acceleration and growth of the alteration layer possibly indicates a greater access of the brine leachant to the glass/salt surface during the latter 6 months. The glass/glass interfaces showed a much reduced rate of growth during the second time interval. In this case, replenishment of the leachant would be limited by the poor diffusion between the polished glass surfaces and by the accumulation of precipitates with time. The appearance of Ni and Cr in the alteration layer suggests corrosion of the Inconel/Incaloy system components (heater, thermocouple sheath).

Solution equilibrium calculations on Mg-containing multi-component brines [9] show that Al, Si additions to solution from dissolution of Al-, Si-based glasses causes the brines to rapidly achieve saturation with respect to Mg-based clay minerals (smectite or sepiolite). Thus, the appearance of a gel-like alteration layer and its subsequent conversion to smectite is not unexpected. Under conditions of rapid solution replenishment (e.g., at the glass/salt interface), this alteration layer does not appear to present a significant barrier to further leachant attack on the underlying glass surface, as indicated by the thick corrosion layer for these contact surfaces. Where solution replenishment is limited by diffusion, the layer may assist in maintaining a solution composition of glass/alteration layer species that inhibits continued dissolution. However, there is no evidence that this layer provides inhibition of dissolution on its own.

Further analysis of the SIMS and SEM results presented here will be possible when results from other characterization studies being conducted on these and other glasses are available. A more detailed examination of the surface interactions and leaching behaviour of the aluminosilicate glass and sphene glass-ceramics for samples removed after two years leaching will be presented in a future publication.

ACKNOWLEDGEMENTS

The authors gratefully acknowledge the assistance of M. Duclos in providing XRD analyses, I. George and C.A. Boase for glass sample preparation and P. Taylor for helpful review and comments on the manuscript.

REFERENCES

1. G.G. Wicks, 1985. "WIIP/SRL In Situ and Laboratory Testing Programs - Part I: MIIT Overview, Nonradioactive Waste Glass Studies", USDOE Report DP-1706, Savannah River Laboratory, E.I. du Pont de Nemours & Co., Aiken, SC.

2. G.G. Wicks and M.A. Molecke, 1986. "WIPP/SRL In Situ Testing Program", Advances in Ceramics, 20, pp. 657-667.

3. G.G. Wicks, M.E. Weinle and M.A. Molecke, 1987. "WIPP/SRL In Situ Tests - Part II: Pictorial History of MIIT and Final MIIT Matrices, Assemblies and Sample Listings", USDOE Report DP-1733, E.I. du Pont de Nemours & Co., Savannah River Laboratory, Aiken, SC.

4. L.D. Tyler, R.V. Matalucci, M.A. Molecke, D.E. Munson, E.J. Nowak, and J.C. Stormont, 1988. "Summary Report for the WIPP Technology Development Program for Isolation of Radioactive Waste", Sandia Report SAND-88-0844.

5. B.K. Zoitos, D.E. Clark and A.R. Lodding. 1989. "Surface Layer Formation under Simulated Burial Conditions". 4th International Symposium on Ceramics in Nuclear Waste Management. American Ceramic Society Meeting, Indianapolis, Indiana, April 23-27. (These proceedings).

6. Workshop on Testing of High-Level Waste Forms under Repository Conditions. Cadarache, France, October 1988. To be published in proceedings, EUR 12 017 EN (1989).

7. C.W. Magee and R.E. Honig, 1982. "Depth Profiling by SIMS - Depth Resolution, Dynamic Range and Sensitivity". Surf. Interface Anal. 4, pp.35-41.

8. D.E. Newbury, 1980. "Methods for Quantitative Analysis in Secondary Ion Mass Spectrometry". Scanning, 3, pp.110-118.

9. P.J. Hayward, W.H. Hocking, S.L. Mitchell and M.A. Stanchell, 1984. "Leaching Studies of Sphene-based Glass-ceramics". Nucl. Chem. Waste Management, 5, pp. 27-37.

10. G. Bart, E.T. Aerne and H.U. Zwicky, 1986. "Experience in SIMS Surface Analysis of Radioactive Waste Glasses". Nucl. Instrum. Methods Phys. Res., B17, pp.127-136.

11. D.G. Watson and F.E. Doern, 1985. "Software for Surface Analysis". Appl. Surf. Sci., 21, pp.304-313.

Section V. Spent Fuel Characterization and Testing

CHARACTERISTICS OF SPENT NUCLEAR FUEL -THE CHARACTERISTICS DATA BASE

Karl J. Notz
Oak Ridge National Laboratory

Introduction

The Office of Civilian Radioactive Waste Management (OCRWM) is responsible for the spent fuels and other wastes that will, or may, eventually be disposed of in a geologic repository. The two major sources of these materials are commercial light-water reactor (LWR) spent fuel and immobilized high-level waste (HLW). Other wastes that may require long-term isolation include non-LWR spent fuels and miscellaneous sources such as activated metals. This paper deals with spent fuels, but for completeness, the other sources are described briefly.

Detailed characterizations are required for all of these potential repository wastes. These characterizations include physical, chemical, and radiological properties. The latter must take into account decay as a function of time. In addition, the present inventories and projected quantities of the various wastes are needed. This information has been extracted from primary data sources, evaluated, and assembled in a Characteristics Data Base which provides data in four formats: hard copy standard reports, menu-driven personal computer (PC) data bases, program-level PC data bases, and mainframe computer files.

The Characteristics Data Base (CDB) is an integral part of the Systems Integration approach being used by OCRWM. In support of this effort, the CDB provides a standard set of self-consistent data to the various areas of responsibility within OCRWM, including systems integration and waste stream analysis, storage, transportation, and geologic disposal. The data will be used for design studies, trade-off studies, and system optimization by OCRWM and supporting contractors.

The CDB was initiated, on a small scale, in 1985 and the first report was published in 1986, including one PC data base.[1] This report gave partial descriptions of BWR and PWR fuel assemblies, and the PC data base gave the radiological properties. The work was expanded in 1986 to cover the full scope of candidate wastes, and the first draft report was issued that year. Subsequently, two additional draft versions were issued and, finally, the formal report was issued in eight volumes and two printings[2]. A summary of this work is given in the following sections, with major emphasis on spent nuclear fuels.

<u>Overview of CDB</u>

The Characteristics Data Base is organized on the basis of four waste stream categories: LWR Spent Fuel, Immobilized High-Level Waste, Non-LWR Spent Fuels, and Miscellaneous wastes. These can be further broken down into eight waste streams (Fig. 1). For HLW, the generator site identities are retained, and the basic units are individual canisters of immobilized HLW. For LWR spent fuel, the basic units are the individual assemblies, and the distinctions between BWRs and PWRs warrant separation on this basis; the generating reactors are also identified. For non-LWR spent fuels and miscellaneous wastes, the as-stored characteristics are the present basis, but as predisposal treatment and packaging concepts are developed, these will be incorporated.

The above four major categories, plus a summary, comprise the five major sections of the hard copy report[2], which was recently issued as an 8-volume set. Six of these volumes consist of 16 appendices that provide supporting information to the other two volumes. There are approximately 50,000 data entries in these volumes, about 1% of all the data in the PC data bases.

Five of the appendices are Users' Guides to as many menu-driven PC data bases. No programming skills are required to use these, which are described in Table 1. Two more such data bases are under development, one with data on LWR fuel pin characteristics and one for LWR assembly serial numbers. These data bases are designed for use on an

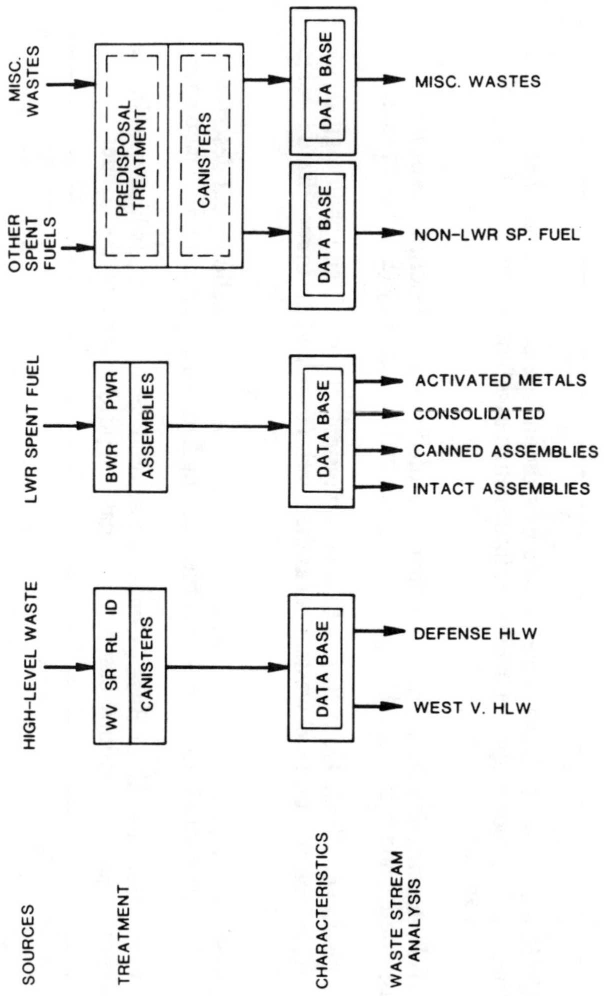

FIG. 1. GENERIC DATA SOURCES FOR THE CHARACTERISTICS DATA BASE

Table 1. Menu-driven PC Data Bases Available from the Characteristics Data Base

LWR Radiological Data Base - Contains radionuclide compositions, heat generation rates, curies, photon spectra, and other information as a function of spent fuel type, burnup, and decay time.

LWR Assemblies Data Base - Contains detailed physical descriptions of fuel assemblies and radiological properties of spent fuel disassembly (SFD) hardware.

High-Level Waste Data Base - Contains physical, chemical, and radiological descriptions of high-level waste, both as the interim form and the immobilized form in canisters.

LWR NFA Hardware Data Base - Contains physical and radiological descriptions of nonfuel assembly hardware; i.e. nonfuel-bearing hardware other than SFD hardware.

LWR Quantities Data Base - Contains data on discharged fuel, as historical inventories and as projected quantities, based on EIA data supplied to them by the utilities.

IBM-PC or PC-compatible hardware. A total of 20 megabytes of disk storage is required for all of them; this represents about 5 million data entries, about 80% of them radiological data and the balance descriptive data and quantities of materials. They are available through the Oak Ridge National Laboratory; ordering information is given in the report[2] on page v.

The above data bases, as well as many of the internal working files, are written in dBASE III PLUS. This permits tabulation of special reports and interactive output. Mainframe files are used to generate some of the above files and some of the hard copy reports. Their use involves SAS and FORTRAN. A generalized flow chart for data processing is shown in Fig. 2.

The radiological characteristics are calculated using ORIGEN2[3], and include:

- quantities of each nuclide (grams or gram atoms);
- radioactivity, total and by nuclide (curies);
- thermal power, total and by nuclide (watts);
- photon energy spectra, by energy group (18 groups);
- neutrons from spontaneous fission (per second);
- neutrons from (alpha, n) reactions (per second); and
- quantity of each element (grams or gram atoms).

The above characteristics are decayed over time, for 24 (or more) time periods from 1-yr to 1-million years. An interpolation function is provided for those users who may need data at intermediate times. For LWR spent fuel, the integral heats (thermal power, integrated over time) are also available, between any two times out of a set of 38 times, also spanning 1- to 1-million years.

LWR Spent Fuel

Data on LWR spent fuel constitutes the largest portion of the Characteristics Data Base, because of the large quantities of material involved and because of the complexity of the data. The data base must deal with 78 (at present) different assembly types or models, over a wide range of fuel burnups (5 to 60 GWd/MTU), both actually discharged and

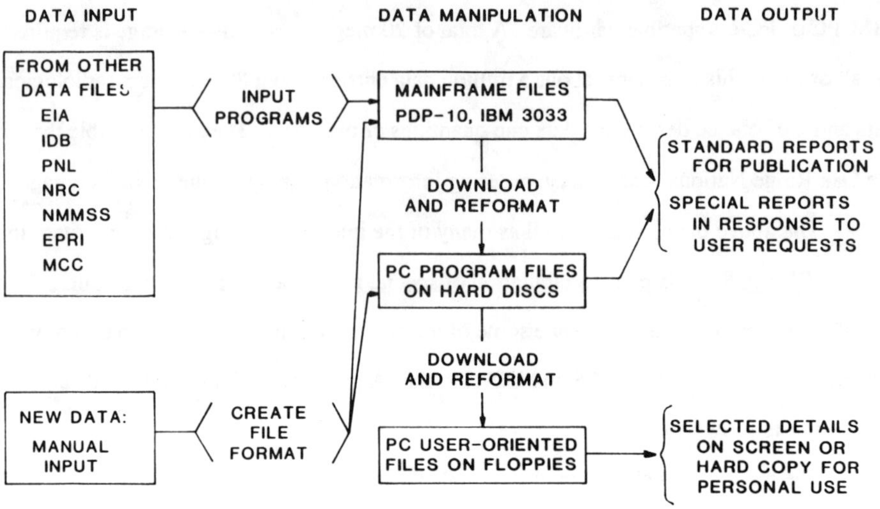

FIG. 2. DATA FLOW IN THE CHARACTERISTICS DATA BASE

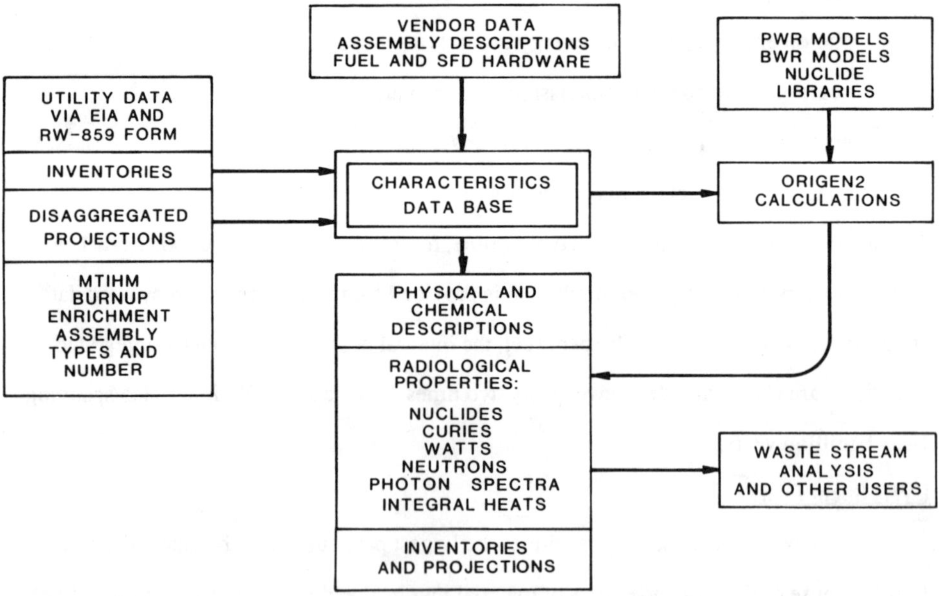

FIG. 3. CHARACTERISTICS DATA FOR LWR SPENT FUEL

Table 2. Annual spent fuel discharges from commercial light-water reactors.

End of calendar year	Mass, MTIHM		Mass, HTIHM		Mass, HTIHM	
	Annual rate	Accumulation	Annual rate	Accumulation	Annual rate	Accumulation
	Boiling-water reactors		Pressurized-water reactors		Totals	
1968	1	1	0	0	1	1
1969	10	10	0	0	10	10
1970	0	10	39	39	39	49
1971	70	81	26	66	97	146
1972	146	226	118	183	264	410
1973	93	320	67	250	160	570
1974	247	567	209	460	456	1,026
1975	196	763	322	781	518	1,545
1976	318	1,081	402	1,184	720	2,265
1977	383	1,464	455	1,639	839	3,103
1978	383	1,848	692	2,331	1,075	4,179
1979	396	2,244	717	3,048	1,113	5,292
1980	620	2,863	626	3,674	1,246	6,538
1981	457	3,321	693	4,367	1,150	7,688
1982	359	3,680	654	5,021	1,013	8,701
1983	498	4,178	767	5,788	1,265	9,966
1984	452	4,630	858	6,646	1,310	11,276
1985	480	5,110	882	7,528	1,362	12,639
1986	467	5,577	1,042	8,571	1,509	14,148
1987	601	6,178	1,154	9,725	1,755	15,902

Table 3. Mass of commercial BWR spent fuel discharged at various ranges of burnup

End of calendar year	Annual mass of discharged spent fuel for various burnup ranges, MTIHM									Total annual mass over all burnup ranges (MTIHM)
	0-5,000	5,001-10,000	10,001-15,000	15,001-20,000	20,001-25,000	25,001-30,000	30,001-35,000	35,001-40,000	40,001-45,000	
1968	0.6									0.6
1969		0.5	1.5	7.4	0.2	0.1				9.7
1970										0.0
1971	47.1	8.8	1.9	10.7	1.9					70.4
1972	71.6	38.0	31.4	4.2	0.5					145.7
1973	9.7	16.7	27.7	36.4	2.3	0.5				93.3
1974		77.9	120.8	44.5	3.4	0.4				247.0
1975	0.3	0.9	44.8	124.4	24.2	1.7				196.3
1976	0.9	65.9	94.3	154.4	2.5					318.1
1977		45.7	42.5	196.9	97.2	0.7				383.1
1978	1.6	33.2	18.1	76.0	239.1	15.4				383.4
1979			18.6	109.5	114.8	152.7	0.3			395.8
1980	14.0	0.4	0.6	95.2	337.0	122.9	9.8			619.9
1981		0.2	0.2	58.1	223.5	164.3	10.3	0.7		457.3
1982		0.2	4.6	19.4	137.2	183.9	13.0	0.6	0.4	359.3
1983			0.9	2.9	90.7	358.5	44.9			497.9
1984		5.4	9.5	0.4	123.8	189.9	122.7		0.4	452.0
1985	16.9	42.5	18.3	36.7	67.7	286.8	11.0		0.2	480.3
1986	50.8	38.9	42.5	67.0	51.4	147.4	69.1			467.1
1987	50.7	48.9	71.7	40.8	116.1	244.8	8.0	20.2		601.1

Table 4. Mass of commercial PWR spent fuel discharged at various ranges of burnup

End of calendar year	Annual mass of discharged spent fuel for various burnup ranges, MTIHM									Total annual mass over all burnup ranges (MTIHM)
	0-10,000	10,000-15,000	15,001-20,000	20,001-25,000	25,001-30,000	30,000-35,000	35,001-40,000	40,001-45,000	45,001-55,000	
1970		1.7	37.3							39.0
1971	4.6			2.5	19.4					26.5
1972		11.9	29.3	31.5	23.2	22.1				117.9
1973			26.2		33.3	7.6				67.1
1974	9.4	88.3	13.2	40.0	57.2	1.1				209.2
1975	45.3	95.0	53.6	79.4	25.3	23.1				321.8
1976		3.7	197.3	83.3	62.5	55.4				402.2
1977		2.8	86.7	132.9	130.5	87.1	15.4			455.5
1978	1.9	47.9	90.3	37.2	335.2	117.8	61.2	0.4		691.9
1979		30.6	107.1	64.0	232.3	236.3	46.5	0.5		717.3
1980		0.9		73.6	237.2	285.8	26.3	2.0		625.9
1981	0.9	17.2	1.9	26.8	232.9	377.4	34.4	1.3		692.7
1982	5.1	1.8	83.0	83.7	57.3	300.6	117.4	2.7	2.6	654.1
1983	5.5	6.4	71.8	51.1	170.1	318.4	138.0	5.4	0.5	767.1
1984		54.2	45.6	53.1	214.8	379.2	107.1	4.1		858.1
1985		20.5	25.9	19.9	223.1	333.9	234.3	24.2	0.4	882.2
1986	0.4	27.1	143.9	11.9	193.3	381.3	246.7	35.0	2.6	1,042.3
1987		27.6	135.1	85.1	126.8	400.6	355.5			1,154.0

projected for future discharge, and the radiological properties over time not only for the fuel itself, but also for the associated hardware. In addition, although most of this fuel is presently in the form of uncanned intact assemblies, some is canned and more of this can be expected (for defective fuel), and future practices may result in spent fuel disassembly and consolidation, which will give rise to activated metals from the hardware components. Figure 3 shows schematically the flow of input and output data for LWR spent fuel. More detailed information on this part of the data base is provided in Chapter 2 of ref. 2.

The LWR spent fuel data in the CDB can be tabulated in many ways. Table 2 shows the annual discharges from BWRs and PWRs, in terms of MTIHM (metric tons initial heavy metal). These data are also available in terms of numbers of assemblies, and by burnup. The latter is summarized in Tables 3 and 4. A graphical summation through 1981 of the burnup data in Tables 3 and 4 is given in Fig. 4. This clearly shows the trend to higher burnups with time, as well as the increasing quantities that were discharged.

Radiological characteristics important to handling and shielding include the thermal power, curies of gamma emissions, and the neutron production rate. Fig. 5 shows the thermal power for standard burnup fuel (about 30,000 MWD/MTIHM) from 5 yrs after discharge to 1000 years. During this time the major sources (Sr-90, Cs-137, Co-60, Cm-244, etc.) decay away, leaving a much lower, but very long-lived, residue of nuclides.

Defective LWR spent fuel is of special interest because it may require special handling procedures, which have not yet been identified, nor have the appropriate waste acceptance criteria been defined. This aspect of the data base is described in Section 2.5 of ref. 2.

Although there are 78 different assembly types listed in the data base, many of these are virtually identical, differing only in minor details or manufacturer, their basic design or configuration being controlled by reactor core design. There are 21 such "assembly classes" (Table 5), plus a number of variants that represent a very limited number of discharged assemblies. A detailed description of this classification scheme, which was

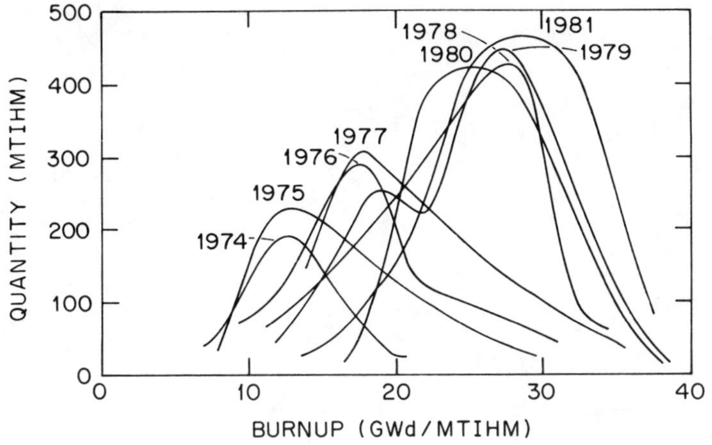

FIG. 4. DISCHARGE HISTORY OF LWR SPENT FUEL

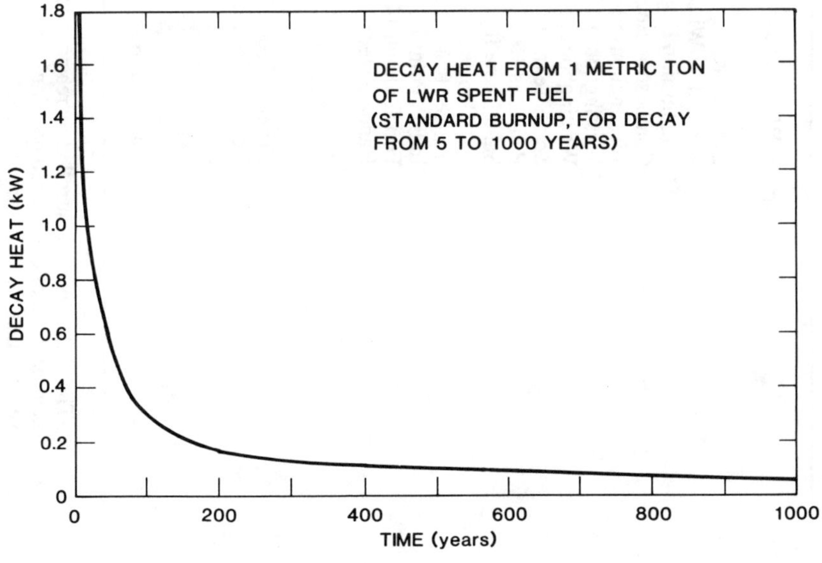

FIG. 5. DECAY HEAT OF LWR SPENT FUEL

Table 5. Major Classes of LWR Assembly Models[a]

Pressurized Water Reactors:	
	Babcock & Wilcox 15 x 15
	Babcock & Wilcox 17 x 17
	Combustion Engineering 14 x 14
	Combustion Engineering 16 x 16
	Westinghouse 14 x 14
	Westinghouse 15 x 15
	Westinghouse 17 x 17
	South Texas 17 x 17 XLR
	Palisades 15 x 15
	Saint Lucie II 16 x 16
	San Onofre I 14 x 14
	Yankee Rowe 15 x 16
	Indian Point I 13 x 14
	Fort Calhoun 14 x 14
	Haddam Neck 15 x 15
Boiling Water Reactors:	
	General Electric BWR/2, 3
	General Electric BWR/4, 5, 6
	Lacrosse
	Big Rock Point
	Dresden I
	Humboldt Bay

[a]These classes are controlled by reactor core configuration. For PWRs, the array size is fixed because of control rod positions within assemblies. For BWRs, the array size is not fixed because the cruciform control blades operate in between assemblies. For both types there are a number of early, one-of-a-kind reactors, identified by reactor name. Later versions are generic. For each class there are several model types manufactured.

developed as part of this work, has been published.[4] A supplemental report on GE BWR assemblies has also been issued.[5]

High-Level Waste

Immobilized HLW will derive from one commercial operation (West Valley) and three defense sites: Savannah River, Hanford, and Idaho. The first three will produce borosilicate glass by a vitrification process while Idaho, which is storing its HLW in the form of a calcine (rather than liquid, sludge, or salt cake), plans to produce a high-density ceramic waste form. All four sites plan to use canisters 10-ft long by 24-in. in diameter. For completeness, and to utilize existing data, the data base also includes the characteristics of the interim waste forms at the four sites, since this will be the feed material for future immobilized HLW. The data base includes detailed information, year-by-year, on the projected number of canisters and their radiological and thermal properties. A condensed summary of this type data is given in Table 6.

Non-LWR Spent Fuel

This category includes HTGR reactors, research and test reactors, and miscellaneous fuels from various reactors now stored at nine locations. Table 7 gives a summary of these fuels. Note that for the HTGRs the quantity is the number of fuel elements, for the FFTF it is the number of assemblies, while the others are in metric tons of heavy metal (MTHM). The Idaho total does not include the Shippingport breeder core, which contains 770 kg of U, mostly U-233, and 47.21 MT of thorium. In addition, the HTGR fuel is also at the Idaho facility. The thermal outputs of these spent fuels are generally quite low. However, the diversity of chemical compositions and physical configurations, coupled with the high enrichments of many of them, may impose some challenging packaging and disposal considerations.

Miscellaneous Wastes

This category includes wastes that are neither spent fuel nor HLW, but which may not qualify for commercial low-level disposal, nor as defense TRU waste. In the absence

Table 6. Characteristics of immobilized HLW

	West Valley	Savannah River	Hanford	Idaho
Weight of canister (kg)	252	500	500	500
Weight of contents (kg)	1895	1682	1650	1825
Total weight (kg)	2147	2182	2150	2325
Kilocuries (per canister)[a]	125	234	416	143
Watts (per canister)[a]	380	700	1150	340
Projected startup (year)	1990[b]	1990[b]	1996[b]	2011
Projected No. of canisters (cumulative) in:				
1992	275	922	0	0
1995	275	2152	0	0
2000	275	4202	653	0
2005	275	5302	1305	0
2010	275	5890	1860	0
2015	275	6350	1860	3800
2020	275	6810	1860	8800

[a]These are maximum values at time of loading. For West Valley, with a fixed input, the average is 10-20% lower. For the defense sites, with future input undefined, the averages may be much lower than the maximum values listed.

[b]Unofficial updates have adjusted these dates to 1992 and 1999, respectively.

Table 7. Summary of Non-LWR Spent Fuels

Reactor/site	Quantities[a]
HTGRs:	
Peach Bottom I	
Core I	804 elements
Core II	804 elements
Fort St. Vrain	725 elements
Research/Test Reactors:	
MTR Plate Fuel	-
TRIGA	-
UO_2-in-Polyethylene	-
PULSTAR	-
FFTF	170 assemblies
Miscellaneous Fuels:	
Argonne	0.31 MTHM
Babcock & Wilcox	0.05 MTHM
Battelle-Columbus	1.50 MTHM
Battelle-PNL	2.25 MTHM
Hanford	0.07 MTHM
Idaho	38.06 MTHM
Los Alamos	0.13 MTHM
Oak Ridge	1.28 MTHM
Savannah River	19.02 MTHM

[a]Data as of December 31,1987.

of intermediate depth disposal or greater-confinement disposal, these materials may be future candidates for consideration for repository disposal. Thus, they are included in the Characteristics Data Base, in order to provide technical data on these materials. Many of them are in the category of Greater-than-Class-C Low-Level Waste, and are not eligible for shallow-land burial without an exemption from the Nuclear Regulatory Commission. Wastes presently included in this section are OCRWM-generated wastes, commercial TRU waste, reactor decommissioning waste, radioisotope capsules, and certain LWR operational wastes.

Future Work

A major revision and update of the entire report and PC data bases is planned for 1989, after comments from current users have been received. This revision will include two additional PC data bases, on LWR Fuel Pin Characteristics and on LWR Assembly Serial Numbers. Descriptive data on recently-introduced LWR assembly models will be added and all inventories and projections will be updated. Improved descriptions of HLW will be incorporated. Radiological calculations for LWR fuel and hardware will be improved by using improved cross section and enrichment data for fuel and newly developed activation parameters for activated hardware.

References

1. J. W. Roddy, et al., "Physical and Decay Characteristics of Commercial LWR Spent Fuel," ORNL/TM-9591, January 1986.

2. "Characteristics of Spent Fuel, High-Level Waste, and Other Radioactive Wastes Which May Require Long-Term Isolation," DOE/RW-0184, Vols. 1-6 (December 1987) and Vols. 7-8 (June 1988):

Volume 1.	Summary, LWR Spent Fuel, and Immobilized High-Level Waste.	
Volume 2.	App. 1A	ORIGEN2 Overview
	App. 1B	ORIGEN2 Library Data
	App. 1C	ORIGEN2 Interpolation Functions
Volume 3	App. 2A	Physical Descriptions of LWR Fuel Assembly
Volume 4.	App. 2B	User's Guide to the LWR Assemblies Data Base
	App. 2C	User's Guide to the LWR Radiological Data Base
	App. 2D	User's Guide to the LWR Quantities Data Base
Volume 5.	App. 2E	Physical Descriptions of LWR NFA Hardware
	App. 2F	User's Guide to the High-Level Waste PC Data Base
Volume 6.	App. 3A	ORIGEN2 Decay Tables for Immobilized High-Level Waste
	App. 3B	Interim High-Level Waste Forms
	App. 3C	User's Guide to the High-Level Waste PC data Base
Volume 7.	Non-LWR Spent Fuels and Miscellaneous Wastes	
Volume 8.	App. 4A	Nuclear Reactors at Educational Institutions in the United States
	App. 4B	Data Sheets for Nuclear Reactors at Educational Institutions
	App. 4C	Supplemental Data for St. Vrain Spent Fuel
	App. 4D	Supplemental Data for Peach Bottom 1 Spent Fuel

3. A. G. Croff, "ORIGEN2 - A Revised and Updated Version of the Oak Ridge Isotope Generation and Depletion Code," ORNL-5621, July 1980, and "A User's Manual for the ORIGEN2 Computer Code," ORNL/TM-7175, July 1980.

4. R. S. Moore, D. A. Williamson, and K. J. Notz, "A Classification Scheme for LWR Fuel Assemblies," ORNL/TM-10901, November 1988.

5. R. S. Moore, "Physical Characteristics of GE Fuel Assemblies," ORNL/TM-10902, April 1989.

ANALYTICAL ELECTRON MICROSCOPY OF LIGHT-WATER REACTOR FUELS

L. E. Thomas and L. A. Charlot
Pacific Northwest Laboratory
Richland, WA

INTRODUCTION

Archetypal spent fuels from commercial light-water reactors (LWRs) are being analyzed by the Materials Characterization Center (MCC) to provide well characterized material for use in spent-fuel testing programs conducted by the U. S. Department of Energy Office of Civilian Radioactive Waste Management Program. Among the characterization methods being used, analytical transmission electron microscopy (AEM) and scanning Auger electron microscopy (SAM) uniquely provide information about the chemical nature and distribution of fission products at sub-micrometer resolution. Burnup-induced changes in microstructure that may affect the oxidation and leaching behavior of LWR fuels are not observable by most other methods.

This paper focuses on AEM and SAM of fission-product aggregates (particles and gas bubbles) observed within UO_2 grains and along grain boundaries. AEM and SAM are complementary methods. AEM reveals the internal structures of thin-foil samples down to the atomic level and provides compositional and crystal structure information from sample areas as small as about 10 nm. SAM is a surface-specific method useful for compositional analyses of selected local areas of grain boundaries and other surfaces exposed by fracturing. Both methods benefit from the use of micro-samples to minimize sample radioactivity for ease of handling and reduced analytical interference. Results from boiling-water and pressurized-water reactor (BWR and PWR) fuels having average burnups from 30 to 48 MWd/kgM and gas releases from <0.2% to nearly 18% are presented.

EXPERIMENTAL DETAILS

Table I lists the fuels examined. Burnup and gas release values were determined by radiochemical analyses of the fuel and analyses of gases released by puncturing fuel rods.[1] Samples for AEM and AES were obtained by chipping 300- to 400-μm-diameter fragments from edge, mid-radius and center locations on transverse sections from peak-power regions.

Table 1. MCC Fuels Examined by Transmission Electron Microscopy

Fuel	Type	Peak Burnup (MWd/kgM)	Gas Release (%)
ATM-101	PWR	32	0.2
ATM-103	PWR	33	0.25
ATM-104	PWR	44	1.1
ATM-105	BWR	31	0.6
ATM-106	PWR	48	11.2 and 17.7

Samples for AEM were prepared by a waterless method described by McCarthy and Thomas.[2] Individual particles were embedded in epoxy within a 3-mm-diameter annular washer, ground to 10 to 15 μm, masked with other glued-on washers, and thinned to perforation by argon ion micromilling. A liquid-nitrogen-cooled holder was used to avoid sample heating during ion milling. This procedure yielded samples suitable for examination in a conventional transmission/scanning transmission electron microscope (TEM/STEM). The 200-KV TEM/STEM was equipped for compositional analysis with energy-dispersive x-ray spectrometry (EDS), and was operated at its maximum accelerating voltage to increase the electron beam intensity and minimize beam broadening during EDS analysis. The smallest embedded second-phase particles on which EDS and electron diffraction analyses were successful were about 10 nm in diameter.

For Auger analysis, a method was developed to fracture fuel microparticles under vacuum in the SAM. Individual particles pressed into a soft indium substrate were crushed in an auxiliary vacuum chamber to expose fresh surfaces. Under slow crushing, the fuel tended to fracture mainly along fuel grain boundaries. Samples were transferred under vacuum between the crushing chamber and the microscope analysis chamber for repeated observation, analysis, and fracturing. Auger electron spectrometry (AES) was done with electron probes as small as 50 nm.

TRANSMISSION ELECTRON MICROSCOPY OBSERVATIONS

Transmission electron microscopy of moderate- to high-burnup fuels with <1% gas release, i.e. ATM-101, -103, -104, and -105, revealed high densities of nm- to µm-size bubbles and particles throughout UO_2 grains and along grain boundaries. Generally, these fission product aggregates were smallest near the pellet edges and coarsened progressively toward the centers in response to the local fuel operating temperatures. Similar radial variations in fission product aggregation were found in all the low-release fuels.

Figure 1 shows a grain boundary near the pellet edge in high-burnup ATM-104 fuel. Prominent features include enlarged (20- to 25-nm) particles associated with the grain boundary, and smaller particles throughout the adjacent grains. Bubble-free zones extended about 0.3 µm from the grain boundary shown in Figure 1, so that no bubbles appear in the figure. Farther from the boundaries, particles and bubbles 2 to 6 µm in size occurred in uniformly distributed particle/bubble pairs. Particles large enough for successful EDS and electron diffraction analysis (>10 nm) were generally the common fission product phase ε-ruthenium, a hcp solid-solution alloy of molybdenum, ruthenium, technetium, palladium and rhodium.

In edge to mid-radius samples, grain boundaries were decorated with small (<1 to 20 nm), closely spaced gas bubbles as shown in Figure 2. Bubble sizes and spacings generally increased from the pellet edges inward. Few bubbles were found near the pellet centers.

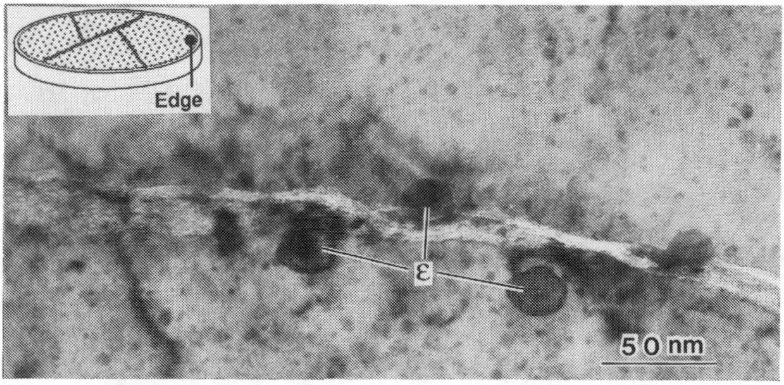

FIGURE 1. Grain Boundary Region Near Pellet Edge in Low-Gas-Release ATM-104 Spent Fuel. Brightfield TEM. Particles along boundary are ε-phase (Mo-Ru-Tc-Pd-Rh alloy).

In a few samples, no bubbles were visible along grain boundaries near the pellet edges. However, the image fine-structure of the boundaries (as in Figure 1) suggested that grain-boundary bubbles too small to resolve individually in the TEM were also present in these samples. The tendency of LWR spent fuel to oxidize and leach along grain boundaries has been attributed to short-circuit penetration of oxygen and water through the grain-boundary microbubbles and is expected to vary with the bubble spacings.[3]

Although the thinning procedure used for preparing TEM samples was controlled to minimize preparation artifacts, localized heating during ion micromilling occasionally produced microstructures with greatly enlarged bubbles (up to 120 nm) and ε-phase particles. The affected microstructures were associated with poorly connected regions in the thin samples, and were avoided during microscopy.

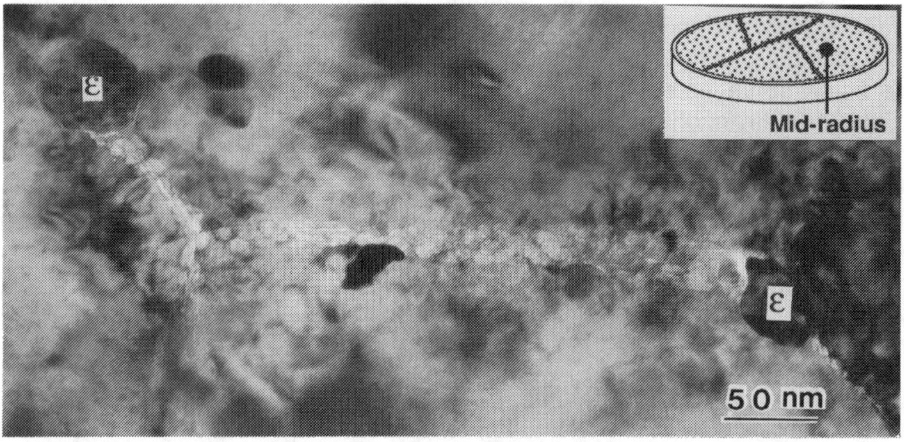

FIGURE 2. Bubble-Coated Grain Boundaries in Low-Release BWR Fuel ATM-105. Brightfield transmission electron micrograph.

A newly discovered phenomenon in LWR fuels is precipitation of fission gases as internally pressurized condensed-phase particles.[4] Near the pellet centers, xenon and krypton form discrete particles associated with ε-phase particles within UO_2 grains. Figure 3 shows these ε–phase/xenon-krypton aggregates surrounded by dislocation tangles in a UO_2 grain. The dislocations are believed to result from a "punching" mechanism of pressure relief by local plastic deformation of the surrounding matrix. Characteristic triangular-appearing pits in the thin AEM samples mark locations where the particles intersected the surfaces, and the fission gas phase was lost. One such pit is marked in Figure 3. Similar xenon/krypton particles were observed near the centers of all the fuels in Table 1.

EDS analyses (Figure 4) readily detected the xenon in the particles. Combined with microstructural imaging and diffraction analyses, EDS showed that the fission gases were concentrated in irregular, noncrystalline particles next to similar-size ε-particles. Compositions of the fission-gas particles determined by EDS were about 95 wt% Xe-5 wt% Kr, approximately the fission yield ratio[1].

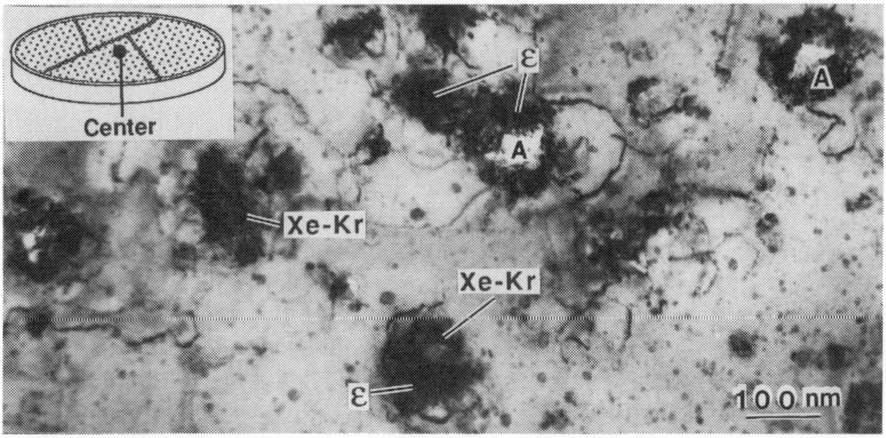

FIGURE 3. Xenon-Containing Particles and Surrounding Dislocation Tangles near Center of Low-Release ATM-103 Spent Fuel. Pits (A) result from gas release at surface-intersected particles.

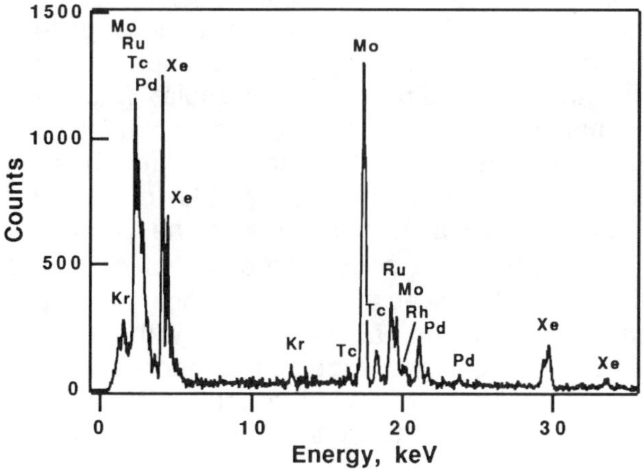

FIGURE 4. X-ray Spectrum from High-Density Xenon-Krypton Particle in ATM-103 Fuel. UO_2 background removed.

The conclusion that the xenon-krypton "particles" observed near the fuel pellet centers are a condensed phase under high pressure is based primarily on EDS analyses. For a suitably thin sample in the AEM, the x-ray intensities determined by EDS are related simply to elemental concentrations and, from measured particle and sample thicknesses, to densities. Analyses[4] of several xenon-krypton particles gave densities up to 3 g/cm^3, nearly that of a solid and about 10 times the density expected in equilibrium gas bubbles of the same sizes.[5] Based on the measured densities, internal pressures near 1 GPa can be inferred from the equation of state for xenon.[6] Similar pressures are needed to produce the observed plastic deformation of the UO_2 surrounding the particles.[7]

The only fission products detected by EDS microbeam analysis in the low-gas-release fuels were the five metals in ε–phase and the fission gases in particles found near the fuel-pellet centers. Other fission products, including cesium, barium, strontium, iodine and tellurium, were not detected and presumably were in solution in the UO_2 matrix. No fission products above detection limits of about 0.3 to 0.5 wt% for EDS were detected at grain boundaries or within particle-free regions of the UO_2 matrix. Based on calculated fission yields[1], concentrations near or below detection limits would be expected if the fission products stayed in solution in the UO_2.

AUGER FRACTOGRAPHY

Grain boundaries in low- and high-gas-release fuels ATM-103, 105, and -106 were analyzed for fission-product segregation in the scanning Auger microscope. Figure 5 shows typical Auger spectra from intergranular and transgranular fracture surfaces near a pellet edge in ATM-105 fuel. An inset scanning electron micrograph shows the fracture surface. The spectrum from the transgranular fracture is characteristic of the UO_2 matrix and contains no detectable peaks from fission products. Although fission-product particles and gas bubbles along grain boundaries were generally not visible in the low-release fuels because of image resolution limitations in the SAM, palladium and the other constituents of ε-phase were detected. It was not possible to detect iodine on UO_2 because of interference by oxygen. However, cesium, barium, and strontium, all detectable without interference from uranium and oxygen, and expected to segregate to grain boundaries, were notably absent from intergranular fracture surfaces near mid-radius and edge regions in the low-release fuels.

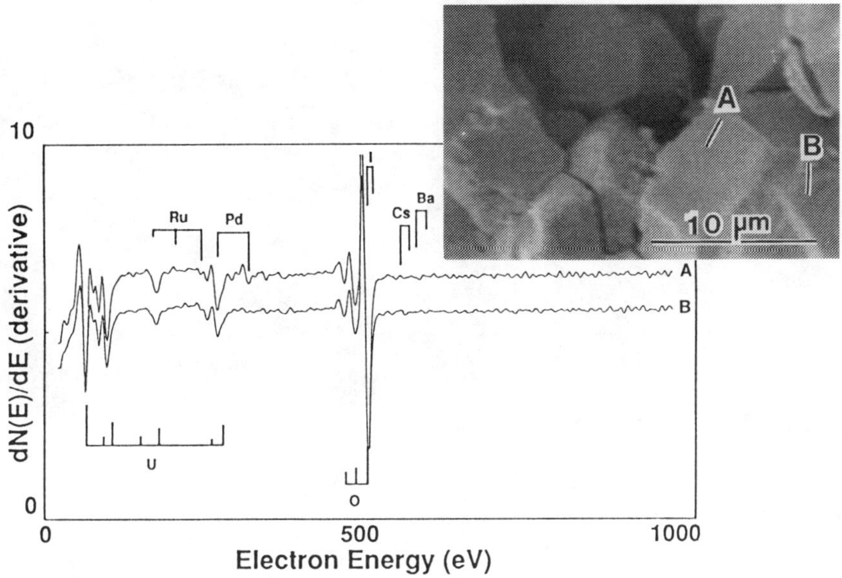

FIGURE 5. Auger Electron Spectrometry of ATM-105 (BWR) Fuel. Inset shows fracture surface. (A) Intergranular fracture. No Cs or Ba segregation. (B) Transgranular fracture.

In high-gas-release (18%) ATM-106 fuel, fracture surfaces near the fuel pellet center (Figure 6a) contained large particles and cavities (assumed to have contained fission gases before fracturing) along the grain boundaries. Transgranular fractures in Figure 6a appear as smooth, terraced surfaces lacking visible particles. The largest particles and cavities, up to several µm, occurred at grain-boundary intersections. EDS and electron diffraction analyses in the AEM identified the particles as ε-phase.

Auger spectrometry of intergranular fractures near the center of ATM-106 fuel showed no detectable concentrations of fission products on particle-free regions of the grain boundaries. However, high concentrations of cesium, tellurium, and sometimes barium were found on the fracture surfaces of the ε-phase particles. The spectrum in Figure 6b, for example, shows cesium and tellurium associated with a 2-µm particle. AES analysis also showed that the particle surfaces were highly enriched in palladium and ruthenium, and additionally contained cadmium, silver, and sometimes tin. The interior compositions of the particles, determined by EDS analysis of thin samples in the AEM, differed greatly from the compositions indicated by AES of the particle/matrix fracture surfaces.

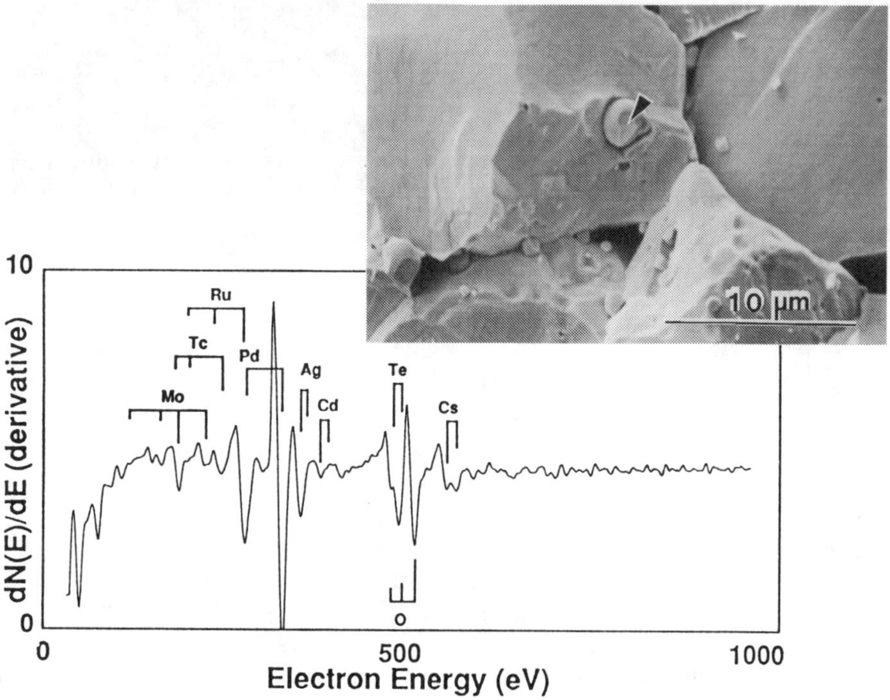

FIGURE 6. Auger Analysis of ε–Phase Particle near Center of 18% Gas-Release Fuel. (a) Fracture surface, showing intergranular particles and cavities. (b) Auger spectrum from particle (arrowed) with surfacial Cs, Te, Pd, Ag, and Cd enrichment.

DISCUSSION

Until recently, there was little direct information about the chemical nature or local distribution of fission products in LWR spent fuels. The sub-micrometer scale of the fission products in LWR fuels is accessible to only a few modern analytical methods, and the radioactivity of spent fuel has generally inhibited application of these methods. Many of the phases predicted in LWR fuels[8] do not seem to exist within the fuels examined in this study. Notable among the missing are the so-called "grey phases"--zirconates, molybdates and uranates that form optically visible particles in the higher temperature fast-breeder and gas-cooled reactors. Fuel temperatures in conventional LWRs may be too low to form these secondary oxides.

Despite the large number of chemically distinct fission products, the low-release LWR fuels contain essentially three phases (four counting the condensed-phase xenon-krypton and the bubbles containing unidentified gases as separate phases). These are the UO_2 matrix, the alloy phase ε-ruthenium, and xenon-krypton. Since the other fission products were not detected in low-release fuels, they probably are dispersed on or near the atomic level in the matrix, at concentrations below the detection limits of the present analytical methods.

From another viewpoint, the question is "Where are the fission products that might contribute to long-term release during leaching tests with water?" The species of greatest interest are cesium, technetium, strontium, and iodine.[9] Although interpretation of leaching tests is outside the scope of this paper, we now know where some of these products are--and are not--found in the fuels. Technetium, for example, exists largely within the relatively insoluble ε–ruthenium phase. Estimates of ε-phase volume fractions and compositions from EDS analyses and measurements of transmission electron micrographs roughly account for the total expected yields of the five metals.

Discovery of a condensed xenon-krypton phase near the centers of the fuel rods was an unexpected result obtained by analyzing spent fuel at high spatial resolution in the AEM. Condensed-phase xenon-krypton was observed in all the fuels listed in Table 1, and probably forms in most LWR fuels. The effects of this phenomenon on fuel behavior remain to be determined.

Auger analysis experiments described in this paper were directed at detecting cesium, strontium, and iodine on grain boundaries. Iodine was not detectable by AES because of interference from oxygen, and the sensitivity for strontium was limited by partial interference from uranium. The absence of detectable cesium on grain boundaries in the low-gas-release fuels was an unexpected result, but indicates that cesium does not significantly segregate to grain boundaries in these fuels. Apparently the cesium remains in the UO_2 matrix. In high-release fuels, electron microprobe analyses indicate that cesium moves from the fuel center toward the edges and accumulates in some outer regions of the fuels.[10] Cesium accumulations in outer regions of high-release fuel were not observed by AEM or SAM, possibly due to localization of deposits and limited sample examinations. The AES analyses complement the electron microprobe results by showing that cesium is retained at ε-phase particles on the grain boundaries near the pellet centers, but is not otherwise concentrated on these boundaries.

CONCLUSIONS

Analytical transmission electron microscopy and Auger electron spectrometry were used to determine the nature and distribution of fission product aggregates in LWR spent fuels having a wide range of burnup and gas-release characteristics. The fission solids and gas bubbles in the fuels ranged in size from <1 nm to several μm, and were dispersed throughout UO_2 grains and along grain boundaries. Although particle/bubble sizes generally increased from the cooler edges to the high-temperature centers of the fuel pellets, only the insoluble metals (Mo-Ru-Tc-Pd-Rh) and fission gases (Xe and Kr) formed separate phases from the UO_2. The absence of secondary oxide phases or detectable grain-boundary segregation in fuels with low gas release indicated that most of the fission products stayed in the UO_2. Cesium was detected on grain boundaries near the center of high-release fuel, but only in association with ε–phase particles on the boundaries. Xenon and krypton form a pressurized, condensed phase near the high-temperature centers of low-gas-release LWR fuels.

REFERENCES

1. R. J. Guenther et al. Characterization of Spent Fuel Approved Testing Material--ATM-103, PNL-5109-103, Pacific Northwest Laboratory, April, 1988.

2. J. M. McCarthy and L. E. Thomas, "Preparation of TEM Specimens from Highly Radioactive Materials," in: EMSA 1985, Ed. G. W. Bailey (San Francisco Press) 184-185.

3. L. E. Thomas, R. E. Woodley and R. E. Einziger, "Microstructural Examination of Oxidized Spent PWR Fuel by Transmission Electron MIcroscopy," J. Nucl. Mater, 166 (1989) 243-251.

4. L. E. Thomas and R. J. Guenther, "Characterization of Low-Gas-Release LWR Fuels by Transmission Electron Microscopy," in Scientific Basis for Nuclear Waste Management XII, Ed W. Lutze, Materials Research Society, 1989, 293-300.

5. D. R. Olander, Fundamental Aspects of Nuclear Reactor Fuel Elements. TID-26711-P1, Technical Information Center, U. S. Department of Commerce, Springfield, VA, 1976, p 203.

6. C. Ronchi, "Extrapolated Equation of State for Rare Gases at High Temperatures and Densities," J. Nucl. Mater. 96 (1981) 314-328.

7. G. W. Greenwood, A. J. E. Foreman and D. E. Rimmer, "The Role of Vacancies and Dislocations in the Nucleation and Growth of Gas Bubbles in Irradiated Fissile Material," J. Nucl Mater. 4 (1959) 306-324.

8. E. H. P. Cordfunke and R. J. M. Konings, "Chemical Interactions in Water-Cooled Nuclear Fuel: A Thermochemical Approach," J. Nucl. Mater. 152 (1988) 301-309.

9. C. N. Wilson and H. F. Shaw, "Experimental Study of the Dissolution of Spent Fuel at 85°C in Natural Ground Water," in Scientific Basis for Nuclear Waste Management X, Eds. J. K. Bates and W. B. Seefeldt, Materials Research Society, 1987, 123-130.

10. T. K. Campbell, R. J. Guenther and E. D. Jenson, "Volatile Fission Product Distributions in LWR Fuel Rods," This Proceedings.

Work sponsored by the U. S. DOE Office of Civilian Radioactive Waste Management. Pacific Northwest Laboratory is operated for the U. S. Department of Energy by Battelle Memorial Institute under Contract DE-AC06-76RLO-1830.

VOLATILE FISSION PRODUCT DISTRIBUTIONS IN LWR SPENT FUEL RODS

T. K. Campbell
R. J. Guenther
E. D. Jenson
Pacific Northwest Laboratory*
P.O. Box 999
Richland, Washington

INTRODUCTION

Results presented here are from spent fuel characterizations being conducted by the Materials Characterization Center at Pacific Northwest Laboratory on a variety of spent commercial power reactor fuels designated as approved testing materials (ATMs). These ATMs have a variety of burnup levels and fission gas releases; they include fuel from both pressurized water and boiling water reactor designs. The purpose of this work is to provide a source of well-characterized spent fuel for testing in the U.S. Department of Energy Office of Civilian Radioactive Waste Management repository programs and, potentially, other programs. Details of these characterization studies for some of the ATMs are available.[1,2,3]

This paper focuses on the radial redistribution of the fission products cesium and iodine in ATM fuel pellets and the apparent axial variation of these radial redistributions along the full length of the rod. The redistribution of these volatiles within a fuel rod is important because of the potential for release of radionuclides to the repository environment. Following breaches of the waste package and the cladding, aqueous dissolution of the radionuclides deposited in the fuel-cladding gap, in the fuel cracks, and possibly along grain boundaries will constitute the initial radionuclide release.

---

* The Pacific Northwest Laboratory is operated by Battelle Memorial Institute for the U.S. Department of Energy under Contract DE-AC06-76RLO 1830.

Gaseous and volatile fission products such as xenon, cesium, and iodine can migrate in uranium oxide fuel during reactor operation. Xenon, the major component of fission gas, is released from the fuel center via diffusion to grain boundaries and eventually to the fuel-cladding gap and fuel rod plenum space if centerline temperatures are sufficiently high.[4] Under similar conditions, cesium and iodine are known to migrate to the grain boundaries and away from the fuel center, depositing either in the cooler regions of the fuel towards the periphery, or on the cladding interior surface.[5]

MATERIALS EXAMINED

A summary of the rods examined is given in Table 1. Gamma scans, fission gas sampling, radiochemical analyses, and metallographic analyses were conducted on a minimum of one rod each from ATM-103, -104, -105, and -106. Rod NBD107 of ATM-106 had a high rod-average fission gas release (indicating high operating temperatures); thus, high levels of volatile fission product movement and, potentially, grain growth were expected for this rod. Rod MLA098 of ATM-103, which had low fission gas release, was expected to have lower levels of fission product movement and little grain growth.

Table 1. Summary of Rods Characterized

ATM	Rod ID	Rod Type	Peak Rod Burnup, MWd/kgM	Measured Rod Avg. Fission Gas Release, %	As-fabricated[(a)] True Grain Size, μm Range	Average ±1σ
103	MLA098	PWR	33.17	0.25	16.6-21.7	18.4±1.5
104	MKP109	PWR	44.34	1.09	9.6-11.3	10.5±0.5
105	ADD2966	BWR	33.94	7.85	14.8-17.6	16.5±1.5
	ADD2974	BWR	31.04	0.59	11.5-13.5	12.3±1.0
106	NBD107	PWR	46.46	11.2	6.4- 9.1	7.6±1.0

(a) From grain sizes at fuel edge. Range and average based on all samples cut from the fuel rod. True grain size = 1.57 x intercept grain size.[6]

PROCEDURES

Each full-length rod was axially gamma scanned for ^{137}Cs and ^{134}Cs using a germanium-lithium gamma ray detector. This provided an axial profile of the relative power/burnup along the length of each rod. Fission gas release was determined by measuring the equilibrium pressure obtained following puncture of the cladding and expansion of the fission gas into a known volume. Small sections (1.3 to 2.5-cm in length) from selected locations along the full length of a rod were then removed for radiochemical analyses (cesium and iodine) of material deposited on the interior cladding, ceramography of the fuel for grain size delineation, center-to-edge microprobe scans, and beta/gamma autoradiography. Each method is discussed in detail below.

The amounts of ^{135}Cs, ^{137}Cs, and ^{129}I released to the interior surface of the cladding were determined by radiochemical methods. After removing the fuel from the cladding, loose particles were removed from the interior cladding surface. Following removal of deposits on the exterior surface of the cladding, the cladding was immersed in an 8N HNO_3 solution to dissolve the interior surface deposits. The ^{137}Cs activity was determined by gamma ray spectrometry; ^{135}Cs by mass spectrometry of a chromographic elution from a cation exchange column. Iodine was separated by distillation and precipitation as AgI, and the ^{129}I activity was determined in a germanium-lithium well detector.

The fuel grain growth provides a relative measure of the temperature profile across the fuel radius and along the fuel rod during reactor operation. Grain sizes were determined using the intercept method[6] on 250x photomicrographs of fuel samples taken at the fuel center, 1/3 and 2/3 the radius, and near the edge. Grain growth was calculated using the near-edge grain size as the initial grain size. In some instances, the maximum grain growth did not occur at the center of the fuel (i.e., grain growth was asymmetric). Thus, reported grain growths are the maximum grain growth, not necessarily the grain growth from the edge to center, unless otherwise noted.

Electron Probe Micro-Analysis (EPMA) was performed on ceramographic sections to obtain chemical concentration data. Analyses were conducted at 200-μm or smaller steps across the fuel radius, providing a detailed radial profile of the relative distribution of the elements examined. Radial profiles of cesium and iodine from peak-power/burnup samples of ATM-103, -104, and -106 were determined by EPMA. Neodymium (a nonmigratory element) concentrations were measured to determine radial burnup profiles; these profiles were used to estimate the radial as-generated concentrations of other elements generated from the fission process.

Autoradiography provides a qualitative radial distribution of the fission products by direct contact of the ceramographic fuel sections on beta/gamma sensitive film. The beta-gamma radiographs reveal relative concentrations of cesium and other fission products. Darker regions on the autoradiograph indicate areas of higher concentration.

AXIAL VARIABILITY OF CESIUM AND IODINE DEPOSITS ON CLADDING INTERIOR

The full-length gamma scans of Rod MLA098 and Rod NBD107, sample locations, radiochemical data, and grain growth are reported in Figures 1 and 2, respectively.

In the low fission gas release Rod MLA098, low grain growth (<30%) was observed in the peak-power/burnup region. Both the cesium and iodine deposits on the cladding interior are consistent with the power/burnup profile of the gamma scan, i.e., the deposits are high in the center of the rod and lower near the rod ends. Deposits at comparable levels of burnup at the top and bottom ends of the rod (i.e., Samples D and DD) were similar. The largest amounts of cesium and iodine release occurred where the grain growth was highest (Sample H). At this location, however, the amounts of elemental cesium and iodine released to the cladding were only 0.2% and 0.04%, respectively, of the amounts of these elements generated in the fuel at the sample burnup.

Evidence of more cesium release was indicated in the ^{137}Cs gamma scan of Rod NBD107, a high fission gas release rod. In contrast to the ATM-103 fuel, spikes of cesium activity (as much as 10% above the nominal activity) were observed in several regions along the fuel rod. This spiking was indicative of high-centerline operating temperatures as later confirmed by grain growth measurements.

Examination of Rod NBD107 of ATM-106 showed cesium and iodine cladding deposits that increased from the ends of the fuel rod towards center region of the fuel rod where higher operating temperatures were expected. The amount of cesium and iodine released to the cladding in the peak-power/burnup region varied from 0.6 to 2.3% and 2.7 to 10.7%, respectively, of the amounts of these elements generated in the fuel at the sample burnups. Thus, the release of cesium and iodine from the ATM-106 fuel was above the levels observed on the low fission gas release rod of ATM-103.

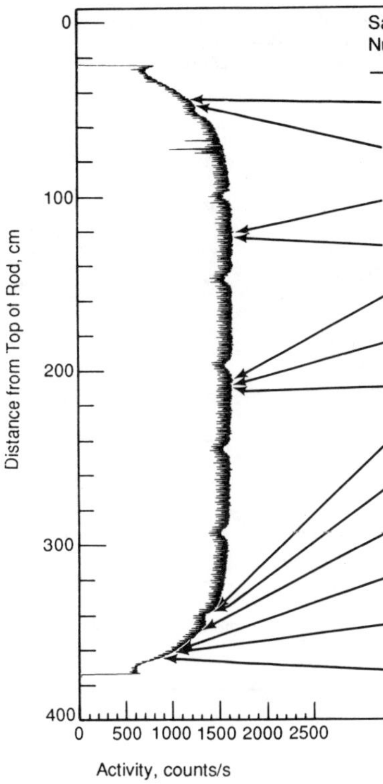

Sample Number	^{135}Cs µCi/cm²	^{137}Cs µCi/cm²	^{129}I µCi/cm²	Max Grain Growth, %
C	—	—	—	~0
D	3.9×10^{-5}	8.3	0.73×10^{-5}	—
H	—	—	—	28
I	214×10^{-5}	354	3.02×10^{-5}	—
N	—	—	—	~0
O	—	—	—	~0
S	7.4×10^{-5}	18.7	2.34×10^{-5}	—
Z	—	—	—	14
AA	—	—	—	11*
DD	4.5×10^{-5}	9.5	0.99×10^{-5}	—
HH	—	—	—	8*
II	—	—	—	6
LL	3.6×10^{-5}	5.8	0.25×10^{-5}	—

Figure 1. Comparison of Axial Variation of ^{135}Cs, ^{137}Cs and ^{129}I Deposits on the Inner Cladding Surface with the ^{137}Cs Gamma Scan and Grain Growth for Rod MLA098 of ATM-103 (Fission Gas Release = 0.25%). Asterisk indicates maximum grain growth was asymmetric.

The axial variability of cesium and iodine deposition in the ATM-106 fuel was consistent with the calculated levels of grain growth observed along the fuel rod. Samples H and BB, which were adjacent to samples with over 100% grain growth, had cesium and iodine deposits on the cladding interior surfaces that were three to four times higher than in Sample U, which had about 78% grain growth. Deposits near the bottom end of Rod NBD107 on Sample PP, which is estimated to have had 60 to 65% grain growth, were approximately two orders of magnitude above those at the top of this rod (Sample B) where little grain growth was observed. This finding was also qualitatively consistent with the higher ^{137}Cs movement observed in the gamma scan of the bottom end of the rod.

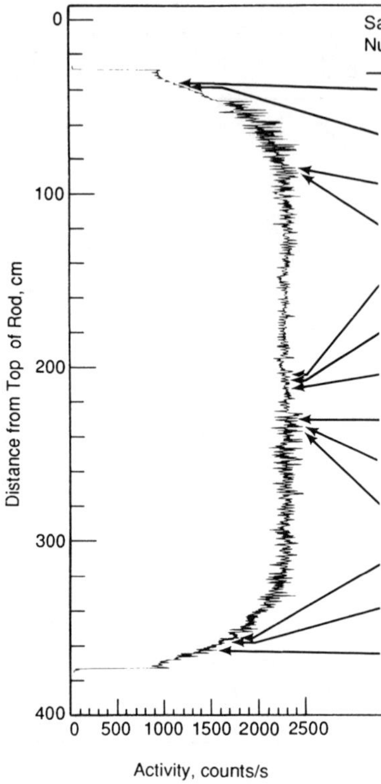

Sample Number	^{135}Cs µCi/cm²	^{137}Cs µCi/cm²	^{129}I µCi/cm²	Max Grain Growth, %
B	11.2x10⁻⁵	19.6	1.26x10⁻⁵	—
C	—	—	—	17*
G	—	—	—	104
H	1800x10⁻⁵	3973	950x10⁻⁵	—
O	—	—	—	75*
P	—	—	—	78
U	640x10⁻⁵	1414	269x10⁻⁵	—
Y	—	—	—	116
Z	—	—	—	112*
BB	2610x10⁻⁵	5811	802x10⁻⁵	—
KK	—	—	—	66
LL	—	—	—	64
PP	1690x10⁻⁵	2707	284x10⁻⁵	—

Figure 2. Comparison of Axial Variation of ^{135}Cs, ^{137}Cs and ^{129}I Deposits on the Inner Cladding Surface with the ^{137}Cs Gamma Scan and Grain Growth for Rod NBD107 of ATM-106 (Fission Gas Release = 11.2%). Asterisk indicates maximum grain growth was asymmetric.

The results of cesium and iodine deposition data for all rods examined to date are summarized in Figures 3 and 4, respectively. The data are expressed in terms of the percentage of the amount of elemental cesium locally generated in the fuel, assuming all isotopes of cesium are released in proportion to their generated values. For both of these volatile elements, the percentage of cesium or iodine released from the fuel and deposited on the interior surface increases with increasing maximum grain growth (a measure of fuel operating temperatures).

For the rods examined, the correlation of volatile fission product deposits on the cladding with grain growth suggests that significant grain growth may have occurred simultaneously with fission

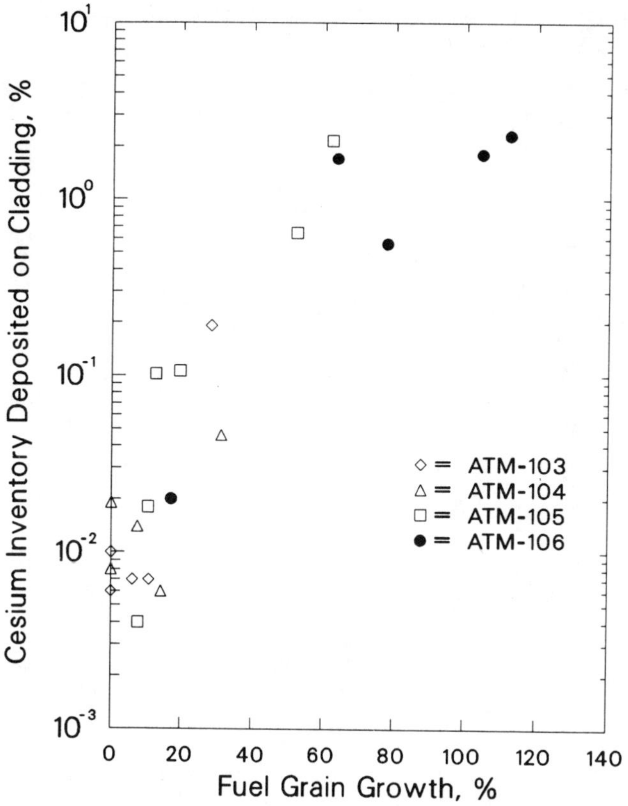

Figure 3. Percentage of the Local Cesium Inventory Deposited on the Inner Fuel Cladding as a Function of Fuel Grain Growth

product release. Fuel temperatures usually are at a maximum early in life when fission product concentrations are low, resulting in grain growth and minimal end of life fission product redistributions. However, under certain conditions later in life, grain growth may occur concurrent with volatile fission product redistribution. Possible short-term increases in the power may have caused simultaneous grain growth and volatile fission product release to the cladding. In addition, release of the fission gas from the fuel into the fuel-cladding gap is known to reduce thermal conductivity, causing an increase in fuel pellet temperatures and additional release of fission gas.[7] Significant release of the fission gas from the fuel at mid-life may have caused an increase in operating temperatures and corresponding increases in grain growth and volatile fission product redistribution.

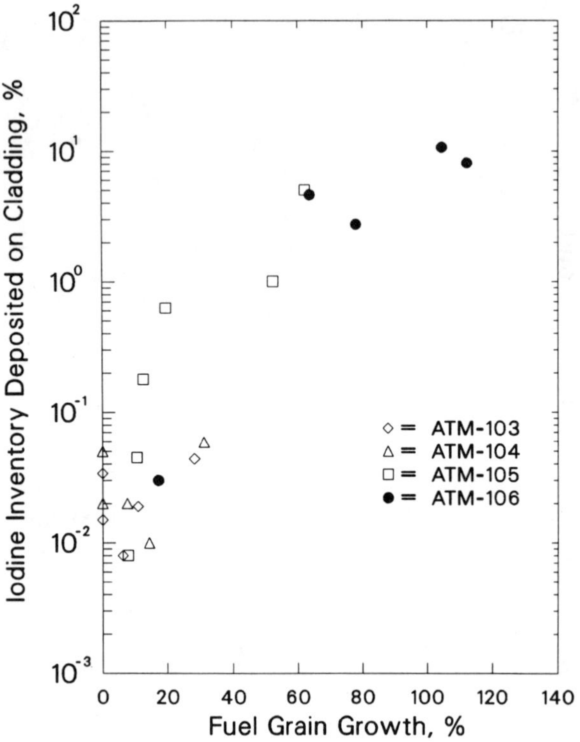

Figure 4. Percentage of the Local Iodine Inventory Deposited on the Inner Fuel Cladding as a Function of Fuel Grain Growth

RADIAL REDISTRIBUTION RESULTS

Distributions of cesium across the fuel radius were determined by EPMA on peak-power/burnup ceramographic samples cut from ATM-103, -104, and -106 rods. Iodine concentrations were too low to provide meaningful results.

To detect migration of the cesium, a comparison of the measured-to-predicted radial profile of cesium concentration was required. Because the measured EPMA data provided a relative distribution of cesium across the fuel radius, the as-measured profile was corrected before comparing it with the predicted curve. Correction was accomplished by normalizing as-measured data points so that a volume integration of the as-measured profile yielded a bulk average cesium concentration that was in agreement with that of the predicted profile. Cylindrical symmetry of both distributions was assumed and adjustment was made in the as-measured data for loss of cesium from the fuel to the cladding.

Neodymium, also measured across the fuel radius, provided the basis for generating the predicted radial profile of cesium concentration. First, local burnups were obtained from the linear relationship of burnup with measured neodymium concentration, as predicted by ORIGEN2.[8,9] Profiles of fission-generated cesium were then obtained using ORIGEN2 estimates of the cesium production with burnup.

A comparison of the measured and predicted profiles of cesium across the fuel radius indicates negligible cesium migration for a sample from the peak-power/burnup region of ATM-104 (Figure 5). This is consistent with the low grain growth and the low fission gas release of 1.1% for this rod. In addition, the relatively uniform exposure of the beta-gamma autoradiograph of this sample (Figure 6) indicates negligible migration of fission products. Similar results were found for a sample from the lower peak-power/burnup region of ATM-103.

In contrast, EPMA examination of a peak-power/burnup sample from high-release ATM-106 indicates that cesium has moved from the center towards the fuel edge. As shown in Figure 7, a net depletion of cesium is observed from the center of the fuel to about 60% of the fuel radius (2840 μm from the center). The residual cesium

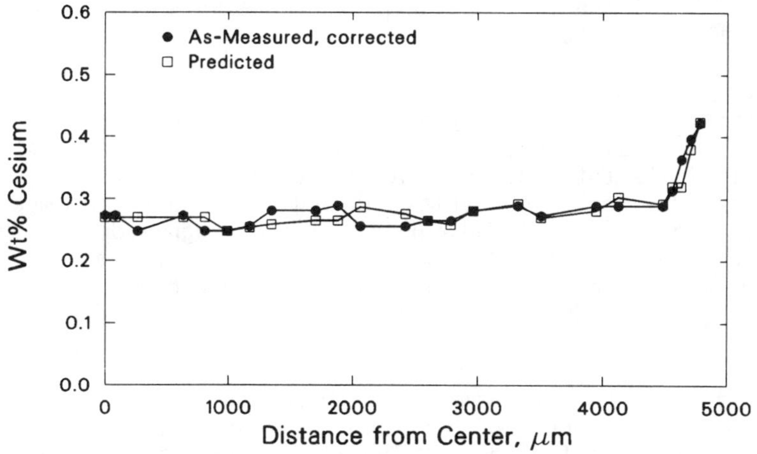

Figure 5. Comparison of Measured EPMA Cesium Radial Profile to ORIGEN2 Predicted Profile of a Peak-Power/Burnup Sample from Low Fission Gas Release Rod MKP109 of ATM-104 (Sample H, Grain Growth = 0%)

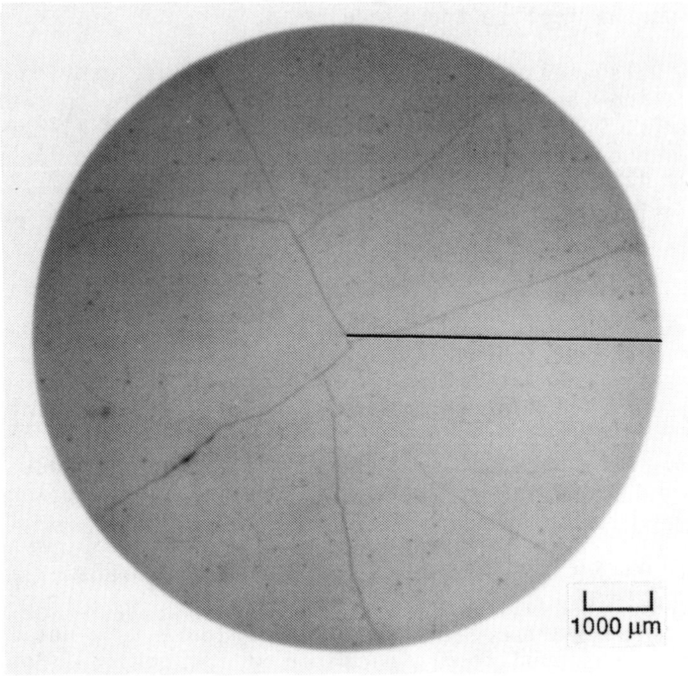

Figure 6. Beta-Gamma Autoradiograph of Sample H from Rod MKP109 of ATM-104. Note: The line on the photo represents the approximate path of the EPMA scan.

concentration remaining in the fuel center indicates that the migration of cesium occurred after the sample had accumulated over one-half of its total burnup. Accumulation of cesium, presumably along the fuel grain boundaries, is indicated where the measured data is above the predicted curve. It is recognized that the single data point near 3000 µm from the fuel center may be an anomaly. Measured edge-to-center grain growth of approximately 100% indicates that the cesium migration was the result of high fuel operating temperatures.

Examination of the beta-gamma autoradiograph of this sample in Figure 8 shows evidence of migration of fission products from the center towards the fuel edge. The lighter central region, which indicates a relatively low concentration of beta-gamma emitters, extends to approximately the same radial distance as the depleted region observed by EPMA in Figure 7. The location of the dark ring immediately following the depleted zone is consistent with the

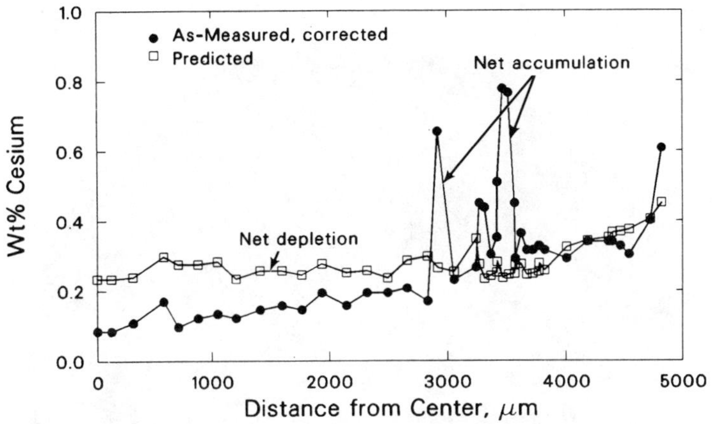

Figure 7. Comparison of Measured EPMA Cesium Radial Profile to ORIGEN2 Predicted Profile of a Peak-Power/Burnup Sample from High Fission Gas Release Rod NBD107 of ATM-106 (Sample Z, edge-to-center grain growth = 93%).

localized high concentrations observed by EPMA. The highest concentrations of fission products are visible near cracks close to the fuel periphery and in the fuel cladding gap.

Examination of a second ATM-106 sample from the same rod with 78% grain growth had qualitatively similar cesium EPMA results. However, the beta-gamma autoradiography revealed a central depleted zone that was not as well defined as the above sample with higher grain growth.

CONCLUSIONS

Cesium and iodine distributions along the length of fuel rods and across the fuel radius have been analyzed at selected locations in LWR spent fuel rods. The rods had rod-average burnup levels of 28 to 43 MWd/kgM and 0.3 to 11% fission gas release.

The axial redistribution of both cesium and iodine indicated a correlation between the degree of release of these elements from the fuel and the level of fuel grain growth, a measure of fuel temperatures. Up to 2% of the cesium and 11% of the iodine were deposited on cladding interior surfaces for samples with grain growths in excess of 100%. High levels of grain growth were associated with fuel rods with high rod-average fission gas release.

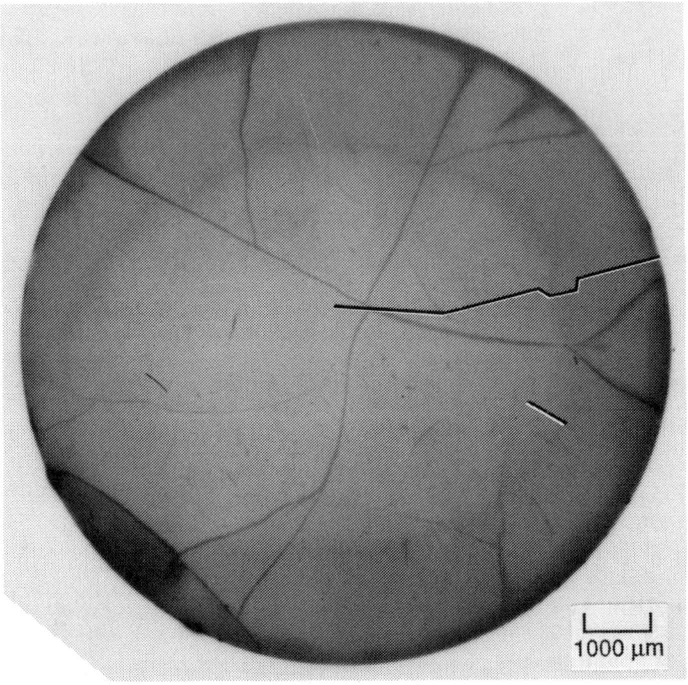

Figure 8. Beta-Gamma Autoradiograph of Sample Z From Rod NBD107 of ATM-106. Note: The lines on the photo represent the approximate paths of the EPMA scan; the shorter scan was conducted at smaller step sizes to verify the apparent cesium accumulation at approximately two-thirds radius.

Radial redistribution of cesium from the center towards the outer edge of the fuel occurred for peak-power/burnup fuels having grain growth in excess of approximately 75% (i.e., high operating temperatures). Fuels with low levels of grain growth (<30%) showed no indication of radial redistribution of cesium.

Thus, of the rods examined, the rods with high levels of grain growth have a significant fraction of their inventory of cesium and iodine deposited at readily releasable locations: in the fuel cladding gap, near peripheral fuel cracks, and presumably along grain boundaries. These types of fuels would be expected to initially release a larger percentage of their cesium and iodine inventory following contact with water than fuels with negligible levels of grain growth.

These investigations are continuing with examination of additional rods from the current stock of spent fuel ATMs. The results will provide characterization data for repository experimenters to interpret the results of performance tests conducted on these fuels.

REFERENCES

[1] J. O. Barner, *Characterization of LWR Spent Fuel MCC-Approved Testing Material - ATM-101*; PNL-5109, Rev. 1, Pacific Northwest Laboratory, Richland, WA, 1985.

[2] R. J. Guenther, D. E. Blahnik, T. K. Campbell, U. P. Jenquin, J. E. Mendel, L. E. Thomas, and C. K. Thornhill, *Characterization of Spent Fuel Approved Testing Material -- ATM-103*; PNL-5109-103, Pacific Northwest Laboratory, Richland, WA, 1988.

[3] R. J. Guenther, D. E. Blahnik, T. K. Campbell, U. P. Jenquin, J. E. Mendel, L. E. Thomas, and C. K. Thornhill, *Characterization of Spent Fuel Approved Testing Material -- ATM-106*; PNL-5109-106, Pacific Northwest Laboratory, Richland, WA, 1988.

[4] W. Hering, "The KWU Fission Gas Release Model for LWR Fuel Rods," *J. Nucl. Mater.*, **114**, 41-49, 1983.

[5] R. Manzel, F. Sontheimer, and R. Würtz, "The Radial Distribution of Fission Gasses and other Fission Products in Irradiated PWR Fuel," *J. Nucl. Mater.*, **126**, 132-143, 1984.

[6] ASTM, ASTM E 112, "Standard Method for Estimating the Average Grain Size of Metals"; pp. 186-220, in *Annual Book of ASTM Standards, Part II*, American Society for Testing and Materials, Philadelphia, PA, 1980.

[7] C. E. Beyer, "An Evaluation of Published High Burnup Fission Gas Release Data"; In *Proceedings of ANS Topical Meeting on LWR Extended Burnup - Fuel Performance and Utilization*, American Nuclear Society, Williamsburg, VA, 1982.

[8] A. G. Croff, *A User's Manual for the ORIGEN2 Computer Code*; ORNL/TM-7175, Oak Ridge National Laboratory, Oak Ridge, TN, 1980.

[9] A. G. Croff, *ORIGEN2 - A Revised and Updated Version of the Oak Ridge Isotope Generation and Depletion Code*; ORNL-5621, Oak Ridge National Laboratory, Oak Ridge, TN, 1980.

STUDIES ON SPENT FUEL DISSOLUTION BEHAVIOR UNDER YUCCA MOUNTAIN REPOSITORY CONDITIONS

C. N. Wilson
Pacific Northwest Laboratory

C. J. Bruton
Lawrence Livermore National Laboratory

ABSTRACT

Nuclide concentrations measured in laboratory tests with PWR spent fuel specimens in Nevada Test Site J-13 well water are compared to equilibrium concentrations calculated using the EQ3/6 geochemical modeling code. Actinide concentrations in the laboratory tests reach steady-state values lower than those required to meet Nuclear Regulatory Commission (NRC) release limits. Differences between measured and calculated actinide concentrations are discussed in terms of the effects of temperature (25°C to 90°C), sample filtration, oxygen fugacity, secondary phase precipitation, and the thermodynamic data in use. The concentrations of fission product radionuclides in the laboratory tests tend to increase continuously with time, in contrast to the behavior of the actinides.

1.0 INTRODUCTION

The Yucca Mountain Project of the U.S. Department of Energy is studying the potential dissolution and radionuclide release behavior of spent fuel in a candidate repository site at Yucca Mountain, Nevada. The repository horizon under study lies in the unsaturated zone 200 to 400 meters above the water table. With the exception of C-14, which may migrate in a vapor phase,[1] and possibly I-129, the majority of long-lived radionuclides present in spent nuclear fuel will be transported from a failed waste package in the repository via dissolution or suspension in water in the absence of a major geological event such as volcanism.

Spent fuel will not be contacted by liquid water infiltrating the rock until several hundred years after disposal when the repository has cooled to below the 95°C boiling temperature of water at the repository elevation. The potential dissolution behavior of spent fuel during the repository post-thermal period is being studied using geochemical models and laboratory tests with actual spent fuel specimens.* Selected initial results from these studies are discussed in the present paper.

2.0 LABORATORY TESTS

Three spent fuel dissolution test series have been conducted in laboratory hot cells using spent fuel specimens of various configurations. Results from the Series 2 and Series 3 tests with bare fuel particles are discussed in the present paper. The Series 2 tests used unsealed fused silica test vessels and were run for five cycles in air at ambient hot cell temperature (25°C). The Series 3 tests used sealed stainless steel vessels and were run for three cycles at 25°C and 85°C. Each test cycle was started in fresh Nevada Test Site J-13 well water and was about six months in duration. Periodic solution samples were taken during each test cycle and the sample volume was replenished with fresh J-13 water. Five bare fuel specimens tested in these two test series are identified in Table 1 and the test configurations are shown in Figure 1. The composition of J-13 well water(2) is given in Table 2. Additional information on the laboratory tests is provided in references 3 and 4.

2.1 Actinide Results

Actinide concentrations (U, Np, Pu, Am and Cm) measured in solution samples rapidly reached maximum levels during the first test cycle and then generally dropped to lower steady-state levels in later test cycles. The concentrations of uranium and the activities of Pu-239+240 and Am-241 measured in 0.4 μm filtered solution samples are plotted in Figure 2. The initial concentration peaks are attributed to dissolution of more readily soluble UO_{2+x} oxidized phases present initially at the fuel particle sur-

---

* This work was performed under the auspices of the U.S. Department of Energy (DOE) by Lawrence Livermore National Laboratory under Contract No. W-7405-Eng-48, and by Pacific Northwest Laboratory operated for the DOE by Battelle Memorial Institute under Contract No. DE-AC06-76RLO-1830.

Table 1. Bare Fuel Test Identification

Identification	Description	Starting Fuel Wt. (g)
HBR-2-25	Series 2, H. B. Robinson Fuel, 25°C	83.10
TP-2-25	Series 2, Turkey Point Fuel, 25°C	27.21
HBR-3-25	Series 3, H. B. Robinson Fuel, 25°C	80.70
HBR-3-85	Series 3, H. B. Robinson Fuel, 85°C	85.55
TP-3-85	Series 3, Turkey Point Fuel, 85°C	86.17

faces, and to kinetic factors limiting the nucleation and growth of secondary phases that may ultimately control actinide concentrations at lower levels.

Uranium (U) concentrations at 25°C were lower in the Series 3 tests than in the Series 2 tests, and with the exception of the Cycle 1 data, U concentrations in the 85°C Series 3 tests were lower than those in the 25°C tests. The very low U concentrations measured during Cycle 1 of the HBR-3-85 test are attributed to a vessel corrosion anomaly. In the later cycles of the Series 2 tests, U concentrations tended to stabilize at steady-state levels of about 1 to 2 μg/ml. In Cycles 2 and 3 of the Series 3 tests, U concentrations stabilized at about 0.3 μg/ml at 25°C and about 0.15 μg/ml at 85°C. Precipitated crystals of the calcium-uranium-silicates, uranophane (Figure 3) and haiweeite, and possibly the uranium-silicate soddyite, were found on filters used to filter cycle termination rinse solutions from both 85°C tests. Phase identifications were based on examinations by X-ray diffraction and microanalysis in the SEM[4]. Secondary phases controlling actinide concentrations other than U were not found.

The 0.4 μm filtered Pu-239+240 solution activities measured in Cycles 2 through 5 of the TP-2-25 test generally ranged from about 100 to 200 pCi/ml (Figure 2). Activities as low as about 20 pCi/ml were measured in the HBR-2-25 test. During Cycles 2 and 3 of the HBR-3-25 test, activities varied from about 60 to 100 pCi/ml. A value of 100 pCi/ml, which corresponds to a Pu concentration of about 4.4×10^{-9} $\underline{M}$ ($\underline{M}$ = molarity), would appear to be a reasonable estimate of steady-state Pu-239+240 activities in 0.4 μm filtered solutions in the 25°C tests. Significantly lower activities on the order of 1 pCi/ml were measured in the 85°C tests. The lower activities at 85°C may result from enhanced nucleation and growth of secondary phases at the higher temperature that limit Pu concentration.

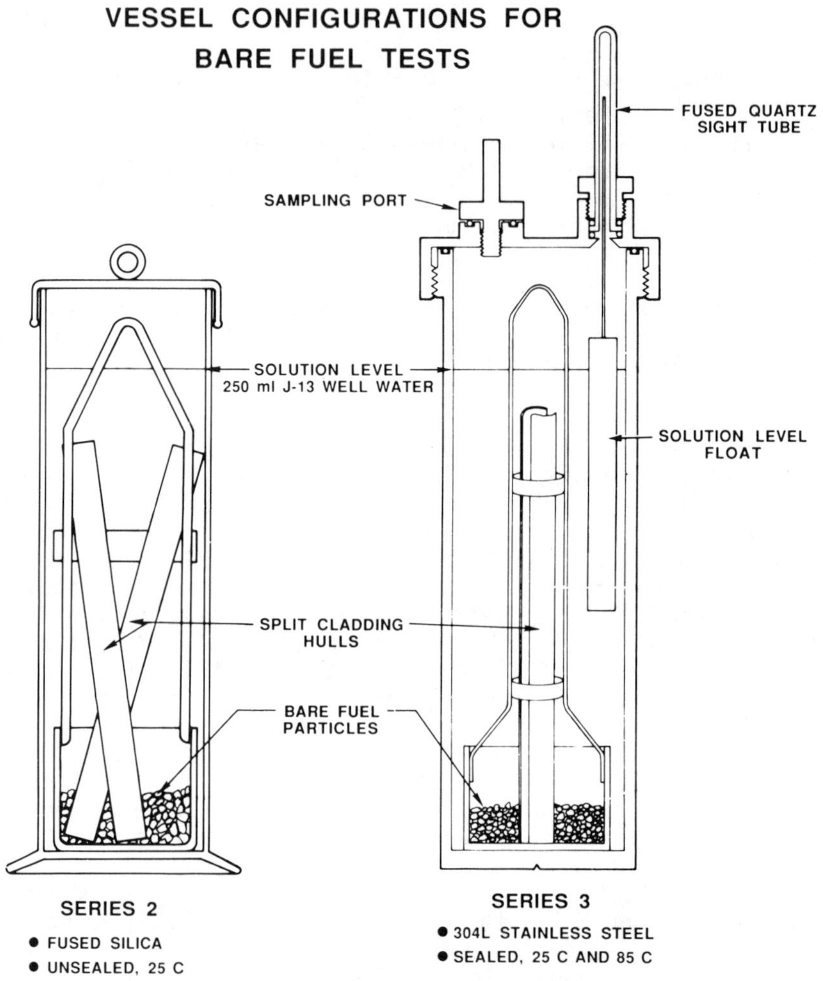

Figure 1. Test Configurations for the Series 2 and Series 3 Bare Fuel Dissolution Tests.

Selected solution samples were centrifuged through membrane filters that provide an estimated filtration size of approximately 2 nm.* Filtering to 2 nm caused Pu-239+240 activities to

---

* Amicon Corporation Model CF-25 centrifuge membrane cone filter.

Table 2. J-13 Well Water Analysis[2]

Component	Concentration ($\mu g/ml$)	Component	Concentration ($\mu g/ml$)
Li	0.042	Si	27.0
Na	43.9	F	2.2
K	5.11	Cl	6.9
Ca	12.5	NO_3	9.6
Mg	1.92	SO_4	18.7
Sr	0.035	HCO_3	125.3
Al	0.012		
Fe	0.006	pH	7.6

decrease by about 20 to 40%. No significant differences between unfiltered and 0.4 μm filtered Pu activities were noted. The 0.4 μm filtered sample data are considered the most significant relative to radionuclide release because larger particles probably would not be transported by water, whereas colloidal particles greater than 2 nm may remain in stable suspension and be transported by water movement.

Steady-state Am-241 activities on the order of 100 pCi/ml, corresponding to Am concentrations of about 1.5 x 10^{-10} M, were measured in 0.4 μm filtered samples during Cycles 2 and 3 of the TP-2-25 and HBR-3-25 tests. The 100 pCi/ml value would appear to be a conservative estimate for Am-241 activity at steady-state and 25°C considering that activities on the order of 10 pCi/ml were measured during Cycles 2, 4 and 5 of the HBR-2-25 test. Much lower 0.4 μm filtered Am-241 activities of about 0.3 pCi/ml were measured during Cycles 2 and 3 of the two 85°C tests. The effects of both 0.4 μm and 2 nm filtration were in general greater for Am-241 than for Pu-239+240. Association of Am with an apparent suspended phase is suggested by unfiltered data from the 85°C tests plotted as dashed lines in Figure 2, and by a relatively large fraction of 0.4 μm filtered Am-241 activity removed by 2 nm filtration (not shown). Cm-244 activity measured in most samples was similar to that measured for Am-241 in each of the tests. However, Cm-244 alpha decays with a 18-year half-life to Pu-240 and will not be present during the repository post-thermal period.

Measured Np-237 activities in most samples were generally not much greater than the detection limit of 0.1 pCi/ml and were below detection limits in several samples. Measured Np-237 activities showed very little dependence on temperature, vessel

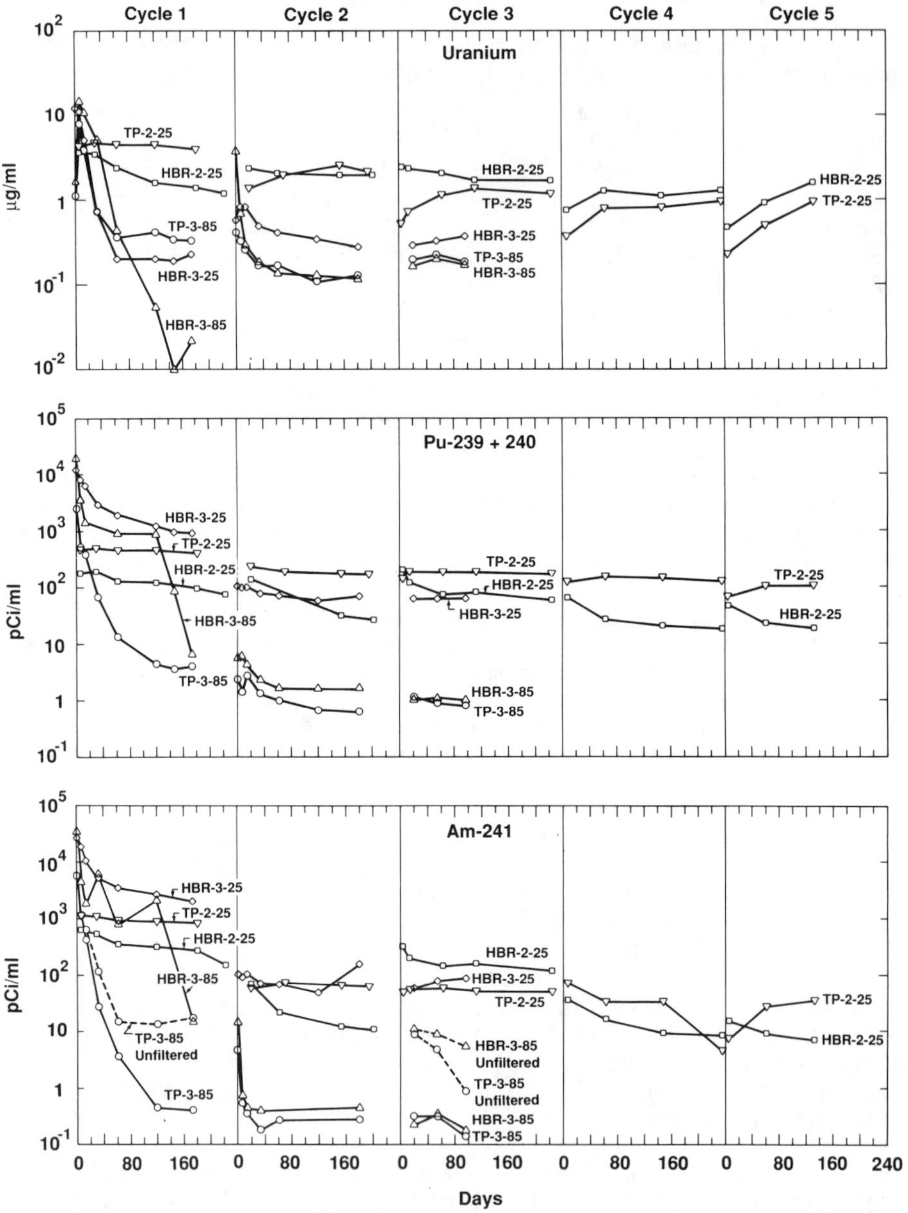

Figure 2. Uranium Concentrations (top), Pu-239+240 Activities (center) and Am-241 Activities (bottom) Measured in 0.4 μm Filtered Solution Samples.

Figure 3. Acicular Crystals of Uranophane Formed on Spent Fuel Grains in the 85°C Series 3 Tests.

type or sample filtration. Following initially higher values at the beginning of Cycle 1, Np-237 activities generally ranged from 0.1 to 0.5 pCi/ml.

2.2 Fission Product Results

Specimen inventory fractions of the fission product radionuclides Cs-137, Sr-90, Tc-99 and I-129 measured in solution are plotted in Figure 4 for the HBR-2-25 and HBR-3-85 tests. Each data point represents the fraction of the ORIGEN-2 calculated specimen inventory in solution on the sample date plus the inventory fraction calculated to have been removed in previous samples from the test cycle. During Cycle 1 of the HBR-3-85 test, Tc-99 fell to below detectable levels as a result of the corrosion anomaly that occurred in this test. Cycle 1 Cs-137 gap inventory release was about 0.7% from the HBR fuel and is therefore off-scale in Figure 4. Sr-90 was not measured during Cycle 1 of the Series 2 tests, and appeared to be limited by association with an unknown precipitated phase in the 85°C tests.

The inventory fractions of Cs-137, Sr-90, Tc-99 and I-129 in solution increased continuously with time, with the exception of the anomalous precipitation of Tc-99 in Cycle 1 of the HBR-3-85

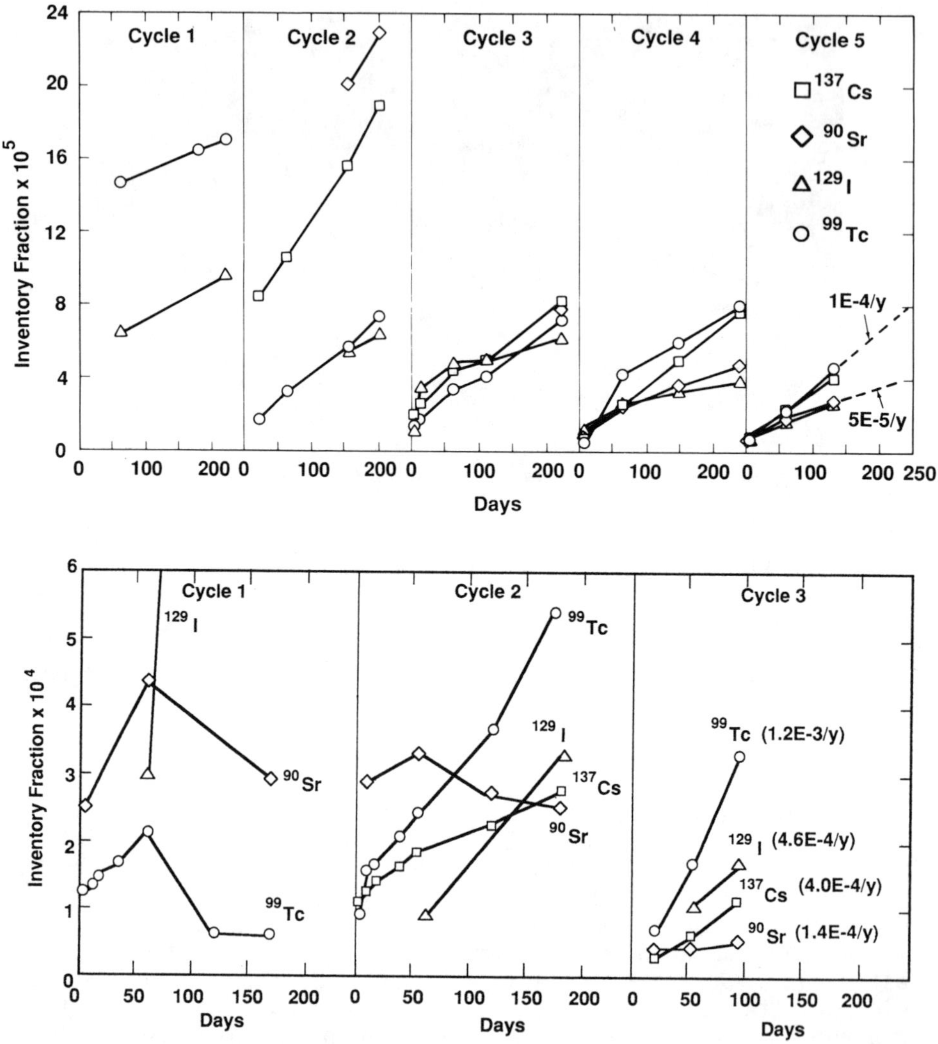

Figure 4. Inventory Fractions of Cs-137, Sr-90, Tc-99 and I-129 Measured in Solution in the HBR-2-25 Test (top) and in the HBR-3-85 Test (bottom). Approximate annual fractional release rates are listed for each nuclide during the last cycle plotted.

test and the limit on Sr-90 activity in solution at 85°C. The continuous release rates of the fission products in units of inventory fraction per year are given in Figure 4 for the final cycle of the two tests. Because the actual quantity of fuel matrix dissolution and precipitation of actinides was not measured, it is not known to what degree the continuous fission product release resulted from preferential leaching of grain boundaries where fission products are thought to concentrate during irradiation. Whether as a result of increased matrix dissolution or increased grain boundary leaching, the soluble fission product release rate is greater in the later test cycles at the higher temperature.

3.0 GEOCHEMICAL MODELING

3.1 Actinide Concentrations in Solution

Spent fuel dissolution in J-13 well water was simulated using the geochemical modeling code EQ3/6[5] to determine whether steady-state actinide concentrations measured in the tests could be related to the precipitation of actinide-bearing solids. Version 3245 of the EQ3/6 code and version 3270R13 of the supporting thermodynamic database were used to simulate spent fuel dissolution at 25°C and 90°C assuming atmospheric CO_2 gas fugacity and two different O_2 gas fugacities of $10^{-0.7}$ (atmospheric) and 10^{-12} bars (see later discussion). The simulation process is described in more detail elsewhere.[6] The computer simulations yield: 1) the sequence of solids that precipitate and sequester elements released during spent fuel dissolution; and 2) the corresponding elemental concentrations in solution. Approximate steady-state actinide concentrations measured at 25°C and 85°C in the Series 3 laboratory tests are compared in Table 3 to concentrations of actinides in equilibrium with the listed solids as calculated in the EQ3/6 simulations. Comparisons of simulation results with experimental results are being used to determine the adequacy of the thermodynamic database and to identify additional aqueous species and minerals for which data are needed.

Uranium (U) concentrations in the simulations vary as a function of the secondary U-bearing precipitates. The following sequence of mineral assemblages are predicted to precipitate and sequester U as increasing amounts of spent fuel dissolve: haiweeite, haiweeite plus soddyite, soddyite, soddyite plus schoepite, and schoepite. The relative compositions of these phases and of U-bearing phases that were observed in residues from the 85°C laboratory tests are shown in Figure 5. Unique, and steadily increasing, concentrations of U in solution are related to each

Table 3. Comparison of Measured and Predicted Actinide Concentrations (log M)

Actinide	Measured[(a)] 25°C	Measured[(a)] 85°C	EQ3/6[(b)] 25°C -0.7	EQ3/6[(b)] 25°C -12.0	EQ3/6[(b)] 90°C -0.7	EQ3/6[(b)] 90°C -12.0	Phase
U	-5.9	-6.2	-7.2/-7.0*	-7.1/-6.9	-8.8/-7.6	-8.5/-7.5	H
			-7.0/-6.9	-6.9/-6.8	-7.6	-7.5	H + S
			-6.9/-4.3	-6.8/-4.2	-7.6/-6.0	-7.5/-5.9	S
			-4.3	-4.2	-6.0	-5.9	S + Sch
			-4.2	-4.1	-6.0/-5.8	-5.8/-5.6	Sch
Np	-8.9	-9.1	-6.2	-9.0	-5.2	-8.0	NpO_2
Pu	-8.4	-10.4	-12.4	-13.8	-11.9	-14.6	PuO_2
			-4.3	-5.7	-4.2	-6.9	$Pu(OH)_4$
Am	-9.8	-12.3	-8.3	-8.3	--	--	$Am(OH)CO_3$
			--	--	-8.4	-8.4	$Am(OH)_3$
Cm	-11.3	-14.3	Cm not in thermodynamic database				

(a) Series 3 tests, 0.4 μm filtered.
(b) At oxygen fugacities log f_{O2} = -0.7 (atmospheric) and log f_{O2} = -12.0 with solubility control by precipitated secondary phases as listed. H = haiweeite; S = soddyite; Sch = schoepite. All phases are in crystalline state except $Pu(OH)_4$ which is amorphous.
* -7.2/-7.0 refers to a range in concentration from -7.2 to -7.0.

mineral assemblage. The concentration of U varies not only as the precipitates vary, but also during the precipitation of a single mineral, such as soddyite, because of changes in the pH and overall chemical characteristics of the fluid. As previously discussed, uranophane, haiweeite, and possibly soddyite were found in the 85°C Series 3 tests. Unfortunately, reliable thermodynamic data for uranophane are not available, which complicates comparison of the laboratory test results to the calculated solubility limits. Haiweeite, a Ca-U-silicate like uranophane, is predicted to precipitate at U concentrations that are lower than the measured steady-state values. In the absence of data for uranophane, the experimental concentrations of U would appear to be consistent with the precipitation of soddyite at both 25°C and 90°C in the simulations.

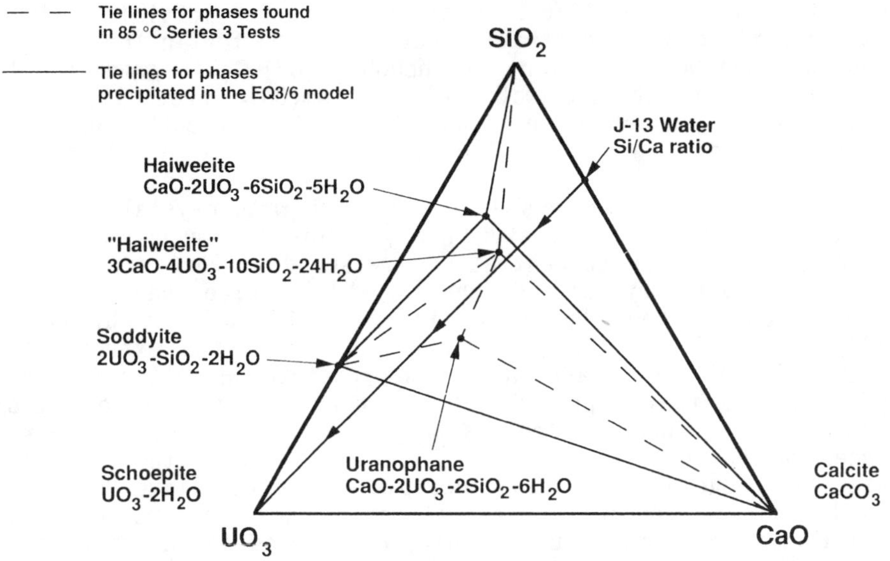

Figure 5. Relative Compositions (mole %) of U-bearing Phases Indicated as Controlling U Concentration in the EQ3/6 Simulation and for which Indications were Observed in the 85°C Series 3 Tests.

Neptunium concentration is controlled by equilibrium with NpO_2 in the simulations. However, the predicted concentration of Np is highly dependent on solution Eh and pH.(7) The O_2 fugacity in the simulations was reduced from $10^{-0.7}$ bars to 10^{-12} bars in order to produce good agreement between the measured and predicted concentrations of Np at 25°C. An O_2 fugacity of 10^{-12} bars may correspond to conditions at the fuel surface in an otherwise oxygenated system (i. e. contains an air cap) that is poorly buffered. Eh was not measured during the laboratory tests, and redox equilibrium may not have been established among the various species and phases within the sealed stainless steel vessels. An oxygen fugacity of 10^{-12} bars over-estimates Np concentration at 90°C, however, because the experimental data do not reflect predicted increases in Np concentration with temperature. The thermodynamic data for Np and other actinides must, consequently, be critically evaluated at elevated temperature.

Significant differences exist between measured and predicted Pu and Am concentrations in Table 3. Measured Am concentrations may have been lower than those predicted because of Am removal from solution by phases such as lanthanide precipitates that were not accounted for in the EQ3/6 simulations. Another possible mechanism controlling Am concentration not accounted for in the simulation may have been sorption. Although $Am(OH)CO_3$ is predicted to control Am concentration at 25°C and $Am(OH)_3$ precipitates at 90°C, the Am concentration in equilibrium with both phases is about the same.

Predicted Pu concentrations in equilibrium with crystalline PuO_2 at both temperatures and oxygen fugacities are much lower than those measured. Pu concentrations measured at 25°C are similar to those reported by Rai and Ryan[8], who measured the solubility of PuO_2 and hydrous $PuO_2 \cdot xH_2O$ in water for periods of up to 1300 days at 25°C. At a pH of 8, which was the extrapolated lower limit of their data and the approximate pH in the Series 2 and 3 tests, they reported that Pu concentrations ranged from about $10^{-7.4}$ $\underline{M}$, where amorphous $PuO_2 \cdot xH_2O$ was thought to control concentration, down to about 10^{-9} $\underline{M}$ where aging of the amorphous material produced a more (but incompletely) crystalline PuO_2 that was thought to control concentration. Concentrations of Pu in equilibrium with amorphous $Pu(OH)_4$, calculated in recognition of the fact that an amorphous or less crystalline phase is more likely to precipitate than crystalline PuO_2, are listed in Table 3. Measured Pu concentrations would be expected to fall between the equilibrium concentrations for PuO_2 and $Pu(OH)_4$, becoming closer to PuO_2 with aging. Equilibrium with amorphous $Pu(OH)_4$ and crystalline PuO_2 at O_2 fugacities of $10^{-0.7}$ and 10^{-12} bars yields predicted Pu concentrations that bracket measured results at both 25°C and 85°C.

3.2 Sources of Discrepancy Between Measured and Predicted Results

Discrepancies between measured and predicted concentrations are to be expected considering database limitations and uncertainty in the interpretation of measured apparent steady-state actinide concentrations. Care must be taken in interpreting the 90°C simulation results because insufficient data exist to accurately calculate the temperature-dependence of the thermodynamic properties of many radionuclide-bearing solids and solution species. The 3270 thermodynamic database is constantly updated through inclusion of new and revised thermodynamic data and the selection of a consistent set of aqueous complexes for each chemical element. Puigdomenech and Bruno[9] have constructed a thermodynamic database for U minerals and aqueous species that they showed to be in reasonable agreement with available experimental solubility data in systems in which U is complexed by OH^- and $CO_3^{=}$. The 3270 database contains many of the same aqueous species and minerals, but Puigdomenech and Bruno have included recent data for aqueous uranyl hydroxides from Lemire[10] which are not yet in the EQ3/6 database. Future plans include a critical evaluation of simulations of spent fuel dissolution made using the Puigdomenech and Bruno U database, and comparison with simulations made using the latest version of the EQ3/6 database. Inclusion of standard Nuclear Energy Agency (NEA) data for U minerals and species will also help to standardize future databases.

Until the U database is better established, calculated U concentrations must be recognized as preliminary and speculative. Simulation results can be used as a vehicle for identifying geochemical trends and studying the interactions between solid precipitation and elemental concentrations in solution. Seemingly small changes in the thermodynamic database can have potentially large impacts on predictions. For example, U concentrations calculated to be in equilibrium with schoepite using version 3270 of the EQ3/6 database are radically lower than those predicted in 1987[6] using an older database. The species $(UO_2)_3(OH)_7^-$ and $(UO_2)_2(OH)_3CO_3^-$ were omitted from version 3270 of the EQ3/6 database because their validity was questioned. $UO_2(CO_3)_2^{-2}$ and $UO_2(CO_3)_3^{-4}$ were left as the only dominant U species in solution throughout the EQ3/6 simulations. U concentrations accordingly remain lower during U mineral precipitation, especially schoepite, than in the previous 1987 calculations. Future work must address the sensitivity of the results to variations in thermodynamic data and the choice of a self-consistent set of aqueous species for elements of interest.

Comparisons between experimental results and predictions in Table 3 are predicated on the assumption that the listed solid phases precipitate from solution and control the solution composition. Except for some U-bearing minerals, no minerals containing radionuclides have been identified in the laboratory tests. Detection and characterization of actinide-bearing secondary phases may be difficult because of the extremely small masses of these actinides involved. Precipitates limiting actinide concentrations in the laboratory tests may also be amorphous, colloidal, or in some other less-than-perfect crystalline state. For instance, Rai and Ryan[8] observed that early Pu precipitates tend to be hydrated oxides which undergo aging to more crystalline solids. The concentrations of the affected actinides would, therefore, gradually decrease as aging progresses.

The chemistry of trivalent Am and Cm can be expected to be almost identical to that of the light lanthanide fission product elements which are present in much greater concentrations in spent fuel than are Am and Cm. Am and Cm may, therefore, be present in dilute solid solution with secondary phases formed by the lanthanides, which would result in lower measured solution concentrations than predicted for Am based on equilibration with $Am(OH)CO_3$ or $Am(OH)_3$. Pu and Np, and possibly Am and Cm, may also have been incorporated at low concentrations in solid solution with the U-bearing precipitates or other secondary phases. Efforts are planned to separate crystals of uranophane from test residues and to perform radiochemical analyses of these crystals to check for incorporation of other radionuclides. Sorption of actinides on colloids or other surfaces such as the fuel or test hardware may also control solution concentrations, but the impact of sorption was not considered in the simulations. Other factors, such as local variations in redox potential, may also contribute to differences between measured and predicted solubilities.

As it is not currently reasonable to expect a geochemical model to predict accurately the effects of all potential concentration-controlling processes over thousands of years, we hope to use modeling predictions to establish upper limits, or conservative estimates, of radionuclide concentrations over time. Lower limits to radionuclide concentrations imposed by solid precipitation are also of interest, however, as a baseline for further calculations, and because radionuclide concentrations may be expected to approach the lower limits over extended time periods. Accordingly, we assume in this paper that the actinide concentrations are controlled by the most stable and insoluble precipitates for which data are available. The consequences of precipitation of progressively less stable precipitates will be explored in future calculations, and upper limits of radionuclide concentrations controlled by solid precipitation will be estimated. In the case of Pu, for example, we have begun to explore the upper limits to Pu concentration as controlled by the precipitation of amorphous $Pu(OH)_4$. Comparison of modeling results with experimental results helps to identify phenomena which may revise our estimates of concentration limits. Processes such as sorption and aging of solids to forms of increasing crystallinity tend to lower element concentrations in solution, and increase the conservative nature of our estimates. However, consideration of colloid formation and colloid migration with the fluid phase may lead to an increase in our estimates of mobile concentrations over those made considering precipitation phenomena alone.

4.0 RADIONUCLIDE RELEASES

Annual actinide releases per failed waste package were calculated assuming that water flowing at a rate of 20 l/yr per waste package transports the actinides at the approximate concentrations measured at steady-state in Cycles 2 and 3 of the HBR-3-25 test. Each waste package was assumed to contain 3140 kg of fuel with an average burnup of approximately 33,000 MWd/MTM. The logarithms of the waste package 1000-year inventory fractions transported annually for each actinide under such conditions is given in Table 4. These releases are at least three orders of magnitude lower than the Nuclear Regulatory Commission (NRC) requirement in 10 CFR 60.113[11] that annual radionuclide releases during the post-containment period shall not exceed one part in 100,000 of the 1000-year inventories. The calculated annual release results would appear to be particularly encouraging for Pu and Am because isotopes of these two actinide elements account for about 98% of the total activity present in spent fuel at 1000 years. These values may be conservative in that they are based on the higher steady-state Pu and Am concentrations measured at 25°C and assume a conservative (high) estimate of the water flux through the repository. The calculated releases do, however, assume maintenance of steady values for actinide concentrations over time, whereas the geochemical simulations suggest that actinide concentrations, and U concentrations in particular, may vary with

Table 4. Annual Actinide Releases as a Fraction of the 1000-Year Inventories Based on HBR-3-25 Test Data

Actinide	Concentration Log(M)	Log(Release)*
U	-5.9	-8.6
Np	-8.9	-8.8
Pu	-8.4	-9.0
Am	-9.8	-9.1

* Assumes water flow rate of 20 l/yr per waste package transporting actinides at the indicated concentrations. Each waste package is assumed to contain 3140 kg of 33,000 MWd/MTM burnup PWR fuel.

time. Confidence in such release predictions will be greatly increased when the chemical mechanisms of solubility control are identified and successfully modeled.

Measured activities of the more soluble fission product radionuclides Cs-137, Sr-90, Tc-99 and I-129 continuously increase in solution at rates generally corresponding to annual release rates in the range of 10^{-4} to 10^{-3} of specimen inventory per year (Figure 4). These release rates imply a problem in meeting the NRC 10^{-5} annual fractional release limit for the more soluble radionuclides if the waste form alone is expected to carry the burden of compliance in the unanticipated case of large quantities of water contacting the waste. However, there are two factors that make these release rates uncertain. First, the degree to which these radionuclides are preferentially released from grain boundaries where they may be concentrated during irradiation has not yet been determined. Preferential release could be expected to provide a lesser contribution over time as exposed grain boundary inventories are depleted and release rates approach the congruent fuel matrix dissolution rate. A second factor is the extent to which the fuel may be degraded over time by exposure to the repository environment. Degradation of the fuel as a result of oxidation to higher oxygen stoichiometries such as U_3O_8, or as a result of preferential grain boundary dissolution, may cause increases in surface area and increased rates of nuclide dissolution from grain boundaries and from the fuel matrix over time.

Flow-through tests in which uranium minerals do not precipitate are being developed to measure the degree to which soluble nuclides are preferentially released during the initial phases of fuel dissolution. Dissolution tests using spent fuel specimens that have been degraded by slow, low-temperature oxidation are also planned. Results from these tests should provide a better understanding of potential long-term releases of the soluble and volatile radionuclides. Additional characterization of potential release of C-14 is important because it is soluble as bicarbonate and could also be released in the vapor phase as CO_2.

5.0 CONCLUSIONS

Laboratory testing and geochemical simulation of the dissolution of spent fuel under conditions selected for relevance to the proposed Yucca Mountain repository have resulted in the following conclusions.

1. Radionuclides of interest in spent fuel appear to fall into three categories of potential release mechanisms: 1) radionuclides whose release appears to be controlled by concentration-limiting mechanisms; 2) more highly soluble radionuclides; and 3) radionuclides that are released in the vapor phase (principally C-14).

2. The principal radionuclides whose releases appear to be controlled by concentration-limiting mechanisms are the actinides U, Np, Pu, Am and Cm. Steady-state concentrations measured for these actinide elements are at least three orders of magnitude lower than those required to meet NRC release limits based on conservative estimates of water fluxes through the repository. This result is of particular significance because isotopes of Pu and Am account for about 98% of the activity in spent fuel at 1000 years. However, results from geochemical modeling suggest that steady-state concentrations may vary significantly with time because of changes in solution composition and the identity of precipitating phases.

3. Good agreement between measured and predicted concentrations was obtained for Np based on equilibration with NpO_2 at 25°C when the oxygen fugacity in the simulation was set at 10^{-12} bars. A broad range of solubilities that bracketed the measured values were predicted for Pu depending upon the assumed oxygen fugacity and solubility-controlling phase. Measured Am concentrations were less than predicted based on data for equilibration with $Am(OH)CO_3$ and $Am(OH)_3$.

4. Dissolution rates for soluble radionuclides (Cs-137, Sr-90, Tc-99 and I-129) exceeding 10^{-5} of specimen inventory per year were measured during the laboratory tests. The implications of these data relative to long-term release of soluble radionuclides from a failed waste package are uncertain. The degree to which these radionuclides were preferentially released from grain boundaries where they may have concentrated during irradiation was not determined. Preferential release could be expected to provide a lesser contribution over time as exposed grain boundary inventories are depleted. However, physical degradation of the fuel over time from exposure to the oxidizing repository environment may result in accelerated release of soluble nuclides.

5. Additional work is required to identify solid phases that control actinide concentrations, and to acquire reliable thermodynamic data on these phases for use in geochemical modeling. In this regard, identification of any stable

suspended phases that can be transported by water movement is also important. In addition, we must better understand the potential release of soluble and volatile radionuclides, which may initially depend on preferential release from gap and grain boundary inventories, but may ultimately depend on the rate of fuel degradation by oxidation or other processes in the post-containment repository environment.

REFERENCES

1. R. A. Van Konynenburg, C. F. Smith, H. W. Culham and C. H. Otto Jr., "Behavior of Carbon-14 in Waste Packages for Spent Fuel in a Repository in Tuff," Scientific Basis for Nuclear Waste Management VIII, C. M. Jantzen, J. A. Stone and R. C. Ewing, Eds., Materials Research Soc., Pittsburgh, PA, Vol. 44, pp 405-412 (1985).

2. J. M. Delany, Reaction of Topopah Spring Tuff with J-13 Water: A Geochemical Modeling Approach Using The EQ3/6 Reaction Path Code, Lawrence Livermore National Laboratory, UCRL-53631 (1985).

3. C. N. Wilson, Results From Cycles 1 and 2 of NNWSI Series 2 Spent Fuel Dissolution Tests, Hanford Engineering Development Laboratory, HEDL-TME 85-22 (1987).

4. C. N. Wilson, "Summary Results From the Series 2 and Series 3 NNWSI Bare Fuel Dissolution Tests", Scientific Basis for Nuclear Waste Management XI, M. J. Apted and R. E. Westerman, Eds., Materials Research Soc., Pittsburgh, PA, Vol. 112, pp 473-483 (1987).

5. T. J. Wolery, Calculation of Chemical Equilibrium Between Aqueous Solution and Minerals: The EQ3/6 Software Package, Lawrence Livermore National Laboratory, UCRL-52658 (1979).

6. C. J. Bruton and H. F. Shaw, "Geochemical Simulation of Reaction Between Spent Fuel Waste Form and J-13 Water at 25°C and 90°C", Scientific Basis for Nuclear Waste Management XI, M. J. Apted and R. E. Westerman, Eds., Materials Research Soc., Pittsburgh, PA, Vol. 112, pp 485-494 (1987).

7. R. J. Lemire, An Assessment of the Thermodynamic Behavior of Neptunium in Water and Model Groundwater from 25°C to 150°C, Atomic Energy of Canada Limited, Whiteshell Nucl. Res. Estbl., AECL-7817 (1984).

8. D. Rai and J. L. Ryan, "Crystallinity and Solubility of Pu(IV) Oxide and Hydrous Oxide in Aged Aqueous Suspensions", Radiochemica Acta, vol 30, pp 213-216 (1982).

9. I. Puigdomenech and J. Bruno, Modelling Uranium Solubilities in Aqueous Solutions: Validation of Thermodynamic Data Base for the EQ3/6 Geochemical Codes, SKB technical report 88-21 (1988).

10. R. J. Lemire, Effects of High Ionic Strength Groundwaters on Calculated Equilibrium Concentrations in the Uranium-Water System, Atomic Energy of Canada Limited, Whiteshell Nucl. Res. Estbl., AECL-9549 (1988).

11. Code of Federal Regulations, "Disposal of High-Level Radioactive Wastes in Geological Repositories - Licensing Procedures," Title 10, Chapter 1, Part 60, section 60.113 (1983).

URANIUM DEPOSITS: ANALOGS FOR SPENT FUEL DISPOSAL

Douglas G. Brookins
University of New Mexico
Department of Geology
Albuquerque, NM 87131

ABSTRACT

Uranium deposits provide useful information on the components of spent fuel in mined geologic repositories. Elements that are in crystal chemical compatibility with U are commonly retained in host pitchblende or coffinite, and include the REE, Nb, Y, Ru, Tc, and the transuranics Np, Pu, Am. The evidence for this retention is supplied by the study of the Oklo natural reactor, Gabon, as well as from other uranium deposits. Study of minerals penecontemporaneous with uranium mineralization shows that clay minerals, and carbonate and sulfate minerals, commonly house Sr, Cs, Rb, Ba, Se, and others. Elements with an affinity for S are found in accessory sulfides and include Mo, Cd, Ag, Sb, and Te. Zr occurs in the pitchblende or in resistate accessories. Retention of Kr and Xe may be approximated by Rn in samples with concordant U-Pb dates. Data are sparse for I and Br. The natural occurrences of U deposits argues that in many cases the fission products and actinides of hypothetically breached spent fuel containers may be effectively isolated by the geologic media used for waste disposal.

INTRODUCTION

This paper is a partial review of sandstone uranium deposits and their possible role as analogs for spent fuel disposal in rocks. It is recognized that the information from these deposits is more applicable to rocks under chemically reducing conditions rather than rocks, such a the tuffs of Yucca Mountain, Nevada, under oxidizing conditions. Nevertheless, the information is of interest and importance to the generic criteria for radwaste immobilization in rocks.

GENERAL GEOLOGY OF THE OKLO DEPOSIT, GABON

Much of Gabon is underlain by Precambrian rocks. A series of ~2.5 Ga granitic massifs are found in the eastern two-thirds of the country; several remnant "windows" of Precambrian sedimentary rocks occur over the massifs. The largest of these is the Francevillian Basin of southeastern Gabon.

The uranium deposits at Oklo are located in the uppermost sandstone unit of the Francevillian sequence known as Couche 1 (C1). There are two principal types of deposits: those located in host sandstones as disseminated, blanket or pod-like deposits, and those filling fracture systems in the sandstones. The first variety has been compared to the Colorado Plateau type in the U.S. by Weber [(1)] and Brookins [(2)]. They are characterized by fairly low-grade uranium with 0.5 to 0.8% uranium oxide typical. The ore consists of pitchblende impregnating organic carbonaceous matter rich sandstone.

The second type of Oklo deposits are located in fracture systems that cut the low grade ore. They occur mainly as rather massive vein-like zones of pitchblende with a paucity of the other constituents of the low-grade ore. Presumably the Eh-pH conditions were sufficient to cause oxidation and dissolution of some of the Oklo low-grade ore, but these conditions were not sufficient to mobilize the V, Mo, Se, etc. [(2)]. This second type of uranium is commonly very rich in grade with several tens of percent uranium oxide. In several places, the reactor ore zones along these fractures had conditions favorable for nuclear fission reactions to take place.

Conditions for nuclear fission reactions at Oklo have been summarized by Naudet [(3)] and Brookins [(4)]. In brief, a combination of high uranium content, high ^{235}U percent, lack of neutron poisons, and a favorable amount of water (as moderator) present were necessary and sufficient criteria for the reactions to occur.

In order to attempt to apply the Oklo physical conditions to man-sited waste repositories, it is necessary to place limits on the parameters for temperature, pressure, depth of burial, porosity, and water availability. The determination of these factors is very complex. What is apparent is that the conditions for radionuclide transport involved temperatures on the order of 200-400°C and pressures in the very general range of 0.5 to 1.5 kilobars (i.e., corresponding to assumed depths of 1500-4500 m). The temperature range thus is consistent with those expected in buried man-made radwaste (250-300°C).

FISSION PRODUCTS, ACTINIDES, AND ACTINIDE DAUGHTERS AT OKLO

In the discussion of individual elements to follow, the following symbols will be used: M_n (naturally occurring element), M_f (fission product element), M_a (actinide element), M_{ad} (actinide daughter element), and M_t (total amounts such as $M_n \pm M_f$ or $\pm M_a$ or $\pm M_{ad}$).

Rubidium, Cesium, Strontium, Barium: From crystal chemical arguments, it is unlikely that Rb_f, Cs_f, Sr_f, and Ba_f should remain in the UO_2 structure. The ionic radii of Rb^+, Cs^+, Sr^{2+}, and Ba^{2+} are much larger than the ionic radius of U^{4+}. Brookins [5] found that Rb_f and Sr_f remained fixed in the local Oklo rocks and did not migrate extensively. Brookins [5] and Brookins et al. [6] note that there exist positive correlations of $^{137}Ba/^{135}Ba$ vs $^{85}Rb/^{87}Rb$ when a fission component is present, thus arguing for local fixation of Ba_f. This correlation implies that there was no Cs_f migration during the reactor's operative lifetime; otherwise, 137/135 ratio would reflect ^{135}Cs loss after all ^{137}Cs had decayed to ^{137}Ba.

Krypton and Xenon: The loss of Kr_f and Xe_f is accepted in the Oklo rocks [3,7]. Kr_f and Xe_f are unstable in the UO_2 structure and quickly diffuse to the surface of grains; from there they are lost to circulating hydrothermal solutions. It is assumed, but not necessarily proved, that they have been totally flushed from the mine site.

Yttrium, Zirconium and Niobium: The retention of Y_f at Oklo has been established [8]. Retention of Zr_f is predicted based on the crystal chemistry alone. In addition, Frejacques et al. [9] have shown that since $^{90}Zr_f$ is not depleted relative to other Zr_f isotopes, its immediate precursor, $^{90}Sr_f$, must have been compatible in the UO_2 structure for a few hundred years as well. Further, fissiogenic Nb_f has not migrated at the Oklo site. Early spark source mass spectrographic data [3,9] show no depletion of Nb_f on calculated fission yield curves.

Molybdenum: Significant amounts of fissiogenic Mo isotopes are produced during nuclear fission, thus making detection of the presence of Mo_f straightforward. I have suggested [2] that the preliminary depletion observed in Mo_f in Oklo ores by Naudet [3] and Frejacques et al. [9] could be explained by the use of Eh-pH diagrams. And, further, that Mo_f should be found with other chalcophile elements suspected of migrating. Curtis et al. [10] found that roughly 80% of Mo_f and 25% of Ru_f had migrated from the reactor zone, but that the majority of the Mo_f and Ru_f were fixed within 1.5 m of the reactor zones in gangue sandstone, primarily in secondary sulfide minerals.

Technetium and Ruthenium: Technetium is a major element of concern in man-made radwaste. The $^{99}Tc_f$ produced has a 2.13 x 10^5 y half-life and thus must be isolated for roughly one million years from the environment. Curtis et al. [11] report that some 730 kg of $^{99}Tc_f$ were produced at Oklo, and that 60 to 85% of this total was retained in the reactor zone pitchblende. Of that which migrated, a major part is found with Ru_f within 10 m or so of the reactor zones. Curtis et al. [11] estimate that 70 to 90% of Ru_f has been retained in the host pitchblende. They have shown, for example, that Ru_f is found with Mo_f and Tc_f (now present as Ru_f) in secondary sulfides near the reactor cores.

Palladium and Rhodium: The retention of Pd_f at Oklo is well documented [12,13]. The Eh-pH diagram shows that much of the stability field of water is covered by solids [14]. The Rh_f produced at Oklo is presumed to have remained within the host pitchblende where formed [2,9].

Silver, Cadmium, Indium, Tin, Antimony: The isotopic systematics of silver in the Oklo ores have been investigated by Delaeter et al. [12], and Loss et al. [13]. These authors argue for limited migration of Ag from the Oklo ores, and this is consistent with crystal chemical arguments [4]. It is noteworthy that the ionic radius of Ag^+ is very large at 1.23 nm; therefore, if any local oxidation occurred, Ag^+ would be extremely metastable in the UO_2 structure. The initial spark source mass spectrographic study of Oklo ores revealed apparent deficiencies of cadmium [8,9], a fact subsequently verified in detail by DeLaeter et al. [12] and Loss et al. [13]. The Eh-pH diagram for Cd species [14] shows that Cd_f would be expected to migrate. Only ^{115}In was produced by fission at Oklo, and then only in very small quantities [9]. From crystal chemical grounds, a strong case can be made for In_f retention at Oklo, and is also consistent with the Eh-pH prediction [14]. Migration of some tin at the Oklo natural reactor has been documented by DeLaeter et al. [12] who estimate losses of Sn_f isotopes in the amounts of 60 to 70% of many samples. The systematics of antimony at Oklo have not been studied. Preliminary spark source mass spectrometric data do not show major Sb_f deficiency, but the data are not high quality [8]. The Eh-pH diagram for Sb species [14] shows that most of the stability field of water is occupied by solids.

Tellurium: Data for Te at Oklo have been reported by DeLaeter et al. [12] and by Loss et al. [13]. Brookins [2,4] has also discussed Te at Oklo in some detail. All isotopes of Te except

^{126}Te fall on calculated fission yield curves. The ^{126}Te depletion is due to migration of ^{126}Sn, the immediate parent to ^{126}Te. From a crystal chemical viewpoint, it is surprising that Te^{4+} is retained in host pitchblende, but the Eh-pH diagram explains this [14]. A large field of native Te occupies a wide expanse of Eh-pH space, and includes the zone probable for pitchblende formation. Even under the slight oxidizing conditions accompanying high grade uranium ore formation, and lasting during the reactor's operative lifetime, native Te would be the likely form of Te_f.

Bromine and Iodine: Bromine and iodine were both produced at Oklo. The production of Br_f with lower fission yields is of less concern. No data have been obtained for Br_f. All data taken to date indicate loss of I_f from Oklo ores.

Rare Earth Elements: The rare earth elements (REE) are ideal for studying many facets of the Oklo natural reactor. Their fission yields are high, they contain strikingly different fissiogenic and natural isotopic compositions, and many REE isotopes have large enough capture cross sections so that their study yields useful nuclear parameters [4]. From a crystal chemical viewpoint, the REE are fairly compatible in the UO_2 structure. For most of the likely pH range at Oklo, the Eh-pH diagrams show that solids occupy most of the stability field of water. This fact, coupled with the favorable crystal chemical characteristics, argues for retention of the REE at Oklo (an argument supported by over ten years of gathering data).

Uranium: Uranium migration at Oklo has occurred, but to what extent is not yet clearly known. Chauvet [15] has shown that there is enrichment of uranium on the up-dip side of 900 Ma dolerite dikes which cut the southern end of the Oklo deposit and which indicate that major amounts of U have been mobilized post-900 Ma. In the reactor zones, Naudet [3] suggests that some limited migration of ^{235}U-depleted U has occurred. Basically, since the fission product-transuranic-uranium budget is fairly constant, and since the mobilities of all the different elements involved are quite different from each other, this budget argues for U retention in the reactor zones. The Eh-pH diagrams for U [14] indicate that for overall chemically reducing conditions, U as UO_2 should not be mobile. For more oxidizing conditions, transport of U (VI) as a carbonate complex is probable.

Neptunium: Neptunium was produced at Oklo in response to the high neutron flux in the highest grade uranium ores. The crystal chemical and Eh-pH diagram for Np_a argues for retention [14].

Plutonium: Plutonium at Oklo was produced by several reactions. The crystal chemistry of Pu is similar to that for U. Further, there is almost ideal solid solution between PuO_2 and UO_2. Consequently, if the UO_2 has remained largely unaffected by post-2 Ga events, the Pu (as PuO_2) in solid solution in the UO_2 should have been retained until destroyed by radioactive decay. The Eh-pH diagram for the Pu_a species agrees with Pu retention at Oklo [14]. The Oklo data show conclusively that Pu_a can be retained under natural conditions, even in rocks of fairly high porosity and in which hydrothermal solutions were circulating.

Americium: Small amounts of Am_a were formed at Oklo, but it is not possible to document any aspects of Am_a retention at Oklo as the yields were very small. From a crystal chemical and Eh-pH approach, Am may be as immobile as Pu, Np, and U at Oklo [4].

THORIUM AND BISMUTH BUDGET AT OKLO: COMPARISON WITH REACTOR FUEL

Thorium and bismuth are detected in the centers of reactor zones in amounts well above presumed background, and Frejacques et al. [9] have argued that these elements owe their presence to alpha decay from plutonium or uranium and neptunium respectively. Thorium is present as ^{232}Th and bismuth as ^{209}Bi. Under the Eh-pH conditions for uranium transport during Oklo reactor ore formation, neither Th nor Bi should be soluble. Thus background Th and Bi in the reactor zones should be very low. In the centers of the reactor zones the fluence was sufficient for formation of Np, Pu, and some Am. The Th and Bi noted in these zones is due to decay from ^{236}U and ^{237}Np respectively. Brookins [4] notes the agreement between calculated and measured ^{232}Th and ^{209}Bi data, and a proportionate amount of excess ^{235}U due to alpha decay from ^{239}Pu is also found in the same zones.

RADIONUCLIDE SORPTION AT OKLO

Curtis et al. [11] have discussed sorption in evaluating radionuclide mobility at Oklo. They have calculated the sorption ratio for Tc_f and Ru_f at 10^4 cm^3/g. However, even with large errors, the K_d values will be positive and greater than 10 in all instances. This calculation is interesting as it predicts Ru_f and Tc_f retention within a few tens of meters of source and is contrary to many laboratory experiments that include man-made waste. A key reason for high K_d values may be due to the extremely low fO_2. Laboratory experiments at such low fO_2 values have not been realized nor maintained [11]. These authors propose that due to temperature drop outside the reactor zone, the transported Ru_f and Tc_f became supersaturated and therefore precipitated.

For those elements retained in the UO_2 the Kd_f, values must be greater than those for Ru_f and Tc_f, and would include REE, Y, Nb, Zr, Pd, Rh, Te, and, perhaps, Sb. Elements with K_d values close to those for Ru_f and Tc_f include Mo_f, Ag_f, Sn_f, and Ba_f. From arguments presented earlier (see Cadmium), Cd_f may have a K_d close to that for Mo_f. The actinides, U, Th, Np, Pu, and probably Am, must all possess K_d values at least on the order of 10^4 cm^3/g.

SANDSTONE URANIUM DEPOSITS: ANALOGS FOR SPENT FUELS

The ore and other minerals of sandstone uranium deposits contain most of the elements found in spent unreprocessed fuel (SF). The actinides present include U and some Th, and even small amounts of Pu have been calculated to be present [2]. Further, a good data base exists for Pb behavior in these deposits. Analog elements for SF fission products include Rb, Sr, Zr, Mo, Nb, Cd, Ag, Y, Sb, Cs, Ba, and the REE (La-Lu). Some data also exist for Br and I. There are few data for the elements Rh, Pd, Ru, Sn, Te, but a combined Eh-pH diagram with sulfide phase study allows assessment of the mobility of these elements. Further, information concerning Tc can be obtained from use of Mo as analog as well as from theoretical considerations [16].

The fission product and actinide analog elements in sandstone uranium deposits are distributed among several different phases, whereas, by definition, these same elements are contained within the UO_2 of the SF. Thus the uranium deposit is like spent fuel where fission elements have migrated to the surface of UO_2 pellets. Further, the accessory and gangue uranium deposit minerals are often as susceptible to alteration and leaching as are the ore minerals, hence the uranium deposit assemblage is essentially a worst case scenario for the SF. If the analog elements of interest were retained in the uranium ore, then the same fission and actinide elements of SF would be more firmly fixed in a similar natural setting

FORMATION OF SANDSTONE URANIUM DEPOSITS

The processes by which sandstone uranium deposits form are fairly well known [1, 17,18]. Solutions containing dissolved oxyions of uranium along with oxyions of V, Se, Mo, and As penetrate permeable sandstone. When chemically reducing conditions are encountered, reduction of the high valence U(VI), Mo(VI), V(V), Se(VI), and As(V) in oxyions occurs with resultant precipitation of U(IV), Mo(IV), Se(O,,-II), V(III,IV), and As(III) solid species. The zone of reduction is marked in sandstones by transition from Fe(III) oxyhydroxide-bearing, organic carbon-free, sulfate-carbonate cemented sandstone of the oxidized zone to pyritiferous, organic carbon-rich, calcite cemented sandstone of the reduced zone.

Certain clay mineral assemblages are also commonly characteristic of the sandstone uranium deposits. In the oxidized rocks, kaolinite is commonly the dominant phase along with some detrital chlorite and re-worked smectites. In the reduced rocks, smectites and illites, authigenic chlorite, and mixed layer varieties are common. In the mineralized rocks, rosette chlorite with a high vanadium content is typical, although vanadiferous illite or smectite sometimes occur instead of chlorite [18].

United States Sandstone Uranium Deposits: In the United States most sandstone U deposits are found in the San Juan Basin of New Mexico-Colorado-Utah, the sedimentary basins of Wyoming, and in the basins of the Gulf Coastal area of Texas. In particular, we have studied the uranium deposits in the well known Grants Mineral Belt (GMB), San Juan Basin [17,19].

BEHAVIOR OF SPECIFIC ELEMENTS IN GRANTS MINERAL BELT ORES

Extensive data for Grants Mineral Belt sandstone U ores have been compiled by the author [17,19] and his students Riese [20], Lee [21], and Della Valle [22].

Rb, Sr, Ba: Most Rb is contained in clay minerals. Work by Lee [21] and Lee and Brookins [23] shows that many clay minerals formed penecontemporaneously with the ore, ~140 Ma, have retained their isotopic and chemical integrity. Primary ores in the Ambrosia Lake subdistrict have measured Rb-Sr ages of 135-145 Ma. In the Laguna subdistrict, early (~140 Ma) ore was mobilized due to extreme Mid-Cretaceous tectonism, and redeposited at about 110-120 Ma. Mobilization was thought to be no more than some tens of meters. The Rb-Sr studies also document the retention of Sr in these same clay minerals. Barium behavior is moderately well known. Ba is contained in accessory barite and, locally, other sulfates in oxidized rocks, in calcite, and partly in clay minerals in the ore zones. Since Ba is less mobile than Sr under most crustal conditions, then the fact that the Rb-Sr studies argue for Sr retention can be used indirectly to argue for Ba retention as well.

Y, Zr, Nb, Mo, Ru, Tc, Pd, Re, Pd, Rh, Cd, Ag, Sb, Sn, Te, I, Cl, Se: The Y is largely contained in detrital phases, although some local Y accumulations occur when phases like magnetite are destroyed by alteration. In which case, secondary oxide coatings contain the Y. No evidence for Y migration is noted and is consistent with findings elsewhere [2]. Zr and Nb are largely contained in detrital phases, and no abnormal abundances of these elements are noted. Data for Mo are abundant [17]. In ore zones, Mo is fixed in gangue minerals such as jordisite or pyrite, or, when these minerals are subjected to mildly oxidizing waters, as ilsemannite. The behavior of Mo is readily explainable in terms of Eh-pH diagrams [14], and it can be readily mobile under strongly oxidizing conditions. Mo is not readily mobile under reducing to mildly oxidizing conditions or in the pH range from about 6-9. Ru is present in such low crustal amounts as to not be readily determinable by INAA and Tc consists only of radioactive isotopes. Theoretical considerations argue that both Ru and Tc should be less mobile than Mo and U [2]. Consequently, since the migration of Mo and U is limited, Ru and Tc should be even more highly retained in the sandstone ores. Analyses of the rocks for Re, as analog for Tc, would be of great value here. The Eh-pH diagrams for Pd and Rh show that the greater part of the stability field of water is covered by solids, hence no migration is likely for these elements. There are limited data for Cd and Ag for sandstone U deposits, but no data for Pd. Theoretical considerations, Eh-pH diagrams mainly, argue for either complete retention of these elements or, at worst, very limited migration [2]. The Cd and Ag are contained in sulfide phases. When the sulfides are altered, Ag will remain immobile while Cd may either be incorporated into newly formed calcite $(Ca,Cd)CO_3$ or an oxyhydroxide phase in most of the pH range 6-9; whereas, at lower pH it will be quite mobile. However at higher temperatures Cd mobility will increase [14]. Data for Sb show this element to be fixed in sulfides in reduced rocks and in oxyhydroxide phases in near-ore oxidized rocks. The Eh-pH diagram shows a field of $Sb(OH)_3$ occupying that part of the central pH range over the sulfide field [14]. Occasionally, Sb haloes in oxidized rocks can also be used for localizing remobilized U [17]. Sb does not appear to be mobile in the environment of the sandstone U deposits. There are few data for Sn and Te. Theoretical considerations again argue for retention of both elements in the environment of the sandstone U deposits [2]. Data are virtually non-existent for I and other halides, although there are some Cl data from fluid inclusion work [17]. It is thus not possible to comment on I mobility or retention, although elsewhere [2] the author has commented on the possibility of I fixation on sulfide phases. Selenium behavior in sandstone uranium deposits is well known. The Eh-pH stability field for Se species includes a large field for native Se. Hence, when rocks mineralized with U, Mo, and Se are oxidized, much of the Se remains fixed in the rocks even though the U and Mo may be mobilized.

The Rare Earth Elements (REE): The data base for the REE is extensive [17,22]. There are some data [24] which suggest mobilization of the REE with U during formation of the sandstone U deposits, but no data suggests mobilization after the ore formed.

The Actinides: Data are available only for U and Th for the Grants samples. For the most part, Th is fixed in detrital phases. There are abundant U data, most of which suggest little if any mobility of U under reducing conditions. In the pH range of 6-9, Brookins [14] suggests that U can only be successfully remobilized under dynamic high oxidizing conditions.

CONCLUSIONS: APPLICATION OF OKLO AND SANDSTONE URANIUM DEPOSIT STUDIES TO USA RADWASTE MANAGEMENT PROGRAM

The Oklo natural reactor is an excellent analog for discussion of behavior of radionuclides in rocks. The UO_2 of the reactor zone is the analog for spent fuel. It is surrounded by ferromagnesian clay minerals (chlorite and illite) that may be taken as analogs for engineered barrier clay material (i.e., should this be a consideration). The surrounding rocks at Oklo include argillaceous sandstone, shale rich in organic carbonaceous matter, and some carbonate-rich zones as well.

The Oklo reactors were operative for roughly 0.1 to 0.8 Ma, where reactor core temperatures were on the order of several hundred degrees (simulating buried man-made radwaste). The reactors allow some constraints to be placed on the near field behavior of radionuclides under conditions of prolonged temperature and time. The data from Oklo support the fact that very little of the radionuclide inventory migrated during the 700,000 or so years of reactor operation. Further, nearly two billion years later, most of the radionuclides and/or their decay products at Oklo have not migrated or have migrated from a few cm to a maximum of a few meters. This fact is a very strong endorsement of the ability of rocks to retain and retard radionuclides under natural settings.

Of critical importance, however, is that even for the measurable elements that have migrated, calculated sorption ratios are on the order of 10^3-10^4 ml/g for species such as Tc and Ru.

Studies of sandstone uranium deposits in the Grants Mineral Belt have proven successful in addressing the behavior of numerous analog elements for fission products and actinides common to spent fuels. In particular, argillaceous rocks, including argillaceous sandstones, retain alkali, alkaline earth, REE, actinide and other elements. Use of such material for engineered backfill, or even repository medium, can be supported in the multi-barrier system. In the Grants Mineral Belt, strong retention properties for Mo, Tc, Sb, REE, U, Cs, Sr, and others are indicated. These studies supplement and compliment the Oklo studies and may be applicable to man-made radioactive waste.

In oxidized rocks such as the tuffaceous rocks at Yucca Mountain, Nevada, there is concern over the mobility of the isotopes ^{99}Tc, ^{79}Se, ^{129}I, and ^{14}C. The studies of uranium deposits formed under chemically reducing conditions do not directly address the behavior of these isotopes, but studies of oxidized sandstone uranium deposits may be useful. Oxidized rocks in the Grants Mineral Belt area typically contain secondary U(VI) minerals, and, often, these same rocks are rich in Se and carbonate minerals. Isotopic studies show that many calcites contain carbon derived from original organic sources and especially in oxidized rocks. This suggests that any ^{14}C released may be fixed in interstitial carbonates. For ^{129}I and ^{99}Tc, the data from oxidized uranium deposits is even less certain, but it would be useful to study the abundance and distribution of Re as analog to Tc and for survey data on Br and I.

References

1. F. Weber, "Une Serie Precambrienne du Gabon: le Francevillien, Sedimentologie, Geochimie, Relations avec les Gites Mineraux," PhD Thesis, University Strasbourg, 328 p, 1968.
2. Brookins, D. G., Geochemical Aspects of Radioactive Waste Disposal, Springer-Verlag, New York, Berlin, Heidelberg, 1984, 347 p.
3. Naudet, R., "Synthese des Donnees Concernant la Stabilite et les Remobilisations de l'Uranium et des Terres Rares," in I A E A, 1978, pp. 643-678.
4. Brookins, D. G., "Radionuclide Behavior at the Oklo Natural Reactor," TM 10501, 1987, Oak Ridge National Laboratory, 183 p.
5. Brookins, D.G., "Alkali and Alkaline Earth Element Studies at Oklo, "Scientific Basis for Nuclear Waste Management, III (J. G. Moore, ed.), Plenum Press, New York, 1981, pp. 275-282.
6. Brookins, D. G. et al., "Search for Fission-produced Rb, Sr, Cs, and Ba at Oklo," in I A E A , 1976, pp. 401-414.

7. Drozd, R. J., Hohenberg, C. M., and Morgan, C. J., "Heavy Rare Gases from Rabbit Lake (Canada) and the Oklo Mine (Gabon): Natural Spontaneous Chain Reactions in Old Uranium Deposits," Earth Planetary Science Letters, Vol. 23, pp. 28-37.
8. Naudet, R., "Summary Report on the Oklo Phenonemon,," Frence Centre Energie Atomique, Report Bulletin Inf. Science Technology., Vol. 193, 1974, pp. 7-35.
9. Frejacques, C. et al., "Conclusios Tirees de l'Etude de la Migration de Produits de Fission," in I A E A, 1976, pp. 509-526.
10. Curtis, D. B. et al., "Transport of Fission Products at the Oklo Natural Reactor," LA-UR-82-1276, 1982, pp.
11. Curtis, D. B., Benjamin, T. M., and Gancarz, A. J., "The Oklo Reactors: Natural Analogs to Nuclear Waste Repositories," LA-UR-81-3/83, 1983, 32 p.
12. DeLaeter, J. R., and Rosman, J. K. R., and Loss, R. D., "Fission Product Studies at WAIT," Australian Physicist, Vol. 22, 1985, 5 p.
13. Loss, R. D., Rosman, J. K. R., and DeLaeter, J. R., "Transport of Symmetric Mass Region Fission Productgs at the Oklo Natural Reactors," Earth Planetary Science Letters, Vol. 68, 1984, pp. 240-248.
14. Brookins, D. G., Eh-pH Diagrams for Geochemistry, Springer-Verlag, New York, Berlin, Heidelberg, 1988, 192 p.
15. Chauvet, R. J., "Description du Gisement d'Uranium d'Oklo," in I A E A, 1976, pp. 53-66.
16. Brookins, D.G., 1983, Migration and retention of elements at the Oklo natural reactor, Env. Geol. 4, 201.
17. Brookins, D.G., 1979, Uranium deposits of the Grants, New Mexico Mineral Belt (II), DOE Report No. BFEC 76-029E.
18. Brookins, D.G., 1982, Geochemistry of clay minerals for uranium exploration in the Grants Mineral Belt, New Mexico, Min. Deposita 17, 37.
19. Brookins, D.G., 1976, Uranium deposits of the Grants, New Mexico Mineral Belt, ERDA Report No. GJBX 76-16.
20. Reise, W.C., 1980, PhD thesis, Univ. New Mexico.
21. Lee, M.J., 1975, PhD thesis, Univ. New Mexico.
22. Della Valle, R.S., 1981, PhD thesis, Univ. New Mexico.
23. Lee, M.J., and Brookins, D.G., 1978, Rb-Sr investigations of the Morrison Formation (late Jurassic), Grants Mineral Belt, New Mexico: minimum ages of sedimentation, uranium mineralization, and provenance, Amer. Assoc. Petr. Geol. Bull. 62, 1673.
24. Della Valle, R.S. and Brookins, D.G., 1983, in The Significance of Trace Elements in Solving Petrogenetic Problems and Controversies, S.S. Augustithius, ed., Theophrastus Publications, Athens, Greece, pp. 793-818.

Section VI. Waste Form Development, Characterization, and Qualification

A MODEL FOR THE VISCOSITY OF MULTICOMPONENT GLASS MELTS

X. FENG, E. SAAD*, AND I. L. PEGG

THE CATHOLIC UNIVERSITY OF AMERICA, VITREOUS STATE LABORATORY, WASHINGTON, D.C. 20064
*Present Address: MOBAY CO., BALTIMORE, MD

ABSTRACT

A simple model for the composition and temperature dependence of the viscosity of multicomponent silicate melts is described. The component-specific parameters are obtained from the heats of formation of the constituent oxides which are then modified by four rules that summarize the different structural roles of particular groups of oxides. The remaining three to six adjustable parameters are obtained by least-squares fitting. The model is tested on nine sets of data from the literature totaling 1674 data points. The standard deviation of log (viscosity/poise) is typically around 0.25 and is considerably lower for the individual data sets.

INTRODUCTION

A number of models have been described that correlate the viscosity of glass melts with their composition [1-5], some of which were motivated by the particular importance of melt viscosity in determining processability in nuclear waste vitrification schemes. The model of Bottinga and Weill [1] covers perhaps the broadest range of compositions. In this approach the logarithm of the viscosity is represented by a linear function of the mole fractions x_i

$$\log n = \Sigma_i \bullet x_i \bullet D_i$$

and the coefficients D_i are determined by least-squares fitting. The components i are a combination of oxides and aluminates. The D_i are, of course, temperature dependent but are also found to depend on the SiO_2 concentration and, in fact, they are reasonably constant only over about a 10% range. As a result, extensive tables of coefficients are required. Application of this model to nuclear waste glasses would require new data fitting to determine D_i tables for such components as U, Th, and Fe(III).

In this paper we describe the performance of a correlation of melt viscosity with composition and temperature that we have developed. The coefficients in this model are simply the literature values of the standard formation energies of the constituent oxides modified by a few simple rules that reflect the effect of specific groups of oxides on the glass network. The remaining parameters, which presently may be as few as three, are determined by least squares fitting. A scheme such as this has the advantage that the coefficients for all oxides are automatically determined, regardless of whether or not they appear in the fitting database. A disadvantage is that one expects the data to be less well represented than by an approach that fits all of the component-specific coefficients. We have found, however, that the model compares surprisingly well with the Bottinga and Weill [1] approach over wide composition ranges.

FORMATION ENERGY MODEL

An early form of this model was presented previously [6] and a full description will be given in a latter publication; only a brief description can be given here. It is well known that, broadly speaking, good glass formers tend to increase melt viscosity while network breakers tend to decrease viscosity. An effective correlation scheme must make this observation quantitative. It is therefore useful to consider other bases for distinguishing "good glass formers" (e.g. SiO_2, Al_2O_3, etc.) from "network breakers" or "modifiers" (e.g. Na_2O, CaO etc).

Sun [7] has presented an interesting ranking of oxides based on estimated X-O bond strengths which gives quite an effective separation that is in good agreement with the observed roles of the oxides. A conspicuous exception is B_2O_3 which has a large B-O bond strength but is generally thought of as an intermediate or modifier.

Similarly the good glass former Al_2O_3, with a large Al-O bond strength, does not always increase the melt viscosity: if there is insufficient alkali for Al to maintain a tetrahedral coordination addition of Al_2O_3 may, in fact, decrease the melt viscosity [8-10]. These examples illustrate that a successful correlation must incorporate some knowledge of the structural roles of the constituent oxides in addition to simple bond strength arguments.

We have used these ideas in constructing our initial model in the following way. We have made use of the simple empirical observation that the bond strengths D(X-O) calculated by Sun are very well represented (to within 15%) by the equation

$$D(X\text{-}O) = \frac{2 \bullet \Delta H_f}{n \cdot Z}$$

where ΔH_f is the standard heat of formation of the oxide X_nO_m and Z is the coordination number of X. Thus, we choose to correlate viscosity data with ΔH_f directly since these data are more extensive, more readily available, and somewhat less ambiguous. We suppose $\Delta H_{f,i}$ to be one contribution to the V_i which are the parameters reflecting the influence of oxide i on viscosity. The V_i are then calculated as follows:

A) Intermediates + all Al_2O_3, and then ZrO_2, in excess (by moles) of total alkalies

$$V_i = \Delta H_{fi},$$

where we have chosen the simplest possible relationship. Note that quantities of alumina and zirconia that cannot be combined with alkali are treated as intermediates.

B) Major Network formers: SiO_2, Al_2O_3, and ZrO2 for which there is sufficient alkali for these to be incorporated into the network

$$V_i = K \bullet \Delta H_{fi}.$$

We would expect K to be greater than one since these components have a greater effect on viscosity than those in (A). In fact, K=2 gives substantial improvement over K=1 and larger values give little further change. Thus, at this point we chose to fix K at the value two; optimization of K will be discussed further in a later publication.

C) Network breakers (alkalies)

$$V_i = \Delta H_{fi} - V_{net} \qquad (1)$$

where V_{net} is the average contribution of a network-former (as in (A)) bond

$$V_{net} = \Sigma x_j \bullet \Delta H_{fj} / \Sigma x_j \bullet n_j \bullet z_j,$$

j = components in (A), and x_j is the mole fraction of oxide j. Equation 1 may then be thought of as the net contribution on replacing a network bond by a "broken" bond.

D) Boron oxide

$$V_{B2O3} = \Delta H_{f,B2O3} - 2 \bullet V_{net}$$

where now each boron atom is supposed to contribute a "broken" bond. Despite the high bond strength of the B-O bond [7] this is in keeping with observations that addition of B_2O_3 reduces the viscosity of silicate melts and also the findings of Krogh-Moe [11]. He argues that the labile trigonal-tetrahedral coordination change of boron facilitates viscous flow and does not require the breaking and reforming of bonds.

With these assignments of the V_i we suppose that the effect of composition on viscosity is through a single parameter V which is simply the mole fraction weighted sum of the individual V_i,

$$V = \Sigma x_i \bullet V_i.$$

The temperature dependence of melt viscosity is usually well represented by the Vogel-Fulcher equation [9, 10]

$$\log n = A + B/(T-T_0) \qquad (2)$$

and frequently the Arrhenius form, $T_0 = 0$, is adequate unless the data extend close to the glass transition temperature.

We will use equation (2) and assume each of the three parameters to be composition dependent and express that relationship as a simple linear dependence on V. Thus,

$$A = a_0 + a_1 \cdot V$$
$$B = b_0 + b_1 \cdot V$$
$$T_0 = t_0 + t_1 \cdot V,$$

where the six parameters are to be determined by a least-squares fit to data. Note that these are the only six adjustable parameters in the model. As we will show below, for many data sets the Arrhenius form gives a satisfactory representation and we find that allowing t_o and t_1 to be nonzero gives a negligible improvement in fit. Thus, in many cases four, and frequently only three, adjustable parameters suffice.

In the following sections we will demonstrate the performance of this model on a number of data sets covering a wide range of compositions.

DATABASE

We have used nine groups of data in our analysis which we will refer to by the acronyms below where we also include a brief description of the composition and temperature ranges covered.

VSL

498 data points measured at Catholic University on 72 glasses around the expected West Valley nuclear waste vitrification process region [12]. Composition: SiO_2 (36-47%), B_2O_3 (7-15%), Na_2O (8-14%), Fe_2O_3 (9-17%), P_2O_5 (0-3.5%), ThO_2 (0-5.5%), UO_2 (0-1.2%), ZrO_2 (0-5.5%), TiO_2 (0-5%), Al_2O_3 (3-14%). Temperature: 850-1150°C.

NBS

47 data points on the three NBS Standard reference materials SRM 710, SRM 711, and SRM 717. Composition: SiO_2 (46-71%), B_2O_3 (0-17%), Na_2O (0-8.7%), CaO (0-12%), PbO (0-45%), Al_2O_3 (0-3%), Li_2O (0-1%). Temperature: 850-1600°C.

WV

80 data points on 10 glasses produced at West Valley

Nuclear Services as part of their cold-operations test runs [13]. Composition: around that of WV205 but with Al_2O_3 content ranging from 2% to 9%. Temperature: 1000-1150°c.

PNL1

88 data points on 29 nuclear waste glass compositions measured at Pacific Northwest Laboratory in the development of the WV205 composition [14]. Composition: SiO_2 (43-54%), Na_2O (7-20%), CaO (0-13%), B_2O_3 (1-16%), Fe_2O_3 (10-17%), Al_2O_3 (1-6%). Temperature: 950-1200°C.

PNL2

172 data points on 91 glasses examined at PNL in the development of the WV205 composition [15]. Composition: SiO_2 (35-60%), B_2O_3 (0-20%), Na_2O (7-30%), CaO (0-15%), Al_2O_3 (0-12%) and Fe_2O_3 (0-24%). Temperature: 750-1800°c.

RED

14 data points on a set of 14 redox glass compositions in the system Na_2O - FeO - Fe_2O_3 - SiO_2 with a Fe(III)/Fe ratio of 0.18 to 1.0 [16]. Temperature: 1200° and 1430°C.

NSNA

41 data points on 6 glasses in the system $Na_2Si_2O_5$ - $Na_4Al_2O_5$ [17]. Composition: from $(Na_2Si_2O_5)_{100}$ - $(Na_4Al_2O_5)_0$ to $(Na_2Si_2O_5)_{60}$ - $(Na_4Al_2O_5)_{40}$. Temperature: 980-1350°C.

B&W

420 data points on 117 glasses collected by Bottinga and Weill [1] on a wide range of silicate melt compositions ranging from simple binary and ternary systems to multicomponent systems of natural molten rocks and minerals. Composition: SiO_2 (28-90%), Al_2O_3 (0-20%), FeO (0-65%), TiO_2 (0-3%), Li_2O (0-33%), Na_2O (0-36%), K_2O (0-35%), MgO (0-42%), CaO (0-45%), SrO (0-64%), and BaO (0-72%). Temperature: 1100-1800°C.

URB

315 data points on 30 glasses in simple binary and ternary silicate systems reported by Urbain et al. [18]. Composition: SiO_2 (17-70%), Al_2O_3 (0-56%), FeO (0-83%), Na_2O (0-12%), K_2O (0-7%), MgO (0-44%), CaO (0-58%), SrO (0-63%), BaO (0-72%) and MnO (0-54%).
Temperature: 1170-2200°C.

ALL

1664 data points on 372 glasses resulting from all of the above data sets. The composition range covered by these data is given in Table I; the table is not exhaustive.

Table I. Composition range, in wt%, covered by "ALL" data set.

SiO_2	B_2O_3	Al_2O_3	FeO	Fe_2O_3
17-90	0-20	0-56	0-83	0-24
ZrO_2	ThO_2	UO_2	Li_2O	Na_2O
0-5.5	0-5.5	0-1	0-33	0-36
MgO	CaO	SrO	BaO	PbO
0-44	0-58	0-64	0-72	0-45
MnO	TiO_2	K_2O	P_2O_5	
0-54	0-5	0-35	0-3.5	

RESULTS

Each data set was first analyzed separately by fitting the model described above using (i) the four parameters a_0, a_1, b_0, b_1 (ii) five parameters; adding t_0, and (iii) six parameters; adding t_1. In all cases the five-parameter fits gave insignificant improvements over the four-parameter fits and will therefore not be considered further here. The results of the four- and six-parameter fits to each data set are listed in columns three and four of Table II where the standard deviation (of the log viscosity) is given for each fit. The only substantial improvement afforded by the six-parameter fit is for the URB data set and a somewhat smaller improvement for the ALL data set. This suggests that the improvement in the fit to the ALL data on addition of t_0 and t_1 is mainly a result of the inclusion of the URB data. This is substantiated by the results of fitting the ALL data set with the URB data removed which gave four- and six-parameter standard deviations of 0.269 and 0.237, respectively, indicating an insignificant improvement. A comparison of measured and calculated values from the six-parameter fit to all 1664 data points is shown in Figure 1.

Table II. Fitting results. Entries are standard deviation of log n for the given number of parameters.

DATA SET	Fits to each data set individually			Parameters fixed at ALL values	
	3	4	6	3	6
VSL	0.161	0.161	0.159	0.204	0.205
NBS	0.133	0.125	0.081	0.527	0.424
WV	0.095	0.096	0.095	0.318	0.304
PNL1	0.161	0.161	0.137	0.193	0.196
PNL2	0.155	0.155	0.153	0.212	0.209
RED	0.137	0.060	0.076	0.128	0.190
NSNA	0.072	0.072	0.144	0.556	0.567
B&W	0.218	0.216	0.198	0.296	0.228
URB	0.418	0.413	0.166	0.566	0.209
ALL	0.339	0.339	0.235	0.339	0.235
ALL (without URB)	0.269	0.269	0.237	0.269	0.237

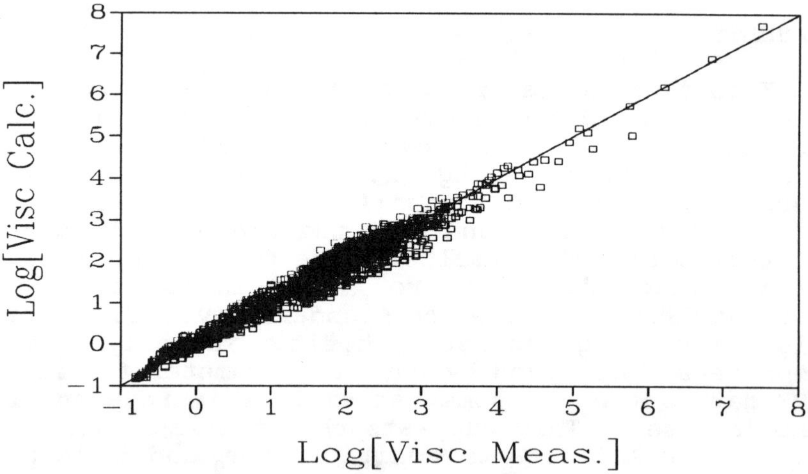

Figure 1. Comparison of measured and calculated viscosities (in poise) from the six-parameter fit to the "ALL" data set. 1664 points.

A review of the results of the four-parameter fits revealed that b_0 was consistently very much smaller than its uncertainty, suggestive of a redundant parameter. Accordingly, column one in Table II shows the standard deviations obtained with 3 parameter fits using only a_0, a_1, and b_1. The fits are generally comparable to or only slightly worse than those obtained with four parameters. This result implies that the apparent activation energy is directly proportional to V (with a proportionality constant of Rb_1, where R is the gas constant). The three parameters obtained by fitting the ALL data set with the URB data removed are

$$a_0 = -4.341\ (0.09)$$
$$a_1 = -0.004740\ (0.0003)$$
$$b_1 = -23.451\ (0.3)$$

where the standard deviations are given in parentheses. The units are those appropriate to viscosity in Poise, temperature in Kelvin and formation energy in kilocalories per mole. The V values obtained from the data fall predominantly in the approximate range -140 to -420 kcal/mole which yields activation energies in the range 14-45 kcal/mole; these values are typical of many glass melts.

Figures 2, 3, and 4 show graphically the fits obtained to three of the data sets, VSL, NBS, and NSNA, respectively. Four-parameter fits were used to generate these figures though, as Table II shows, the three parameter fits gave comparable results. The NSNA data are particularly interesting since Dingwell [17] observed a maximum and minimum in viscosity at 1300°C as a function of mole % $Na_4Al_2O_5$ and examined the ability of the Bottinga and Weill model to reproduce those features. Figure 4 shows these results together with the corresponding results obtained from the model described in this paper. The quality of the fit is quite good and it is also encouraging that the present model does exhibit a maximum and minimum over this composition range as observed.

Finally, if we now consider the parameters to be <u>fixed</u> at their values determined by the fits to the full data sets, we can examine the standard deviation of each

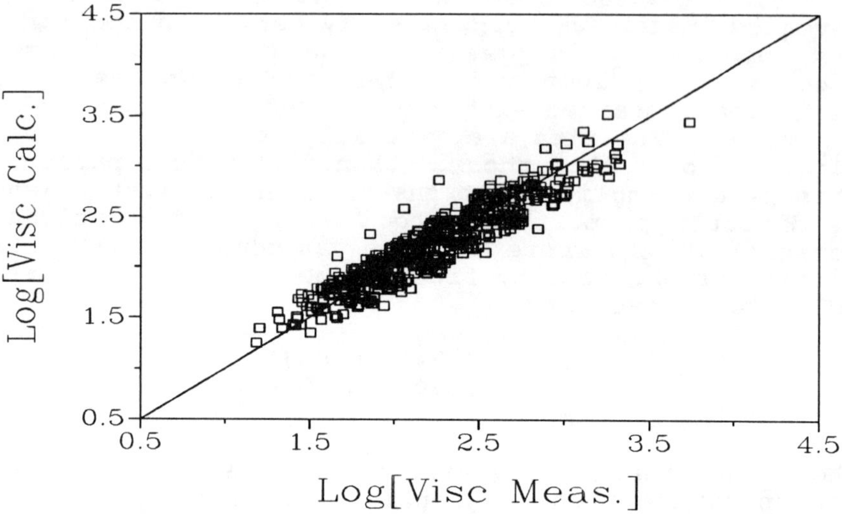

Figure 2 As Fig.1 but four-parameter fit to VSL data. 498 points.

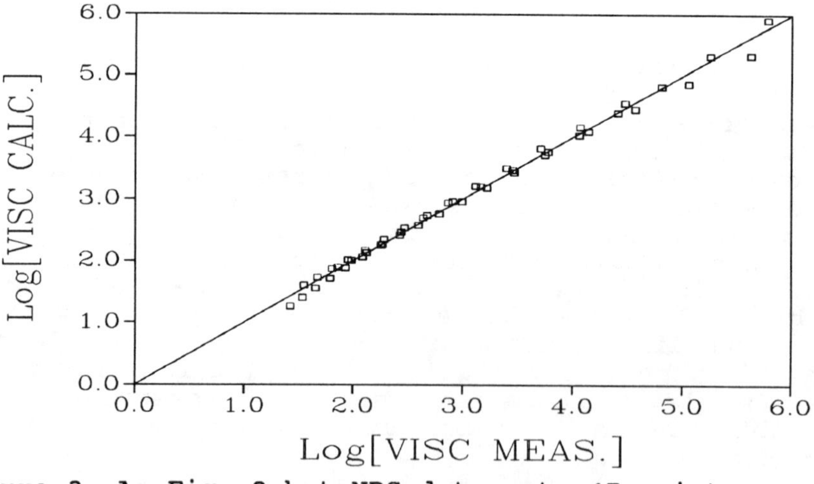

Figure 3. As Fig. 2 but NBS data set. 47 points.

individual data set from the values calculated from the model. The results are shown in columns five and six in Table II where we have used both the three-parameter fit, with the parameters listed above, as well as the six-parameter fit to the ALL data set. The standard deviations of each individual data set are now, of course, larger than those obtained on fitting each data set alone since representing this wider range of compositions results in poorer fits to some of the details in the individual data sets. However, the results are encouraging and together suggest that the model gives reasonable representations of data covering both limited and extended composition ranges.

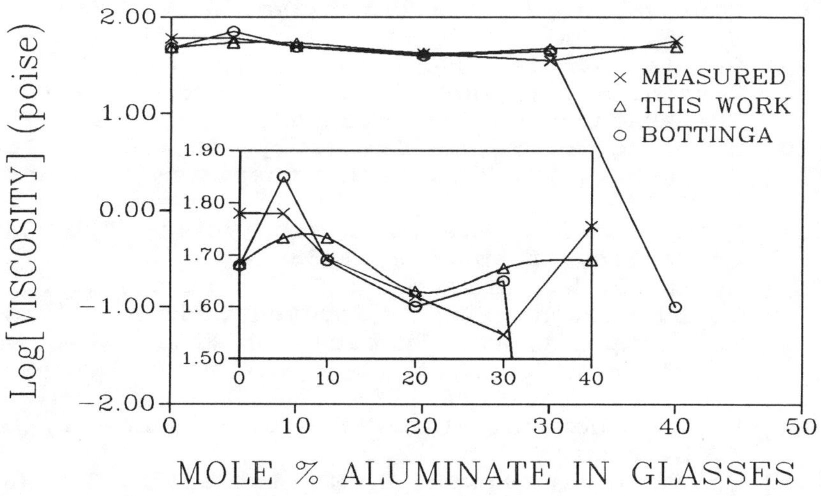

Figure 4. Data from Dingwell [17] at 1300°C compared with Bottinga and Weill model [1] and this work.

ACKNOWLEDGMENTS

We are grateful to L. R. Bunnell of PNL for reference [14] and to the Computer Center at CUA for computer time. This work is supported in part by West Valley Nuclear Services Inc. under contract number DE-AC07-81-NE 44139.

REFERENCES

[1] Bottinga, Y., Weill, D. F., Am. J. Sci., 272, 438 (1972)
[2] Jantzen, C. M., and Plodinec, M. J., report no. DPST-862.-372, Savannah River Lab, 1986
[3] Urbain, G., Cambier, F., Deletter, M., and Anseau, M. R., Trans. J. Br. Ceram. Soc. 80, 139 (1981)
[4] Turkdogan, E. T. Bills, P. M., Am. Ceram. Soc. Bull. 39, 682 (1960)
[5] Ilhan, A., and Hrma, P., Presented at 89th. annual meeting of the Amer. Ceram. Soc., Abstract # 153-G-87, Pittsburg Pa, April 26-30, 1987
[6] Xiangdong Feng, Ph. D. Dissertation, The Catholic University of America, 1988
[7] Sun, K. H., J. Am. Ceram. Soc, 277 (1947)
[8] Bobkova, N. M., Gorodetstkaya, O. G., Yankovskaya, S. A., and Tizhovka, Th. S., Fizikai Khimiya Stekla, 9, 414 (1983)
[9] Rossin, R., Bersan, J., and Urbain, G., Rev. Hautes Temperatures et Refractaires, 1, 159 (1964)
[10 Riebling, E. F., J. Chemistry, 42, 2811 (1964)
[11] Krogh-Moe, J., Physics and Chemistry of Glasses, 1, 26 (1960)
[12] Saad, E. E., Ph. D. Dissertation, The Catholic University of America, 1988
[13] Bowan, B. W., Armstrong, S. C., Bauer, B. P., Dean, E. J., Eaton, W. C., Goodyear, M. A., "Viscosity and Compositional Summary for FACTS RUNS SF-1 to SF-9", HK: 87:0093, West Valley Nuclear Services Co. Inc., TA#75, 5/19/87
[14] Private communication from L. R. Bunnell, PNL, to I. L. Pegg
[15] Chick, L. A., Bowen, W. M., Lokken, R. O., Wald, J. W., Bunnell, L. R., Strachan, D. M., Report No. PNL-4992, 1984
[16] Dingwell, D. B., and Virgo, D., Geochim. Cosmochim. Acta, 51, 195 (1987)
[17] Dingwell, D. B., Geochim. Cosmochim. Acta, 50, 1261 (1986)
[18] Urbain, G., Bottinga, Y. and Ritchet, P., Geochim, Cosmochim. Acta, 46, 1061 (1982)

CHARACTERIZATION OF HIGH LEVEL WASTE GLASS

K. KAWAMURA, T. TAKAHASHI, M. HORIE, N. TSUNODA
Power Reactor and Nuclear Fuel Development Corporation, Tokai Works, Muramatsu, Tokai-mura, Ibaraki-ken, Japan

ABSTRACT

Sensitivity studies were carried out using simulated vitrified waste to assess the effects of variations in waste composition on major glass properties : density, viscosity, electrical resistivity, characteristic temperatures (glass transition temperature, softening point) , thermal stability, and chemical durability. The developed glass has enough flexibility for changes in the process parameters. The effects of noble-metal on viscosity and electrical resistivity were elucidated.

INTRODUCTION

As of April of 1989, about 350m^3 of high-level liquid waste (HLLW) generated from the reprocessing of spent LWR fuels are being stored at the Tokai Reprocessing Plant (TRP) of the Power Reactor and Nuclear Fuel Development Corporation (PNC).

PNC is planning to solidify this HLLW into a borosilicate glass waste form at the Tokai Vitrification Facility (TVF).[1-5] The construction of TVF has already started in 1988 and is scheduled to begin hot operation in 1992.

PNC has been developing vitrification technology since 1975. In particular, full-scale cold mock-up vitrification tests and laboratory-scale hot vitrification tests for the process development and waste glass characterization have been performed since 1981 and 1982, respectively. This paper describes the development of the TVF glass in cold tests.

GLASS COMPOSITION

Extensive development work on TVF waste glass was carried out, and about 700 simulated waste glasses have been evaluated against the following criteria.

① Homogeneity of waste elements in the glass.
② Sufficient viscosity at the process temperature range.
③ Durability of the glass.

As a result of the work, an optimum composition of borosilicate waste glass has been developed. The composition of the glass additive for TVF is shown in Table I. The glassfiber form of the additive has been developed to reduce the amount of particles released from the melter to the off-gas line.[4, 5] Besides the components of glass additive, sodium is added to HLLW in TVF to adjust the sodium oxide content in the waste glass to 10 wt% in order to get the sufficient viscosity on melting and pouring the glass.

The standard HLLW shown in Table II was fixed on the basis of analytical values, process parameters in TRP, and ORIGEN calculation. Average burn-up of the spent fuel is 28,000 MWd/t, and cooling time before vitrification is 5.5 years. HLLW contains fission products, unextracted uranium and plutonium, transuranic elements, corrosion products and additives in reprocessing process.

The standard waste glass contains 25 wt% of waste oxides as shown in Table Ⅲ. In this case, the fission products content is about 10 wt%. The 300kg of waste glass is produced from one metric ton of heavy metals of spent fuel.

SOLUBILITY OF ELEMENTS IN GLASS

The composition of HLLW is affected by the kind of fuel, process condition, the amount of corrosion products, and the entrainment of the precipitates from the HLLW storage tanks. The solubility of the waste components was evaluated using the glasses melted at 1150°C and guenched. The soluble upper limit in the glass is indicated in Table Ⅳ. When the contents exceed these values, the crystals in the third column are observed in the glass under microscope or micro probe analysis.

In order to keep the homogeneity and chemical durability of the waste glass, the waste loading rate is controlled. In particular, "yellow phase" which mainly consisted of caesium-sodium molybdates easily dissolves in water, therfore the amount of molybdenum which was main component of yellow phase should be adjusted within the limit. On the other hand, noble metals do not affect on the chemical durability of the glass, therefore it is not required that they dissolve in the glass.

Sensitivity tests were carried out in order to assess the effects of variations in waste composition on glass properties. The composition of the glasses used in the sensitivity tests is shown in Table V.

The curves of leach rate vs. content of waste oxide and sodium oxide are shown in Figure 1. The leach rate increased with waste and sodium concentration. The curve of waste concentration departed from linearity above 30 wt% waste oxide.

Further, the curves of density vs. content of waste oxide and sodium oxide are shown in Figure 2. The effect of changes in waste concentration on density was greater than changes in sodium concentration, therefore density can be used for roughly assessing the waste content in the glass.

GLASS PROPERTIES RELATED TO MELTING PROCESS

The important properties to the melter operation are viscosity and electrical resistivity of the molten glass. The temperature dependence of them are shown in Figure 3 and 4. Both results indicate that the effect of changes in waste concentration was similar to the effect of changes in sodium concentration under ±20% variance from each standard concentration.

GLASS PROPERTIES RELATED TO STORAGE

Thermal stability of the waste glass was evaluated by using the Time-Temperature-Transformation (T-T-T) diagram shown in Figure 5. The yield of the crystalization was detected by X-ray diffractometer (detection limit of the detected crystals : 0.5wt%). The crystals were cerium oxide and calcium-cerium molybdate (Figure 6). The glass held at 550°C for one year contained 1. 2wt% of the crystals, but the glasses held below the glass transition temperature (500°C) did not contain crystal except ruthenium oxide which existed in the untreated glass. Maximum crystalization can be expected if a glass is held at an intermediate temperature (750°C) between transition temperature and liquidus temperature. The leach rate of the glasses heat-treated for one year at 300 to 550°C was almost same as the leach rate of the untreated glass. The result is shown in Figure 7.

The maximum temperature of the waste glass in storage is kept under the glass transition temperature. Figure 8 shows that characteristic temperature (Tg : glass transition temperature, Ts : softening point) decreased with the content of waste oxide and sodium oxide.

Based on the thermal conductivity of the canistered glass, the maximum temperature of the glass in storage at TVF is expected to be about 300°C, which is enough below the glass transition temperature.

GLASS PROPERTIES RELATED TO DISPOSAL

Leaching studies under predicted disposal conditions are carried out in PNC [6,7] to evaluate a source term for the multiple package system and long-term stability of waste glass. The results of static leach tests with the bentonite/water weight ratio of 1/2 to 1/200 at 98°C for one year were 4×10^{-4} to 4×10^{-2} kg/m^2 of weight loss and 0.3 to 3μm in thickness of depleted layer. The studies of leaching mechanism have been carried out on the basis of the tests with the various media.

EFFECTS OF NOBLE METAL ON GLASS PROPERTIES

Noble-metals with low solubility in the glass agglomerate at the first stage of the vitrification reaction, then precipitate to the bottom of the crucible. The aggromerates of noble-metals gradually compact and form a noble-metals rich layer. The oxidation state of noble-metals is variable according to the redox condition and temperature of the molten glass. The typical micro structure of noble-metal layer is shown in Figure 9. Rod-shaped RuO_2 particles 3 to 70 μm in length and round Pd-Te particles 10 to 30 μm in diameter were observed.

The electrical resistivity of the layer containing 16 wt% noble-metals (RuO_2/Pd, 13/3wt%) was one order lower than that of the standard waste glass (RuO_2/Rh/Pd, 0.7/0.1/0.3wt%), as in Figure 10. RuO_2 is one of the most conductive metal oxides, therfore the resistivity of the layer indicates combined properties of waste glass and noble metals.

The viscosity of the layer containing 16 wt% noble-metals (RuO_2/Pd, 13/3wt%)was measured by a rotation viscometer. The viscosity decreased as the rate of shear increased, as shown in Figure 11. The fluid behaves as pseudo-plastic. In order to increase the fluidity, it is useful to increase the amount of shear.

PNC has also been developing the recovery technology of noble-metals from HLLW or molten glass for the purpose of utilization as resources.

ACKNOWLEDGEMENT

The authors are favored to have the assistance of Mr. K. Terao, Mr. K. Seki, Mr. Y. Kanazawa, Mr. I. Nagai, and Mr. M. Honma who attributed their experimental skill.

REFERENCES

[1] K. Uematsu, "Design of the Vitrification Plant for HLLW Generated from the Tokai Reprocessing Plant" ; pp. 427-433 in Waste Management' 86 ; Vol. 2. Edited by R. G. Post, University of Arizona Press, Tucson, 1986.

[2] N. Sasaki, M. Karino, H. Okamoto, H. Kashihara, and M. Yamamoto, "Solidification of the High-level Liquid Waste from the Tokai Reprocessing Plant" ; pp. 147-160 in Proc. Fuel Reprocessing and Waste Management, Jackson, Wyoming, August 26-29, 1984, Vol. 1, American Nuclear Society, 1984.

[3] T. Tsuboya, N. Tsunoda, "The Japanese Vitrification Program" ; pp. 181-188 in Waste Management' 88 ; Vol. 2. Edited by R. G. Post, University of Arizona Press, Tucson, 1988.

[4] N. Sasaki, S. Torata, H. Igarashi, H. Kashihara, and M. Yamamoto, "Advances in Vitrification Techniques in Japan" ; pp. 737-744 in Spectrum' 86 ; Vol. I. Edited by J. M. Pope, I. M. Leonard, and E. J. Mayer, American Nuclear Society, 1986.

[5] S. Torata, N. Sasaki, M. Horie, and N. Tsunoda, "Development of Glass Melter for the PNC Tokai Vitrification Facility" ; pp. 82-84 in Spectrum' 88 ; Proc. Nuclear and Hazardous Waste Management International Topical Meeting, Pasco, Washington, September 11-15, 1988, American Nuclear Society, 1988.

[6] K. Ishiguro, N. Sasaki, H. Kashihara, and M. Yamamoto, "Effects of Rocks and Backfill Materials on Waste Glass Leaching" ; pp. 247-254 in

Scientific Basis for Nuclear Waste Management IX, Edited by L. O. Werme, Materials Research Society, 1985.

[7] K. Ishiguro, K. Kawamura, H. Sonobe, M. Nodaka, T. Arai, Y. Yusa, and N. Tsunoda, "Experimental Approach to Performance of Engineered Barriers under Repository Conditions" ; presented at Conference Seminar 12 of 9th Structural Mechanics in Reactor Technology, Lausanne, Switzerland, August 17-21, 1987.

Table I. Glass Additive Composition

Component(wt%)	
SiO_2	62.3
B_2O_3	19.0
$A\ell_2O_3$	6.7
Li_2O	4.0
CaO	4.0
ZnO	4.0

Table II. Standard Waste Composition
(0.5 m^3/MTU, 28,000MWd/MTU)

Fission Products (g/ℓ)				Actinides (g/ℓ)		Corrosion Products (g/ℓ)		Additives (g/ℓ)	
Se	0.1	Cs	4.4	U	11.6	Fe	8.4	Na	44.5
Rb	0.6	Ba	2.9	Np	0.8	Ni	2.2	P	0.8
Sr	1.6	La	2.2	Pu	0.1	Cr	2.2		
Y	0.9	Ce	4.2	Am	0.1				
Zr	6.6	Pr	2.0						
Mo	5.9	Nd	7.1						
Tc	1.5	Pm	0.1						
Ru	3.5	Sm	1.5						
Rh	0.8	Eu	0.2						
Pd	1.9	Gd	0.1						
Te	0.7								

Table III. Composition of Standard Waste Glass

Component		(wt%)
Glass Additive	75.0	
SiO_2		46.7
B_2O_3		14.3
$A\ell_2O_3$		5.0
Li_2O		3.0
CaO		3.0
ZnO		3.0
Waste	25.0	
Na_2O		9.6
P_2O_5		0.3
Fe_2O_3		1.9
NiO		0.5
Cr_2O_3		0.5
fission products		9.8
actinides		2.4

Table IV. Solubility of Waste Components in Glass Melted at 1,150°C

Component	Solubility(wt%)	Crystalline Phases
P_2O_5	2	(P,Ca)oxide, (P,Ce,Nd)oxide
Fe_2O_3	16	Fe_2O_3
NiO	7	(Ni,Fe,Zn)oxide
Cr_2O_3	1	(Cr,Fe,Zn)oxide
ZrO_2	10	$ZrSiO_4$, ZrO_2
MoO_3	3	(Mo,Cr)yellow phase
PdO	0.05	Pd-Te
RuO_2	<0.1	RuO_2
Gd_2O_3	20	$Ca_2Gd_8(SiO_4)_6O_2$

Table V. Composition of Sensitivity Test Glass

Na_2O (wt%)	CONTENT (wt%)				WASTE (wt%)	CONTENT (wt%)			
NOTE	Na_2O	others	WASTE	glass additive	NOTE	Na_2O	others	WASTE	glass additive
6	6	16	22	78	15	6	9	15	85
8	8	16	24	76	20	8	12	20	80
10	10	15	25	75	25	10	15	25	75
12	12	15	27	73	30	12	18	30	70
14	14	15	29	71	35	14	21	35	65

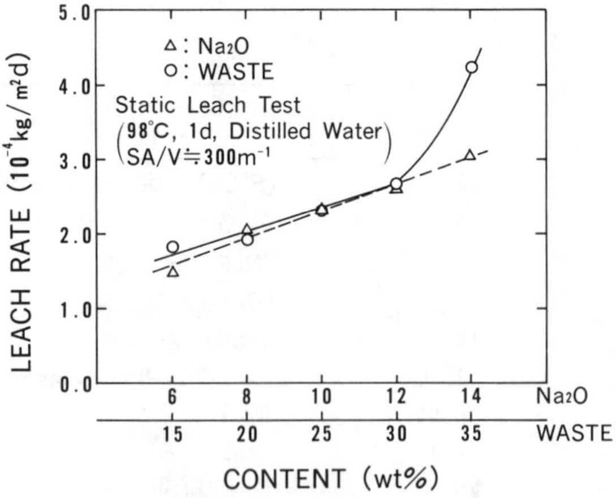

Figure 1. Effect of Na_2O, waste content on leach rate.

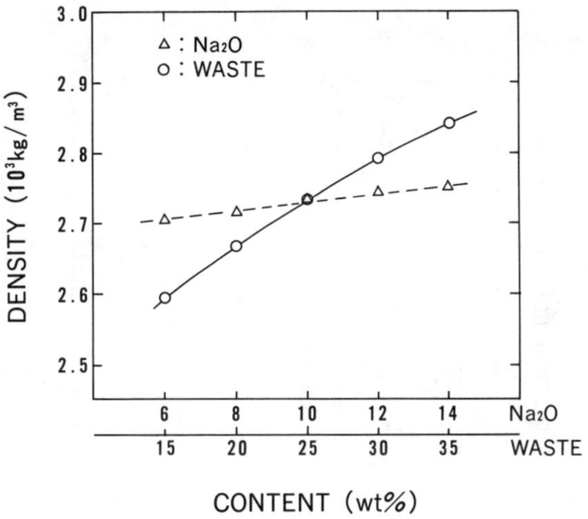

Figure 2. Effect of Na_2O, waste content on density at 20°C.

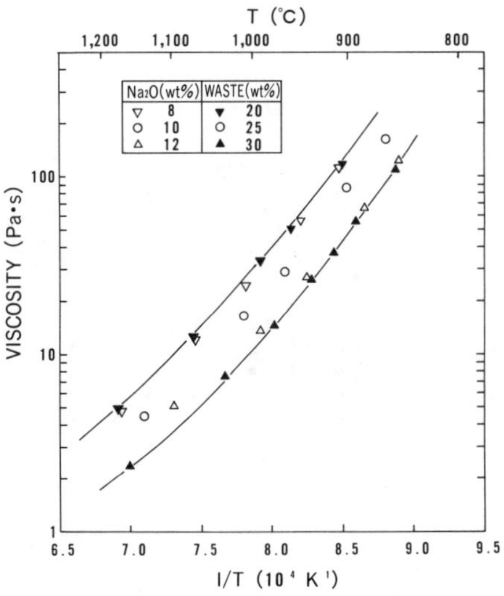

Figure 3. Effect of Na_2O, waste content on viscosity.

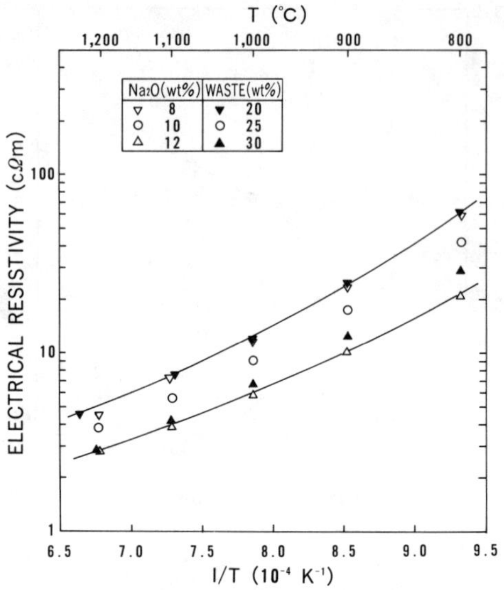

Figure 4 . Effect of Na_2O, waste content on electrical resistivity.

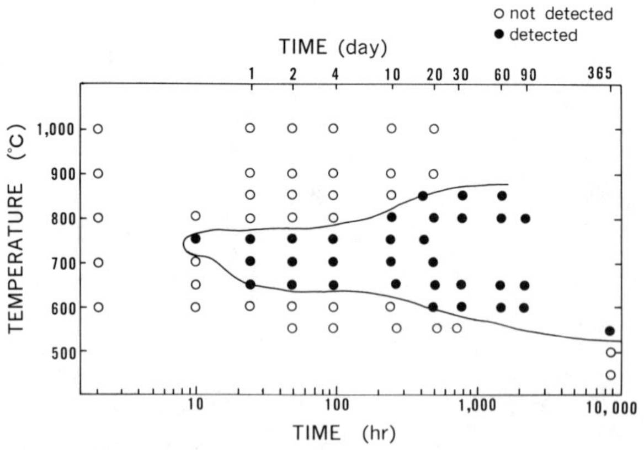

Figure 5 . T-T-T diagram of standard waste glass (crystalline phase detected by XRD).

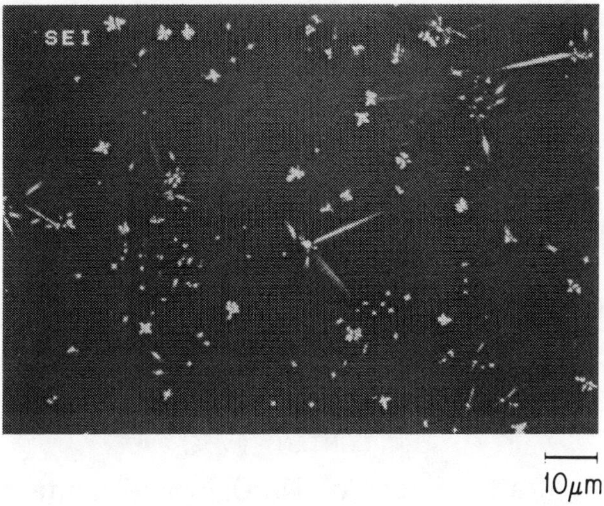

Figure 6 . SEM micrograph of the glass heated for one year at 550°C.

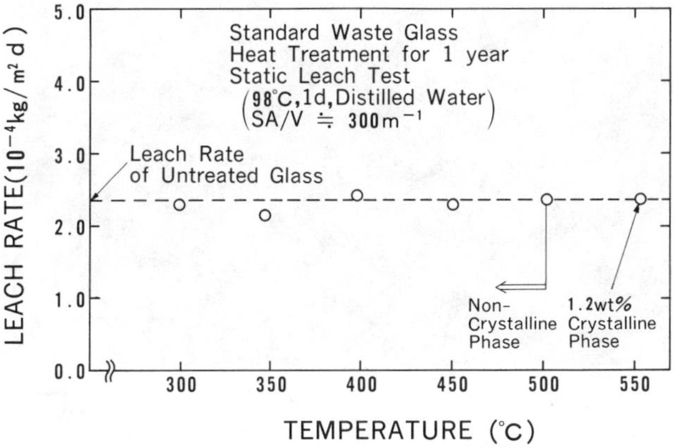

Figure 7 . Leach rate of waste glass after one year heat treatment at various temperature.

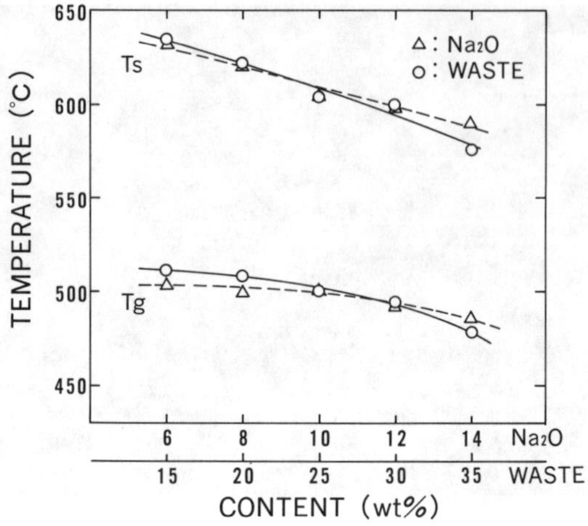

Figure 8 . Effect of Na_2O, waste content on characteristic temperature (Ts: softening point, Tg: glass transition temperature).

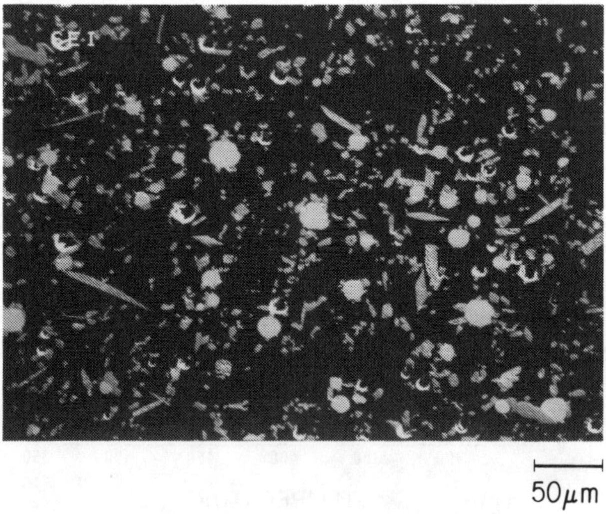

Figure 9 . SEM micrograph of noble-metal layer.

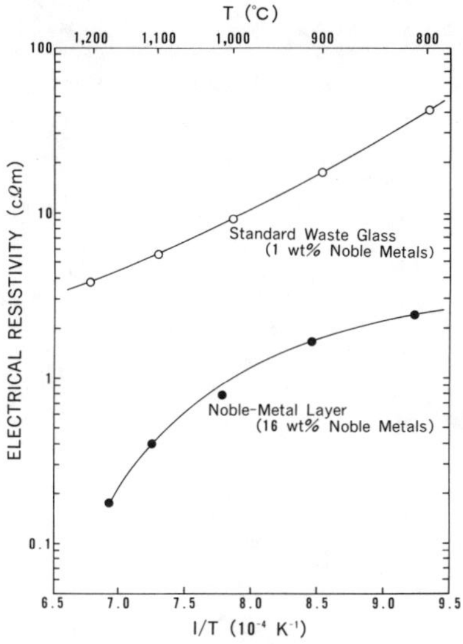

Figure10. Electrical resistivity of noble-metal layer.

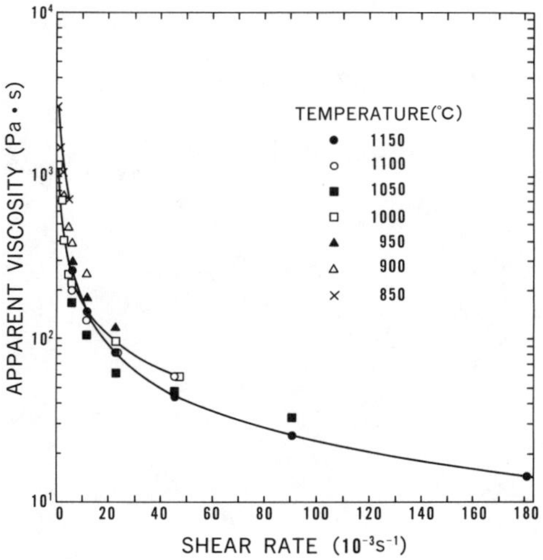

Figure 11. Apparent viscosity of noble-metal layer. (16 wt% noble metals)

STATISTICAL ASPECTS OF COMPLIANCE WITH THE WAPS RADIONUCLIDE RELEASE SPECIFICATION

G.F. Piepel and G.B. Mellinger, Materials Characterization Center, Pacific Northwest Laboratory*, P.O. Box 999, Richland, Washington 99352

ABSTRACT

The paper discusses various statistical aspects of complying with the Waste Acceptance Preliminary Specifications (WAPS) radionuclide release specification, including: (1) interpretation of the 95%/95% criterion in WAPS 1.3.2, (2) use of 95%/95% upper tolerance limits for demonstrating compliance, (3) the level of sampling and leach testing of the production waste form required to demonstrate compliance with WAPS 1.3.2 using an optimal strategy, and (4) alternative approaches for demonstrating compliance that do not require extensive sampling and leach testing of the final production waste form.

INTRODUCTION

Waste Acceptance Preliminary Specification (WAPS) 1.3 (Rev. 1, April 1988) discusses requirements for controlling radionuclide release properties of a waste form and a statistical criterion for demonstrating compliance with the requirements. Specifically, WAPS 1.3.1 requires that a waste form be capable of limiting normalized elemental leach rates for sodium, silicon, boron, cesium-137, and uranium-238 to less than 1.0 g/m^2-day averaged over a 28-day MCC-1 leach test. [Release rates in grams per day per square meter of test specimen surface area are "normalized" by dividing the release rate for a particular glass component (e.g., B_2O_3) by the fraction of the glass that component comprises.] WAPS 1.3.2 requires that sufficient samples of the production waste form be taken and leach tested

---

* Pacific Northwest Laboratory is operated by Battelle Memorial Institute for the United States Department of Energy under Contract DE-AC06-76RLO 1830.

so as to be able to demonstrate with 95% confidence that 95% of the production waste form would yield leach test results that conform to the 1.0 g/m^2-day specification. Finally, WAPS 1.3.3 provides for the use of alternative approaches for demonstrating compliance provided that they can be satisfactorily related to the WAPS 1.3.1 and 1.3.2 requirements.

The purpose of this paper is to discuss the statistical aspects of compliance with the WAPS release specification. Demonstrating compliance is discussed in terms of the 1.0 g/m^2-day MCC-1 specification of WAPS 1.3.1, but the same statistical principles would apply to specifications involving other leach tests and/or other acceptable release values. The topics discussed include:

1. Interpretation of the 95%/95% criterion in WAPS 1.3.2.

2. The use of 95%/95% upper tolerance limits to satisfy the 95%/95% criterion for demonstrating that waste form release properties are acceptable.

3. The level of sampling and leach testing of the production waste form required in order to demonstrate compliance with WAPS 1.3.1 and 1.3.2 using an optimal sampling, leach testing, and statistical strategy.

4. Statistical issues to be considered in developing alternative approaches for demonstrating compliance with the radionuclide release specification, as provided for by WAPS 1.3.3.

An important distinction is made between the approach selected to demonstrate that waste form release properties are acceptable and the strategy chosen to implement the approach. For example, the "approach" specified by WAPS 1.3.2 calls for sampling the production waste form from a position near the canister mouth and leach testing with the 28-day MCC-1 test. However, there are many "strategies" for implementing this approach, often with tradeoffs between the amount of sampling and leach testing performed and the likelihood that the 95%/95% criterion is met. Similarly for the alternative approaches allowed by WAPS 1.3.3, there are many strategies for implementing any given approach.

There are many statistical issues that should be considered when selecting an approach for demonstrating the acceptability of waste form release properties and a strategy for implementing the approach. However, first we discuss the interpretation of the 95%/95% criterion in the WAPS 1.3.2 specification and the use of 95%/95% upper tolerance limits for demonstrating compliance.

INTERPRETATION OF THE 95%/95% CRITERION

Statistical 95%/95% acceptance criteria analogous to that of WAPS 1.3.2 are used routinely in industry and are discussed in statistics books (e.g., Ref. 1). The standard interpretation of such a criterion is that it involves a 95% confidence statement about 95% of the distribution of true property values for all possible population members. In the WAPS case, the property is normalized release in g/m^2-day (for Na, Si, B, Cs-137, and U-238), while the population members include all possible samples of production waste form that could be taken. Thus, the goal of the criterion is to state with 95% confidence that the true 28-day MCC-1 normalized releases of 95% of the production waste form material are below 1.0 g/m^2-day.

In contrast to the above, it is possible to misinterpret the criterion to mean that the glass producer will leach test a suitable number of waste form samples, and deem acceptable those that have normalized release rates for the specified elements and radionuclides less than 1.0 g/m^2-day. Then, if 95% of the samples tested are deemed acceptable, compliance with the WAPS 1.3.2 specification will have been demonstrated (assuming that enough samples were tested to yield 95% confidence).

The main difference between the above two interpretations is that the standard interpretation assumes that the 95%/95% criterion refers to true normalized releases, while the misinterpretation assumes that the criterion refers to observed normalized releases. This distinction is important because the observed releases (that is, the test data) will be affected by a number of sources of uncertainty including those associated with chemically analyzing the glass, performing the leach test, and analyzing the resulting leachate. Each of these sources of uncertainty are subject to short- and long-term within-lab variation as well as lab-to-lab variation. These long-term within-lab and lab-to-lab variations can be significant (see Ref. 2). Sampling variation (waste form inhomogeneity) also contributes to the uncertainty in normalized release rates.

By assuming that the 95%/95% criterion refers to observed rather than true release rates, the uncertainty associated with the data is incorrectly ignored. Rather, the data (which are subject to uncertainty) must be used to make a 95% confidence statement about 95% of the true normalized release values.

The final statistical issue relative to interpreting the 95%/95% criterion is whether it is intended to apply simultaneously or individually to Na, Si, B, Cs-137, and U-238. In the individual case, 95% confidence statements are made about the 95th percentiles of each of the five true release rate distributions. In the simultaneous case, a single 95% confidence statement is made about the 95% envelope of the five-variable true release rate distribution (a distribution in five dimensions). WAPS 1.3.2 needs to be explicit about which of these two interpretations is intended. For brevity, the scope of the remaining discussion is limited to the individual interpretation, although most of the issues raised also apply to the simultaneous interpretation.

95%/95% UPPER TOLERANCE LIMIT

A 95% confidence statement about the 95th percentile of a (single-variable) distribution is known as a 95%/95% upper tolerance limit. If the distribution is normal (Gaussian), the 95%/95% upper tolerance limit is given by the formula

$$TL = m + Ks, \qquad [1]$$

where m and s are, respectively, estimates of the mean and standard deviation of the distribution. K is a 95%/95% upper tolerance limit factor whose value depends on how m and s are estimated (see Ref. 3). A different approach must be taken if the release rates are not normally distributed (see pp. 117-121 of Ref. 4).

Provided the waste form production process is monitored and shown to be in a state of statistical control, compliance with WAPS 1.3.1 and 1.3.2 can be demonstrated by computing a 95%/95% upper tolerance limit and showing that its value is less than 1.0 g/m^2-day. A number of factors affect the magnitude of the 95%/95% upper tolerance limit. These include:

- The mean and standard deviation of the "true" normalized release rate distribution (for each element or radionuclide). The larger the mean or standard deviation of the distribution, the larger the 95%/95% upper tolerance limit.

- The magnitude of sampling and leach testing variations. Leach test results are subject to sampling, short-term within-lab, long-term within-lab, and lab-to-lab variation (the latter two variations affect the results even if leach tests are only performed at one time at one lab). All of these sources of variation must be accounted for, either by incorporating them in the standard deviation s of Eq. [1], or by "removing" them via the sampling, leach testing, and statistical strategy used. The larger the magnitudes of these sources of variation and the more of them included in s, the larger the 95%/95% upper tolerance limit.

- The sampling and leach testing strategy used. Some sampling and leach testing strategies that could be chosen would necessitate including all of the sampling and leach testing sources of variation (discussed above) in s. Other leach testing strategies provide for "removing" the uncertainties due to many of these sources of variation so that they do not have to be included in s. Such strategies yield smaller 95%/95% upper tolerance limits that are more likely to be below the 1.0 g/m^2-day specification.

- The number of production samples and the number of times each is analyzed for composition and leach tested. The magnitude of the 95%/95% upper tolerance limit decreases as the number of samples and the number of times each is analyzed and leach tested increases.

It is clear that a sampling, leach testing, and statistical strategy should be chosen so as to minimize the 95%/95% upper tolerance limit. Doing so maximizes the chance that the tolerance limit will be below 1.0 g/m^2-day. Such a strategy is discussed in the following section.

A STATISTICAL, SAMPLING, AND LEACH TESTING STRATEGY FOR DEMONSTRATING COMPLIANCE WITH WAPS 1.3.1 AND 1.3.2

As written, WAPS 1.3.2 requires sampling and leach testing the production waste form material to demonstrate compliance with the 1.0 g/m^2-day specification via a 95%/95% criterion. A sampling, leach testing, and statistical strategy that minimizes the 95%/95% upper tolerance limit for the WAPS 1.3.2 approach is discussed in this section. Other approaches (provided for by WAPS 1.3.3) for indirectly demonstrating compliance without sampling and leach testing the final product are discussed in the following section.

An Optimal Strategy

Ideally, only the variation in production waste form "true" normalized releases should be included in the standard deviation s used in computing the 95%/95% upper tolerance limit. In practice this usually cannot be done. However, the sampling and leach testing can be performed in such a way that statistical techniques can be used to remove from consideration many of the "nuisance" sources of variation (sampling, short-term within-lab, long-term within-lab, and lab-to-lab). Such a sampling, leach testing, and statistical strategy is outlined in the following steps.

Step 1. Sample the production waste form material "r" times during the span of production for which compliance is to be demonstrated. At each of the "r" times, "m" samples should be collected, analyzed for composition, leach tested, and leachates analyzed. Normalized release rates are then calculated for each element.

Step 2. Concurrently at each of the "r" times, "m" samples of an "appropriate certified" glass should be analyzed for composition, leach tested, leachates analyzed, and normalized releases calculated. The definition of an "appropriate certified" glass is given below.

Step 3. For each element or radionuclide covered by WAPS 1.3.1, and for each of the "r" times, average the "m" compositional analyses and "m" normalized release rates found for the certified glass samples. Then, "bias correct" the "m" production sample normalized release rates using (as appropriate) a ratio or difference correction based on the m-sample averages and certified composition and release rates for the certified glass.

Step 4. Use the "bias corrected" production sample normalized release rates to compute the value of the 95%/95% upper tolerance limit. The details of this calculation (which removes the short-term within-lab uncertainties from consideration) are beyond the scope of this paper, but will be detailed in a forthcoming report.

Step 2 refers to the use of an "appropriate certified" glass. By "appropriate", we mean that the glass is representative of the waste form being produced (e.g., it could be the target composition). In this way, the glass will have similar composition, and hence leach behavior, as the production samples. By "certified", we mean that the normalized release rates and the short-term within-lab, long-term within-lab, and lab-to-lab standard deviations of the normalized release rates are well-estimated (e.g., from a round robin involving at least 7-8 labs, where each lab obtains replicate normalized release rates over both the short- and long-term).

The purpose of analyzing for composition and leach testing the "appropriate certified" glass along with production samples is that it provides for removing long-term within-lab and lab-to-lab variation from the normalized release rates. In a side-by-side chemical analysis, leach testing, and leachate analysis of production and certified glass samples, long-term within-lab variation and lab-to-lab variation will bias the normalized release rates for both glasses. The difference between the certified value and the result obtained from testing the certified glass (for each element of interest) can then be used to "bias correct" (Step 3) the production sample results. Of course, these are only estimated corrections, so the uncertainty in the corrections must be included in computing the 95%/95% upper tolerance limit. Still, the uncertainty in the corrections will usually be less than that due to long-term within-lab and lab-to-lab variation. The end result will be a smaller 95%/95% upper tolerance limit than would be obtained by including long-term within-lab and lab-to-lab sources of variation in s.

Sampling and Leach Testing Requirements

Table 1 presents the widths of 95%/95% upper tolerance limits on normalized release rates for various combinations of "r", "m", s_u and s_p (the notation is defined in Table 1). It is assumed in Table 1 that the short-term within-lab standard deviation for the certified glass normalized release rates is equal to the production sample value. Further, the results in Table 1 are based on a sampling standard deviation (s_s) of 0.03 g/m^2-day and a certified value standard deviation (s_o) of 0.05 g/m^2-day. Three other combinations of these two standard deviations were also investigated [(s_s,s_o) = (.01,.03), (.01,.05), and (.03,.03)] but the tolerance limit widths were not much different than the values given in Table 1.

The values listed in the body of Table 1 indicate how close the mean of the bias-corrected normalized release rates for a given element or radionuclide can be to 1.0 g/m^2-day and still satisfy the WAPS 1.3.1 and 1.3.2 specification. For example, the table value of 0.346 for r = 10, m = 2, s_p = 0.05, and s_u = 0.10 indicates that the means of the bias-corrected normalized release rates for B, Na, Si, Cs-137, and U-238 would have to be 1 - 0.346 = 0.654 g/m^2-day or less in order to demonstrate compliance with the WAPS 1.3.1 and 1.3.2 release specification.

Table 1 can be used to determine "sample sizes" needed in order to demonstrate compliance for particular situations. Values of "r" and "m" necessary to satisfy WAPS 1.3.1 and 1.3.2 can be determined from Table 1 (using anticipated values for s_u and s_p) by choosing the tolerance limit width corresponding to the anticipated distance between the "true" normalized release mean and 1.0 g/m^2-day. For example, suppose that values of s_u = 0.10 and s_p = 0.025 g/m^2-day are anticipated. If a normalized release rate mean of 0.74 g/m^2-day is expected, then Table 1 indicates that r = 20 and m = 3 or 4 would provide for satisfying the WAPS 1.3.1 and 1.3.2 specification. Similarly, if a normalized release rate mean of 0.57 g/m^2-day is expected, Table 1 indicates that r = 5 and m = 2 would be sufficient.

It is clear from Table 1 that the major determinant of the width of a 95%/95% upper tolerance limit is s_u, the standard deviation of the distribution of "true" normalized release rates for the production waste form. This is expected, since the table was constructed based on a strategy that minimizes the "nuisance"

TABLE 1. Widths[(a)] of 95%/95% Upper Tolerance Limits on Normalized Release Rates (in units of g/m^2-day)

	s_p	.025	.025	.025	.05	.05	.05	.075	.075	.075
r	m / s_u	.05	.10	.15	.05	.10	.15	.075	.10	.15
5	2	.253	.426	.634	(b)	.519	.695	(b)	(b)	(b)
5	5	.224	.403	.614	.255	.430	.636	.400	.481	.670
10	2	.199	.311	.448	.256	.346	.472	(b)	.414	.513
10	5	.185	.301	.440	.200	.313	.449	.277	.334	.464
20	2	.173	.261	.370	.200	.276	.381	.281	.309	.400
20	5	.166	.257	.367	.173	.262	.371	.225	.271	.378
50	2	.156	.230	.323	.168	.237	.328	.215	.250	.336
50	5	.153	.228	.321	.157	.230	.323	.194	.235	.326
100	2	.150	.217	.304	.156	.221	.306	.194	.229	.312
100	5	.149	.216	.303	.150	.217	.304	.183	.220	.306

r = The number of times samples are to be collected.

m = The number of samples to be leach tested at each of the r sampling times.

s_p = The short-term within-lab standard deviation of normalized release rates for the production waste form material. In this table, the same value is used for the short-term within-lab standard deviation of normalized release rates for the certified glass.

s_u = The standard deviation of the distribution of "true" normalized release rates.

(a) These "widths" are the smallest amounts below 1.0 g/m^2-day the bias-corrected mean normalized releases can be and still demonstrate compliance with the 95%/95% criterion.

(b) The method used to compute the tolerance limit widths was not applicable for this case (Ref. 5).

uncertainty. A strategy that does not provide for removing any of the nuisance uncertainties would yield considerably larger upper tolerance limit widths. For example, consider the situation with s_u = 0.10 g/m^2-day and sampling (s_s), short-term within-lab (s_p), long-term within-lab, and lab-to-lab nuisance uncertainty standard deviations of 0.03, 0.05, 0.08, and 0.08 g/m^2-day, respectively. In this example situation, the nuisance uncertainties would inflate the 95%/95% upper tolerance limit widths by approximately 45%. Hence, it is seen that strategies for removing nuisance uncertainties are of considerable benefit in reducing the magnitude of 95%/95% upper tolerance limits.

ALTERNATIVE APPROACHES FOR DEMONSTRATING COMPLIANCE

The WAPS 1.3.1 and 1.3.2 specifications require demonstrating compliance with the 1.0 g/m^2-day MCC-1 28-day limit based on sampling and leach testing the production waste form. WAPS 1.3.3, on the other hand, permits the use of other approaches for demonstrating that waste form release properties are acceptable, provided that the results can be related to the requirements of WAPS 1.3.1 and 1.3.2.

A number of alternative approaches might be considered. One alternative involves sampling the production glass but, rather than leach testing the samples, predicting their release rates using a product model that relates glass composition (and possibly other attributes such as crystallinity or redox state) to leaching behavior in the MCC-1 test. The model predictions would then be used to demonstrate compliance with WAPS 1.3. Another alternative is to sample the melter feed instead of the final waste form and use a process model (essentially a melter mass balance) to predict the composition of the glass being produced. A product model would then be used to predict MCC-1 release rates, which would be used to demonstrate compliance with WAPS 1.3. A third alternative would be to leach test the production glass using a leach test other than MCC-1. A model that relates the results of this alternate leach test to MCC-1 results would then be used to predict MCC-1 behavior, and these predictions used to demonstrate compliance with WAPS 1.3.

It is assumed for any alternative approach that the 95%/95% criterion is still the basis for demonstrating acceptable release properties, and that 95%/95% upper tolerance limits falling below the specified release limit is still the way to demonstrate acceptability. The general formula for a 95%/95% upper tolerance limit given earlier in Eq. [1] still applies, but the alternative approach adopted and the statistical strategy for implementing it will affect the magnitude of the tolerance limits. The alternative approaches are more complicated than the approach of WAPS 1.3.2, and have sources of uncertainty associated with process and/or product models in addition to the other sources discussed earlier. Just as with the WAPS 1.3.2 approach, the choice of implementation strategy determines whether the sources of uncertainty can be removed from consideration or must be included in the standard deviation s of Eq. [1]. In selecting an approach and implementation strategy for demonstrating compliance with WAPS 1.3, a producer will have to consider the tradeoffs between minimizing the magnitude of the 95%/95% upper tolerance limit and the type and amount of sampling, chemical analyses, and leach testing that must be performed.

PRIOR TO AND DURING WASTE FORM PRODUCTION

With any approach and strategy for demonstrating acceptability of waste form release properties, a waste form producer must convince himself prior to production that the selected approach and strategy will be successful. Relative to the 95%/95% criterion, the producer must have very good estimates of the mean, standard deviation, and shape of the distribution of "true" normalized release rates for each element or radionuclide to be controlled. This information might initially be obtained by simulating the waste form composition with an adequate process model and then applying a product model (e.g., a 28-day MCC-1 model) that predicts release rates as a function of composition (and possibly other attributes). This initial work should be verified and updated using data from test or cold runs of the process. A waste form producer must also have good estimates of the nuisance uncertainty standard deviations that must be included in the 95%/95% upper tolerance limit because of the approach and strategy selected. The details of how to obtain these estimates are beyond the scope of this paper. However, it is suggested that a statistician be involved in the planning of experimental work and the development of process and product models so that there is adequate data and support for estimating the required nuisance standard deviations.

The 95%/95% upper tolerance limit technique for demonstrating that waste form release properties are acceptable requires that the waste form production process remain in a state of statistical control. Statistical control can be demonstrated during production by sampling at various stages of the process and maintaining control charts or monitoring process parameters. Prior to production, these activities can be defined and explored with data from simulated process runs and cold runs.

SUMMARY

WAPS 1.3 delineates the requirements for demonstrating the acceptability of waste form radionuclide release properties. WAPS 1.3.1 and 1.3.2 call for sampling the production waste form and performing 28-day MCC-1 leach tests so as to be able to state with 95% confidence that 95% of the waste form material produced meets the release rate limit of 1.0 g/m^2-day for sodium, silicon, boron, cesium-137, and uranium-238. WAPS 1.3.3 provides for demonstrating the acceptability of waste form release properties via alternative approaches, provided they can be related to the 1.3.1 and 1.3.2 approach. Presumably the 95%/95% criterion also applies to alternative approaches.

The important statistical issues regarding the WAPS 1.3.1 and 1.3.2 approach and alternative approaches permitted by WAPS 1.3.3 are as follows:

- The 95%/95% criterion should be interpreted as requiring a 95% confidence statement about 95% of the distribution of "true" release rates over the production period for which compliance is to be demonstrated.

- WAPS 1.3 needs to explicitly state whether the 95%/95% criterion applies to Na, Si, B, Cs-137, and U-238 individually or simultaneously. Simultaneous confidence of 95% is more restrictive than individual confidences of 95%, and requires the use of multivariate rather than univariate statistical techniques.

- A 95%/95% upper tolerance limit (TL) is the statistical tool that provides for demonstrating that the release properties (individually) are acceptable according to the 95%/95% criterion (provided the waste form production process is monitored and shown to be in a state of statistical control). If the TL for each element and radionuclide is below the release rate specification (e.g., 1.0 g/m^2-day), then acceptability of the waste form release properties has been demonstrated.

- Release rate data are subject to sampling, short-term within-lab, long-term within-lab, and lab-to-lab uncertainties. If an alternative approach (to that of WAPS 1.3.1 and 1.3.2) is adopted and it involves the use of process and product models, then the release rate "data" will also be subject to model uncertainties. All of these are "nuisance" uncertainties relative to the goal of making a 95%/95% statement about the distribution of true release rates.

- The approach and implementation strategy chosen for demonstrating compliance with WAPS 1.3 determines which of the "nuisance" uncertainties can be removed from consideration and which must be included in the calculation of TL. Ideally, the standard deviation s used to compute TL should only contain the variation in "true" release rates over the period of waste form production for which compliance is to be demonstrated. However, no approach and implementation strategy can remove all "nuisance" sources of uncertainty, and there may be very little benefit (in terms of reducing TL) to removing negligible sources of uncertainty. The approach and strategy selected should be as simple as possible and still remove from consideration the largest sources of nuisance uncertainty.

Many of the statistical issues raised in this paper are not well-recognized in the waste management community, and yet it is important that they be considered both in future revisions of WAPS and by waste form producers developing approaches (and strategies for implementing them) for demonstrating compliance with the waste form release properties specification.

REFERENCES

1. Schilling, E.G. (1982), Acceptance Sampling in Quality Control, Marcel Dekker, New York.

2. Reimus, M.A.H., G.F. Piepel, G.B. Mellinger, and L.R. Bunnell (1988), West Valley Glass Product Durability Studies, FY 1987 -1988: Effects of Composition, Redox State, Thermal History, and Groundwater, PNL-6723, November 1988, Pacific Northwest Laboratory, Richland, Washington.

3. Owen, D.B. (1963), Factors for One-Sided Tolerance Limits and for Variables Sampling Plans, SCR-607, Sandia Corporation.

4. Conover, W.J. (1980), Practical Nonparametric Statistics, Second Edition, John Wiley & Sons, New York.

5. Gaylor, D.W. and F.N. Hopper (1969), "Estimating the Degrees of Freedom for Linear Combinations of Mean Squares by Satterthwaite's Formula," Technometrics, Vol. 11, No. 4, 691-706.

SAMPLING AND ANALYSIS STRATEGIES TO SUPPORT WASTE FORM QUALIFICATION

J. H. Westsik, Jr., B. A. Pulsipher, D. L. Eggett, and W. L. Kuhn
Pacific Northwest Laboratory(*), Richland, WA 99352

ABSTRACT

As part of the waste acceptance process, waste form producers will be required to 1) demonstrate that their glass waste form will meet minimum specifications, 2) show that the process can be controlled to consistently produce an acceptable waste form, and 3) provide documentation that the waste form produced meets specifications. Key to the success of these endeavors is adequate sampling and chemical and radiochemical analyses of the waste streams from the waste tanks through the process to the final glass product. This paper suggests sampling and analysis strategies for meeting specific statistical objectives of 1) detection of compositions outside specification limits, 2) prediction of final glass product composition, and 3) estimation of composition in process vessels for both reporting and guiding succeeding process steps.

INTRODUCTION

The U.S. Department of Energy established the Waste Acceptance Process (WAP) to assure that high-level radioactive waste forms prepared from existing radioactive wastes will be acceptable for disposal in a deep geologic repository. As part of the WAP, waste form specifications are developed[1]. Each waste form producer must 1) demonstrate that the waste form that they intend to produce will meet the specifications, 2) demonstrate that the process to be used to make the waste form can be controlled such that an acceptable waste form will consistently be produced, and 3) provide the appropriate documentation that the production waste form meets specifications.

Key to the success of these endeavors are adequate sampling and chemical and radiochemical analyses of the waste streams from the

---

(*)The Pacific Northwest Laboratory is operated by Battelle Memorial Institute for the United States Department of Energy under Contract DE-AC06-76RLO 1830.

waste tanks through to the final glass product. Sampling and analyses of the process solutions and slurries within Hanford Waste Vitrification Project (HWVP) process vessels are necessary in order to make appropriate chemical additions to the solutions/slurries, to detect when the melter feed slurry is out of specification, and to report the composition of the final glass product. This paper suggests sampling and analysis strategies for meeting these objectives. The strategies are based on reasonable assumptions for the compositions' tolerances.

In recommending a sampling and analysis strategy to support waste form qualification, four specific sampling objectives are defined and are assigned to key process vessels. The approach for determining an appropriate sampling/analysis strategy for three of the objectives is then outlined.

SAMPLING OBJECTIVES

Figure 1 shows the HWVP vitrification process flowsheet. When determining an appropriate sampling/analysis scheme for the several tanks in the HWVP vitrification process, it is necessary to define the statistical objectives of sampling for each tank. The objectives can be classified into four broad categories: estimation, process control, prediction, and out-of-specification detection.

The objective is **estimation** if sampling is to provide an estimate of the true average composition of the material to within a specified level of precision. For example, the major purpose for sampling the Waste Recycle Collection Tank (WRCT) is to estimate the composition in a particular batch. The required precision of such an estimate will drive the required number of samples and analyses. Although estimation will be a sampling and analysis objective for all of the HWVP process tanks; detection, prediction, and/or process control objectives may drive the sampling and analysis schemes employed for any particular tank.

The objective is **detection** if sampling is conducted in order to detect whether the true composition of the material falls in or out of predefined specification limits. Samples will be obtained from the Slurry Mix Evaporator (SME) to demonstrate that the slurry constituents are within some acceptable limits that reflect the compositional variability that the process can accommodate while still producing an acceptable glass. If the true concentration for a constituent is outside the specification limit such that an acceptable glass would not be produced, the sampling scheme should detect that situation. Thus, the sampling/analysis scheme should be set up to allow detection of out-of-specification occurrences.

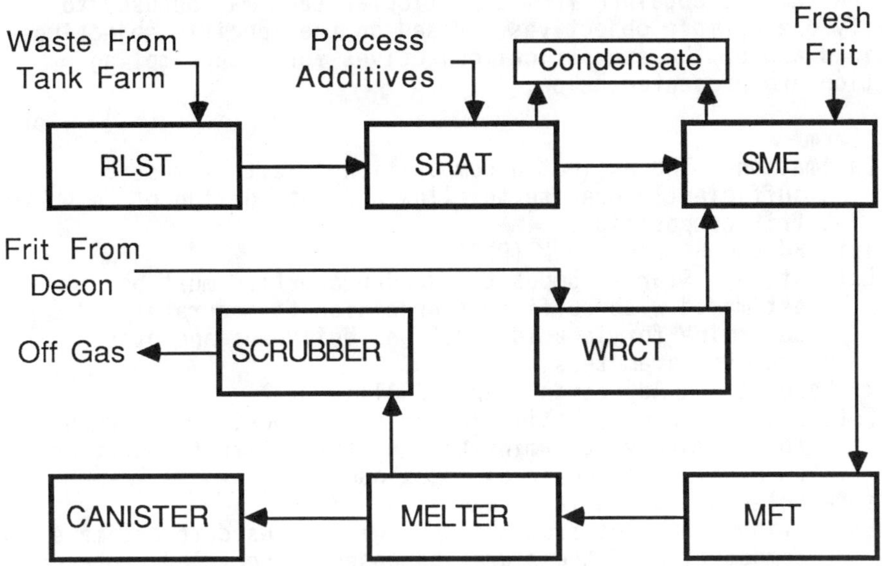

Figure 1. Schematic of HWVP Vitrification Process Flowsheet (Acronyms are defined on next page)

The objective is **process control** if it is important to detect a change in the process or a given process parameter, even though the shift may not immediately produce an out-of-specification product. Process control refers to statistical quality control methods. Although it is important to detect compositions in the SME that are outside of the acceptable operations envelope, it may be equally important to detect trends or shifts in the process that could affect operations or product quality in the long term if not corrected. Additionally, the reliability of predicted glass and feed compositions further along in the process depends on having a process that is in statistical control. Sampling and analytical schemes should provide the ability to detect process shifts and undesirable trends before serious processing difficulties are encountered.

The objective is **prediction** if the samples will be used to predict the composition later in the process. For example, the Melter Feed Tank (MFT) samples can be used to predict the glass composition in the canisters. The required precision of estimates of the glass composition in the canisters would then drive the sampling and analytical schemes required in the MFT.

The samples obtained from a particular tank may be used to satisfy one or more objectives. Based on the specific objectives outlined above, the statistical objectives for each sampling location are presented below.

Tank Farm

Estimation - The estimated composition in a tank must be sufficiently precise to allow accurate design of the HWVP frit composition.

Receipt and Lag Storage Tank (RLST)

Estimation - Slurry composition and properties must be estimated with sufficient precision to accurately determine formic acid addition, boildown, and other process parameters.

Slurry Receipt and Adjustment Tank (SRAT)

Estimation - Precise estimation of the composition is required to accurately determine the quantity of frit and other chemicals (as required) to be added.

Waste Recycle Collection Tank (WRCT)

Estimation - Samples from the WRCT will be used to estimate the composition of the slurry composed of recycled process streams and spent frit from canister decontamination activities.

Slurry Mix Evaporator (SME)

Process Control - Trends or changes in the process beyond those that occur under normal, in-control operations should be detected and corrected.

Detection - If major glass-forming constituents are not within the predefined acceptable composition envelope, this must be detected.

Estimation - If major glass-forming constituents have been detected to be outside acceptable limits, then precise estimation of the composition is needed to accurately determine the chemical adjustments to be made.

Prediction - SME samples can be used to estimate MFT and glass composition.

Melter Feed Tank (MFT)

Prediction - MFT samples can be used to estimate the glass composition.

Process Control - Trends and process shifts should be detected to apprise operators of possible problems upstream that can be corrected to avoid future compromise of product quality or operations.

Glass (Canister)

Estimation - Glass samples will be analyzed to estimate the composition of the glass in the canister.

REQUIRED SAMPLING TO MEET THE PREDICTION OBJECTIVE

The prediction objective involves predicting the glass composition from the SME and MFT samples. A propagation of errors method coupled with a mass balance process model was used to determine the required sampling scheme for the SME and MFT under various conditions[2]. Several parameters were evaluated, including analytical uncertainty (RSDA), batch-to-batch variation (RSDB), and the tolerable degree of uncertainty about the estimated glass composition. The recommended sampling schemes assuming a 2.5% relative standard deviation between replicate samples (RSDS) are shown in Table 1.

A conservative sampling scheme for prediction purposes could be determined by assuming a relative standard deviation between replicate analyses (RSDA) of 30% and a batch-to-batch variability of 5%. This batch-to-batch relative standard deviation (RSDB) is

Table 1. Best Sampling Schemes for the Prediction Objective (SME[a] and MFT) Assuming a 2.5% Sampling Relative Standard Deviation.

	RSDB=1.5%							
	HALFWIDTH=15%				HALFWIDTH=20%			
RSDA	NSSME[b]	NSMFT[c]	NAMFT[d]	TOT[e]	NSSME	NSMFT	NAMFT	TOT
5%	1	1	1	3	1	1	1	3
10%	1	1	1	3	1	1	1	3
15%	1	3	1	5	1	1	1	3
20%	1	6	1	8	1	3	1	5
30%	2	7	2	17	2	6	1	9
40%	7	7	3	29	3	6	2	16

	RSDB=5%							
	HALFWIDTH=15%				HALFWIDTH=20%			
RSDA	NSSME	NSMFT	NAMFT	TOT	NSSME	NSMFT	NAMFT	TOT
5%	1	1	1	3	1	1	1	3
10%	1	3	1	5	1	1	1	3
15%	1	4	2	10	1	2	1	4
20%	1	7	2	16	1	4	1	6
30%	-	-	-	--	2	5	2	13
40%	-	-	-	--	4	6	3	23

(a) - Only one analysis per SME sample is recommended.
(b) - NSSME is the number of SME samples required.
(c) - NSMFT is the number of MFT samples required.
(d) - NAMFT is the number of analyses per MFT sample required.
(e) - TOT is the total number of analyses including the analysis of the one glass sample.

the variability between the MFT batches after final frit and chemical additions and adjustments and does not refer to the batch-to-batch differences in the incoming waste batches. From Table 1, if 2 samples from the SME are analyzed once each and 5 samples are obtained from the MFT and analyzed twice each, then one can be approximately 95% confident that the true elemental concentration will be within ±20% of the reported concentration provided: 1) the feed information is combined with the glass sample estimates, 2) the process remains in statistical control, and 3) the mass balance model relating feed compositions to the final glass composition is correct (the melter is assumed to be a single well-mixed tank).

REQUIRED SAMPLING TO MEET THE DETECTION OBJECTIVE

When detection is the major objective, information from sampling will be used to detect an out-of-specification occurrence. The ability to detect out-of-specification values is dependent on the number of samples (NS) obtained, the number of analyses (NA) per sample, and the sampling and analytical variations. The probability (P) of detecting a true concentration that is outside the specification limit is dependent on the magnitude of the relative difference (D) between the true concentration and the specification limit. Assuming that the RSDS and RSDA are known and that the spread of the distribution (assumed to be normal) is small relative to the range between the upper and lower specification limits, the relative detectable difference can be expressed as

$$D = Z_p * \sqrt{\frac{RSDS^2}{NS} + \frac{RSDA^2}{NS*NA}}$$

where Z_p is obtained from the normal distribution table and RSDS and RSDA are the sampling and analytical relative standard deviations, respectively.

To determine the appropriate sampling scheme, various combinations of the values for RSDS, RSDA, NS, NA, and P were used to calculate the detectable difference. The quantities explored are shown below.

Factor	Levels
RSDS	2.5%, 5%, 10%
RSDA	5%, 10%, 15%
NS	1, 2, 3, 4, 5, 6, 7
NA	1, 2, 3
P	0.90, 0.95

For each combination of the factor levels, the detectable difference was calculated. The calculated detectable differences were categorized into one of five categories: <2.5%, <5%, <7.5%, <10%, and <15%. For each combination, the sampling scheme that minimized the total number of analyses (NS*NA) for a particular category is presented in Table 2. Where no sampling scheme is shown, the required number of samples and analyses would be greater than 7 and 3, respectively.

To illustrate the use of Table 2, suppose the sampling relative standard deviation is 5% and the analytical relative standard deviation is 15%. If seven samples are obtained with only one analysis per sample, one would have at least a 95% chance of concluding that the mean value was out of specification if the true mean value was at least 10% greater than the upper specification limit or 10% less than the lower specification limit. If only a

Table 2. Best Sampling Schemes for the Detection Objective.

			Detectable Difference (D)									
			2.5		5.0		7.5		10.0		15.0	
RSDS[a]	RSDA[b]	P[c]	NS[d]	NA[e]	NS	NA	NS	NA	NS	NA	NS	NA
2.5	5	0.95	7[f]	3	4	1	1	2	1	1	1	1
2.5	5	0.90	5	2	1	3	1	1	1	1	1	1
2.5	10	0.95			7	2	2	3	3	1	1	2
2.5	10	0.90			7	1	2	2	1	2	1	1
2.5	15	0.95					4	3	7	1	1	3
2.5	15	0.90			6	3	7	1	2	2	1	2
5.0	5	0.95			6	1	3	1	2	1	1	1
5.0	5	0.90			4	1	2	1	1	1	1	1
5.0	10	0.95			7	3	7	1	4	1	1	2
5.0	10	0.90			5	2	4	1	1	3	1	1
5.0	15	0.95					7	2	7	1	2	2
5.0	15	0.90			7	3	3	3	5	1	2	1
10.0	5	0.95					7	1	4	1	2	1
10.0	5	0.90					4	1	3	1	1	1
10.0	10	0.95					7	3	6	1	3	1
10.0	10	0.90					6	1	4	1	2	1
10.0	15	0.95							6	2	4	1
10.0	15	0.90					7	2	6	1	3	1

(a) - RSDS is the sampling relative standard deviation.
(b) - RSDA is the analytical relative standard deviation.
(c) - P is the probability of concluding that the value is out of specification when the true value is D% greater or less than the specification limit.
(d) - NS refers to the number of samples.
(e) - NA refers to the number of analyses per sample.
(f) - Blanks indicate more than 7 samples and 3 analyses needed.

90% chance of detecting a true difference of 10% were required, then only five samples are needed with only one analysis per sample.

REQUIRED SAMPLING TO MEET THE ESTIMATION OBJECTIVE

Estimation is the primary objective when the composition of the contents of a tank must be known with some precision. The degree of precision required for a given estimate depends on the end use of the estimate. If a slight imprecision in an estimate could have a major effect on the product quality, a greater number of samples or analyses would be required. If imprecise estimates do not greatly affect the product quality, then fewer number of samples or analyses may be required.

Determining the sampling scheme for the estimation objective is similar to that for the detection objective. Instead of a detectable difference, the required precision is defined as the halfwidth of a statistical confidence interval. Assuming that the RSDS and the RSDA are known from previous experience, the halfwidth of an $1-\alpha/2$ statistical confidence interval is computed as:

$$H = Z_{1-\alpha/2} * \sqrt{\frac{RSDS^2}{NS} + \frac{RSDA^2}{NS*NA}}$$

where $Z_{1-\alpha/2}$ is obtained from the normal distribution table, NS is the number of samples, and NA is the number of analyses per sample.

Reasonable estimates for the RSDS and RSDA for the major constituents are 2.5% and 10%, respectively. Given these estimates, the halfwidths of 95% confidence intervals were computed for several combinations of the number of samples and analyses and are shown in Table 3. For example, if two samples were obtained and two analyses were performed on each sample, then one would be 95% confident that the true constituent concentration was within ± 10.4% of the estimated concentration. If ± 10.4% of the estimated concentration is an acceptable degree of precision, then this sampling scheme would be appropriate.

Estimation is a major WFQ objective for the samples obtained from the SRAT and the WRCT. The contents of these tanks must be known well enough to determine the appropriate amount of frit to add in the SME. The imprecision in the determinations of these tank compositions will affect the ability to make an acceptable glass that has near-optimal durability properties. Currently the required precision for the SRAT and WRCT estimates of composition are not defined. Such precision requirements could be determined through process simulations which examine the propagation of the errors through the process to the variations in the glass composition[2]. Information from such process simulations coupled with the outcome of the glass composition variability studies could be used to determine the required degree of precision in the SRAT and WRCT estimates for producing an acceptable product.

REQUIRED SAMPLING TO MEET THE PROCESS CONTROL OBJECTIVE

The required sampling schemes for meeting the process control objectives remain to be examined through process simulations using the waste form qualification simulation model coupled with statistical process control methods. Although a shift or change in the process may not immediately jeopardize the operations or quality of the product, it could signal the start of a trend or develop into a long-term problem. Moreover, the process control system's ability to detect certain changes and shifts in the process due to mechanical failure, instrument failure, or operator errors may be important. Demonstrating that a process is in statistical control is fundamental in assuring reliable predictions of the glass composition from SME and MFT feed samples.

SUMMARY AND RECOMMENDATIONS

Samples from various HWVP process tanks will be collected to satisfy several objectives related to waste form qualification. The recommended sampling schemes depend on the purpose of the

Table 3. Confidence Interval Halfwidths for the Estimation Objective Assuming RSDS=2.5% and RSDA=10%.

Number of Samples	Number of Analyses/Sample	Total Analyses	95% Conf. Interval Halfwidth
2	1	2	14.3%
2	2	4	10.4%
2	3	6	8.7%
3	1	3	11.7%
3	2	6	8.5%
3	3	9	7.1%
4	1	4	10.1%
4	2	8	7.4%
4	3	12	6.2%
5	1	5	9.0%
5	2	10	6.6%
5	3	15	5.5%

samples. For each process tank, the purpose of sampling has been identified and classified into four statistical objectives: detection of compositions outside specification limits, prediction of final glass product composition, process control, and estimation of composition in process vessels for reporting and for guiding succeeding process steps. Sampling schemes which meet the detection, prediction, and estimation objectives have been identified for each process tank. Development of sampling schemes to support process control objectives will require additional evaluation.

Experience suggests that the analytical relative standard deviation (RSDA) is less than 10% for the major constituents (Si, B, Na, Fe, Zr, Li, and Al) and less than 30% for minor constituents, that the sampling relative standard deviation (RSDS) is approximately 2.5%, and that the batch-to-batch relative standard deviation (RSDB) in the glass (corresponding to the MFT batches) is conservatively 5%. For these conditions, the sampling schemes for each tank are shown below.

Tank	Number of Samples	Analyses per Sample
SRAT	4	2
WRCT	4	2
SME	4	2
MFT	5	2
Glass	1	3

This sampling/analytical scheme will assure that 1) for the estimation objective, one is 95% confident that the true elemental concentration is within ±7.5% of the estimated concentration, 2) for the detection objective, there is a 95% chance of detecting elemental concentrations that are at least 7.5% above/below a predefined upper/lower specification limit, and 3) for the prediction objective, a ± 20% halfwidth for a 95% confidence interval on the estimated glass elemental concentration will be achieved given that the SME and MFT sample estimates are combined with the glass sample estimates and that the melter can be modeled as a well-mixed tank. The estimation objective requires more total analyses than the detection objective such that the estimation objective drives the sampling strategy outlined for the SRAT, WRCT, and SME. The sampling scheme shown above will meet the estimation objective and will improve the detection objective capability. More stringent requirements on detection would require more samples and analyses while less stringent requirements would require fewer samples and analyses.

These sampling schemes are appropriate for the detection, prediction, and estimation objectives but may not be appropriate for the process control objectives, particularly for the SRAT and SME tanks. Further study on the required sampling and analysis schemes for process control objectives is recommended.

The sampling and analytical strategies are based on assumed values for the required precision and accuracy of the results (i.e., ±20% halfwidth for a 95% confidence interval, 95% chance of detecting concentrations 7.5% beyond a limit, 95% confidence that the true composition is within ±7.5% of the estimated composition). Required precisions and accuracies have not been defined but will be developed as glass variability studies are conducted and as process limits are defined.

In determining the sampling schemes presented here, it has been assumed that the uncertainties in measurements are known and are normally distributed. As uncertainty data become available from testing prior to production, these assumptions should be reviewed. If the RSDS and RSDA values are not well defined and must therefore be estimated from samples taken during production, then tabled values from a t-distribution should be used rather than those from a normal distribution.

The sampling and analysis strategies outlined in this report should be reviewed as the requirements develop. The estimates of sampling and analytical precision include several components of variation (long- and short-term instrument, long- and short-term procedural, etc.). It may be possible to eliminate some of the sources of sampling and analytical variations by modifying procedures and instrument parameters or by conducting concurrent analyses on well-characterized reference materials in order to make bias corrections. The required sampling schemes should be updated as the sampling and analytical precisions change.

REFERENCES

1. DOE-OCRWM. 1986. Waste Acceptance Preliminary Specifications for the Defense Waste Processing Facility High-Level Waste Form, OGR/B-8, U. S. Department of Energy-Office of Civilian Radioactive Waste Management, Washington, D.C.

2. J. H. Westsik, Jr., W. L. Kuhn, B. A. Pulsipher, and J. L. Nelson, 1988, "Evaluation of Strategies for Controlling Hanford Waste Vitrification Plant Glass by Process Simulation," Spectrum '88, American Nuclear Society.

COMPOSITIONAL ANALYSES: WHERE IS THE UNCERTAINTY COMING FROM?

Brent A. Pulsipher and Dennis L. Eggett
Pacific Northwest Laboratory[a]
P. O. Box 999
Richland, Washington 99352

Charles W. McVay, John H. Marlow, and Coleen L. Gerwitz
West Valley Nuclear Services
P. O. Box 191
West Valley, New York 14171-0191

ABSTRACT

The West Valley vitrification process is designed to solidify the high-level radioactive waste solids stored at West Valley, New York. As specified in the Waste Acceptance Preliminary Specifications, an estimate of the chemical composition of the waste glass is required and an "estimate of precision and accuracy, and the basis for the estimate of the precision shall be reported in the Waste Compliance Plan."

Results from a statistically designed experiment to quantify 1) short-term instrument variations, 2) long-term instrument variations, 3) short-term preparation variations, 4) long-term preparation variations, and 5) variations between replicate samples of the ground powder from a single glass sample are presented for ICP and AA analyses at West Valley. Besides determining the overall analytical precision, statistical analyses identified which major contributors to the uncertainty should be targeted for future improvements.

INTRODUCTION

The West Valley vitrification process is designed to solidify the high-level radioactive waste solids stored in underground storage tanks at West Valley, New York. The resulting glass

(a) Pacific Northwest Laboratory is operated for the U.S. Department of Energy by Battelle Memorial Institute under Contract DE-AC06-76RLO 1830.

product will be poured into canisters that will be sent to a deep-geological repository. The U.S. Department of Energy established the Waste Acceptance Process to ensure that high-level radioactive waste forms prepared from existing radioactive wastes will be acceptable for disposal. A waste acceptance committee prepared Waste Acceptance Preliminary Specifications (WAPS) as part of the waste acceptance process. As specified in the WAPS (Specification 1.1.2), an estimate of the chemical composition of the waste glass is required and an "estimate of precision and accuracy, and the basis for the estimate of the precision shall be reported in the Waste Compliance Plan."

Numerous factors could affect the precision of the estimates of chemical composition. Sampling methods may not always provide similar samples that are representative of the true composition. Variations between replicate samples may also be due to analytical uncertainties. It is important to quantify all major components of variation and provide a statistically defensible estimate of analytical precision and accuracy for waste form qualification purposes.

In addition to the waste form qualification objective, characterization of the analytical variation is necessary to ensure proper statistical designs for comparative studies currently being conducted at West Valley. If long-term analytical variations exist, then these studies must be designed to eliminate any confounding of the effects of interest with the inherent long-term analytical variations. Such considerations must be taken into account in order to ensure that the effects of long-term analytical variations are not mistakenly concluded to be manifestations of changes in experimental parameters.

In an effort to quantify the analytical precision associated with glass chemical composition determinations and isolate the several major sources of variation affecting the precision, a statistically designed test plan was developed. The test plan provides for estimating the analytical uncertainties from 1) grinding and sampling of the ground glass, 2) variations in preparation of the ground glass for analysis over the short term, 3) variations in preparation over the long term, 4) short-term variations in the analytical instrument, and 5) long-term instrument variations. The test plan also allows sample-to-sample differences to be and reported separately from the analytical precision. By isolating the major components of analytical uncertainty, not only can the overall analytical precision be defined, but direction will be provided for defining which step(s) in the analytical process should be targeted for future improvements. A program to ensure maintenance of current precision levels or to improve the precision can then be implemented.

POSSIBLE SOURCES OF VARIATION

This study focuses only on the Inductively Coupled Plasma (ICP) and Atomic Absorption (AA) methods for determining elemental composition. There are many steps in the ICP and AA analytical processes. For glass pieces that enter the analytical laboratory for ICP or AA analysis, one of the first preparation steps is the grinding of the piece into a fine powder. Potentially, the grinding process could be slightly different from piece to piece resulting in particle size differences between samples or slight cross contamination from "left over" material on the mortar and pestle. Such factors could introduce some variability (probably minor).

Next, an approximately 50-mg sample of the powder is obtained and weighed. The imprecision of the weighing device may have a noticeable effect on the precision of the analytical results. Another source of variation is the powder sample-to-sample differences that could exist. An analysis of only one sample from the powdered material may not be entirely representative of the true composition. A better representation may require an averaging of the analyses of replicate samples from the powdered material.

After weighing, the sample is transferred into a digestion vessel where nitric acid is added followed by an addition of hydrofluoric acid. The accuracy of the amounts and ratios of the acid additions, the transfer of the sample from the weighing boat into the digestion vessel, and the purity of the acids could be possible sources of variation.

Vessels are then heated in a microwave oven for approximately three minutes. Subsequent to this "microwave digestion" the vessels are removed, cooled, examined visually for complete dissolution, and transferred to a 100-mL volumetric flask. A scandium spike is added and the solution is further diluted with deionized water. In these final preparation steps, several sources could contribute to the analytical imprecision. The microwave time may have some effect, the visual examination may not be consistent across laboratory personnel or over time, the transfer of material from the vessels to the flask may not be entirely uniform, and further additions and dilutions may not be exact.

Finally the prepared solutions are analyzed by the ICP or AA instrument. Several factors may affect the precision of the ICP instrument. Changes in room temperature or humidity, nebulizer parameters, flame characteristics, flow parameters, personnel, standards, etc., may all have a potential effect on instrument precision. Likewise, other operating parameters for the AA method can affect the precision of the AA results.

Some of the possible sources of variation outlined above may have a negligible effect on the overall analytical precision. However, before passing judgement on the contribution to the overall uncertainty of these specific factors, the contributions of more generalized clusters of factors should be examined. An experiment to isolate and quantify all of the specific components of variation would be inefficient and infeasible. If most of the analytical uncertainty could be attributed to the instrument variations, then it would be pointless to focus effort on estimating the specific components of variance associated with the preparation steps. Likewise, if the analytical uncertainty is attributable to the preparation steps, efforts towards estimating the specific components of variance affecting the instrument precision would be wasted. Thus, the test matrix was designed to permit quantifying the general sources of variation that may affect the overall analytical precision.

The above specific possible sources of variation have been categorized into more general factors. The preparation steps have been split into two general steps: 1) grinding, and 2) digestion which includes sampling of the ground material, weighing, and all further steps to arrive at a sample ready for instrument analysis. The factors affecting the instrument precision have been combined into an overall instrument factor.

It is possible that some components of variation may only be exhibited over an extended period. For example, the analytical instrument may produce very reproducible results when analyses are conducted on the same day but significant variability may be present when similar aliquots are analyzed on different days. These long-term variations must be included in the overall estimate of analytical precision. Thus, the test matrix was designed to ensure some measure of long-term preparation and instrument variation.

Finally, differences between glass samples must be separable from the pure analytical uncertainties. The test design discussed in this paper allows the variations between glass samples to be extracted from the total uncertainty of the reported results. Additionally, to minimize this source of variation, a nearly "homogeneous" reference glass was used (MCC WV205 reference glass, similar to the MCC ATM 10 glass except nonradioactive).

EXPERIMENTAL TEST MATRIX

As outlined above, several underlying sources of uncertainty exist which may contribute to the overall analytical uncertainty. To improve the analytical precision, efforts should be focused on reducing the major underlying sources of uncertainty. Thus, estimates of the variation contributed by these underlying components of variance must be derived.

An overall estimate of analytical precision could be derived from the analysis of replicate grinds on similar material. However, such an estimate would be a conglomeration of subsampling, digestion, and instrument variation. To separately estimate each major component of variation, replication at appropriate places in the analytical process is necessary. The test matrix shown in Figure 1 provides the structure for estimating the overall analytical precision, as well as, isolating the major contributors to the uncertainty.

Sixteen days of testing over a 2.5-month period (approximately once every third work day) were proposed. On each day of testing, one glass sample is divided into two subsamples and both are submitted to the analytical laboratory. Each subsample is ground and two powder samples are obtained from each grind (four total samples per day). These four powder samples are then prepared for ICP (or AA) analysis. After the digestion step is complete, each solution is divided into two aliquots and the aliquots are analyzed (resulting in eight analyses per day).

The short-term instrument precision can be estimated from the analyses of the replicate aliquots from each solution. The short-term digestion step variation can be estimated from the replicate preparations per glass subsample. The grinding/ subsampling variation can be estimated from the replicate subsamples of glass. However, the glass sample-to-sample variations are confounded with the long-term instrument, digestion, and subsampling fluctuations.

To separate the long-term analytical uncertainties from the glass sample variations, additional tests as shown in Figure 1 are required. Before any tests began, a preparation day (day 0) in the analytical laboratory was required. One of the

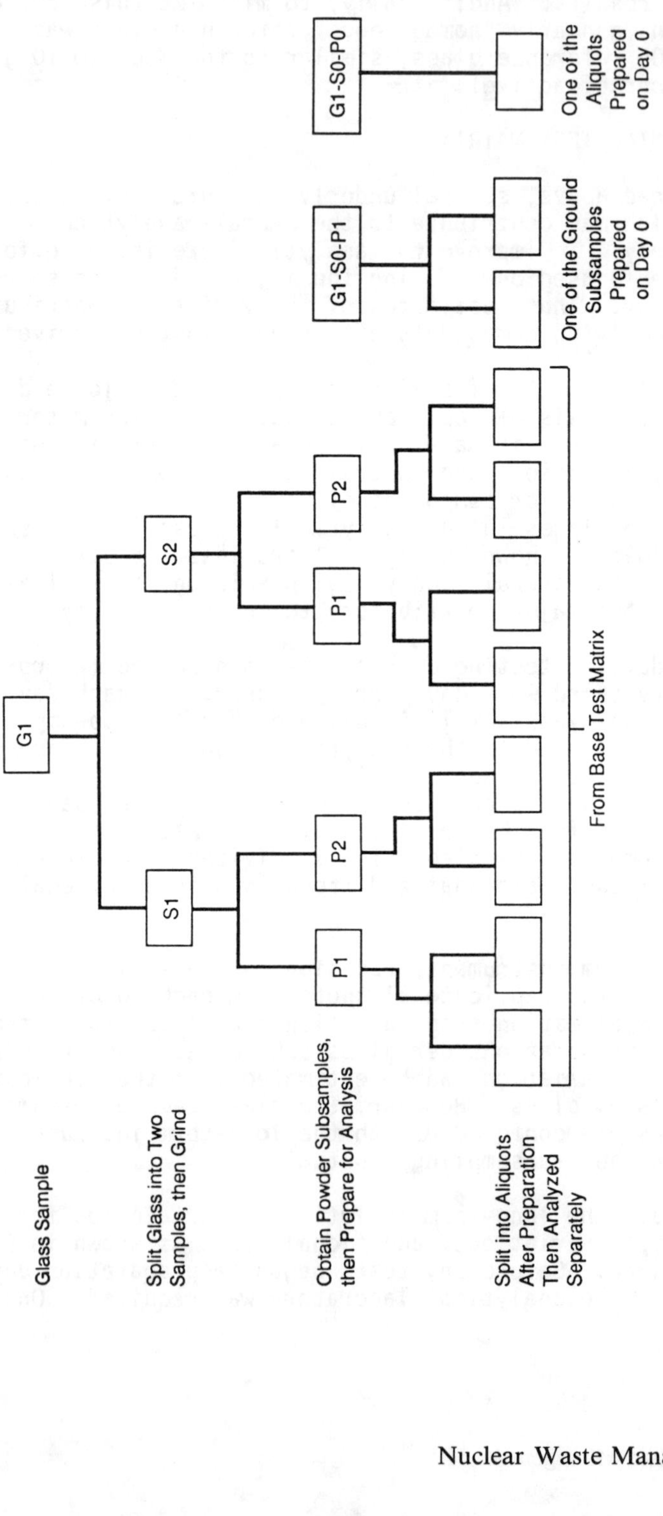

Figure 1. Complete Test Matrix for 1 Day

glass samples was ground in the analytical laboratory and divided into 17 subsamples. Sixteen of those subsamples were set aside for future digestion and analysis. The remaining one subsample was further prepared for analysis. Once the digestion and dilution steps were complete for that one subsample, the resulting solution was further divided into 16 aliquots, which were set aside for future analysis.

On each of the 16 days of testing, the digestion and dilution steps for one of the 16 powder subsamples that was set aside on day 0 was completed. After digestion and dilution, it was split into two aliquots and analyzed with the rest of the aliquots prepared on that day. Additionally, one of the 16 aliquots that was prepared on day 0 was analyzed for composition. The full test matrix for a single test day is shown in Figure 1.

STATISTICAL EVALUATION OF THE EXPERIMENTAL RESULTS

At the time of this paper, twelve of the sixteen test days were completed. A preliminary statistical analysis of the data compiled to date was conducted and the results are presented herein.

The statistical techniques of analysis of variance and variance component estimation (Ref. 1 and 2) were used to obtain the results shown in Tables 1 and 2. In Table 1, the results obtained using normalized data (weight percent oxides normalized to 100%) are presented and in Table 2 the results obtained using PPM data (concentration in parts per million divided by 10000) are shown. The overall average composition and total relative standard deviation (RSD) estimates are shown for each constituent. Only the precision of the analytical laboratory will be addressed in this paper. The accuracy will be addressed in a later report.

Because of the statistical design of the test matrix, several of the major components contributing to the overall analytical uncertainty could be isolated and quantified. The components of variation that are estimable are: 1) short-term instrument variations, 2) long-term instrument variations, 3) short-term preparation variations, 4) long-term preparation variations, 5) variations between halves of the same glass sample, and 6) variations between glass samples. The variations attributable to short-term preparation include the differences between powder subsamples obtained from the same ground glass subsample. The estimates of these components of variation are shown in Tables 1 and 2 as percentages of the total variation.

Table 1. Analytical Laboratory Uncertainty Estimates for the Normalized Results

ELEMENT	Mean Oxide Wt%	Total RSD[a]	Percent of Total Variance Due To: Short[b] Term Instr.	Long[c] Term Instr.	Short[d] Term Prep.	Long[e] Term Prep.	Split[f]	Glass[g]
Si	46.74	1.13	16.1	60.8	17.4	0.0	0.0	5.7
Fe	11.52	1.34	23.4	41.4	11.8	19.0	0.0	4.4
Na	10.47	2.82	17.1	75.9	7.0	0.0	0.0	0.0
B	9.95	1.91	11.1	62.3	8.6	0.0	0.0	18.0
Al	6.44	5.13	11.8	7.4	28.0	46.1	0.0	6.7
K	3.13	11.32	3.0	95.9	0.5	0.0	0.6	0.0
Li	2.78	2.76	37.2	60.1	2.3	0.0	0.5	0.0
P	2.28	3.89	27.0	69.0	2.9	0.0	0.0	1.1
Zr	1.44	1.23	28.0	50.3	9.6	0.0	12.1	0.0
Mg	1.20	11.22	10.1	0.0	49.9	40.0	0.0	0.0
Mn	1.02	2.47	4.5	77.8	11.2	3.8	0.0	2.7
Ti	0.92	1.81	13.4	33.2	16.8	35.6	1.0	0.0
Ca	0.66	30.88	0.3	0.0	99.7	0.0	0.0	0.0
Ni	0.31	8.26	6.1	75.0	0.2	18.0	0.7	0.0
S	0.28	33.57	12.0	87.9	0.0	0.0	0.0	0.0
Ce	0.24	25.48	30.7	54.5	3.9	6.2	4.7	0.0
Cr	0.24	23.00	20.7	54.5	0.0	22.4	2.5	0.0
Nd	0.19	27.28	49.6	12.6	0.0	36.2	1.5	0.0
Cs	0.07	19.64	21.9	58.1	12.7	7.3	0.0	0.0
Ba	0.05	13.51	30.0	67.5	2.5	0.0	0.0	0.0
Sr	0.03	5.37	41.2	42.0	11.5	0.0	5.3	0.0

(a) Total RSD is the total standard deviation relative to the mean wt% (100*s/mean) including all components of variation. This is the uncertainty that would be expected for an individual glass sample analyzed once on any given day.

(b) The short-term instrument variation is the percent of the total variation that is due to variations between replicate analyses on the same day.

(c) The long-term instrument variation is the percent of the total variation that is due to variations between replicate analyses of the same digestion sample conducted over different days.

(d) The short-term preparation variation is the percent of the total variation that is due to variations in sample preparation within the same day.

(e) The long-term preparation variation is the percent of the total variation that is due to variations in sample preparation that occur from day-to-day.

(f) The split variation is the percent of the total variation that is due to differences between the two glass subsamples (glass halves).

(g) The glass variation is the percent of the total variation that is due to differences between the actual glass samples.

Table 2. Analytical Laboratory Uncertainty Estimates for the PPM Results

ELEMENT	Mean PPM	Total RSD[a]	Short[b] Term Instr.	Long[c] Term Instr.	Short[d] Term Prep.	Long[e] Term Prep.	Split[f]	Glass[g]
			Percent of Total Variance Due To:					
Si	21.52	3.24	7.6	43.5	15.6	22.8	7.7	2.7
Fe	7.94	2.62	11.7	26.7	21.6	25.8	14.2	0.0
Na	7.59	3.79	11.6	44.7	7.5	21.9	6.3	7.9
Al	3.36	6.72	6.9	17.3	19.7	41.6	5.9	8.6
B	3.05	3.35	9.1	73.5	10.9	0.0	6.5	0.0
K	2.55	11.61	2.6	96.2	1.2	0.0	0.0	0.0
Li	1.28	4.39	11.3	58.2	3.9	22.5	4.0	0.0
Zr	1.06	3.00	2.5	40.6	14.5	30.1	12.3	0.0
P	0.98	3.82	35.9	46.4	17.7	0.0	0.0	0.0
Mn	0.78	2.33	14.8	36.5	39.5	0.0	9.2	0.0
Mg	0.71	11.56	9.8	0.0	47.2	37.2	5.7	0.0
Ti	0.54	2.32	2.9	9.4	34.1	18.2	35.4	0.0
Ca	0.47	31.23	0.2	0.0	99.8	0.0	0.0	0.0
Ni	0.24	9.28	6.8	50.0	1.2	39.4	2.5	0.0
Ce	0.20	25.52	32.4	47.2	3.0	11.0	6.4	0.0
Cr	0.16	22.10	24.1	47.3	0.0	23.9	4.8	0.0
S	0.11	34.47	12.2	86.6	0.0	0.0	1.2	0.0
Nd	0.11	27.79	50.2	7.2	0.0	39.8	2.9	0.0
Cs	0.07	20.54	19.5	47.8	11.3	21.4	0.0	0.0
Ba	0.05	13.81	32.2	59.9	0.5	7.4	0.0	0.0
Sr	0.02	6.18	33.7	16.3	12.4	27.7	9.9	0.0

(a) Total RSD is the total standard deviation relative to the mean ppm (100*s/mean) including all components of variation. This is the uncertainty that would be expected for an individual glass sample analyzed once on any given day.
(b) The short-term instrument variation is the percent of the total variation that is due to variations between replicate analyses on the same day.
(c) The long-term instrument variation is the percent of the total variation that is due to variations between replicate analyses of the same digestion sample conducted over different days.
(d) The short-term preparation variation is the percent of the total variation that is due to variations in sample preparation within the same day.
(e) The long-term preparation variation is the percent of the total variation that is due to variations in sample preparation that occur from day-to-day.
(f) The split variation is the percent of the total variation that is due to differences between the two glass subsamples (glass halves).
(g) The glass variation is the percent of the total variation that is due to differences between the actual glass samples.

In general, most of the analytical variation is due to instrument variations. This suggests that the laboratory personnel are doing an excellent job in sample preparation and handling but are greatly hampered by the lack of precision inherent in the analytical instrument. Particularly, the long-term instrument variation is a major contributor to the overall uncertainty of the results presented in this paper. The long-term instrument variation can be caused by several factors. It may be that the instrument is extremely sensitive to changes in environmental conditions. If a number of laboratory personnel use the instrument, different instrument settings may be used at different times, causing long-term instrument variation. Recalibration models and schedules could also be a possible source of long-term instrument variations. Another possible source of long-term instrument variation is the method of blank-correction. The effect of blank-correction will not be addressed in this paper but will be examined later by comparing the precision estimates with and without blank correction.

Significant preparation variations were apparent for some of the elemental determinations. For the normalized wt% results, the preparation variation was large (>20%) for Fe_2O_3, Al_2O_3, MgO, TiO_2, CaO, Cr_2O_3, and Nd_2O_3. For the PPM results, the preparation variation was significant for more of the elements. Thus, normalizing to 100% appeared to smooth out the preparation variations for many of the constituents.

Overall, the variations between glass samples and halves of the same glass sample (splits) were negligible. This suggests that the reference bar of glass from which samples were obtained was very homogeneous.

CONCLUSIONS AND RECOMMENDATIONS

The total analytical precision for ICP and AA glass sample analyses was determined based on data obtained using current laboratory practices and methods at West Valley. The total relative standard deviation estimates are less than 5% for the major constituents (normalized >1.5 wt%) in the glass (with the exception of potassium and possibly aluminum) and less than 30% for the minor constituents (with the exception of sulfur).

In general, instrument variation accounted for the majority of the total analytical uncertainty. Long-term instrument variations were a major contributor to the instrument variations. If necessary, future efforts to reduce analytical uncertainty should be focused on identifying and eliminating factors that affect long-term instrument variations. Because it may prove difficult to identify and eliminate all factors causing long-term variations, it may prove beneficial to concurrently analyze a reference sample of glass with the samples of interest and bias-correct using the reference sample analysis result. This may be feasible given the fact that the statistical analysis indicated that the reference glass used in this experiment was very homogeneous. Such a reference material would need to have composition similar to the samples of interest and would require careful characterization.

Experimenters must account for the long-term analytical instrument variations that exist when planning and conducting their experiments. Frequently, experimenters have ignored such variations. They often confound the effects of long-term instrument variations with the effects of varying the experimental parameters. Such practices can cause erroneous conclusions because the manifested effects are interpreted as effects resulting from changing the experimental parameters when, in fact, they may be due only to inherent long-term analytical variations.

The precision estimates presented herein are applicable only to glass sample analyses of glasses with compositions similar to the WV205 glass analyzed at the WVNS analytical laboratory. The estimates of precision are not transferrable to other laboratories. A similar study is planned to characterize the analytical precision of feed slurry samples. Because there are additional preparation steps required for analyzing the feed samples, one might expect greater variability result from the preparation.

This paper does not address the accuracy (bias) of the West Valley Nuclear Services (WVNS) analytical laboratory. Although certain components of variation may not have a significant effect on the analytical precision, they could have an effect on the accuracy. The estimated components of variation reported in this paper pertain only to the precision (not accuracy) of the WVNS analytical laboratory.

REFERENCES

1. Ostle, B., and R. W. Mensing. 1975. Statistics in Research. Iowa State University Press, Ames, Iowa.

2. Bowen, W. M., and C. A. Bennett. 1988. Statistical Methods for Nuclear Material Management. NUREG/CR-4604, Office of Nuclear Regulatory Research, Washington, DC.

Section VII. Glass Waste Form Processing

WASTE MANAGEMENT STRATEGY AT THE WEST VALLEY DEMONSTRATION PROJECT

Eli Maestas
West Valley Project Office
U. S. Department of Energy
West Valley, NY 14171-0191

Henry F. Walter
Office of Remedial Action and Waste Technology
U. S. Department of Energy
Washington, DC 20545

ABSTRACT

High-level radioactive waste stored at West Valley (New York) is to be immobilized in a durable borosilicate glass for later emplacement in a federal repository. Technical choices concerning waste forms and solidification processes for both high-level and low-level waste and the selected options will be discussed.

HOW WEST VALLEY CAME TO BE

In 1958 the Saint Lawrence Seaway opened and Nelson Rockefeller was elected governor of New York. Western New York, which had been a rail-water trans-shipping place for grain and minerals, was adversely impacted by the Seaway because ocean going ships from abroad could sail directly to other ports on the Great Lakes. Sensing the need for new industry in the area, believing that nuclear technology was the wave of the future, and encouraged by a federal promise of shared technology, New York State planned an integrated nuclear regional industrial park, beginning with a facility which would recover fissionable elements and radioisotopes from spent fuel and turn them into useful products for industry and medicine.

The Western New York Nuclear Service Center, owned by the New York State Energy Research and Development Authority, was sited at a geologically favorable location near West Valley, New

York, some 40 miles south of Buffalo. An operating contractor was selected, the plant was completed in 1966, operated until 1972, and then shut down to expand its capacity. The proposed expansion was never completed. The reasons are many and diverse: determination that expansion required an Environmental Impact Statement, questions about the seismic resistance of the buildings, changes in public perception of nuclear power, falling prices for uranium which brought the economics of reprocessing into question, and finally the 1977 federal policy decision to halt reprocessing of spent nuclear fuel by private industry. As a result, the operating contractor announced its intention not to renew the lease when it expired in 1980.

Over the next two years, several studies examined the technical options for dealing with the waste. The West Valley Demonstration Project Act (PL 96-368; October 1, 1980) established within the U. S. Department of Energy a demonstration program with five specific tasks: (1) to solidify the liquid high-level waste, (2) to develop suitable containers, (3) to transport the solidified waste to a federal site, (4) to decontaminate and decommission the tanks and facilities used in the solidification, and (5) to dispose of project low-level and transuranic waste.

The project is directed by the Department of Energy and funded jointly by the Federal government (90%) and New York State (10%). (Planning began in 1978 for a similar facility for solidifying defense wastes, the Defense Waste Processing Facility now nearing completion at the Savannah River Plant, South Carolina. Further, the West Valley Act predated, by two years, the Nuclear Waste Policy Act which established the Federal program for permanent disposal of high-level waste in a geologic repository.)

The operating contractor for the Project is West Valley Nuclear Services Company, Inc., a wholly owned subsidiary of Westinghouse Electric Corporation. WVNS is supported by Pacific Northwest Laboratory, Dames and Moore, Ebasco Services, Catholic University of America, and Alfred University.

WEST VALLEY WASTE

The chemical separation process used at West Valley reflected then conventional methodology for reprocessing spent nuclear fuel: dissolution of fuel in hot nitric acid, recovery of fissile elements by solvent extraction, neutralization of the waste with caustic, and storage in carbon steel underground tanks. During six years of operation, processing of 640 metric tons of fuel yielded about 2.2 million litres of alkaline liquid high-level waste now stored in two underground tanks. Table 1 gives details of the 31 million Curie inventory. (This estimate is based upon plant records and calculations. Analysis of samples of waste suggest there is about 5% less activity.) Almost all the radioactivity is due to Sr-90/Y-90 and Cs-137/Ba-137m; mixed fission products and transuranic elements contribute less than 2%.

Table 1. Estimated Radioactivity of West Valley Waste (1987)

Fission Products		Transuranic Elements	
Nuclide	Mega Curies	Nuclide	Mega Curies
Cs	8.9	Pu	0.0734
Ba	8.4	Cm	0.0088
Sr	6.7	Am	0.0206
Y	6.7	Np	0.0002
Sm	0.2	U	0.0000
Eu	0.1		
Pm	0.1		
Others *	0.0		
TOTAL FISSION	31.1	TOTAL TRU	0.1030

* Sb, Te, Tc, H, Ru, Rh, Zr, Nb, Ce, Pr

BOROSILICATE GLASS HIGH-LEVEL WASTE FORM

In 1983, the Department selected borosilicate glass (25% nominal waste loading) as the waste form for the West Valley high-level waste. Borosilicate glass is durable, able to immobilize a wide variety of metal ions, and is processable remotely at atmospheric pressure and accessible temperatures using current technology. This decision relied on an extensive data base developed by the Department, beginning in the early 1970s for power reactor wastes and later for existing defense wastes. This data base was first used formally for selecting a waste form for the Defense Waste Processing Facility.

The waste glass contains silica as its basic ingredient with additives to give it particular properties: alumina to increase durability, boron for easier processing, alkalies to control viscosity, and the waste which is to be immobilized. Desirable properties which had to be balanced against each other included operating temperature, chemical durability, and waste loading. Higher temperature favors homogeneity and throughput but shorter lifetime for melter and refractory; higher alumina increases durability but decreases ease of processing; higher loading means fewer canisters but a smaller compositional region of acceptable durability.

Glass formulation is currently in the refinement stage at West Valley. The reference formula, given in Table 2, has been finalized. Small adjustments to composition and operating parameters are being made so as to identify the region where quality product can be made routinely under stable and controllable operating conditions.

Table 2. Composition of West Valley Reference Borosilicate Glass Wasteform

	Weight Percent Concentration	
Oxide of	Radioactive	Nonradioactive
Silicon	41.24	43.35
Iron	12.12	12.12
Sodium	11.42	11.42
Boron	10.26	10.26
Aluminum	6.54	6.54
Potassium	3.73	3.73
Lithium	3.15	3.15
Phosphorous	2.36	2.36
Thorium	3.61	0.00
Fission Products	0.58	0.58 *
Transuranic	0.61	0.00
Zirconium	0.19	2.30 **
Balance	4.19	4.19
TOTAL	100.00	100.00

* Nonradioactive isotopes used.
** In nonradioactive test glass, zirconium is a surrogate for thorium and transuranic elements.

LIQUID FED CERAMIC MELTER VITRIFICATION PROCESS

In 1983, the Department selected the liquid fed joule heated ceramic melter, developed at the Pacific Northwest Laboratory, as the vitrification process. The ceramic melter dissipates the joule heat efficiently to the glass and induces strong internal stirring. he same melter process was also selected for the Defense Waste Processing Facility. By controlling the relative flow of energy between the one bottom and two top electrodes, heat is applied to melt the feed, to assure melt turnover, to establish homogeneity in the melt, to boil off water, and to decompose nitrates and carbonates.

Since 1984, the Project has operated a full sized ceramic melter for testing and training purposes. Currently, the integrated system is being tested in nonradioactive operation to confirm that the melter and its associated components for feed adjustment, feed rate control, off-gas treatment, and process control will meet the operating requirements. Also, activities are underway to convert the test facility into a shielded cell for radioactive production operations, which are currently scheduled to begin in the mid-1990s.

The Department of Energy has established a formal process to assure coordination among interested parties of activities related to determining and verifying waste form quality. Preliminary specifications for the waste form have been defined and reference product quality tests identified. Demonstration that the production waste form has the required qualities will involve a combination of theoretical and empirical modeling of the melter system and the glass product and also comparison of nonradioactive glass produced under reference conditions with radioactive glass produced under the same conditions. Eventually, production records will need to show continuing compliance of the system with specifications.

CEMENT LOW-LEVEL WASTE FORM

Because of former operating practices, the West Valley waste is relatively high in sulfate ion. Incorporating the liquid waste directly into glass is constrained by the limited solubility of sulfate ion in the glass. Removing the bulk of the nonradioactive sodium salts (sulfate, nitrate, nitrite, and hydroxide ions) shrinks the number of glass canisters from 1500 to 300 but creates additional low-level waste of about 4000 cubic meters. This waste is being immobilized in a cement waste form. Figure 1 illustrates the details of this strategy for handling the two waste streams.

Removal of the excess supernatant liquid and salts began in May 1988 and will continue for about 2 years. Radiocesium, the only radionuclide present in significant amounts in the supernate, is retained on a tailored zeolite ion exchange medium, to be combined at a later time with the washed sludge and glass forming chemicals and sent to the melter. The decontaminated salt solution is adjusted to 39% (wt/wt) concentration, blended with cement and chemical additives in a high-shear mixer, poured into square drums of 350 litre

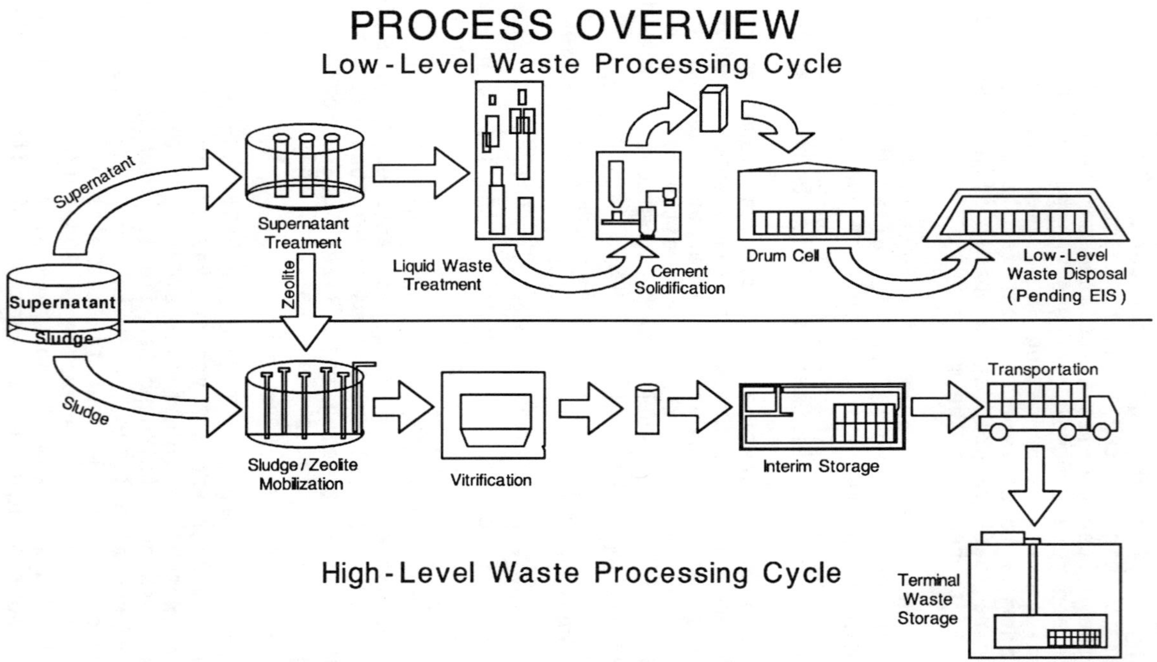

Figure 1. Processing strategy for West Valley waste. The liquid waste is being separated into a low-level waste stream and a high-level waste stream. The low-level stream, which contains sodium salts and small amounts of radiocesium, is being solidified in a cement waste form. The high-level stream containing the radionuclides will be mixed with glass forming chemicals and processed to a borosilicate glass waste form.

capacity, allowed to set, and then moved to a shielded storage facility. The final product contains about 18% (wt/wt) of salt. About 12,000 drums will be produced. Staff of the Nuclear Regulatory Commission have reviewed and approved the qualification data which demonstrates that this waste form exceeds the technical requirements for land disposal of low-level waste given in 10 CFR 61. A sampling program will confirm continued adherence to specifications.

OTHER PROJECT ACTIVITIES

Decontamination and decommissioning activities in the early years of the project were aimed at recovering space for project reuse. Later, after the high level waste is vitrified, all the facilities used in the project are to be decontaminated and decommissioned to levels to be established by the Nuclear Regulatory Commission.

Some of the contaminated equipment is sufficiently radioactive that it will have to be handled by remote equipment and behind shielding. It is planned to use the space of the former spent fuel storage pool for size reduction of large process vessels, piping, and equipment. As a first step to doing this, 625 spent fuel assemblies were returned to the owning utilities 1986 . The remaining 125 assemblies of spent fuel will be shipped by the end of 1990 to the Idaho National Engineering Laboratory in a demonstration of the feasibility of using a single cask for both transportation and storage.

In 1982, the Department issued a Final Environmental Impact Statement for the high-level waste solidification portion of the project. In 1988, the Department began preparation, jointly with New York State, of an Environmental Impact Statement for disposal of project low-level waste and project closure by the Department and for site closure by New York State.

Ultimately, the canisters of high-level waste will be sent to the Federal geological repository for permanent disposal. Responsibility for site long-term maintenance and disposition remains with New York State.

CORRELATION OF LABORATORY TESTING AND ACTUAL OPERATIONS FOR THE WEST VALLEY SUPERNATANT TREATMENT SYSTEM[a]

D.E. Kurath, L.A. Bray, W.A. Ross
Pacific Northwest Laboratory, Richland, WA 99352 (509) 376-8309

D.K. Ploetz
West Valley Nuclear Services Co., Inc., West Valley, NY 14171-0191
(716) 942-4276

ABSTRACT

An ion exchange process which removes Cs-137 from the supernatant liquid associated with neutralized PUREX high-level waste at the West Valley Demonstration Project was developed at the Pacific Northwest Laboratory. West Valley has now achieved initial operation of the supernatant treatment system, and results have confirmed the laboratory studies. This paper addresses the development of the predictions of system operations and the initial results of the system.

INTRODUCTION

West Valley Nuclear Services (WVNS) has begun processing the 2.1 million L of supernatant (Table I) associated with neutralized PUREX high-level waste at the West Valley Demonstration Project. The process employs a zeolite ion exchange process to remove the Cs-137 from the supernatant. The decontaminated supernatant is being solidified as cement, and the zeolite will be vitrified along with the PUREX sludge and THOREX waste, other HLW, and glass-forming materials.

The main objective of the supernatant treatment system (STS) is to remove at least 99.9% of the radioactive cesium while using less than 47,000 kg (all zeolite masses are given on an anhydrous basis) of the zeolite (IE-96) ion exchange material. This limit is necessary to avoid increasing the amount of glass produced.

(a) Work supported by the U.S. Department of Energy.

Table I. Major Components of the Supernatant

Ion	West Valley Supernatant	Synthetic Supernatant
Na^+	6.9	≈7.0
K^+	1.9 E-1	1.6 E-1
Rb^+	1.1 E-4	1.1 E-4
Cs^+	1.4 E-3	1.1 E-3
NO_3^-	3.5	3.4
NO_2^-	2.1	2.2
$CO_3^=$	3.4 E-1	3.9 E-1
$SO_4^=$	3.0 E-1	2.5 E-1
pH	10 to 11	10 to 11
Cs-137	3.38 Ci/L	Trace

The ion exchange process[1] was developed at the Pacific Northwest Laboratory using a combination of laboratory testing with synthetic supernatant and mathematical modeling. This paper combines these laboratory results with the operating experience to date to more accurately project the zeolite requirement.

SUPERNATANT TREATMENT SYSTEM PROCESS DESCRIPTION

Process Flow Diagram

Figure 1 is a simplified schematic of the STS flow diagram. The supernatant in Tank 8D2 is separated from the sludge with a combination of decanting and filtration and pumped to the feed

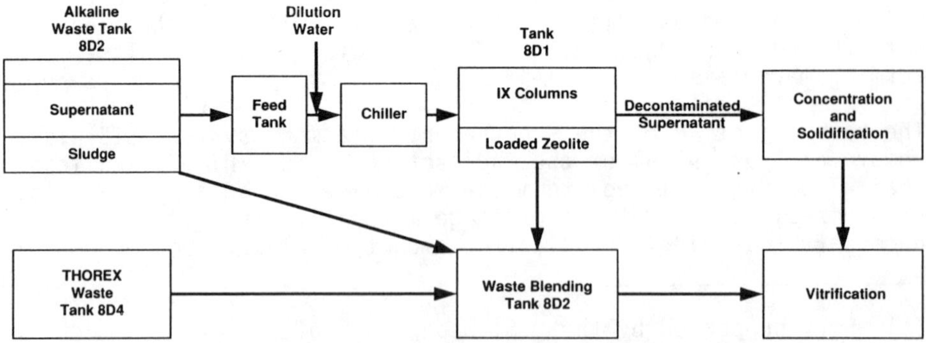

Figure 1. Process Flow Diagram of the Supernatant Treatment System

tank. Dilution water may be added to the supernatant to improve the cesium capacity of the zeolite, and the solution is then chilled to a temperature of 10°C to 12°C before it enters the ion exchange columns located in Tank 8D1. The cesium-loaded zeolite is dumped to Tank 8D1, and the decontaminated supernatant is concentrated and solidified for final disposal. When the bulk of the supernatant in Tank 8D2 has been processed, the sludge will be washed to remove interstitial liquid and soluble materials. The wash solution will also be processed through the ion exchange columns. Ultimately, the washed sludge, the loaded zeolite, and the acidic thorex waste in Tank 8D4 will be blended for vitrification and final disposal in a U.S. DOE repository.

Ion Exchange Column Description/Operation

The STS uses a series of four ion exchange columns for contacting the supernatant with the zeolite. Each column is ≈4 m long, with an ID of 0.93 m, and is typically filled with 1300 kg of zeolite IE-96. The volume of this zeolite is about 1700 L. The four columns are connected in series with the chilled supernatant flowing down through each column at a rate of 12 to 25 L/min. The supernatant is collected near the bottom of each column, and the effluent is routed to the top of the downstream column. The flow of feed is maintained until the Cs-137 breakthrough is greater than 95% from the lead column while maintaining a Cs-137 decontamination factor (DF) >1500. The DF is defined as the concentration of Cs-137 in the feed/concentration of Cs-137 in the effluent. At this point the columns are flushed with demineralized water and the cesium loaded zeolite in the lead column is dumped to Tank 8D-1. The now empty lead column is placed at the end of the column series and filled with fresh zeolite for the next loading cycle.

ION EXCHANGE BACKGROUND

In a column operation such as the STS, the exchangeable ion (cesium) is continuously removed and accumulates on the exchanger (zeolite). Initially all cesium is removed in the top portion of the zeolite bed. As processing continues, the zeolite becomes loaded with cesium and the cesium level in the effluent starts to rise. Eventually the zeolite would be completely loaded and the cesium in the effluent would be the same as the concentration in the feed. This behavior results in the characteristic S-shaped breakthrough curve shown in Figure 2. A given effluent concentration (C) is typically denoted as a percent of the inlet feed concentration (Co):

$$\% \text{ breakthrough} = C/C_o$$

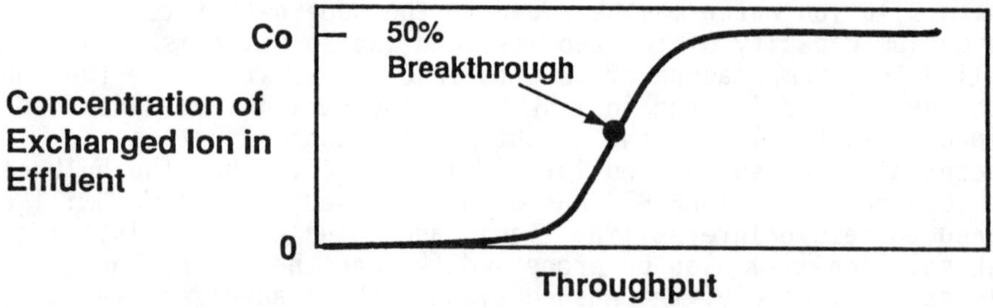

Figure 2. Typical Breakthrough Curve for Ion Exchange

The column loading behavior of zeolite materials can be approximated by the following equation (Buckingham, 1967).

$$C/Co = 1/2[1 + P_i(A \ln N/No)]$$

where $P_i(x)$ = normal probability integral
A = slope of the line on log-probability paper, dimensionless
N = bed volumes of feed solution
No = bed volumes of feed at 50% breakthrough.

This equation indicates that the breakthrough curves for multiple loading cycles in operation of the STS should give a sequence of straight lines on log-probability paper.

The capacity of an ion exchange material is usually given as the volumetric distribution coefficient (Kd). The cesium distribution coefficient, which is conveniently determined by simple batch experiments, is defined as:

$$Kd = \frac{\text{Cesium concentration on the zeolite}}{\text{Cesium concentration in the solution}} * (\rho_b) \; [=] \; \text{Dimensionless}$$

where ρ_b is the anhydrous density of the zeolite.

The point No on the breakthrough curve is related to Kd by the Equation (2)

$$Kd = No[\exp(\ 1/2A^2\ )]$$

In most applications, A is large enough that Kd ≈ No so that Kd is simply the column volumes (cv) of supernatant that can be processed per column volume of zeolite.

EXPERIMENTAL TEST EQUIPMENT/PROCEDURE

Batch Testing

Numerous simple batch tests were performed to determine the equilibrium behavior of the zeolite under various conditions. The batch procedure involved adding a known quantity (≈ 0.1 g) of unloaded zeolite to 10 mL of synthetic supernatant. After 48 hr of contact time with periodic agitation, the amount of Cs-137 remaining in solution was measured and the amount of Cs-137 adsorbed on the zeolite was determined by the difference. This allowed a calculation of Kd. The anhydrous density of the zeolite IE-96 was found to be 0.73 g/mL.

Laboratory-Scale Column Testing

The PNL laboratory-scale column system shown in Figure 3 was used to simulate the full-scale process conditions. The test equipment consisted of four glass columns, 1 in. in ID, connected in series. The temperature of the columns was controlled by circulating water through the jackets. Each column was packed with 200 mL (146 g anhydrous) of the zeolite IE-96. Supernatant was fed to the

Figure 3. Laboratory Ion Exchange Columns

columns by means of a pressurized (3 psig) feed bottle and was controlled using a needle valve. The flow rate was controlled to 0.6 cv/hr, which was the expected process-scale flow rate.

RESULTS

Temperature Effect

The batch distribution coefficients were determined at several temperatures and are shown in the form of the Arrhenius equation (Figure 4). The plot shows that the distribution coefficient increases exponentially with a decrease in temperature. This demonstrates the importance of operating at low temperature since a higher distribution coefficient leads to a higher cesium loading on the zeolite and consequently a reduced zeolite requirement.

Dilution Effect

It was also found that supernatant dilution will increase the cesium capacity of the zeolite. It is thought that this is a

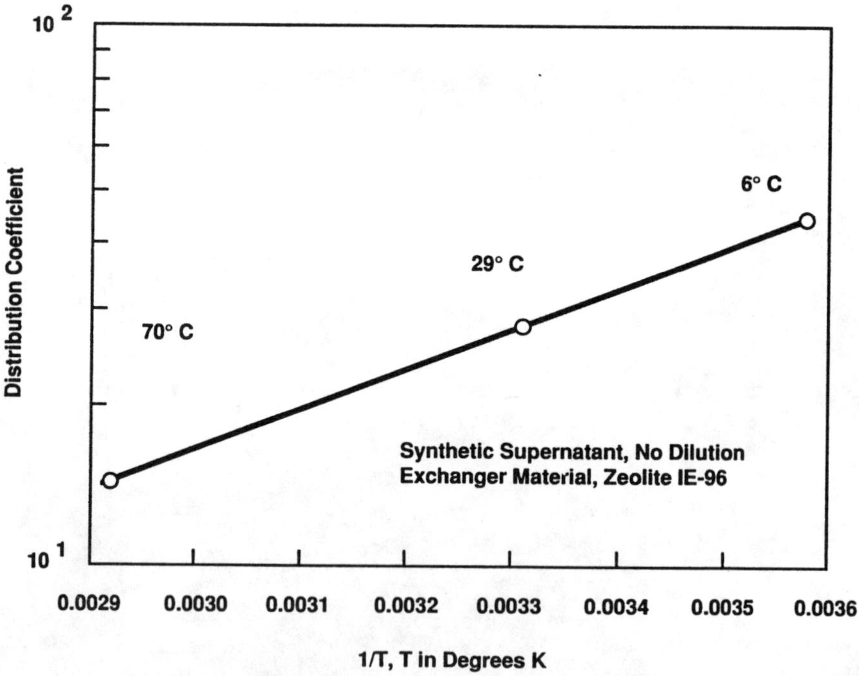

Figure 4. Cesium Distribution Coefficient as a Function of Temperature

result of the non-ideal behavior of the activity coefficients that determine the equilibrium of the Cs-137 between the supernatant and the zeolite. The laboratory data at 6°C indicated that the Kd with no dilution was 45 and the Kd with solution diluted 2:1 was 65. Both Kd's are given as the volume of undiluted solution/volume of zeolite. These data project an increase in the zeolite capacity of about 45% in going from undiluted to supernatant diluted 2:1.

Column Testing

Figure 5 compares the breakthrough curves for the first two supernatant processing runs to the curves obtained with the laboratory columns. The fact that the curves are nearly identical probably results from compensating differences in operation between the two

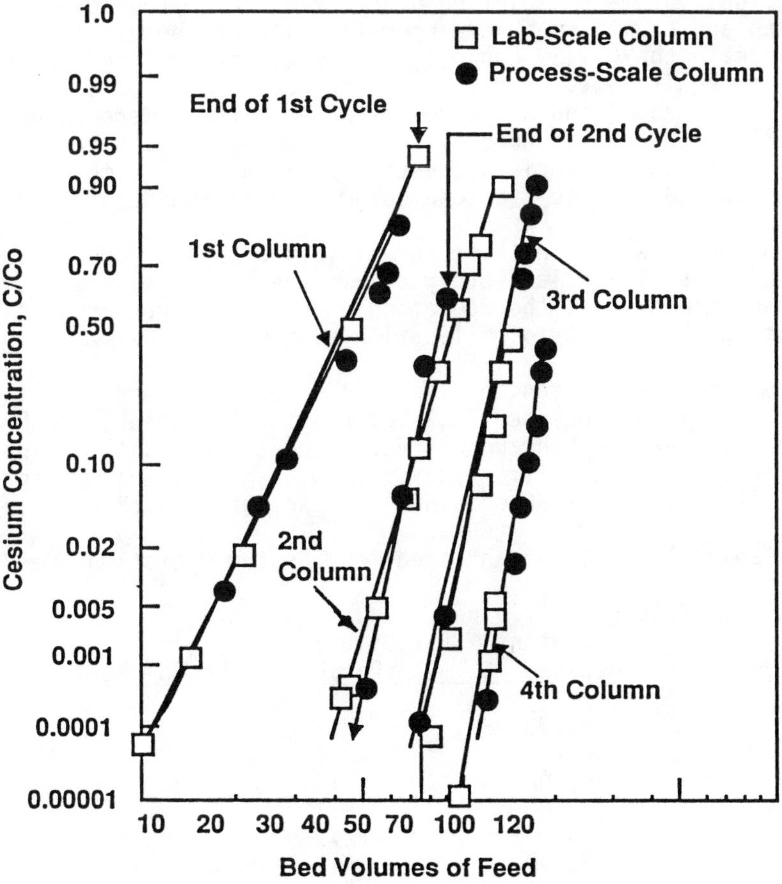

Figure 5. Comparison of Breakthrough Curves

sets of columns. The process columns were operated at a higher temperature (18°C) than the lab columns (6°C), which would decrease the zeolite capacity by ~20%. The dilution of the supernatant was ~20%, which offset this effect.

Supernatant Treatment System Operation

The results of the STS operation to date are summarized in Table II. Before the start of processing, the supernatant was partially diluted by adding 380,000 L of condensate from Tank 8D1, and another 380,000 L was added before the start of run 6. The volumes processed during each run are given as the volume of original undiluted supernatant. A dilution factor (volume of dilution water:volume of original supernatant) is given to indicate the actual volume of solution processed through the columns. The seventh and eighth runs were operated with deliberate feed dilution to increase the Cs-137 capacity of the zeolite and reduce the zeolite requirement. Portions of the decontaminated supernatant from runs 1 and 4 and all supernatant from run 5 were returned to Tank 8D2 because of inadequate cesium removal. Data from runs 5 and 6 were not available because of difficulty in removing all the cesium-loaded zeolite from one column during run 5.

Most of the DFs are in the range of 10,000 to 100,000 as predicted on the basis of the laboratory column tests. This is significant because it minimizes the radiation exposure of the personnel operating the cement solidification system.

The cesium distribution coefficients were determined from the bed volumes of feed processed (No) between the 50% breakthrough point for successive breakthrough curves. The cesium distribution coefficients for the first four runs averaged 35 which compares well with a value of 36 obtained from an interpolation of the

Table II. Supernatant Treatment System Operation Summary

Run	Volume of Original Supernatant Processed, L	Dilution Ratio	Cesium Kd	DF
1	75,700	0.2:1	34	25,000
2	54,700	0.2:1	28	180,000
3	80,500	0.2:1	38	40,000
4	43,700	0.2:1	42	4,300
5	NA	0.2:1	NA	NA
6	49,600	0.4:1	NA	11,100
7	93,800	2.0:1	60	75,000
8	100,000	2.0:1	57	45,350

laboratory data. The average value of the cesium distribution coefficient for runs 7 and 8 (58) is somewhat more than the 55 predicted from an interpolation of laboratory data, but within the range of estimated error for the laboratory and process data.

Estimate of Zeolite Requirement

Based on the STS operational experience to date, the total zeolite requirement for the supernatant processing is estimated to be 42,500 kg. Some details of this estimate are shown in Table III. The number of remaining supernatant processing runs was projected from the volume of feed processed between the 50% breakthrough point in runs 6,7 and 8. It is estimated that four column volumes of zeolite will be required to process the sludge washing solutions and that there will be two partially loaded columns at the end of processing. The total zeolite requirement is thus estimated to be 4500 kg (more than three column volumes) less than the goal of 47,000 kg.

Table III. Projected Zeolite Requirement

	Column Volumes of Zeolite	Zeolite IE-96, kg, anhydrous
Cold testing	--	900
Runs 1-8	9	11,700
Remaining runs	17	22,100
Sludge washing	4	5,200
Remaining after processing	2	2,600
Total		42,500

CONCLUSIONS

WVNS operations have demonstrated the effectiveness of a multiple-column zeolite ion exchange system for removing cesium from the supernatant associated with neutralized PUREX high-level waste. Nearly all the DFs are in the range of 10,000 to 100,000, as predicted.

Based on the laboratory data and the existing STS operational data, decontamination of the supernatant should be completed using 42,500 kg (anhydrous basis) of zeolite IE-96. This will meet the processing goal of using less than 47,000 kg and provides a contingency of more than three column volumes (4500 kg, anhydrous).

The zeolite requirement can be minimized by continuing to dilute the supernatant and by operating the ion exchange columns at reduced temperature.

The usefulness of laboratory development of the ion exchange process has been demonstrated by the correlation between the laboratory data and the process-scale results.

REFERENCES

1. L. A. Bray, L. K. Holton, Jr., T. R. Myers, G. M. Richardson, B. M. Wise, Experimental Data Developed To Support the Selection of a Treatment Process for West Valley Alkaline Supernatant, PNL-4969, January 1984, Pacific Northwest Laboratory, Richland, WA 99352.

2. J. S. Buckingham, ed., Waste Management Technical Manual, ISO-100, Isochem Inc., Richland, WA 99352.

SLURRY FEED VARIABILITY IN WEST VALLEY'S MELTER FEED TANK AND SAMPLING SYSTEM

C. L. Fow, D. E. Kurath, B. A. Pulsipher
Pacific Northwest Laboratory*, Richland, WA 99352
(509) 376-8309

B. P. Bauer
West Valley Nuclear Services, West Valley, NY 14171
(716) 942-3235

ABSTRACT

The present plan for disposal of high-level wastes at West Valley is to vitrify the wastes for disposal in a deep geologic repository. The vitrification process involves mixing the high-level wastes with glass-forming chemicals and feeding the resulting slurry to a liquid-fed ceramic melter. Maintaining the quality of the glass product and proficient melter operation depends on the ability of the melter feed system to produce and maintain a homogeneous mixture of waste and glass-former materials.

To investigate the mixing properties of the melter feed preparation system at West Valley, a statistically designed experiment was conducted using synthetic melter feed slurry over a range of concentrations. On the basis of the statistical data analysis, it was found that 1) a homogeneous slurry is produced in the melter feed tank, 2) the liquid-sampling system provides slurry samples that are statistically different from the slurry in the tank, and 3) analytical measurements are the major source of variability. A statistical quality control program for the analytical laboratory and a characterization test of the actual sampling system is recommended.

---

* Pacific Northwest Laboratory is operated for the U.S. Department of Energy by Battelle Memorial Institute under Contract DE-AC06-76RLO 1830.

BACKGROUND AND OBJECTIVES

High-level radioactive liquid wastes are stored in underground tanks at several Department of Energy facilities in the United States, including the Savannah River Plant (SRP) in Aiken, South Carolina; the West Valley Demonstration Project (West Valley) in West Valley, New York; and the Hanford Site near Richland, Washington. The current plan for managing high-level radioactive wastes is to convert the waste to a durable borosilicate glass, and isolate the glass in a deep geologic repository.

The process at West Valley for converting the wastes to glass involves mixing the high-level wastes with glass-forming chemicals. The resulting slurry is then transferred to a melter feed tank and processed through a high-temperature glass furnace known as a slurry-fed ceramic melter. Maintaining a consistent quality of the glass product and safe, efficient operation of the melter depends on the ability of the melter slurry preparation and feed systems to produce and maintain a homogeneous mixture of waste and glass-former materials. A schematic of a typical preparation and feed system is presented in Figure 1.

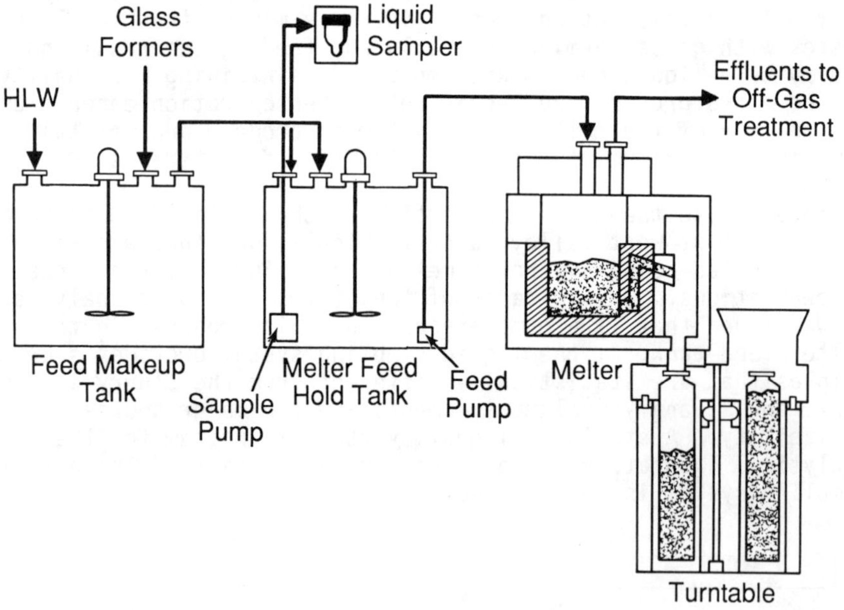

Figure 1. Typical Melter Feed Preparation and Feed System

In FY 1987, Pacific Northwest Laboratory and West Valley conducted full-scale tests at West Valley using their melter feed hold tank (MFHT). The objectives of these tests were to 1) quantify the homogeneity of the melter slurry in the MFHT, 2) determine the representativeness of the slurry sample obtained by the liquid-sampling system in the MFHT, and 3) determine the time required to resuspend the solids and homogenize the melter slurry in the MFHT after a simulated, unexpected power outage.

TEST APPROACH

Two types of tests were performed at West Valley using three concentrations of simulated, nonradioactive melter feed slurry. The tests are identified as the homogeneity/sampling system test and the resuspension test. The melter slurry concentrations that were tested included the reference melter feed (42 wt% total solids), the reference melter feed concentrated to 75 vol% (concentrated feed - 46 wt% total solids), and the reference melter feed diluted to 125 vol% (dilute feed - 34 wt% total solids).

The homogeneity/sampling system tests were performed to characterize the melter feed system. Characterization included quantifying the homogeneity of the melter feed and determining if the liquid-sampling system obtains representative samples of the slurry. The homogeneity of the slurry in the MFHT was measured by comparing the average wt% total solids and selected elemental concentrations of slurry samples taken from six sampling locations within the tank, as shown in Figure 2. The representativeness of the sampling system was determined by comparing the average wt% solids and selected elemental concentrations between the samples taken at six tank sampling locations and the sampling system.

The resuspension test was performed to determine the time required to resuspend the solids and homogenize the melter slurry in the MFHT after an unexpected power outage. For this test, the agitator was turned off for approximately 4 days. Samples were taken from the tank at the six specified locations approximately 2, 6, and 24 hours after the agitator was restarted. The homogeneity of the tank was determined by comparing only wt% total solids.

The statistical design of the homogeneity/sampling system tests required 21 samples at each sampling location to attain a 95% confidence level of detecting a difference of greater than 4.5 elemental wt% between any two sampling locations. The statistical design of the resuspension tests required six samples at each

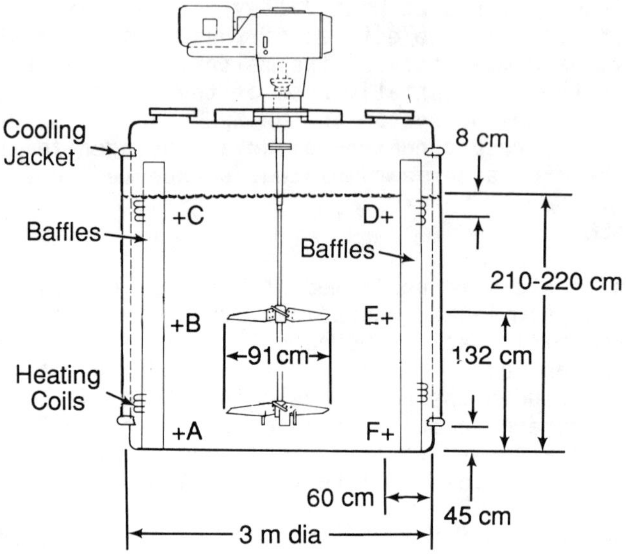

Figure 2. Schematic of West Valley MFHT with Sample Point Locations

sampling location. Samples were taken from the tank and analyzed in a random order for total solids and elemental concentrations.

All sample analyses were conducted at West Valley. For a more detailed discussion of the experimental statistical design, see Section 4.1 in Fow, Kurath, and Pulsipher (1989).[1]

TEST EQUIPMENT

The equipment used included the MFHT at West Valley, a mechanical liquid-sampling system, and a bottle-and-rod tank sampler. This section describes the major equipment used for this mixing study.

West Valley Melter Feed Hold Tank

The MFHT is the hold tank from which an air-displacement slurry (ADS) pump is used to feed the radioactive liquid waste and glass formers slurry (melter feed) to the ceramic melter. The stainless steel tank (see Figure 2) is 3 m (10 ft) in diameter and approximately 3 m (10 ft) high, with a capacity of 19,000 L (5,000 gal). The MFHT is equipped with an agitator, four equally spaced standard baffles, heating coils, and a cooling jacket. The heating coils are positioned inside the tank near the tank wall. The agitator motor is 11,000 W (15 Hp), single speed (155 rpm), with a 460 volt, 3-phase DC drive. The agitator is equipped with two 91-cm (36-in.)-dia., high-efficiency turbine impellers. Each impeller has three blades.

Mechanical Liquid-Sampling System

The purpose of the liquid-sampling system is to obtain a representative sample of melter feed from the MFHT. During actual melter operations, the sample will be used for process control and to support qualification of the resulting glass for disposal in a repository. The liquid-sampling system consists of 1) the ADS pump, and 2) the Hydragard Series 2000® liquid sampler.

<u>Air-Displacement Slurry Pump</u>

The basic operating sequence for the ADS pump is portrayed in Figure 3. The first three steps position the valve and allow the chamber to fill with the liquid slurry. The fourth step is the

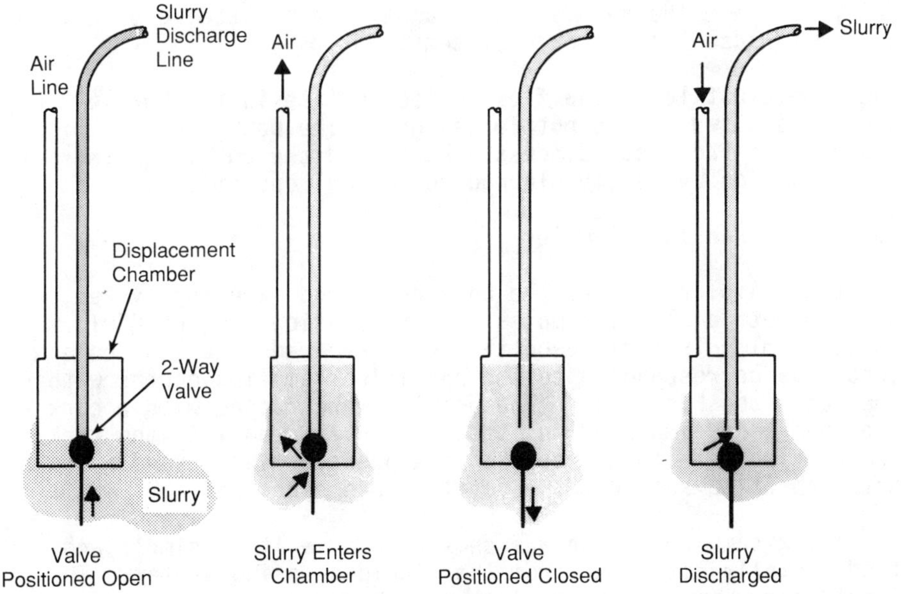

Figure 3. Operating Cycles of ADS Pump

® Registered tradename of Hinds International, Inc., Hillsboro, Oregon.

discharge step, where compressed air is valved into the chamber, thus forcing the slurry out the discharge line. The action of the air is like a reciprocating piston. By controlling the air pressure and cycle time, the flow rate can be controlled. This cycle creates a pulsating flow.

The slurry pump should be capable of pumping the melter feed in the turbulent flow regime, which should prevent solids from settling in the piping and either plugging the line and/or not providing a representative sample to the in-line liquid sampler. The ADS pump is capable of pumping the melter feed slurry at flow rates between 50 and 250 L/hr. An average flow rate of 250 L/hr was used for these tests. This flow rate corresponds to an average velocity of 0.55 m/s (1.8 ft/s) in a 1.27-cm I.D. pipe (0.5-in. I.D.).

Hydragard Series 2000 Liquid Sampler

The Hydragard sampler (Figure 4) was developed for the SRP and is designed to provide positive extraction and collection of samples from in-line process flows that are toxic, radioactive, or chemically severe. The sampler is a manually activated, closed loop system for use with robotic systems.

The sample bottles allow flow to recirculate in the bottle. Figure 5 shows two sample bottle designs. The bottle on the right was used in the tests discussed here, and the bottle on the left is the design currently planned for actual operations.

Bottle-and-Rod Tank Sampler

The tank sampler, called the bottle-and-rod sampling system, was used to obtain slurry samples at six locations in the MFHT. Three 60-ml polyethylene bottles were secured to a long pole at locations corresponding to the specified sampling depths within the tank (see Figure 2). The bottles were capped with a cork that had a rod imbedded in it, and then lowered into the tank. The bottles were then uncorked in a predetermined, random order. After about 10 seconds, the bottles were recorked.

This tank sampler design was chosen because it is simple and fundamentally different from the liquid-sampling system. The liquid-sampling system consists of a pipeline from the ADS pump to the sampler, and solids could settle in that line and/or in the sample bottle. The tank sampler collects samples directly from the tank. Comparing data from the two sampling systems provided information on possible deficiencies associated with the ADS pump/Hydragard sampling system.

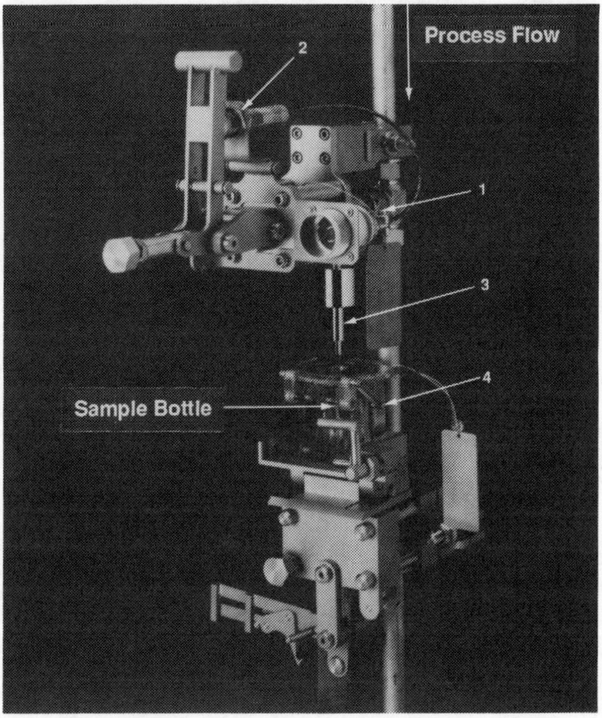

---

(1) Sampler Subassembly - Extracts sample from process line on operator or robot command.
(2) Limit Switch Subassembly - Provides system status verification signals for the valve "open" and "closed" positions.
(3) Concentric Needle Filler Subassembly - Transfers the sample to the sample container and returns excess sample to the process flow.
(4) Sample Container Shroud Subassembly - Positions the sample container in the sampling system, and provides a backup containment seal. Material: optically clear polycarbonate.

Figure 4. Hydragard Series 2000 Liquid-Sampler

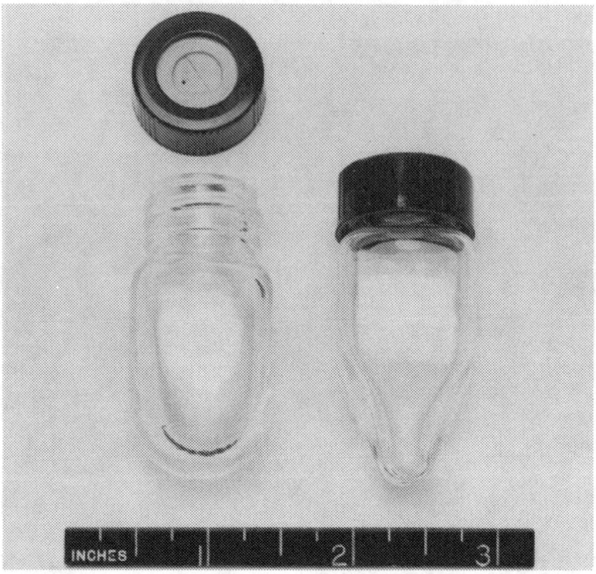

Figure 5. Sample Bottles for Hydragard Liquid-Sampler

EXPERIMENTAL RESULTS AND DISCUSSION

Statistical data analyses were performed on wt% total solids and as-analyzed and normalized values of the elemental concentrations. Results from the elemental concentration analyses on samples taken during the homogeneity/sampling system tests are summarized in Table 1. The ranges shown in Table 1 represent the minimum and maximum biases and relative standard deviations (RSD) determined from the statistical analyses of the seven major elemental constituents (Al, B, Ca, Fe, Li, Na, and Si). Four significant conclusions can be drawn from this data:

1) Bias ranges for the "as-analyzed" elemental concentrations are significantly larger than the ranges for the "normalized" elemental concentrations. Analysis of normalized values can sometimes mask effects. It can be deceptive when certain conditions affect some elements but not others, or affect all elemental constituents similarly. The effects of interactions can have significant impact during feed makeup.

2) The 11,000 W (15 Hp), single-speed (155 rpm) agitator installed in the MFHT produces a homogeneous mixture for the three concentrations of simulated melter feed slurry

Table 1. Results from Statistical Analysis of Elemental Concentration Data

Slurry Concentration	Sampler Bias Range[a]	Tank Inhomogeneity RSD[b] Range[c]	Analytical RSD Range[d]
	As-Analyzed Elemental Concentration		
Concentrated	0.3% - 5.7%	0.0% - 1.3%	1.4% - 7.8%
Reference	0.4% - 12.2%	0.0% - 0.0%	2.4% - 12.0%
Dilute	2.1% - 29.2%	0.0% - 2.3%	2.0% - 20.7%
	Normalized Elemental Concentration		
Concentrated	0.1% - 1.7%	0.0% - 0.7%	2.6% - 5.3%
Reference	0.8% - 6.9%	0.0% - 0.0%	3.2% - 9.6%
Dilute	0.2% - 13.9%	0.0% - 0.8%	4.2% - 16.1%

(a) Sampler bias is the difference in elemental concentration between the average of the six tank sampling locations and the liquid-sampling system.
(b) RSD is defined as relative standard deviation (SD*100/mean).
(c) Tank inhomogeneity range is the variability between the six sampling locations (see Figure 2).
(d) Analytical range is the variability in the analytical results for a particular sampling location.

used in these tests. This conclusion is reflected in the statistically insignificant ranges of RSD in Column 2 of the table ("Tank Inhomogeneity"). Slurries with significantly greater settling rates may produce different results.

3) The liquid-sampling system used in these tests obtains samples that are significantly different in certain elements (note the large bias range in Column 1 - "Sampler Bias Range"). This may be because of the flow rate of the slurry in the sampling system, and/or the type of sample bottle used, and/or the number of ADS pump cycles used to fill the sample bottle.

4) The major source affecting precision of the results is the variability in replicate analytical measurements. (The RSD ranges presented in Column 3, "Analytical RSD Range," are significantly higher than the ranges presented in Column 2.)

Other significant conclusions were drawn from analyses of the data from both types of tests. These conclusions are as follows:

- After 4 days of settling in the MFHT (simulating an unscheduled power outage), less than 2 hours of agitation was required to resuspend and uniformly mix the slurries tested.

- The maximum flow rate of the ADS pump used for the sampling system does not achieve turbulent flow in a 1.3-cm (0.5-in.)-(inside)-dia. pipe, which may result in a non-representative sample due to solids settling in the lines.

- The mixing Reynolds numbers calculated for the three concentrations of slurry tested were 1 to 2 orders of magnitude greater than the Reynolds number at which fully turbulent mixing is defined. This difference implies that the speed of the agitator could be reduced without affecting the homogeneity of the feed in the MFHT, if slurry settling rates are similar to the tested slurries.

The following are recommendations for improving the sampling system and reducing the number of required analyses per sample.

- Establish a statistical quality control program in the analytical laboratory to reduce analytical uncertainties due to long-term effects (e.g., lab conditions, calibration procedures). A statistical quality control program would isolate and eliminate some of the major sources of variation. Reduced analytical uncertainty minimizes the number of required analyses per sample, thereby reducing cost.

- Perform characterization tests on the actual liquid-sampling system to determine procedures that will provide a more representative sample. These tests should include an evaluation of sample bottle design, sampling cycle times, minimum flow rates, and flushing requirements.

REFERENCE

1. Fow, C. L., D. E. Kurath, B. A. Pulsipher. Evaluation of the Mixing System for the West Valley Melter Feed Hold Tank. PNL-6724, Pacific Northwest Laboratory, Richland, Washington, 1989.

LIQUID-FED CERAMIC MELTER TESTING FOR THE HANFORD WASTE VITRIFICATION PLANT

J. M. Perez, Jr., R. W. Goles, R. K. Nakaoka,
G. J. Sevigny and D. E. Larson
Pacific Northwest Laboratory(*)
P.O. Box 999
Richland, Washington 99352

O. L. Kruger
Westinghouse Hanford Company
Richland, WA, 99352

INTRODUCTION

A fundamental goal of the Defense Waste Management Plan is to end the 40-plus years of interim storage of defense wastes and to provide for their permanent immobilization and disposal. Pursuant to this national goal the Hanford Defense Waste Program has embarked on an integrated strategy for permanent disposal of all Hanford defense wastes. Defense high-level wastes (HLW) and transuranic wastes currently stored in underground double-shell tanks are to be immobilized before shipment to a geologic repository. Subject to satisfactory completion of the appropriate National Environmental Policy Act steps, the Hanford Waste Vitrification Plant (HWVP) is being designed to accomplish the objective of waste immobilization. The HWVP will process these pretreated wastes into a borosilicate glass that will be cast into stainless steel canisters. The prime Hanford contractor, Westinghouse Hanford Company, has lead responsibility for the design, construction, and operation of the HWVP. The Hanford research and development contractor, Pacific Northwest Laboratory (PNL), is responsible for development and testing in support of the HWVP.

(*) Operated for the U.S. Department of Energy by Battelle Memorial Institute under contract DE-AC06-76RLO 1830.

The HWVP Project is proceeding aggressively on simultaneous design and development paths. The major development goals are intended to support design and process definition, waste form qualification, and site and permitting activities. In achieving these goals the HLW glass producer shares many of the challenges of the commercial glass industry. A reliable feedstream with consistent composition and physical properties is required to maximize throughput. The prevention of secondary phases such as spinels, insoluble alkali salts, and reduced metals is crucial to maintain a high on-stream factor and long melter life. The treatment of process off gases to maximize process recycle and eliminate environmentally harmful gases and particulate matter is a necessity. Finally, by thoroughly understanding the process and its control, a high-quality glass can be produced that will meet disposal requirements. This paper focuses on liquid-fed, ceramic-lined melter and off-gas treatment equipment testing to date. Following the process description and related testing requirements, the results of melter testing is presented with emphasis on FY 1988 test results.

PROCESS DESCRIPTION

The first HLW to be processed in the HWVP will be a pretreated alkaline waste resulting from fuel reprocessing activities. Metals resulting from the dissolution of fuel cladding and equipment corrosion, together with chemicals added in the separations and neutralization steps, comprise the majority of the waste solids to be vitrified. This waste, defined as neutralized current acid waste (NCAW), is currently stored in underground million-gallon tanks. The waste contains over half of the elements in the periodic table as exhibited in Figure 1. Although many of the elements are present in very low concentrations, elements such as the noble metals, halides and sulfur can have a significant impact on processing.

Before processing within the HWVP, the NCAW will be pretreated to remove a significant fraction of the soluble and nonradioactive salts (see Figure 2). This will minimize the amount of glass product (which is very expensive to dispose of in the geologic repository). Upon retrieval from the tank farm, the sludge will be transported to B-Plant for pretreatment. The treated supernatant and wash water will be processed as a low-level waste by the Hanford Grout Treatment Facility.[1] The treated sludge (still containing >99% of the transuranics) will then be stored in an underground storage tank. Periodic batches will be transported from the storage tank to the HWVP for processing.

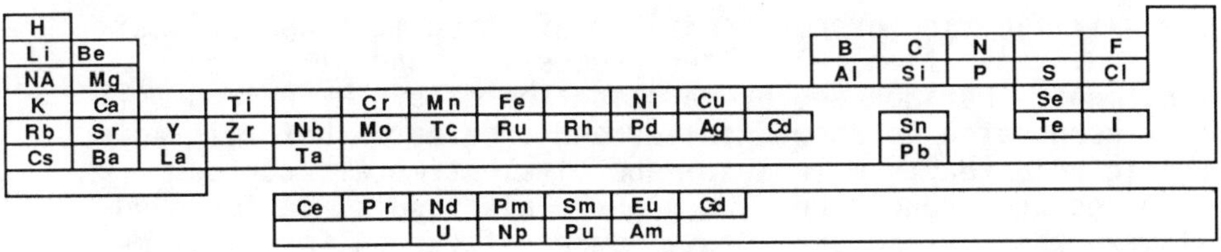

Figure 1. Periodic Table of the Elements Represented in NCAW.

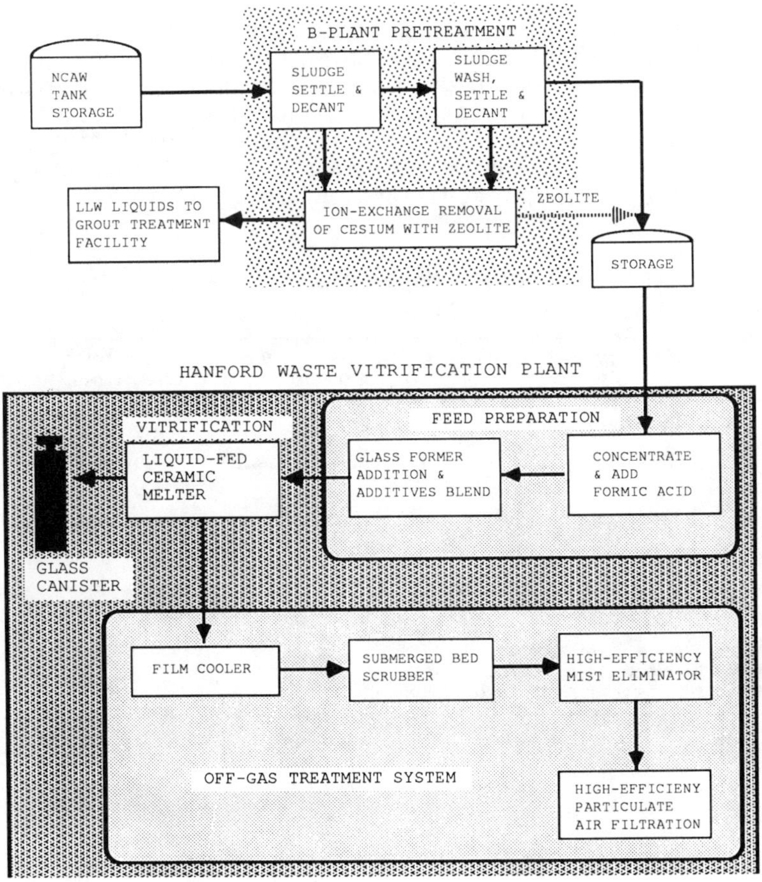

Figure 2. NCAW Process Flowsheet

The vitrification process consists of three main operations: 1) feed preparation, 2) melter processing, and 3) off-gas treatment. During feed preparation the dilute HWVP feed will be concentrated to 30 wt% solids and treated with formic acid. Formic acid reacts with mono- and divalent metals to form formates and reduce the slurry pH. This results in improved slurry rheology and provides a source of carbon for glass melt stability (i.e., prevents glass foaming or "reboil"). Glass forming frit and any additional reductant or shim chemicals are added next. The slurry is fed continuously to the melter while the glass can be transferred continuously or in batches into stainless steel canisters. The off gas is treated by the melter off-gas treatment system to remove water vapor, particulates, and hazardous gases.

The melter design employs a matrix ceramic design. The philosophy of the design is to back up successive refractories with a material that has higher thermal insulating properties and is typically less resistant to molten glass corrosion. Ceramics employed primarily for their thermal insulating properties do not contact the molten glass. Typical refractory materials used in melter construction are given in Table 1.

Table 1. Melter Refractory Materials

Refractory	Typical Composition, wt% Al_2O_3	SiO_2	Cr_2O_3	ZrO	Fe_2O_3	MgO	CaO	Density, g/cc	Thermal[a] Conductance, W/m•°C
Monofrax K-3[b]	60.4	1.8	27.3	--	4.2	6.1	--	3.9	2.9 (@1100 °C)
Monofrax E[b]	4.0	1.6	77.7	--	7.9	8.7	--	4.2	5.0 (@1100 °C)
Zirmul[c]	70.0	10.2	--	19.5	--	--	--	3.1	1.9 (@900 °C)
Alfrax 57[b]	94.8	0.4	--	--	0.1	0.1	4.7	1.2	0.7 (@550 °C)
Alfrax 66[b]	95.9	--	--	--	0.1	--	4.0	2.7	1.4 (@550 °C)
Fiberfrax Duraboard[b]	56.0	39.0	--	--	--	--	--	0.3	0.07 (@300 °C)

[a] 1 BTU•in./h•ft^2•°F = 0.144 W/m•°C
[b] A registered tradename of the Carborundum Co.
[c] A registered tradename of Chas. Taylor & Sons Co.

The reference off-gas system for the HWVP is based on a system designed for the West Valley Demonstration Project (WVDP).[2] The system consists of a film cooler, submerged bed scrubber (SBS), high-efficiency mist eliminator (HEME), and high efficiency particulate air (HEPA) filters (see Figure 2). The film cooler maintains a clear and unobstructed line between the melter and the SBS by injecting air and steam into the off-gas line. The SBS consists of a packed bed submerged in a water tank. The SBS condenses the process stream and captures gross particulates (>1μm). The HEME is a high efficiency filter designed to capture aerosols in the submicron size range. The HEPA filtration at the end of the off-gas treatment train ensures capture of virtually any remaining aerosols that escape previous filtration. Each of these equipment pieces is described in detail by Scott, Goles, and Peters (1985).[3] Significant strengths of the design are its simplicity and minimization of mechanical systems. The use of pumps or valves that would be exposed to the harsh chemical and radioactive environment is minimized. As a result, reliability is enhanced.

PROCESS TESTING REQUIREMENTS

The HWVP will make maximum use of vitrification technology developed in support of the Defense Waste Processing Facility at the Savannah River Plant, the WVDP, and numerous programs at the PNL. Adaptation of process technology requires process verification testing where site differences in waste composition could affect performance. The major areas of testing are composition definition, process/equipment design verification, waste form/process qualification, and environmental permitting. These major requirements for testing are discussed below.

Composition Definition

Significant properties and characteristics of the glass and composition are as follows:

- Waste Loading - Maximize within variability constraints
- Glass Former Composition - Select frit vs bulk chemicals
- Viscosity - 100 Poise in temperature range 1075-1150°C,]
- Electrical Conductivity - 0.18 to 0.5 $(\text{ohm}\cdot\text{cm})^{-1}$ at 100 Poise temperature
- Secondary Phase Formation - Prevent or minimize
- Glass Melt Redox Level - 0.005 to 0.3 ferrous/ferric ratio to control reboil
- Slurry Oxide Loading - Maximize and establish range
- Waste Variability - Establish limits based on glass and process limitations
- Glass Durability - Maximize leach resistance.

The criteria are generally established to ensure that the composition is compatible with the as-designed melter, will fall within process limits, and that the melter glass product will meet repository acceptance requirements.

Process/Equipment Design Verification

The key process equipment has been specified based on previous Department of Energy-funded development. Therefore, the emphasis of the work is to determine process performance with Hanford-specific wastes. The major issues and their criteria are:

- Melter Process Rate - Minimum of 45 L-slurry/h·m^2 glass surface area
- Glass Production Rate - Minimum of 100 kg/h
- Process Conditions - Stable feed pile for surface coverage of 50 to 100%
- Secondary Phases - No significant molten salts, crystalline phases, or precipitates
- Off-Gas Equipment - Achieve required decontamination treatment factors
- Process Control - Establish operating limits of equipment
- Unsteady State - Characterize equipment responses.

Waste Form Qualification

To allow shipment of the waste form to the federal repository, the waste producer must meet waste form specifications, demonstrate process control, and document the production of the qualified waste form. The pre-operation data will be obtained through a series of testing and modeling activities. These include melter and process modeling; determination of sources of error such as sampling, analytical, and process variations; canister thermal analysis; and product testing. In addition, small-scale radioactive testing will provide much-needed verification data. Before radioactive startup of the HWVP, cold testing for final qualification of the plant will be carried out.

Environmental Permitting

Demonstrating that the melter off-gas emissions can be treated to prevent environmental impact requires testing under conditions that are representative of plant operations. Such testing will provide data for designers to properly size and specify the off-gas treatment equipment. Testing will be done under steady-state, melter idling, and unsteady-state conditions.

PILOT-PLANT TESTING EQUIPMENT

Vitrification testing at PNL has been performed in two different melter systems: an engineering-scale melter, identified as the high-bay ceramic melter (HBCM), and the pilot-scale ceramic melter (PSCM). The HBCM and PSCM are shown in Figures 3 and 4, respectively. Table 2 summarizes the specifications of the two melters. The HBCM is employed for initial process and product investigations, waste variability testing, and for obtaining rudimentary off-gas results. Testing in the

PILOT-PLANT TEST RESULTS

The NCAW flowsheet testing has been ongoing at PNL since 1985. Since that time two PSCM and eight HBCM experiments have been conducted. Details of the experiments are given in Table 3. The most recent test was conducted in 1988 in which a 20-day PSCM experiment was conducted. Significant results to date are presented in the following paragraphs. Emphasis is placed on the most recent test, HWVP-15 in the PSCM in 1988.

Basic Flowsheet Testing

The first melter test of a melter feed based on NCAW was the HBCM run HWVP-11.[4] A waste loading of 25 wt% (glass oxide basis) and slurry oxide loading of 400 g/L were selected based on previous experience and favorable laboratory results. Glass formers composed of a premade glass frit (HWVP-11a) and bulk, unreacted chemicals (HWVP-11b) were each tested. The use of frit as the sole glass former was found to result in a 25% process increase (40 $L/h{\cdot}m^2$ vs 32 $L/h{\cdot}m^2$). Both feeds exhibited acceptable melter processing characteristics, i.e., the cold caps were flexible and maintained good contact with the glass, cap thickness was 2 to 7 cm, the solids melted easily into the glass, the slurries readily spread across the cold caps, and no secondary phases were evident. Based on the higher process rates achieved with using frit, frit was selected for subsequent testing.

Figure 3. High Bay Ceramic Melter

Figure 4. Pilot-Scale Ceramic Melter

Table 2. PNL Nonradioactive Test Melter Specifications

Specification	HBCM	PSCM
Glass Surface Area, m^2	0.25	0.76
Glass Tank Depth, cm	21.6	38
Glass Holdup Volume, L	55	288
Nominal Design Process Rate, L/h	15-20	35-45
Nominal Design Glass Production Rate, kg/h	7.5-10[a]	17.5-23[a]
Process Temperature, °C	1150	1150
Plenum Vacuum, cm H_2O	12.7	12.7

[a]Based on a feed slurry oxide content of 500 g/L.

Table 3. Summary of NCAW Melter Tests

Melter Test	Date	Test Length, h	Slurry Processed, L	Glass Produce, kg
HWVP-11a/HBCM	April'85	72	687	272
HWVP-11b/HBCM	April'85	48	407	149
HWVP-12/PSCM	July'85	384	15,950	6,131
HWVP-13a,b,c/ HBCM	May'86	149	2,243	1,093
HWVP-14a,b,c/ HBCM	July'86	148	2,510	1,255
HWVP-15/PSCM	June'88	471	15,000	7,450

PSCM supports process and flowsheet evaluations, off-gas treatment system testing, and waste form qualification model verification. In addition, full-scale canister experiments will be performed using the PSCM.

The first PSCM run, HWVP-12, established the large-scale process characteristics of the NCAW flowsheet. An equipment on-stream factor of 99% demonstrated the excellent processability of the simulated HWVP melter feed. A nominal slurry process rate of 62 L/h·m^2 was achieved. This rate exceeded design criteria process and glass production rates by about 33%. The average bulk glass temperature was maintained between 1150 and 1200°C. With a cold cap coverage over the glass melt of 60 to 75%, the plenum temperature above the melt surface ranged from 200 to 400°C. Melter off-gas losses and off-gas treatment equipment performance were found to be similar to previous experimental results in which formated slurries were processed. Ruthenium volatility loss from the melter was only 2% because of reducing conditions in the melt.

Optimum Slurry Oxide Loading

In 1986 the HWVP-13 tests in the HBCM determined the optimum slurry oxide loading. The melter is a poor evaporator compared with standard evaporators; and the largest energy input requirement is the evaporation of water. As a result, any water that can be eliminated from the melter feed substantially increases the glass production rate. However, if the slurry drys prematurely before flowing out across the molten glass, the feed piles up. The increased feed pile decreases heat transfer and reduced process rate results. Slurry oxide loadings of 415, 515, and 654 g/L were tested. The results of the HWVP-13 tests are shown in Figure 5. Nominal rates (i.e., approx. 75% coverage) exceeded design criteria by 61 to 93%. Based on these results, a nominal slurry oxide loading of approximately 500 g/L was recommended as a new process criteria. This corresponds to a solids loading of about 42 wt% in the melter feed.

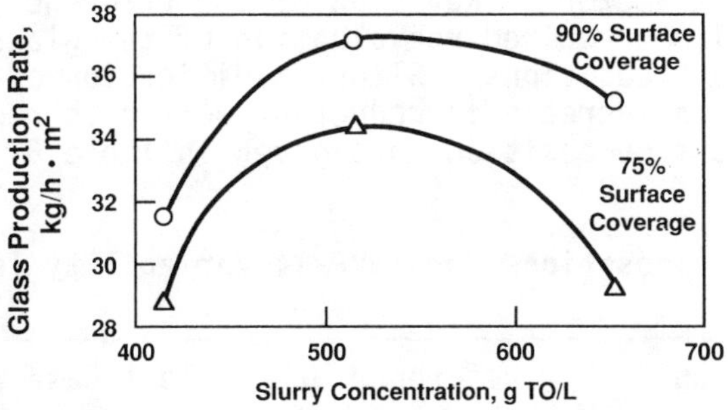

Figure 5. HWVP-13 Test Results Demonstrating Optimum Slurry Oxide Loading

Composition Variability

The HWVP will be required to accommodate a level of variability in the NCAW during processing. To ensure that the glass product will still be acceptable, an extensive laboratory testing program is mapping the composition field to determine the "process envelope." Extreme composition variations are to be tested in a melter system to determine that processing characteristics such as melt rate and cold cap condition are acceptable and that no secondary phases will form. The three composition variations tested to date are shown in Table 4.

The processing characteristics for all three segments of HWVP-14 test were acceptable. Cold cap characteristics for all three segments of the tests were similar and closely matched those of the previous HWVP melter runs. Cold cap coverage, thickness, flexibility, and venting were essentially identical. No creation or accumulation of spinel or molten salt phases was observed for any of the three feeds tested. Nominal processing (i.e., feeding) rates for the high sodium oxide, low iron oxide, and high zirconium oxide feeds were 89, 61, and 80 $L/h \cdot m^2$, respectively. These values were 132, 90, and 118%, respectively, of the nominal rate achieved with the reference feed slurry during HWVP-13.

The HWVP-15 experiment conducted in the PSCM was a 20-day run. The prime objectives of the test were to evaluate the melter processing characteristics and develop off-gas treatment data for design. The test consisted of a 15-day feeding period followed by a 5-day idling period. Table 5 lists the pertinent melter process parameters. Revision of the reference NCAW composition in 1988 required reevaluation of the glass former frit and reductant additions. Slight modification of the frit composition and an increase in reductant were both required. The feed and glass compositions are given in Table 6.

Table 4. NCAW Compositions for HWVP-14 Variability Testing

Condition	Reference Wt%	Test Case Wt%
High Sodium as Na_2O	10.4	13.2
Low Iron as Fe_2O_3	11.2	5.5
High Zirconium as ZrO_2	0.6	10.0

1988 Testing Results

Table 5. HWVP-15 Average PSCM Test Parameters

Parameter	Value
Bulk Glass Temperature	1156°C
Melter Feed Rate	42 L/h
Plenum Temperature	528°C
Plenum Pressure	-2.2 in. WC
Power Consumption	61.4 kW
Electrode Resistance	0.13 ohms
Electrode Current	706 amps

Processing Summary

The feed rate was adjusted until an average cold cap coverage of approximately 80% was established. Once the cold cap was stabilized, the average thickness was determined and the flexibility and venting behavior were evaluated. Average glass temperature, plenum temperature, electrode power, and melter pressure were monitored to establish the stability of the process. On-stream efficiency was excellent -- better than 98%. The feed rate to the melter was consistent at 43 to 44 L/h [~57 $L/h \cdot m^2$ (1.4 $gal/h \cdot ft^2$)]. This rate is quite comparable to the HWVP-12 rates. During HWVP-12 nominal rates of 51 $L/h \cdot m^2$ (formated feed with 3.5 g/L sugar added) to 62 $L/h \cdot m^2$ (formated feed without sugar added) were achieved and are in excess of the design basis value of 45 $L/h \cdot m^2$ used for the HWVP melter. Combined with the oxide loading in the slurry, the nominal glass production rate achieved was 58% higher than the design criteria of 18 $kg/h \cdot m^2$.

During the first 8 h of startup some degassing of the original glass inventory occurred. This was expected, given the initial oxidized condition of the glass batch. The degassing resulted in a light foam accumulation forming over the open glass surface. It dispersed as the HWVP-15 glass gradually displaced the original glass tank contents.

Throughout the test the cold cap characteristics remained essentially constant. The cold cap coverage of the glass surface typically ranged from 60 to 80%. The cold cap remained flexible throughout the run period. Venting was routinely observed, and the cap thickness appeared to remain consistent at about 2 to 5 cm (1 to 2 in.). At the termination of

Table 6. HWVP-15 Melter Feed Composition

Chemicals Used[a] In Makeup	Concentration g/L	Assumed State After Reaction	Concentration in Glass, wt%[b]
$Al(OH)_3$	18.11	Al_2O_3	2.36
H_3BO_3	0.23	B_2O_3	0.03
$BaSO_4$	0.80	BaO	0.10
$CaHPO_4*H_2O$	1.21	CaO	0.08
$Ce(OH)_3$	0.87	CeO_2	0.16
$Cr(NO_3)_3*9H_2O$	2.06	Cr_2O_3	0.13
$CsNO_3$	1.09	Cs_2O	0.16
$CuSO_4*5H_2O$	2.47	CuO	0.16
$Fe(OH)_3$	49.31	Fe_2O_3	7.36
LaF_3	1.97	La_2O_3	1.29
$La/Nd(NO_3)_3*6H_2O$	17.25	MgO	0.09
$MgSO_4$	1.23	MnO_2	0.16
$Mg(NO_3)_2*6H_2O$	0.30	MoO_3	0.31
$Mn(NO_3)_2*4H_2O$	2.27	Na_2O	4.60
MoO_3	1.58	Nd_2O_3	0.67
NaCl	0.63	NiO	0.61
$Na_2Cr_2O_7*2H_2O$	0.53	P_2O_5	0.10
$Na_2C_2O_4$	13.95	Pr_6O_{11}	0.10
NaF	0.79	SiO_2	1.05
$NaNO_3$	19.58	Sm_2O_3	0.05
NaOH	10.46	SrO	0.10
$Na_2HPO_4*7H_2O$	0.10	Y_2O_3	0.05
Na_2SO_4	0.65	ZnO	0.50
NdF_3	2.02	ZrO_2	3.94
$Ni(OH)_2$	3.76	F	0.30
$Pr(OH)_3$	0.59	Cl	0.08
SiO_2	5.26	SO_3	0.45
$Sm(OH)_3$	0.30	NO_3	35.00 g/L[c]
$Sr(NO_3)_2$	1.07	TOC	2.50 g/L[c]
$Y(OH)_3$	0.33		
ZnO	2.50		
$Zr(OH)_4$	25.51		
Subtotals:	188.8 g/L		25.0 wt%
HW-39-4 Frit Composition[d]		SiO_2	52.50
		B_2O_3	10.50
		Na_2O	6.75
		Li_2O	3.75
		CaO	0.75
		Mg	0.75
		Subtotal	75.0 wt%
Reductant Added		HCOOH[e]	28.7 g/L[f]
		Sugar	10.0 g/L

[a]Before formic acid treatment.
[b]Values assume no volatility losses.
[c]Value is not included in subtotal values.
[d]Size distribution of frit was -80/+200 mesh with a maximum of 10 wt% -200 mesh.
[e]Value given is for a 90 wt% solution of HCOOH.
[f]Value includes 90 wt% HCOOH added with frit, equivalent to 1 wt% of frit added, or 4.2 g/L.

feeding, the cold cap melted into the glass quickly (i.e., <30 min). No refractory or molten salt accumulations were observed.

Off-Gas Treatment System Performance

PSCM off-gas studies were conducted to characterize melter emissions under steady-state processing and non-processing (idling) conditions and characterize off-gas treatment equipment performance. To characterize individual component, as well as integrated off-gas system behavior, effluents were sampled before and after passing through each major piece of off-gas process equipment (i.e., the SBS, HEME and HEPA).

The melter effluent source was found to be typical of previous investigations in that most melter off-gas particulate losses were associated with aerosol escape. The average size distribution of effluent aerosols was bi-modal; the larger sized particles (>1 μm in diameter) were predominantly non-volatiles, and the small-diameter component of the aerosol was composed of semi-volatile feed components. Gross particulate entrainment losses were very low during steady-state melter feeding conditions; well over half of the airborne particulate mass leaving the melter was associated with aerosols having aerodynamic diameters less than or equal to 1 μm. Semi-volatile emissions were also responsible for most melter off-gas losses during the idling, or nonfeeding portion of the run. Semi-volatile emission rates during idling did not exceed emission rates observed during the feeding portion of the run. The concentration of combustible gases, CO and H_2, resulting from decomposition of the organic reductants were very low and did not present an off-gas flammability hazard under any of the operating conditions investigated during the test.

As an off-gas system quencher, the SBS successfully reduced the temperature of the off-gas stream to below 30°C and condensed 84% of the process steam. As an effluent scrubber, it essentially removed all effluent aerosols with diameters larger than 1 μm, and provided significant semi-volatile off-gas decontamination over the range of flows tested (3,100 to 4,500 slpm [110 to 160 scfm]). Melter idling was not found to increase effluent concentrations in the SBS exhaust. Although its gas scrubbing capabilities have not been fully evaluated, SBS NO_x removal efficiency was ~1%. This low efficiency was due to both low NO_x concentrations and high $NO:NO_2$ ratios.

The HEME exhibited very high aerosol removal efficiencies under all off-gas flow and inlet water loading conditions used during the test. The HEME effectively removed all off-gas aerosols having diameters greater than 0.5 μm, and reduced aerosol concentrations to values so low that HEPA service life under these conditions was estimated to exceed the required service life of 1 yr by a factor of 5. Aerosols captured in the HEME were effectively purged from the filter element by collection and run off of the normal SBS mist carryover (3.5 mg/slpm). Increases in pressure resulting from particulate buildup were not observed under any of the operating conditions of the test. The use of a 15 L/h water spray at the inlet to the HEME did cause a 50% increase in the normal operating pressure drop across the HEME filter element.

Based on the off-gas system testing results and previous studies, recommended design-basis decontamination factor (DF) values for the feed components were recommended as shown in Table 7. They are felt to be conservative yet realistic such that they do not unduly restrict design and equipment requirements.

Table 7. HWVP Off-Gas System Decontamination Factors Recommended for Design

Material	Decontamination Factors		
	Melter	SBS	HEME
Nonvolatiles	500	40	400
Semi-Volatiles	500	5	50
Volatiles	500	5	400
B	100	5	50
C	1	1	1
Cd	7	3	400
Cl	2	5	40
Cs	14	3	400
F	4	5	40
H_3	1	5	1
I	1	1	1
NO_3	1	1	1
Pb	(100)[a]	(5)[a]	(50)[a]
Ru	40	5	400
S	5	5	5
Se	10	5	50
Tc	14	5	50
Te	10	5	50

[a]Projected based on chemical properties.

SUMMARY

Flowsheet and equipment testing has been conducted at PNL in support of the HWVP project since 1985. A significant number of engineering design, process, and qualification issues have been addressed or are in the process of being resolved. To date the NCAW flowsheet has been amenable to applying existing U.S. Department of Energy technology to plant design and process approach. The melter and off-gas treatment systems have demonstrated production rates and operational performance in excess of current design criteria.

REFERENCES

1. R. H. Guymon, L. D. Vanselow, and G. D. Campbell, "The Grout Treatment Facility--Processing Facilities for Low-Level Waste Immobilization and Disposal." pp. 95-98 in *SPECTRUM'88: Proceedings of the Nuclear and Hazardous Waste Management International Topical Meeting, Pasco, Washington*, Pacific Northwest Laboratory, Richland, WA, 1988.

2. S. M. Barnes, J. M. Pope, C. C. Chapman, "Three Year's Progress of the West Valley Demonstration Project Titrification System." pp. 195-202, in *Proc. of the Symposium on Waste Management'88, Vol. II, Tucson, AZ*, 1988.

3. P. A. Scott, R. W. Goles, and R. D. Peters, *Technology of Off-Gas Treatment for Liquid-Fed Ceramic Melters*; PNL-5446, Pacific Northwest Laboratory, Richland, WA, 1985.

4. J. M. Perez, Jr., and R. K. Nakaoka, "Vitrification Testing of Simulated High-Level Radioactive Waste at Hanford;" pp. 495-505 in *Proc. of the Symposium on Waste Management '86, Vol. II*. ed. R. G. Post and M. E. Wacks, University of Arizona, Tucson, AZ, 1986.

PROPOSED NOBLE METALS-COMPATIBLE MELTER: PRELIMINARY MODELING RESULTS[a]

R. D. Peters, M. L. Elliott, and C. C. Chapman
Pacific Northwest Laboratory
Richland, Washington 99352 - (509)376-4579

ABSTRACT

The noble metals palladium, rhodium, and ruthenium are present in waste to be vitrified at Hanford, and they can disrupt operations by precipitating and settling in the glass melter. This paper reports ongoing work to evaluate and better understand the characteristics of noble metals in support of the Hanford Waste Vitrification Plant. In this context, a sloped melter bottom, a bottom drain freeze valve, and three electrodes to enhance convection are being evaluated. This paper discusses the strategy for accommodating noble metals, describes theoretical modeling of freeze valve operation, and describes progress on physical modeling of the motion of glass and precipitates in the melter cavity. The investigation focuses on a sloped-bottom, three-electrode melter design, which is a promising approach for accommodating noble metals.

INTRODUCTION

Stored high-level waste (HLW) at Hanford, Washington, is to be converted to glass in the Hanford Waste Vitrification Plant (HWVP). The characteristics of the waste and a description of the HWVP process are described by Gurley et al.[1] A liquid-fed ceramic melter (LFCM) will be used to process the HLW slurry. The waste originates from processing of defense production reactor fuel. A wide variety of fission products are generated during fuel burn-up, including the noble metals palladium, rhodium, and ruthenium. While the HLW slurry is predominantly compounds of iron, aluminum, sodium, silicon, and zirconium which

---

(a) Work supported by the U.S. Department of Energy under Contract DE-AC06-76RLO 1830.

are glass-soluble, there is a small yet significant quantity of noble metals in the waste. These noble metals are not soluble in the LFCM glass melt and could form settled precipitates. If such precipitates are allowed to accumulate, then melter operation could be disrupted. The metals are conductive and could "short circuit" the resistance-heating through the glass.

With a properly designed and operated LFCM it is believed that noble metals will remain suspended and will be removed with routine glass discharging. If the noble metals tend to accumulate at the bottom of the melter, the LFCM design should also incorporate features that promote their resuspension. This paper describes the strategy for accommodating noble metals, provides results of mathematical modeling of a bottom drain freeze valve, and gives preliminary results from physical modeling of the LFCM molten glass cavity. During the course of the work several models will be evaluated. Currently results from one model are available. The work described here is being performed at Pacific Northwest Laboratory (PNL) for technical support of the HWVP project, which is managed by Westinghouse Hanford Company.

STRATEGY FOR NOBLE METALS-COMPATIBLE MELTER

The main features of the LFCM design to accommodate noble metals are a sloped melter bottom, a bottom drain freeze valve, and three electrodes to enhance molten glass convection stirring. Figure 1 is a conceptual sketch of a melter cavity with these features. The bottom is sloped 45° to promote transfer of settled particles to the bottom drain. A freeze valve is used to control glass discharging through a drain in the melter floor. A three-electrode configuration provides uniform heating and promotes glass circulation. When current is passed between the bottom and side electrodes, the glass between these regions is joule-heated. Buoyancy drives heated glass upwards, promoting formation of convection cells. Such fluid motion may be necessary to keep insoluble material suspended in the molten glass so it can be discharged during canister filling. The three-electrode, sloped-bottom design was proposed by Chapman(2) and has been deployed in the LFCM at the West Valley Demonstration Project in New York. The design has also been used in Japan at the Tokai Vitrification Facility(3) and the Ishikawajima-Harima Heavy Industries Plant.(4)

FREEZE VALVE PERFORMANCE MODELING

The bottom drain is based on a freeze valve mechanism that consists of a nickel-alloy pipe surrounded by an induction coil

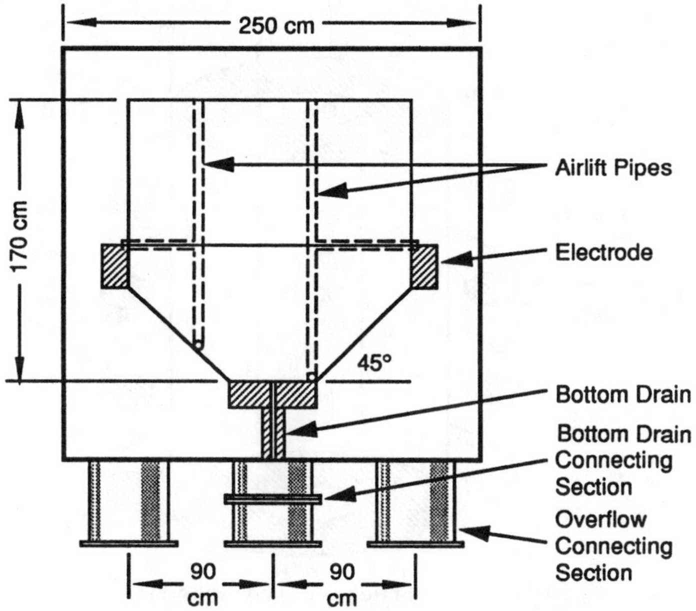

Figure 1. Proposed Melter Concept

(Figure 2). The flow of glass through the valve is initiated by heating a frozen plug in the nickel-alloy pipe with the induction coil. Draining is terminated by cutting power to the induction coil, which allows the flowing glass to refreeze. Cooling water that circulates through the induction coil accelerates the freezing of the glass. Theoretical modeling has been done to describe the performance of the valve and to guide selection of design parameters.

The valve must be sized to achieve the desired glass discharge rate yet still be able to stop flow of glass when necessary. To select a valve diameter, an analysis is performed assuming that glass flow (Q) through the pipe is laminar, steady and isothermal. Molten glass is further assumed to be Newtonian, and then both the head (H) in the melter and drain valve length (L) are fixed. The following relationship is obtained:

$$Q = \frac{\pi \rho g (H+L) R^4}{8 \mu L} \tag{1}$$

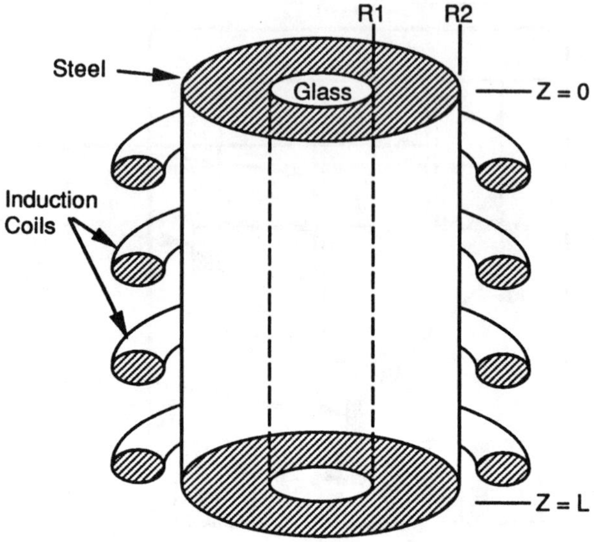

Figure 2. Freeze Valve Schematic

(Symbols for all equations are defined at the end of this paper.) To drain glass at a required minimum of 225 kg/h, the required valve diameter was found to be 16 mm.

The second requirement for modeling the performance of a freeze valve is to determine the time required to stop the flow of glass or "close" the valve. This requires solving the governing equations for heat and momentum transfer in the valve. A heat balance is established for the pipe wall because this is the pathway for removing heat from the molten glass stream. A heat and momentum balance is also written for the glass stream. Boundary conditions are established and the two systems of partial differential equations are solved numerically.

Heat Balance for Freeze Valve Pipe

The energy equation for unsteady heat transfer in the walls of the freeze valve pipe is:

$$\frac{\partial T}{\partial t} = \alpha \frac{\partial^2 T}{\partial r^2} + \frac{\alpha}{r} \frac{\partial T}{\partial r} + \alpha \frac{\partial^2 T}{\partial z^2} \qquad (2)$$

There are five initial and boundary conditions. Initially the pipe is at a uniform temperature, T_{start}:

$$\text{for } t = 0,\ 0 \le z \le L, \text{ and } R_1 \le r \le R_2 : \quad T = T_{start} \tag{3}$$

Heat leaving the glass enters the inside wall of the freeze valve pipe:

$$\text{for } 0 \le t \le \infty,\ 0 \le z \le L, \text{ and } r = R_1: -k \left.\frac{\partial T}{\partial r}\right|_{metal} = -k \left.\frac{\partial T}{\partial r}\right|_{glass} \tag{4}$$

Heat loss at the outside pipe wall is determined by convection to air and radiant loss to the water-cooled coil.

$$\text{for } 0 \le t \le \infty,\ 0 \le z \le L, \text{ and } r = R_2 :$$

$$-k \frac{\partial T}{\partial r} = h_c(T - T_{air}) + \sigma F_{2-3} (T^4 - T_W{}^4) \tag{5}$$

Finally it is assumed that no heat loss occurs from either end of the solid portion of the freeze valve pipe:

$$\text{for } 0 \le t \le \infty,\ z = 0, \text{ and } R_1 \le r \le R_2 : \quad \frac{\partial T}{\partial z} = 0 \tag{6}$$

$$\text{for } 0 \le t \le \infty,\ z = L, \text{ and } R_1 \le r \le R_2 : \quad \frac{\partial T}{\partial z} = 0 \tag{7}$$

Heat Balance for Flowing Glass

The energy equation for flowing glass in the freeze valve is:

$$\frac{\partial T}{\partial t} + v_Z \frac{\partial T}{\partial z} = \alpha \frac{\partial^2 T}{\partial r^2} + \frac{\alpha}{r} \frac{\partial T}{\partial r} \tag{8}$$

Velocity components in radial and angular directions are neglected. Axial heat conduction is assumed to be negligible compared to convective heat transfer and radial conduction. Initially glass is flowing at a uniform temperature:

$$\text{for } t = 0,\ 0 \le z \le L, \text{ and } 0 \le r \le R_1 : \quad T = T_{start} \tag{9}$$

Radial heat conduction at the center of the cylindrical glass stream is zero:

$$\text{for } 0 \le t \le \infty,\ 0 \le z \le L, \text{ and } r = 0: \quad \frac{\partial T}{\partial r} = 0 \tag{10}$$

At the inside pipe wall, or outside glass stream boundary, the temperature of the glass and metal is matched:

$$\text{for } 0 \leq t \leq \infty,\ 0 \leq z \leq L, \text{ and } r = R_1:\quad T_{glass} = T_{metal} \tag{11}$$

The glass temperature at the inlet to the freeze valve pipe is constant:

$$\text{for } 0 \leq t \leq \infty,\ z = 0,\ 0 \leq r \leq R_1:\quad T = T_{start} \tag{12}$$

Momentum Balance for Flowing Glass

The equation of motion for the flowing glass in the freeze valve is as follows:

$$\rho \frac{\partial v_Z}{\partial t} = \frac{P_o - P_I}{L} + \frac{\mu}{r}\frac{\partial v_Z}{\partial r} + \frac{\partial v_Z}{\partial r}\frac{\partial \mu}{\partial r} + \mu \frac{\partial^2 v_Z}{\partial r^2} \tag{13}$$

This is derived by assuming constant glass density; also the angular and radial velocity vectors in the glass are zero. The continuity equation then reduces to $\partial v_Z/\partial z = 0$. Note in Equation (13) that viscosity is not constant but varies with radial position. This is important to consider because the radial temperature gradients result in sharp radial viscosity gradients. The following equation for glass viscosity as a function of temperature was used:

$$\mu(\text{poise}) = 4.4 \times 10^{-5} \exp(19{,}880/T(^\circ K)) \tag{14}$$

This relation was derived from regression analysis of waste glass viscosity data collected in this Laboratory. The initial and boundary conditions for Equation (13) are described in Equations (15) through (17). Initially glass flow is laminar so that the radial velocity profile is parabolic:

for $t=0,\ 0 \leq r \leq R_1$, and $0 \leq z \leq L$:

$$v_Z = \frac{(P_o - P_I)R^2}{4\mu L}\left\{1 - \left(\frac{r}{R_1}\right)^2\right\} \tag{15}$$

Axial symmetry about the center of the cylinder is imposed:

$$\text{for } 0 \leq t \leq \infty,\ r = 0, \text{ and } 0 \leq z \leq L:\quad \frac{\partial v_Z}{\partial r} = \frac{\partial \mu}{\partial r} = 0 \tag{16}$$

Finally, a no-slip boundary condition is applied at the pipe wall:

$$\text{for } 0 \leq t \leq \infty,\ r = R_1, \text{ and } 0 \leq z \leq L: \quad v_z = 0 \tag{17}$$

Solution Method

Equations (2), (8), and (13) were simultaneously solved subject to their respective initial and boundary conditions by finite difference methods implemented on a FORTRAN computer code. The two energy Equations (2) and (8) were solved using an explicit finite difference method. The momentum Equation (13) was solved using an implicit, central finite-difference method because of instabilities at small times with the explicit method. The implicit method resulted in a tri-diagonal matrix which was solved at each time step. The matrix was solved using the method of Thomas[5] to avoid an entire matrix inversion at each time step. The algorithm worked by solving the axial and radial temperature distribution using the velocity gradient acquired in the previous time step. These temperatures were then used to calculate the new viscosity and velocity gradients and the process was repeated.

An example of the program output is shown in Figure 3. As the FORTRAN program that generated these data has not yet been formally verified, the results should be considered preliminary. The glass flow rate versus time is shown for a starting glass temperature of 1150°C with valve dimensions of: R_1 = 10 mm, R_2 = 32 mm and L = 400 mm. A 2-mm corrosion allowance is assumed, so the calculation is for a 20-mm rather than the 16-mm-dia pipe specified by Equation (1). Initially heat is removed from the wall of the pipe and the glass flow rate is unchanged. This condition is followed by a decrease in flow rate as the glass begins to freeze. Glass flow is predicted to decelerate more slowly in the final minutes because of heat transfer limitations through the frozen glass surrounding the molten stream. The flow effectively terminates after about 7 min, which is adequate for controlling glass discharge.

An effort is now under way to install a freeze valve, with the dimensions derived from modeling (R_1, R_2, and L), on a nonradioactive experimental LFCM. The performance and operational characteristics of the freeze valve will be tested. Data on glass flow termination will be collected and compared with model results.

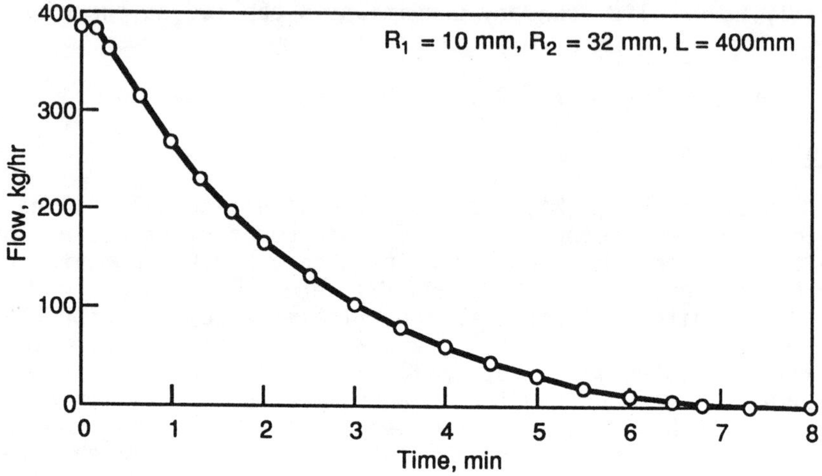

Figure 3. Freeze Valve Shut-Off

PHYSICAL MODELING

The objective of physical modeling is to predict the behavior of noble metal precipitates as a function of particle size and of melter operating parameters. The empirical technique has long been used to characterize and predict processes in commercial glass production furnaces.[6] Several investigators have used the technique to study the LFCM for HLW treatment.

Quigley and Kreid[7] from this Laboratory performed early studies on a flat-bottom cavity design with side electrodes. They found that the upper portions of the cavity were well-mixed and had significant stirring. However, the lowest fluid velocities and poorest mixing occurred in the bottom quarter of the model. Skarda et al.[8] evaluated two configurations, one with a flat bottom and one with a 5° sloped bottom; both had side electrodes. Mixing was enhanced through the use of bubblers, which induced two counter-rotating eddies symmetric about the model's vertical center. Such bubbling appeared necessary to prevent temperature stratification in the two-electrode melter design.

Routt[9] performed physical modeling using a cylindrical melter cavity with four symmetrically spaced bayonet-type electrodes. This represented a proposed design for vitrification of HLW at Savannah River. Voltage gradients were measured, thus providing an indication of power density. Highest power densities occurred on a diagonal path between opposing electrodes.

The bottom of the cavity was poorly heated, and, consistent with this, the velocity of the liquid there was slow compared to higher locations.

Asano et al.[10] reported physical modeling of three different cavity geometries, which had bottom slopes of 0°, 30°, and 45°. All had a three-electrode design, with two electrodes mounted on side walls and one on the bottom. The distribution of heating between electrode pairs was controlled sequentially by time-sharing. When only the upper two electrodes were fired, temperature was stratified, with the lowest temperatures at the bottom. This situation was remedied by firing from the bottom electrode in the two models with sloped bottoms. In the flat-bottom model, temperature stratification persisted even while all three electrodes were fired. Flow cells in the sloped-bottom models were sufficient to maintain a simulated insoluble phase (powdered copper) in suspension. In the case of the flat-bottom model, copper tended to accumulate on the bottom of the cavity even when all three electrodes were being fired.

The purpose of the ongoing modeling effort is to evaluate additional sloped-bottom cavity geometries each with two side electrodes and a bottom electrode. The heating method will differ from Asano's[10] in that all three electrodes will be fired simultaneously with varying proportions of power. Simultaneous firing has been used in the West Valley LFCM. The objective is to provide information on thermal-hydraulic conditions in the melter. Criteria for a good design are: well-mixed fluid with turbulent conditions in the cavity and ability to keep noble metals in suspension. The literature described above suggests that a sloped bottom design has advantages over the earlier flat-bottom configurations with respect to these criteria.

<u>Model Scaling Relationships</u>

To properly scale the physical model and select an appropriate model liquid, the similarity analysis method as described by Quigley and Kreid[7] and Skarda et al.[8] has been used. The following dimensionless groups are equated for the model system and the full-scale melter system:

$$\text{Rayleigh No.} = \frac{g\mathcal{L}^3\beta\Delta T}{\nu\alpha} \qquad (18)$$

$$\text{Peclet No.} = \frac{U\mathcal{L}}{\alpha} \qquad (19)$$

$$\text{Power No.} = \frac{\phi s \mathcal{L}^2}{k \Delta T} \tag{20}$$

$$\text{Nusselt No.} = \frac{h \mathcal{L}}{k} \tag{21}$$

The models in this study are 1/4 scale (based on linear dimensions) of a melter designed for 45 kg/hr glass production rate, the current reference HWVP capacity. The criteria of the dimensionless groups in Equations (18) - (21) require selection of a modeling liquid with the appropriate physical properties. Stanek(11) describes properties of a variety of modeling liquids, and best agreement is obtained with 7.5 wt% LiCl in glycerin prepared by heating at 180°C for 1 hr. The nominal operating temperature of the HWVP melter is 1150°C, and the corresponding model liquid temperature is 55°C. The physical properties of these liquids are given in Table I.

Description of Experiments

A schematic of the experimental system is shown in Figure 4. The model cavities to be tested have three electrodes and a sloped bottom. (Although results are currently available from only one model, several models will be evaluated during the course of the work.) Three single-phase power supply and control systems are used for three possible firing pairs: A - B, B - C, and C - A. Constant-current silicon-controlled rectifiers (SCRs) are manually regulated to achieve desired power levels. An outer water-filled cooling jacket removes heat from the side walls and bottom of the cavity. A separate cooling coil of 5/16-in.-OD vinyl tubing removes heat from the top glycerin liquid surface. This surface simulates the "cold cap" of unmelted feed material

Table I. Properties of Waste Glass and Modeling Liquid

Property	HW-39 Waste Glass at 1150°C	Glycerin + 7.5 wt% LiCl at 55°C
Density, g/cm^3	2.17	1.26
Thermal expansion coefficient, $°C^{-1}$	1.06×10^{-4}	5.3×10^{-4}
Resistivity, ohm-cm	2.86	688.
Viscosity, g/cm-s	45.00	2.8
Heat capacity, cal/g°C	0.60	0.59
Thermal conductivity, cal/cm-s°C	9.11×10^{-3}	7.69×10^{-4}
Kinematic viscosity, cm^2/s	20.74	2.22
Thermal diffusivity, cm^2/s	7.0×10^{-3}	1.0×10^{-3}

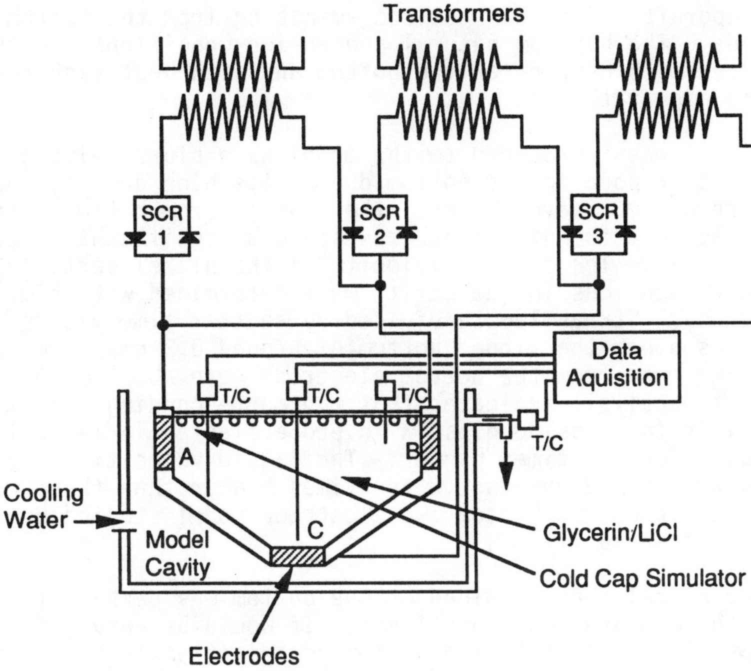

Figure 4. Schematic of Model Melter System

in an actual melter. Stainless steel-sheathed thermocouples have been rubber-insulated and placed at various locations in the cavity. Terminal velocity scaling was used to select particles to simulate noble metals. Powdered nickel in the size range of 12 to 44 μm represents noble metals of specific gravity 12 (such as palladium or rhodium) and with diameter ranging from 50 to 200 μm. Asano et al.(10) used copper particles of the same size (density similar to that of nickel) to represent noble metals.

Preliminary Results and Discussion

A model of the cavity shown in Figure 1 was operated over a 4-hr period at a nominal temperature of 55°C and a power level of 550 W. About 80% of the heat was generated between the bottom-to-side electrodes, and the other 20% between the side electrodes. A heat balance shows that the cold cap removed 40 to 50% of the heat generated in the model liquid, with the balance being removed through side walls to the outer cooling jacket. The liquid near the cold cap was typically 5 to 10°C cooler than the bulk glycerin. The cold liquid could be seen moving downward, and an updraft effect was noticed emanating from the bottom electrode. The highest natural convection conditions are thus associated with heating at the bottom and with heat exchange to the cold cap at the top.

Nickel was introduced to the model as a slurry with glycerin. It dropped to the bottom due to its high density and unique rheology. Nevertheless, the updraft from bottom electrode firing was able to entrain and resuspend a significant fraction of the nickel. The general velocity of the nickel particles at different locations in the cavity were determined with time-lapse photography. Streak length divided by shutter time yields velocities along the sloped bottom of around 0.2 cm/s. Updraft velocities away from the bottom electrode were about 0.15 cm/s. Similarity analysis indicates that the corresponding velocities in a melter (processing glass with properties as in Table I), would be about 1.7 times higher. The liquid velocities measured in this work are about two to ten times higher than those measured in the flat-bottom two-electrode model studied by Quigley and Kreid.(7)

The nickel that remained at the bottom was perhaps trapped inside the momentum boundary layer. It would be very difficult to remove this material without forced agitation such as with a bubbler. It may be important to keep particles from settling in the first place. When the melter is idling, molten glass tends to become quiescent, and a means to promote glass circulation during this period may be prudent. Future testing is planned and will include different cavity geometries, power distributions, and use of a bubbler.

SUMMARY AND CONCLUSIONS

A physical model for simulating the thermal-hydraulic behavior of the proposed LFCM design has been constructed and preliminary results obtained. With proper electrode firing, 10-40 μm nickel particles can be kept suspended. These simulate 50-200 μm-sized palladium or rhodium noble metals in an LFCM. The operation of a bottom drain for glass has been modeled based on the governing equations for heat transfer and fluid dynamics. Calculations show that a valve pipe 20 mm ID by 400 mm long can deliver adequate glass flow for filling canisters and can be shut off in about 7 min. A freeze valve has been designed for performance testing with critical dimensions derived from the modeling. The three-electrode, sloped-bottom cavity design with a freeze valve drain can be considered a promising approach for accommodating trace noble metals in the HLW at Hanford to be vitrified in the LFCM.

SYMBOLS

F_{2-3} = radiant view factor from induction coil to freeze valve
g = acceleration due to gravity
h = heat transfer coefficient out of melter or model
H = glass head in the melter
h_c = convective heat transfer coefficient out of freeze valve
k = thermal conductivity
L = freeze valve length
$\mathcal{L}$ = characteristic dimension or length for physical model
P_o = pressure component (at $z = 0$) = ρgH
P_l = pressure component (at $z = L$) = $-\rho gL$
Q = mass flow rate
r = radial direction
R = radius of freeze valve
R_1 = radius of glass channel in freeze valve
R_2 = outside radius of freeze valve pipe
t = time
T = temperature
T_W = cooling water temperature
ΔT = temperature difference in liquid
U = liquid velocity in model or melter cavity
v_Z = molten glass velocity in axial direction in freeze valve
z = axial direction
α = thermal diffusivity
β = thermal expansion coefficient of liquid
ν = kinematic viscosity
ϕ_S = volumetric heat source due to electrical resistance
μ = viscosity
ρ = density
σ = Boltzman constant.

REFERENCES

(1) Gurley, R. N., et al. 1986. "Hanford Waste Vitrification Plant Project - An Overview." Proceedings of American Nuclear Society International Topical Meeting, September 14-18, 1986. Niagara Falls, New York, pp. 719-728.

(2) Chapman, C. C. 1983. "Design Preferences for a Slurry-Fed Ceramic Melter Suitable for Vitrifying West Valley Wastes." Ceramics in Nuclear Waste Management, Vol. 8, ed. by G. G. Wicks and W. A. Ross, American Ceramic Society.

(3) Tsuboya, T., and N. Tsunoda. 1988. "The Japanese Vitrification Program." Proceedings of Waste Management '88. Board of Regents, University of Arizona, Tucson, Arizona.

(4) Asano, H., et al. 1988. "LFCM Mock-up Test For Commercial Scale Vitrification." Waste Management '88, Board of Regents, University of Arizona, Tucson, Arizona, pp. 229-233.

(5) Davis, M. E. 1984. Numerical Methods and Modeling for Chemical Engineers. John Wiley & Sons.

(6) Bansal, B. 1978. "Physical Modeling of All-Electric Furnaces." IEEE Transactions - Industry Applications, Vol. IA, Issue 1, pp. 55-61.

(7) Quigley, M. S., and D. K. Kreid. 1979. Physical Modeling of Joule-Heated Ceramic Glass Melters for High-Level Waste Immobilization. PNL-2809, Pacific Northwest Laboratory, Richland, Washington.

(8) Skarda, R. J., et al. 1985. Preliminary Evaluation of PSCM and BIPP Melter Design and Operating Conditions Using Physical Modeling. PNL-5498, Pacific Northwest Laboratory, Richland, Washington.

(9) Routt, K. R. 1983. "Physical Modeling of a Glass Melter Designed for Vitrification of Defense Wastes." Proceedings of the American Ceramic Society Annual Meeting. Chicago, Illinois, April 23-27, 1983.

(10) Asano, H., et al. 1986. "Physical Modeling Test of Joule-Heated Ceramic Melter System for High Level Waste Vitrification." Proceedings of American Nuclear Society International Topical Meeting, September 14-18, 1986. Niagara Falls, New York, pp. 942-951.

(11) Stanek, J. 1977. Electric Melting of Glass. Elsevier, Amsterdam.

IMMOBILIZATION OF HIGH-LEVEL NUCLEAR WASTE IN GLASS: PROCESSING CONSTRAINTS IMPOSED BY IRON-NICKEL-SULFUR INTERACTIONS

Henry D. Schreiber
Samuel J. Kozak
Douglas G. Wetmore
Charlotte W. Schreiber
J. Sloan Downey

Center for Glass Chemistry
Virginia Military Institute
Lexington, Virginia 24450

ABSTRACT

Due to the precipitation of metal sulfides from glass-forming systems, iron-nickel-sulfur chemistry in borosilicate melts establishes the lower limit with respect to oxygen fugacity for the processing of high-level nuclear waste. The solubilities of iron sulfide and of nickel sulfide were characterized for the reference melt SRL-131. Nickel sulfide precipitates at oxygen fugacities lower than about 10^{-4} Pa {10^{-9} atm} at 1150°C, while iron sulfide and the respective metals form at even lower oxygen fugacities. Iron (Fe^{3+}) and sulfur (S^{2-}) interact via an electron exchange reaction to enhance the solubility of iron sulfide over the range of oxygen fugacities from 10^{-4} to 10^{-6} Pa {10^{-9} to 10^{-11} atm} in this reference system.

INTRODUCTION

The fundamental strategy for the disposal of high-level nuclear waste is to immobilize the waste in a borosilicate glass for eventual storage in underground repositories. Several constraints on the incorporation of the waste into the glass-forming system have been established on the basis of the redox chemistry of the borosilicate solvent [1,2]. For example, the processing of the waste plus borosilicate frit dictates that reducing agents be added to prevent the system from foaming. However, if the system becomes too reducing, metals and metal sulfides precipitate from the glass-forming solution under typical melter conditions. Accumulation of these materials in a layer on the bottom of the melter can lead to the shorting of the electrodes and/or accelerated corrosion of the melter floor [2]. Conditions necessary to form the metal sulfides specify the lower limit for the prevailing oxygen fugacity during the operation of the melter. This lower limit has been defined by an imposed oxygen fugacity of 10^{-4} Pa {10^{-9} atm} at 1150°C [3], or by an internal Fe^{2+}/Fe^{3+} ratio of no more than 0.5 in the melt [3,4].

OBJECTIVE

The goal of this study is to ascertain the solubility of nickel sulfide and iron sulfide in the reference borosilicate glass SRL-131 (frit #131 of the Savannah River Laboratory). This glass has been previously used as a model composition for nuclear waste immobilization and has been extensively characterized with respect to its redox chemistry [3,5,6].

EXPERIMENTAL PROCEDURES

Base composition SRL-131 was supplied as a homogeneous glass powder by the Savannah River Laboratory. The composition of SRL-131 is that of an alkali borosilicate: 57.9 wt% SiO_2, 1.0 % TiO_2, 0.5 % ZrO_2, 14.7 % B_2O_3, 0.5 % La_2O_3, 2.0 % MgO, and 17.7 % Na_2O. SRL-131 glass containing a certain percentage of nickel or of iron was prepared by mixing weighed quantities of the frit and the metal oxide, melting at 1150°C for 24 hours, quenching to a glass, and grinding to a fine powder. Mixes of the metal sulfides with the SRL-131 frits were prepared on a weight percent basis.

Individual samples were synthesized in alumina crucibles (of 2 mL capacity) suspended in furnaces whose temperature was regulated at 1150 $\pm$ 2 °C. The sample atmospheres were $CO_2/CO/SO_2$ gas mixtures, which simultaneously controlled the oxygen fugacities and the sulfur fugacities. The imposed oxygen fugacities were restricted to be between 10^{-4} and 10^{-6} Pa {10^{-9} and 10^{-11} atm} in a gas mixture containing 15 vol% SO_2, a sufficiently reducing atmosphere to stabilize the sulfide ion in SRL-131 melts but yet not too reducing to produce significant metal in the melt [3,7,8]. Equilibration times of 4 to 8 hours were deemed adequate to establish steady-state partitioning between the immiscible borosilicate and sulfide phases, as shown in figure 1.

Several samples were cut the entire vertical length of the container in order to expose both borosilicate and sulfide layers. Stepwise concentration profiles of Na, Si, S, and Ni (or Fe) from top to bottom were obtained with a scanning electron microscope equipped for simultaneous elemental analyses. Several representative specimens of borosilicate glass from each sample were then powdered.

Duplicate 0.1-0.2 gram aliquots of the borosilicate glass were analyzed for total sulfur content by a procedure that entailed the dissolution of the glass, reduction of the sulfur contained therein to hydrogen sulfide, distillation and collection of the sulfide in a standard anti-oxidant buffer, and measurement of concentration by a selective ion electrode [7,9].

For the samples containing iron, duplicate 10 mg aliquots of the borosilicate glass powder were dissolved under an inert atmosphere to prevent oxidation and were analyzed colorimetrically for Fe^{2+} content by complexation with o-phenanthroline in a buffered solution [9]. The total iron

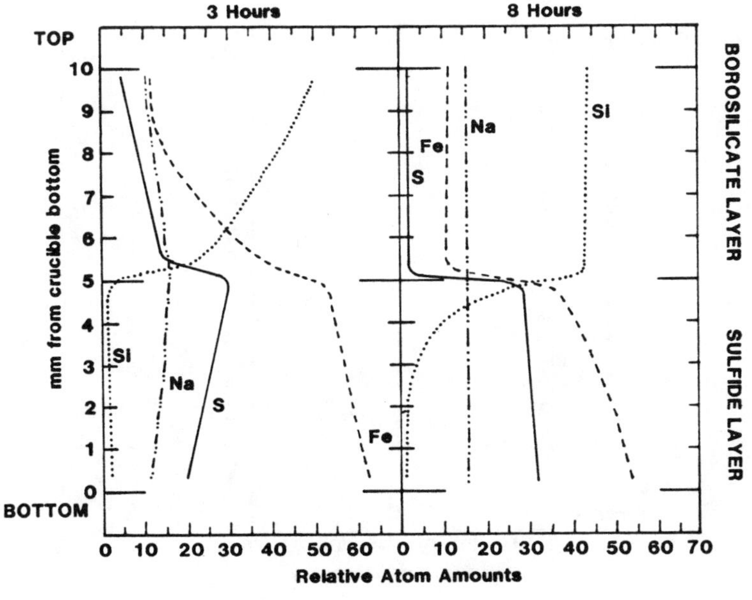

Figure 1. Elemental equilibration between borosilicate and iron sulfide phases as a function of equilibration time within a 1 cm deep sample. System was SRL-131 with 50 wt% FeS melted at 1150^{0}C in an atmosphere whose -log fO_2 {atm} = 9.2 with 15 volume % SO_2.

content was also determined in the same aliquot by reduction of all the iron to the Fe^{2+} state, followed by colorimetric analysis [9].

For the samples containing nickel, polished sections of the borosilicate glass were prepared. Spectrophotometric analysis of the 440 nm absorbance, characteristic of Ni^{2+} in this glass matrix, was employed as a monitor of Ni^{2+} concentration in this phase [8,9]. The absorptivity index (net absorbance divided by glass thickness) of this 440 nm peak had been previously calibrated to absolute Ni^{2+} concentration [8].

RESULTS

Figure 1 displays the distribution of elements between the borosilicate phase (SRL-131) and iron sulfide. The results indicate that the sulfide layer is not simply composed of FeS, but also contains appreciable amounts of alkali metals. The borosilicate/sulfide partition coefficient for sodium is near

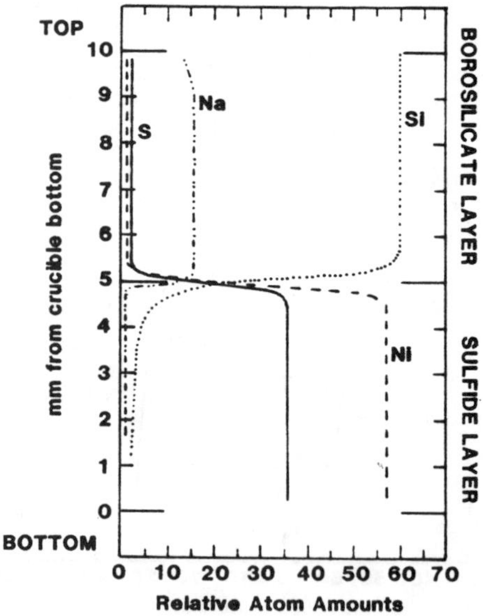

Figure 2. Elemental equilibration between borosilicate and nickel sulfide phases within a 1 cm deep charge. System was SRL-131 with 50 wt% NiS melted for four hours at 1150°C in an atmosphere whose -log fO_2 {atm} = 9.2 with 15 volume % SO_2.

unity in this system. Such results indicate that alkali ferrosulfates (for example; $NaFeS_2$) and oxysulfides are present in the sulfide layer coexisting with the borosilicate liquid [10].

A typical elemental distribution profile between SRL-131 and nickel sulfide is shown in figure 2. Nickel sulfide is much less soluble than iron sulfide in SRL-131 melts. The alkalis and silicon are effectively excluded from the nickel sulfide phase, unlike the iron sulfide system. The nickel sulfide in equilibrium with the borosilicate melt is stoichiometric Ni_3S_2, in agreement with nickel sulfides precipitated from similar silicate solvents [2] as well as from chloride melts [11].

The solubilities of iron sulfide and of nickel sulfide in the model borosilicate melt are both complicated by the fact that neither FeS nor NiS exists as a pure entity in equilibrium with SRL-131. The iron sulfide phase contains not only appreciable sodium but also possesses an Fe/S atomic ratio greater than one. The nickel sulfide phase excludes sodium and has a composition of Ni_3S_2.

The solubility of iron sulfide in SRL-131 melts at 1150°C as a function of the imposed oxygen fugacity is summarized in figure 3. The total sulfur content in the borosilicate layer in equilibrium with the iron sulfide phase is elevated over that observed in saturated SRL-131 melts or in saturated iron-containing SRL-131 melts. All iron in the borosilicate liquid coexisting with the sulfide layer was analyzed as Fe^{2+}, even though about 10 to 30% of the iron should have been present as Fe^{3+} over this oxygen fugacity range [3,9] in the absence of the sulfide phase. The iron content of the borosilicate liquid decreased and the sulfur capacity increased, as the oxygen fugacity was lowered. Solubility measurements were made on systems initially composed of SRL-131 + 50 wt% FeS, SRL-131 + 25 wt% FeS, 5 wt% Fe in SRL-131 + 50 wt% FeS, and 10 wt% Fe in SRL-131 + 50 wt% FeS. That essentially the same solubilities were determined in all systems, regardless of whether excess iron was initially in the borosilicate or the sulfide phase, was further evidence for the establishment of equilibrium between the two layers. To summarize the results of figure 3; the solubility of sulfur in the borosilicate melt seems to be controlled by the presence of the iron sulfide phase and not by the sulfur content of the atmosphere, and the presence of the coexisting iron sulfide phase controls the redox state of iron as exclusively Fe^{2+} in the borosilicate melt.

The solubility of nickel sulfide in SRL-131 melts at 1150°C as a function of the imposed oxygen fugacity is displayed in figure 4. The concentration of dissolved nickel in the SRL-131 melt coexisting with the sulfide phase is always about 1 wt%, independent of the imposed oxygen fugacity. This nickel solubility was attained, regardless whether the SRL-131 initially mixed with the nickel sulfide contained no nickel or excess nickel. Even though in the absence of nickel sulfide about 14-65 % of the nickel should have been reduced to nickel metal dispersed through the melt under these synthesis conditions [8,9], all the nickel in the borosilicate phase was identified as Ni^{2+} with no dispersed metal particulates. As in the iron/sulfur system, the sulfide phase buffers the redox state of the metal species dissolved in the borosilicate layer. The dissolved

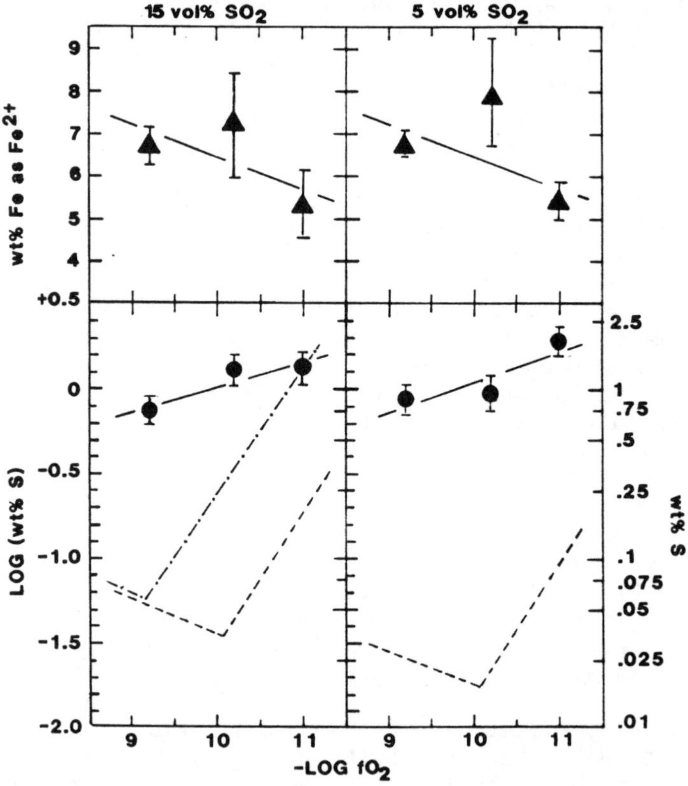

Figure 3. Solubility of iron sulfide in SRL-131 melt at 1150°C as a function of the imposed oxygen fugacity {atm} for two different sulfur contents of the atmosphere. Solubility was monitored by the iron and the sulfur content of the melt in the reference system SRL-131 mixed with 50 wt % FeS. Dashed lines represent the solubility of sulfur in SRL-131 melts, while the dashed-dotted line represents the solubility of sulfur in a SRL-131 melt containing 10 wt% iron for reference.

sulfur contents in SRL-131 melts in equilibrium with nickel sulfide are quite low. In essence, the sulfur content of the melt is controlled by the sulfur fugacity in the atmosphere and does not depend on the presence (or absence) of the nickel sulfide phase. The solubility of nickel sulfide in SRL-131 melts, as monitored by the resulting Ni^{2+} content of the melt, was essentially the same at a melt temperature of 1050°C as for 1150°C.

Nickel sulfide is much less soluble in the reference borosilicate liquid than is iron sulfide. This is in agreement with prior identifications that iron sulfide is poorly soluble in silicate glasses whereas nickel sulfide is virtually insoluble [10].

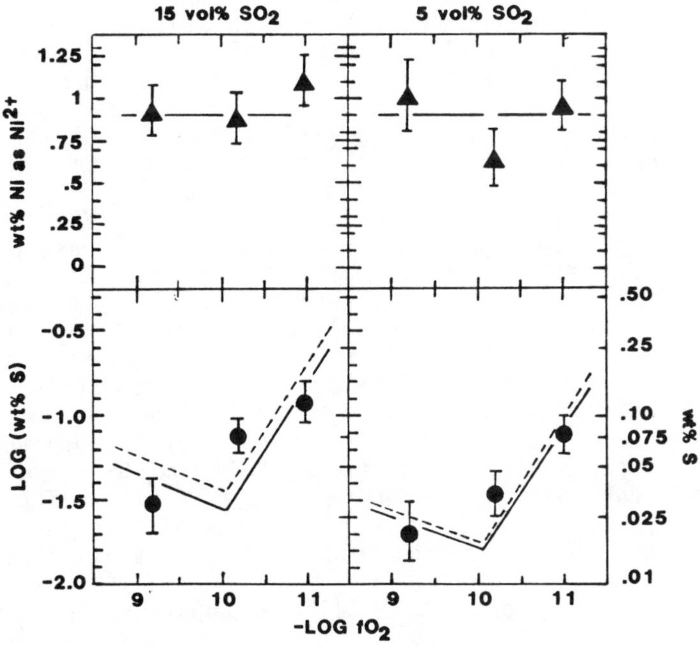

Figure 4. Solubility of nickel sulfide in SRL-131 melt at 1150°C as a function of the imposed oxygen fugacity {atm} for two different sulfur contents of the atmosphere. Solubility was monitored by the nickel and the sulfur content of the melt in the reference system SRL-131 mixed with 50 wt% NiS. Dashed lines represent the solubility of sulfur in SRL-131 melts.

DISCUSSION

The solubility product constant (K_{sp}) represents quantitatively the solubility of an ionic solid in aqueous solutions. For example, the solubility products for iron sulfide (FeS) and for nickel sulfide (NiS) in water are defined and reported as,

$$K_{sp} = [Fe^{2+}][S^{2-}] = 6.3 \times 10^{-18}$$

$$K_{sp} = [Ni^{2+}][S^{2-}] = 3.2 \times 10^{-19}$$

respectively at 25°C, where the ion concentrations are expressed in units of molarity. Thus, iron sulfide is about an order of magnitude more soluble than nickel sulfide in water. Similar solubility product constants have been employed to represent the solubilities of these ionic sulfides in other solvents. Table 1 compiles the solubility products of iron sulfide and of nickel sulfide in a eutectic LiCl/KCl melt at 450°C [11], as well as in the reference SRL-131 melt at 1150°C. It is, of course, realized that such expressions are only approximate, since the iron sulfide is not pure FeS and the nickel sulfide is actually Ni_3S_2. Nevertheless, table 1 clearly illustrates that iron sulfide is more soluble in SRL-131 melt than is nickel sulfide.

The redox systems of $Fe^{3+}/Fe^{2+}/SO_4{}^{2-}$, $Fe^{2+}/Fe^{0}/S^{2-}$, and $Ni^{2+}/Ni^{0}/S^{2-}$ in the reference melt SRL-131 represent stable entities with respect to mutual interactions; that is, the sulfur redox state present will neither oxidize nor reduce the metal redox ions [9]. The only electron exchange reaction between the iron/sulfur or nickel/sulfur redox states is oxidation of S^{2-} by Fe^{3+} to form $S_2{}^{-}$ (if the iron content of the melt is 1 wt% or less) or to form polysulfide species such as $S_x{}^{2-}$ (if excess iron is present in the melt). However, the simultaneous presence of interacting Fe^{3+} and S^{2-} ions in the melt only occurs over a very narrow range of oxygen fugacities. Nevertheless, such interactions are responsible for the enhanced incorporation of sulfur in iron-containing melts from oxygen

Table 1. Solubility product equilibrium constants, K_{sp}, for NiS and FeS in two solvent systems. Solubility products are calculated on the basis of ion fractions.

Solvent	K_{sp} for NiS	K_{sp} for FeS
LiCl/KCl eutectic melt; T = 450°C [11]	2.9×10^{-14}	2.3×10^{-12}
SRL-131 borosilicate melt; T = 1150°C	$9(\pm 7) \times 10^{-7}$	$1.5(\pm .5) \times 10^{-4}$

fugacities of 10^{-4} to 10^{-6} Pa {10^{-9} to 10^{-11} atm} at 1150°C in SRL-131 [9].

The enhanced solubility of iron sulfide versus that of nickel sulfide over the oxygen fugacity range of 10^{-9} to 10^{-11} atm may be related to the internal oxidation of the sulfide ion to polysulfide species by the Fe^{3+} ion. As seen in figure 3, the more Fe^{3+} initially present in the melt, the greater the enhancement of the sulfur solubility. In essence, all the Fe^{3+} is reduced by the presence of the sulfide layer. No such mechanism exists for the dissolution of nickel sulfide, since Ni^{2+} and S^{2-} do not interact by electron exchange. In this case, the sulfur content of the melt is controlled not by the nickel sulfide but by the sulfur gases in the atmosphere.

Several samples of metal sulfide partitioning into SRL-131 melts were synthesized under uncontrolled atmospheric conditions, that is air, at 1150°C. The iron or nickel content of the resulting borosilicate layer was the same as that determined for the samples synthesized at the lower oxygen fugacity and 15 volume% SO_2 in the atmosphere. Even under these quite oxidizing conditions, where the sulfide phase is at best a metastable solid, the presence of the iron sulfide layer was sufficient to buffer the iron redox state to exist exclusively in the Fe^{2+} state in the SRL-131 melt.

CONSTRAINTS ON NUCLEAR WASTE IMMOBILIZATION

The lower limit for oxygen fugacity during the processing of nuclear waste in the melt at a reference temperature of 1150°C was previously established as 10^{-4} Pa {10^{-9} atm} for the first sign of sulfide in the melt and as 10^{-7} Pa {10^{-12} atm} as the appearance of significant metal precipitation [3]. However, just because sulfide ions exist in the melt does not necessarily mean that metal sulfides will precipitate, since sufficient sulfide is required to exceed the solubility product constant. Use of 8.50 wt % Fe, 0.25 wt % Ni, and 0.10 wt % S in a model wasteform (glass components plus nuclear waste) [3] and the experimentally determined solubility constants would imply that precipitation of nickel sulfide at an oxygen fugacity of 10^{-4} Pa {10^{-9} atm} is marginal (but it should definitely form by $10^{-5.5}$ Pa {$10^{-10.5}$ atm}), and that any sulfide phase that forms would be primarily nickel sulfide. Although nucleation kinetics, synergistic relations, and the presence of other crystals may affect the overall scheme; nickel sulfide would precipitate before iron sulfide, iron metal or nickel metal in the nuclear waste melter. In order to preclude the precipitation of nickel sulfide during the processing of the nuclear waste, the lower limit of the prevailing oxygen fugacity should be about 10^{-3} Pa {10^{-8} atm} in order to include a margin of safety; however, this is still well within the redox scale defined by an Fe^{2+}/Fe^{3+} ratio of less than 0.5 in the resulting glass [3,4].

ACKNOWLEDGEMENT

This research was supported through the Savannah River Laboratory, operated by E. I. DuPont for the Department of Energy, and by the Virginia Center for Innovative Technology.

REFERENCES

[1] M. J. Plodinec, "Vitrification Chemistry and Nuclear Waste", J. Non-Cryst. Sol. **84**, 206-214 (1986).

[2] D. F. Bickford, R. B. Diemer, and D. C. Iverson, "Redox Control of Electric Melters with Complex Feed Compositions. II. Preliminary Limits for Radioactive Waste Melters", J. Non-Cryst. Sol. **84**, 285-291 (1986).

[3] H. D. Schreiber and A. L. Hockman, "Redox Chemistry in Candidate Glasses for Nuclear Waste Immobilization", J. Am. Ceram. Soc. **70**, 591-594 (1987).

[4] D. S. Goldman and D. W. Brite, "Redox Characterization of Simulated Nuclear Waste Glass", J. Am. Ceram. Soc. **69**, 411-413 (1986).

[5] H. D. Schreiber, "An Electrochemical Series of Redox Couples in Silicate Melts. A Review and Applications to Geochemistry", J. Geophys. Res. **92**, 9233-9245 (1987).

[6] M. J. Plodinec, "Improved Glass Compositions for Immobilization of SRP Waste", pp. 223-229 in Scientific Basis for Nuclear Waste Management, **2**, edited by C. J. M. Northrup, Plenum Press, 1980.

[7] H. D. Schreiber, S. J. Kozak, P. G. Leonhard, and K. K. McManus, "Sulfur Chemistry in a Borosilicate Melt. Part 1. Redox Equilibria and Solubility", Glastech. Ber. **60**, 389-398 (1987).

[8] H. D. Schreiber, B. E. Carpenter, L. M. Minnix, G. B. Balazs, and T. N. Solberg, "Precipitation of Iron, Cobalt, and Nickel Metal from a Borosilicate Melt", Glastech. Ber. **56K**, 1017-1022 (1983).

[9] H. D. Schreiber, S. J. Kozak, C. W. Schreiber, D. G. Wetmore, and M. W. Riethmiller, "Sulfur Chemistry in a Borosilicate Melt. Part 3. Iron-Sulfur Interactions and the Amber Chromophore", Glastech. Ber., submitted.

[10] M. B. Volf, Chemical Approach to Glass. Elsevier, Amsterdam (1984).

[11] J. R. Selman and M. L. Saboungi, "Electrochemistry of Sulfur in Halide Melts", pp. 81-150 in The Sulfur Electrode, Fused Salts and Solid Electrolytes, edited by R. P. Tischer, Academic Press, 1983.

INDEX